HANDBOOK OF MODERN FERROMAGNETIC MATERIALS

THE KLUWER INTERNATIONAL SERIES
IN ENGINEERING AND COMPUTER SCIENCE

HANDBOOK OF MODERN FERROMAGNETIC MATERIALS

Alex Goldman, B.S., A.M., Ph.D.
Ferrite Technology Worldwide

Kluwer Academic Publishers
Boston/Dordrecht/London

Distributors for North, Central and South America:
Kluwer Academic Publishers
101 Philip Drive
Assinippi Park
Norwell, Massachusetts 02061 USA
Tel: 781-871-6600
Fax: 781-871-6528
E-mail: kluwer@wkap.com

Distributors for all other countries:
Kluwer Academic Publishers Group
Distribution Centre
Post Office Box 322
3300 AH Dordrecht, THE NETHERLANDS
Tel: 31 78 6392 392
Fax: 31 78 6546 474
E-mail: orderdept@wkap.nl

 Electronic Services: http://www.wkap.nl

Library of Congress Cataloging-in-Publication Data

Goldman, Alex.
  Handbook of modern ferromagnetic materials / Alex Goldman.
    p. cm. -- (The Kluwer international series in engineering and computer science ; SECS 505)
  Includes bibliographical references.
  ISBN: 0-412-14661-4
  1. Magnetic materials. 2. Electronics--Materials. 3. Ferrites (Magnetic materials) I. Title. II. Series.
TK7871.15.M3G65   1999
621.381--dc21   99-20757
                                                                         CIP

Copyright © 1999 by Kluwer Academic Publishers.

All rights reserved. No part of this publication may be reproduced, stored in a retrieval system or transmitted in any form or by any means, mechanical, photo-copying, recording, or otherwise, without the prior written permission of the publisher, Kluwer Academic Publishers, 101 Philip Drive, Assinippi Park, Norwell, Massachusetts 02061

Printed on acid-free paper.

Printed in the United States of America

Dedication

This book is dedicated to the memory of three men who greatly influenced the scientific life of the author; Professor Takeshi Takei, Dr. Richard M. Bozorth and Dr. Gilbert Y. Chin. Professor Takei, to whom I dedicated my first book and whose preface to that book follows this dedication, was a greatly loved friend and teacher of mine. He passed away on March 12, 1992. He will be remembered as a founding father of ferrites, a great teacher and the organizer of ICF(International Conference on Ferrites).. Since the present book includes chapters on metallic magnetic materials, the contributions of both Dr. Richard M. Bozorth and Dr. Gilbert Y. Chin are also remembered. Richard M. Bozorth was head of the Magnetics Research Department at the Bell Telephone Laboratories in Murray Hill, New Jersey for many years. His book, "Ferromagnetism" is probably the most important reference book on the subject. Dr. Gilbert Y. Chin was a dear friend of mine whose untimely death was a great blow to me and the world of magnetics. He was responsible for my trip to the Peoples Republic of China to discuss magnetic materials processing. The scientific community has been well served by the lives of these three men.

TABLE OF CONTENTS

Foreword by Takeshi Takei	xv
Preface	xvii
Acknowledgements	xix

Chapter 1-Applications and Functions of Ferromagnetic Material — 1

What are Ferromagnetic Materials?	1
History of Ferromagnetic Materials	1
General Categories of Magnetic Materials	2
Applications of Magnetic Materials at DC	4
Power Applications at Low Frequencies	5
Low Frequency Transformers	6
Entertainment Applications	9
High Frequency Power Supplies	11
Microwave Applications	12
Magnetic Recording Applications	13
Miscellaneous Applications	14

Chapter 2-Basics of Magnetism-Source of Magnetic Effect — 17

Magnetic Fields	17
Electromagnetism	21
Atomic Megnatism	21
Paramagnetism and Diamagnetism	26
Paramagnetism of Metals	27
Ferromagnetism	29
Magnetic Moments in Ferromagnetic Metals and Alloys	32
Antiferromagnetism	34
Ferrimagnetism	39

Chapter 3-The Magnetization in Domains and Bulk Materials — 41

The Nature of Domains	41
Proof of the Existence of Domains	47
The Dynamic Behavior of Domains	47
Bulk Material Magnetization	48
Hysteresis Loops	53
Permeability	54
Magnetocrystalline Anisotrophy Constants	55
Magnetostriction	57
Important Properties for Hard Magnetic Materials	57

Chapter 4-AC Properties of Magnetic Materials   59

AC Hysteresis Loops   59
Eddy Current Losses   59
Permeability   62
Disaccomodation   66
Core Loss   68
Microwave Properties   69
Microwave Precessional Modes   71
Logic and Switching Properties of Ferrites   72
Properties of Recording Media   73

Chapter 5-Materials for Permanent Magnet Applications   75

History of Permanent Magnet   75
General Properties of Permanent Magnets   78
Commercial Permanent Magnet Materials-Properties   81
Oxidic or Hard Ferrite Materials   86
Commercial Oriented and Non-Oriented Hard Ferrites   86
Rare-Earth-Cobalt Permanent Magnet Materials   90
Neodymium-Iron-Boron Permanent Magnet Materials   92
Iron-Chromium-Cobalt Magnet Materials   95
Ductile Permanent Magnet Alloys   95
Miscellaneous Permanent Magnet Materials   95
Criteria for Choosing a Permanent Magnet Material   95
Stabilization of Permanent Magnets   100
Cost Considerations in Permanent Magnet Materials   100
Cost of Finished Magnets   101
Calculations and Design of Magnets   102
Optimum Shapes of Ferrite and Metal Magnets   106
Recoil Lines-Operating Load Lines   106

Chapter 6-DC and Low Frequency Applications   107

Material Requirements for DC and Low Frequencies   107
Metallic Materials for DC or Low Frequencies   108
Soft Iron as a Magnetic Material   110
Low Carbon Steels   112
Silicon Steels   115
Iron-based Amorphous Materials   122
Other Low-frequency Magnetic Alloys   127
Nanocrystalline Materials   130
Sintered Soft Iron for Magnetic Applications   130

Chapter 7-Soft Cobalt-Iron Alloys   137

History of Iron-Cobalt Alloys   137
Chemistry and Structure of Cobalt-Iron Alloys   137

TABLE OF CONTENTS

Commercial Availability of Soft Cobalt-Iron Alloys	142
Applications of Soft Cobalt-Iron Alloys	143
Magnetic Components Using Cobalt-Iron Alloys	144

Chapter 8-Metallic Materials for Magnetic Shielding Applications. 145

Shielding Factor and Attenuation	145
Materials for Magnetic Shielding	147
Processing of Shield Materials	149
Measurement of Shield Material	149
Commercial Forms of Supply	151
Shielding Factor as a Function of Frequency	151

Chapter 9-High Permeability-High Frequency Metal Strip 155

History of Nickel-Iron Alloys	156
High Permeability Nickel-Iron Alloys	157
Binary Nickel-Iron Alloys in the 80% Ni Range	158
Nickel-Iron Alloys with 65% Nickel	168
Nickel-Iron Alloys in the 36% Nickel Range	170
Amorphous Materials-High Perm-HighFrequency	170
Nanocrystalline Materials-High Frequency Operation	176

Chapter 10-Metal Powder Cores for Telecommunications 183

History of Powder Cores for Telecommunications	184
Iron Powder Cores for Audio and RF Applications	185
Processing of Iron Powder Cores	186
Moly-Perm Powder Cores for Telecommunications	191
Properties of Moly-Perm Powder Cores	196

Chapter 11-Crystal Structure of Ferrites 207

Classes of Crystal Structures in Ferrites	207
Normal Spinels	212
Inverse Spinels	213
Magnetic Moments in Inverse Spinels	213
Mixed Zinc Ferrites	216
Sublattice Magnetizations	217
Hexagonal Ferrites	220
Magnetic Rare-Earth Garnets	222
Miscellaneous Structures	226

Chapter 12-Chemical Aspects of Ferrites 229

Intrinsic and Extrinsic Properties of Ferrites	229
Mixed Ferrites for Property Optimization	230
Permeability Dependence on Chemistry	230

Effect of Iron Content on Permeability	232
Effect of Divalent Ion Variation	236
Permeability Dependence on Zinc	238
Effect of Cobalt on Permeability	239
Oxygen Stoichiometry	240
Effect of Purity on Permeability	244
Effect of Foreign Ions on Permeability	245
Temperature Dependence of Initial Permeability	247
Time Dependence of Permeability (Disaccomodation)	250
Chemistry Dependence of Low Field Losses Loss Factor	251
Tetravalent and Pentavalent Oxide Substitution	251
Losses at High Power Levels	256
Chemistry Considerations for Hard Ferrites	259
Saturation Induction of Microwave Ferrites and Garnets	260
Chemistry Dependence of Microwave Properties	261
Ferrites for Memory and Recording Applications	262

Chapter 13-Microstructural Aspects of Ferrites — 265

Microstructural Engineering for Desired Properties	265
The Effects of Grain Size on Permeability	266
Exaggerated Grain Growth in Ferrites	268
Duplex Grain Structures	271
Effect of Porosity on Permeability	271
Separation of Grain Size and Porosity Effects	273
Effect of Grain Size and Porosity on B-H Loop Parameters	276
Grain Boundary Considerations	279
Early Studies of Grain Boundaries	280
High Frequency Materials	284
Considerations of Microstructure for Microwave Ferrites	291
Microstructural Considerations of Hard Ferrites	292
Resistive and Capacitive Effects in Grain Boundaries	293
Phase Transformation and Oxidation	299

Chapter 14-Ferrite Processing — 305

Powder Preparation-Materials Selection	305
Weighing and Blending	309
Calcining	310
Milling	313
Granulation or Spray Drying	313
Pressing	314
Non-conventional Processing	317
Powder Preparation of Microwave Ferrites	325
Hard Ferrite Powder Preparation	325
Sintering Spinel Ferrites	326
Sintering Manganese-Zinc Ferrites	328

TABLE OF CONTENTS

Hold or Soak Temperature and Duration	330
Atmosphere Effects	332
Sintering of MnZn Ferrites for Specific Applications	333
Power Ferrites	335
Microstructural Development	336
Ferrite Kilns	339
Firing of Microwave Ferrites and Garnets	341
Firing of Hard Ferrites	342
Finishing Operations for Ferrites	342
Ferrite and Garnet Films	343
Single Crystal Ferrites	350

Chapter 15-Ferrite Inductors and Transformers for Low Power — 361

Inductance	362
Effective Magnetic Parameters	364
Measurement of Effective Permeability	365
Magnetic Considerations:Low Level Applications	366
Pot Core Assembly	370
Pot Core Shapes and Sizes	373
Designing for Inductance in Pot Cores	375
Designing a Pot Core Inductor for Maximum Q	377
Designing Ferrite Inductors for Stability Requirements	378
Flux Density Limitations in Ferrite Inductor Design	379
DC Bias Effects in Ferrite Inductor Design	381
Surface-Mount Design for Pot Cores	381
Low Level Transformers	381
Ferrite Pulse Transformers	385
Ferrites for Low-Level Digital Applications	385
ISDN Components and Materials	387
Multilayer Chip Inductors and LC Filters	389

Chapter 16-Soft Magnetic Materials for EMI Suppression — 391

The Need for EMI Suppression Devices	391
Materials for EMI Suppression	396
Frequency Characteristics of EMI Materials	399
Mechanism of EMI Suppression	401
Components for EMI Suppression	402
Common-Mode Filters	403
Differential Mode Filters	407
Metal Powder Cores for EMI Suppression	408
Amorphous-Nanocrystalline Materials for EMI Suppression	420

Chapter 17-Ferrites for Entertainment Applications — 425

TABLE OF CONTENTS xiii

Magnetic Recording Head Properties	541
Audio and Video Magnetic Recording	542
Magnetic Recording Media	543
Magnetic Recording Heads	544
Magnetoresistive Heads	545
Magneto-optic Recording	553
Outlook for Areal Densities in Magnetic Recording	555

Chapter 20-Ferrites for Microwave Applications — 559

The Need for Ferrite Microwave Components	559
Ferrite Microwave Components	560
Commercially-Available Microwave Materials	568

Chapter 21-Miscellaneous Magnetic Material Applications — 571

Magnetostrictive Transducers	571
Sensors	572
Copier Powders	576
Ferrofluids	576
Electrodes	576
Delay Lines	577
Ferrite Tiles for Anechoic Chambers	577
Reed Switches	580

Chapter 22-Physical-Thermal Aspects of Magnetic Materials — 581

Densities of Ferrites	581
Mechanical Properties of Ferrites	581
Thermal Properties	583
Pressure Effects on Ferrites	585
Radiation Resistance	587

Chapter 23-Magnetic Measurements-Materials and Components — 589

Measurement of Magnetic Field Strength	589
Measurement of Magnetization	591
Magnetization Curves and Hysteresis Loops	594
Measurements on Hard Ferrites	595
Magnetocrystalline Anisotropy	596
Magnetostriction	597
Inductance and Permeability Measurements	599
Loss Factor	602
Q Factor	602
High Frequency Measurements	602
Permeability Measurements on Ferrite Components	603
Loss Separation at Low Flux Density	607

Losses at High Power Levels-Core Losses 608
Pulse Measurements 615
Amplitude Permeability 615

Bibliography 623

Appendix 1-Abbreviations and Symbols 627
Appendix 2-List of World's Major Ferrite Suppliers 633
Appendix 3-Units conversion from CGS to MKS(SI) System 641

Index 643

FOREWORD

Below is a copy of Professor Takeshi Takei's original preface that he wrote for my first book, Modern Ferrite Technology. I was proud to receive this preface and include it here with pride and affection. We were saddened to learn of his death at 92 on March 12, 1992.

Preface

It is now some 50 years since ferrites debuted as an important new category of magnetic materials. They were prized for a range of properties that had no equivalents in existing metal magnetic materials, and it was not long before full-fledged research and development efforts were underway. Today, ferrites are employed in a truly wide range of applications, and the efforts of the many men and women working in the field are yielding many highly intriguing results. New, high-performance products are appearing one after another, and it would seem we have only scratched the surface of the hidden possibilities of these fascinating materials.

Dr. Alex Goldman is well qualified to talk about the state of the art in ferrites. For many years Dr. Goldman has been heavily involved in the field as director of the research and development division of Spang & Co. and other enterprises. This book, *Modern Ferrite Technology*, based in part on his own experiences, presents a valuable overview of the field. It is testimony to his commitment and bountiful knowledge about one of today's most intriguing areas of technology.

In the first part of his book, Dr. Goldman discusses the static characteristics of ferrites based on the concept of ferrimagnetism. He then considers their dynamic properties in high-frequency magnetic fields.

Dr. Goldman follows this up with a more detailed look at some of these characteristics. In a section on power materials he examines the need to use chemical adjustment and microstructural optimization to attain high saturation and low core loss at high frequencies. He uses anisotropy and fine particle concepts to describe permanent magnetic ferrites and reviews how gyromagnetic properties help explain the actions of microwave ferrites.

In a section on applications, he introduces such production technologies—some conventional, some unconventional—as coprecipitation, spray roasting and single crystal preparation. He also discusses some of the special difficulties that ferrites can impose from a design point of view in actual applications. Turning to the subject of magnetic recording, Dr. Goldman discusses in detail the impressive strides being made in magnetic media and magnetic head applications.

The book is rounded out with valuable appendices, including a list of the latest physical, chemical and magnetic data available on ferrites and listings of world ferrite suppliers.

Modern Ferrite Technology presents the reader with the latest thinking on ferrites by the scientists, researchers and technicians actually involved in their development, leavened with the rich experiences over many years of theauthor himself. It is a work of great interest not only to those researching new ferrite materials and applications, but to all in science and industry who use ferrites in their work.

April 1987

Takeshi Takei

Preface

It has been almost fifty years since Dr. Richard M. Bozorth wrote his landmark book, "Ferromagnetism". This book has been considered by many (including the author) to be the "Bible" of ferromagnetic materials. Evidence for its popularity has been shown by the fact that the IEEE in 1993, some 40 years after its original publication, reissued the book. In my present book, I have liberally referenced from Dr. Bozorth's book for several reasons. First, many of the basic concepts on ferromagnetism have not changed much since then. A number of the materials discussed in Bozorth's book such as 4-79 Permalloy strip and 2-81 Moly-perm powder cores are still being used in pretty much the same form. Second, Bozorth's book presents a bench mark or comparison for so many of the newer materials including ferrites, amorphous materials and of course, the newest nanocrystalline materials.

In my earlier book, Modern Ferrite Technology, I remarked that ferrites were "the new kids on the block". Well, by now, the kids have turned into teenagers and, as such, are making their presence felt. However, they are being challenged by a new breed of metallic materials including the non-crystalline amorphous alloys and the magnetic nanocrystalline materials. How could Bozorth have dreamed of such revolutionary materials and such amazing techniques such as multilayer films so important in magnetic recording? As a matter of fact, in Bozorth's book, ferrites barely got six pages in his volume of nearly 1000 pages.

In expanding my earlier book, Modern Ferrite Technology, I am attempting to accomplish several objectives;

1. To update the information on ferrites to include the advances since the original work including the major features of ICF6 and ICF7.
2. To provide a method of comparison of the main features of all the commercially available ferromagnetic materials (and some that are not available such as the Colossal Magnetoresistive Materials).
3. To add chapters on magnetic materials of expanding importance, such as those for EMI suppression and for entertainment applications
4. To introduce the magnetic materials and components for new technologies such as ISDN, planar magnetics, surface-mount techniques, magnetoresistive recording and integrated magnetics.

As I wrote in the preface in my first book, the attempt here is to describe in as non-technical language as possible, the material, magnetic, and stability characteristics of a wide variety of materials. These materials are introduced in chapters according to frequency. In this scheme, we would then start with the low-frequency metallic power materials, continue to the higher permeability and higher frequency strip materials, to the metal powder cores and finally to ferrites at the highest frequencies including microwaves.

The book is aimed at several different audiences. The first chapter takes a broad view of all the applications and functions of magnetic materials. This chapter would appeal to somebody just getting his or her feet wet in the Sea of Magnetism without getting into the materials themselves. The next three chapters are also introductory in nature and deal with the basic concepts of origins of magnetic phenomena, domain and magnetization behavior, magnetic and electrical units and finally to the the action of magnetic materials under ac drive. Then come the sections on the materials themselves, starting with the DC applications and specifically the permanent magnet materials. The first is one of the few chapters in which metallic and non-metallic materials are discussed together for comparison purposes. The next chapter involves the metallic strip materials for line or mains (50-60 Hz.) power including a discussion of some of the new sintered iron alloys. Slightly higher (400-1200 Hz.) frequencies are dealt with in the next chapter on the cobalt-iron alloys. The next chapter deals with materials for magnetic shielding for protection of sensitive devices. Some of the same materials appear in the next chapter dealing with the high permeability and higher frequency materials, including the nickel-iron alloys and the amorphous metals. We then leave the metal strip materials and go to the powdered metal materials which are useful at the higher frequencies. We then shift to the ferrite materials and the next three chapters deal with the material properties including crystal structure, chemistry and microstructure. These sections are then followed by a chapter on processing of ferrites. The first part of this book (14 chapters) has dealt mainly with the material science aspects of ferromagnetism. Most of the rest of the book is involved with the specific applications. First, we will try to make use of the special properties of the materials we have studied in the low power or telecommunication area, then the associated EMI and entertainment uses. Finally we come to the very large and important chapter on high frequency power materials and components. Although many of the materials in these chapters are ferrites, in these cases, competing materials are introduced to enable the design engineer to make the best choice based on quality, stability and cost. The next chapter discusses the materials for magnetic recording. The sections on magnetoresistive and magneto-optical recording have been expanded to include the new developments. The next chapter deals with the materials for microwave applications and the last application section is on miscellaneous materials that includes materials for theft deterrence and anechoic chamber tiles. The final chapters discuss the mechanical and thermal aspects ending with a chapter on magnetic measurements. A listing of the IEC and ASTM documents appear at the end of this chapter. The appendices include abbreviations and symbols, listing of major ferrite suppliers of the world and units conversion from cgs to MKS. As I described in the preface to the earlier book, I would hope that this book also provides a bridge between the material scientist and the electrical design engineer so that some common ground can be established and an understanding of the problems of each can be gained.

Alex Goldman

ACKNOWLEDGEMENTS xix

I would like to thank Joseph F. Huth III and Harry Savisky of the Magnetics Div. of Spang and Co. for providing me with some of the photos, photomicrographs and figures in the book. Mr. John Knight of Fair-Rite products also provided me with important information on permanent magnet materials. I would also like to thank the people at Fair-Rite Products, Steward, Allied Signal, Vacuumschmelze, Arnold Engineering, Magnetics, Micrometals, TDK, MMG-North America, Cartech Pyroferric and FDK for their catalog data. I would like thank Prof.M. Sugimoto for the material from ICF6 and 7. Finally, I would like to thank my wife, Adele, who put up with my endless hours at the computer with great love and understanding and also to my children, Drs. Mark, Beth and Karen for lots of pride and inspiration that they have given to me.

1 APPLICATIONS AND FUNCTIONS OF FERROMAGNETIC MATERIALS

INTRODUCTION
Magnetism was probably the first natural force discovered by man but it has only been in the last century that any large usage of magnetic materials has been made. Much of the glamour of modern electronics has been centered on the semiconductor (transistor and IC) industry but many of the devices using these new concepts would not be practical without the accompanying magnetic components. The frequencies of application of magnetic materials range from DC(Direct Current) to the highest ones at which any electronic device can function. The emergence of many new technologies driven by differing requirements, in turn, has led to a large variety of magnetic materials supplied in many different shapes and sizes. This chapter will consider the various applications for magnetic materials and the functions performed by the magnetic components in these applications.

WHAT ARE FERROMAGNETIC MATERIALS?
Ferromagnetic materials are those to which we commonly refer as "magnetic materials". While all materials are magnetic to some degree, many of them fall under the heading of paramagnetic or diamagnetic materials possessing only a very weak and relatively undetectable magnetic property. As such, they are not used to any extent in any practical devices but have great scientific importance. The cooperative effect called "ferromagnetism" leads to materials that have magnetic forces many orders of magnitude larger than the aforementioned materials. We should also distinguish between ferromagnetic materials and superconducting materials that have assumed such prominence in recent years. The mechanism of superconductivity is different than the phenomenon of ferromagnetism. Their one common aspect is the generation of magnetic fields by circulating currents.

A popular misconception by the lay person is that a magnetic material is one that has only permanent magnet properties. As we shall see as we expand the subject, the term also includes soft magnetic materials. In fact, the latter group has many more varieties than the permanent magnet group.

Ferromagnetic materials may exist as conductors (metals), insulators (ceramics) or as semiconductors. The choice of material to be used is based on the operating conditions of the device.

HISTORY OF FERROMAGNETIC MATERIALS

The earliest magnetic material known to man is magnetite, a naturally occurring magnetic ceramic (ferrite). Pieces of this mineral were found to exert attractive or repulsive forces when brought close to another piece of the same material. Later, when iron objects became available, they were attracted to these earlier magnetic materials. When iron needles were rubbed against the magnetite, the needles also became imbued with this new bipolar property, giving the same attraction or repulsion as the original magnetite pieces. The subject of frictional magnetism may be one of the oldest questions about magnetism and still remains one of the least understood phenomena. These magnetized needles served as the first applications of magnetic materials since they functioned as compasses and allowed mariners to find North without the use of the stars.

It was not until the advent of electromagnetism that practical electrical devices using magnetic materials were developed. The properties of repulsion and attraction found in magnetite could be reproduced in iron by passing an electric current through a wire surrounding the iron. The magnetic forces between sets of poles could then converted to mechanical action and thus be used to create motors. By reversing the effect, electrical generators could be made.

Another series of events prompting progress in magnetic materials development was the invention of the telegraph followed by the introduction of the telephone which required new magnetic devices such as speakers, loading coils, channel filters and transformers. Wireless radio and later, television all produced a need for new and better magnetic materials. Radar for flight and space communications required new special magnetic materials. Finally, the memory, recording and computer explosion brought even greater reliance on new magnetic materials. Although some newer technologies such as fiber optical communication may eliminate the need for some of the older magnetic materials, there are still many new emerging areas that are replacing the previous ones.

With the rapidly changing turnover of magnetic materials, it is important that we be aware of the present state of the art in magnetic materials.

GENERAL CATEGORIES OF MAGNETIC MATERIALS

Magnetic materials can be categorized in several different manners. First, they can be classified by market.

1. Consumer- radio, television, video recorder, small motors, transformers.
2. Heavy industrial- Generators, large motors, large transformers
3. Light industrial- circuit components, power supplies
4. Specialty and Custom- aircraft, microwave devices, recording heads

The type of market aimed at will usually determine the cost of the magnetic material or component. The cost is lowest for the first category and successively higher for the remaining ones.

Still another form of categorizing is by function and this may be, for our purposes, the best way.

1. Mechanical- Lifting, separating, suspending
2. Shielding
3. Electromechanical transducers- Motors, speakers, vibrators, relays, ultrasonic generators
4. Mechanico-electric transducers-electrical generators, phonograph needles
5. Voltage and current multipliers-Transformers
6. Impedance Matching
7. Inductor in LC circuit
8. Filter to remove any unwanted frequencies- Wide band transformer, channel filter, emi suppression
9. Output choke- remove ac component from D.C
10. Bistable element in a binary memory device- recording media
11. Magnetic head- Write or read data on tape or disk
12. Microwave devices
13. Delay lines

Although this book is primarily aimed at electrical and electronic applications, it may be useful to enumerate the uses of magnetic materials in which the application is actually mechanical but a magnetic force is used to effect the operation.

These applications include;

1. Lifting magnets
2. Magnetic repulsion
3. Magnetic separation
4. Refrigerator door seals
5. Magnetic chucks and holding devices
6. Magnetic couplings and pumps
7. Removal of foreign bodies from eyes
8. Toys

Another means of classification is according to frequency;

1. D.C.-Permanent magnets and D.C. motors, generators and other D.C. devices
2. Line frequencies-50-60 Hz.
3. Aircraft frequencies-400 Hz.
4. Audio Frequencies-to 20,000 Hz.
5. High frequency power- 25000-100000 Hz. and climbing
6. High frequency telecommunications-100,000 Hz. To 100 MHz.
7. Microwave and Radar-1 GHz. And beyond

In general, the frequency used is also an indication of the size of the component. The lower the frequency, the larger the size while conversely, high frequency components tend to be smaller. In general, the materials description part of this book will be arranged according to frequency starting with D.C. and proceeding with applications at increasing frequencies.

APPLICATIONS OF MAGNETIC MATERIALS AT DC

DC was the first type of electrical current discovered so it is natural that the first practical use of magnetic materials was made in DC devices. The main functions were the conversion of electrical power to mechanical power or the converse. These applications include;

1. DC motors and generators driven by electromagnetic forces created in wound stators and rotors or in permanent magnets.
2. Electro-mechanical devices such as relays, switches, armatures and pole pieces for opening and closing electrical circuits, doorbells, buzzers
3. Plungers, brakes and transducers for controlling flow, pressure switches
4. Reed switches

Permanent magnets are used in electronic magneto-acoustical devices such as telephone diaphragms, loudspeakers, microphones, TV picture tube ion traps and phonograph pickups.

Other DC applications involve the construction of large electromagnets for measurements and scientific applications. Large electromagnets are also used as magnetizing yokes to magnetize permanent magnets.

Another use of magnetic materials at DC involves the case where a permanent magnet biases an electrical or magnetic circuit. Examples of this can be found in bubble domain memory devices, biased magnetic cores, in transverse biasing of microwave devices, such as in TWT (traveling wave tubes) and in magnetrons.

Most applications for magnetic materials at DC involve the production of a DC magnetic field by means of an electromagnet or a permanent magnet. While the two differ in materials used, the overall construction is about the same. The source of the effect is a circulating charge producing a current flow. The field is produced either by electron flow or current through a coil in an electromagnet or, in the case of the permanent magnet, by the circulating electron charge in the atoms or ions present. While the permanent magnet requires no external energy to produce the magnetic field and is smaller, the electromagnet has the advantage of controllability.

An application that also operates at DC (as well as ac) is one in which the magnetic material does not generate a magnetic field but, rather provides shielding against one. Ironically, the only method of shielding a sensitive device from an external magnetic field is with a magnetic material itself. Shielding is one of the few magnetic applications where the magnetic material is not wire-wound to excite it. Here, the purpose is to interact with the external magnetic field and prevent it from penetrating the internal space where it could interfere with the device operation.

APPLICATIONS-FUNCTIONS-FERROMAGNETIC MATERIALS

Devices using shielding are cathode ray tubes, color TV tubes, sensitive instruments, and transformers. In addition, whole rooms can be shielded for elimination of the earth's magnetic field effect in critical experiments.

POWER APPLICATIONS AT LOW FREQUENCIES

Low frequency power generation and conditioning are probably the applications that consume the largest tonnage of magnetic materials but not the largest dollar volume. This application is mostly involved in the production and distribution of 50 and 60 Hertz line or mains frequencies. The magnetic equipment consists mainly of generators, motors and transformers. As we shall see, cost is the main consideration although in an effort to obtain lower losses, there is a range of properties and costs within the overall product range. For very large generators, there is a compromise between higher operating efficiency and the initial cost of the magnetic material, which is a very large part of the generator cost. For smaller motors and appliances aimed at consumer applications, cost is the determining factor at the expense of efficiency. The types of applications in this category include;

1. Large rotating machines- utility generators and very large motors.
2. General use motors-industrial and commercial uses.
3. Small consumer use motors and intermittent service motors
4. Large power transformers
5. Distribution transformers
6. Reactors and magnetic amplifiers
7. Current and Potential transformers
8. Audio transformers
9. Ballasts for fluorescent lighting
10. Welding transformers
11. Magnetic switch cores

We have said that since DC was the first type of electric current discovered, the first practical motors were DC motors. However, when AC current was made available, it was soon discovered that AC. motors were more efficient and easier to construct. As a matter of fact, this was one of the reasons for the early acceptance of AC power. The specific rpm values of AC motors have become standard. On the other hand, DC motors do offer simple electrical speed control.

Another rotating machine application often lumped with 50-60 Hz. line frequency power is the 400 Hz. special equipment most frequently used in airborne applications. The increased frequency is used to reduce the volume and weight so obviously important in this case. Motors and generators for 400 Hz. use special, more costly materials that will be described later. In some deep submergence applications, 800 Hz. power has also been used whereas somewhat higher frequencies (2000-3000 Hz.) power applications have been used in induction melting apparatus or in high frequency fluorescent lighting. The higher frequency can be obtained from an initial 50-60 Hz. source followed by the use of non-magnetic devices such as solid state frequency multipliers.

6   HANDBOOK OF MODERN FERROMAGNETIC MATERIALS

LOW FREQUENCY TRANSFORMERS

Large Power Transformers
After the 50-60 Hz. power is generated, it is usually transmitted at high voltage (for reasons of economy) especially when the distances of transmission are very long. As the need for local power to cities and neighborhoods is required, the high voltage is transformed to an intermediate voltage level in utility substations and transformer banks. From these substations, overhead utility lines then carry the power to the eventual consumers.

Distribution Transformers
Before the power is brought into the homes, the voltages must be reduced for the sake of safety to lower levels (110 V. in the U.S. and 220V. in many European countries). Voltages for industrial facilities are usually somewhat higher (440V.) This change to lower voltages involves the use of a distribution transformer. These are the transformers often seen on utility poles along city streets and rural roads. Because this application has such a common and widespread usage, it is probably the single largest consumer of magnetic materials in terms of tonnage.

Other Low Frequency Transformers
In addition to the distribution transformers, many electrical and electronic devices in the home as well as in industry require voltages other than that brought in from the line or mains. Hence, there is a large demand for small consumer-oriented inexpensive transformers. Often, these transformers are built into the device itself. In addition to transformation of voltages and currents, transformers are also used without any transformation for isolation of the device from the mains. Others provide constant voltages for sensitive instruments. Safety and protection of sensitive circuits are purposes of the latter devices.

DC Power Supply Applications
In many electronic devices, especially those using transistor circuitry, there is a need for a well-regulated DC power at moderately low voltages (5-15 V.). There are two main types of devices for this purpose. They are known as the linear power supply and the switching power supply. The linear power supply consists of a 50-60 Hz. transformer, a rectifier to convert to DC and an output choke to reduce the residual ac ripple present. Both the transformer and the inductor portion of the choke involve magnetic materials of the low frequency variety. The other type of power supply, namely, the switching power supply, converts the 50-60 Hz. ac to a high frequency square wave through the use of a transistor or similar solid state switching device, transforms it to the desired voltage at high frequency. This lowered ac voltage is then rectified and the ac ripple removed. Since the transformer and choke operate at the high frequency, magnetic materials other than those of used in the

linear power supplies are needed. This subject will be discussed later in the section on high frequency magnetic materials.

Ground Fault Interrupter (GFI) Applications
Another application recently developed for use in the 50-60 Hz. frequency range is the ground fault interrupter (GFI) sometimes called the earth fault interrupter. This device is installed strictly for safety considerations. The magnetic core here functions as a flux sensor and amplifier in contrast to the previously cited magnetic functions. The magnetic flux that is sensed is that caused by the difference in the currents flowing in the two lines or mains wires. Normally, the currents in these two wires are 180 degrees out of phase to each other so that the net flux produced by a magnetic core encircling both wires is zero. However, if a person accidentally comes into contact with one wire, there is a current leakage to ground creating a current and flux difference that is sensed in the magnetic core. The voltage induced by the flux created is amplified and can trip a circuit breaker before the person involved is harmed. At present, these GFI systems are required in the United States only in vulnerable areas such as swimming pools. However, there is a growing trend to mandate them in a large number of homes and industrial sites.

Power Control Devices
For the control of electrical and electronic equipment such as motors, furnaces and such, it is necessary to be able to regulate the power input. At present, much of this control is done by SCR(silicon control rectifiers) and similar solid state devices. Before the advent of these devices, the control of industrial power was accomplished by saturable core reactors and magnetic amplifiers, both devices using magnetic materials. In some special cases, these latter magnetic control methods are still used and require special types of magnetic materials.

In applications using very high currents such as welding power supplies, there is a need for special type of transformer capable of handling these currents. Another control application is that of a protective choke to absorb the high currents experienced when a large piece of electrical equipment is started. If this surge is not suppressed, the very costly solid state control equipment could be severely damaged or destroyed.

For the control of power to 50-60 Hz. fluorescent lamps, electronic ballasts are required for current control at start up. Again, when the excitation is of higher frequency as in portable fluorescent lamps, a special high frequency material is used.

Miscellaneous Line Frequency Devices
To complete the types of electrical equipment using primarily line or mains frequencies (50-60 Hz.), there is a whole series of consumer devices. These include the electrically activated switches, latches, relays, solenoids and such. Here as in other consumer products, the primary concern is price with little attention given to elec-

trical losses and product life expectancy. Thus, the magnetic materials used are often the ones with the lowest costs.

Audio Frequency Applications
Moving up from the line frequency applications that we have been discussing, the next higher frequency range is that of the audio frequencies (20-20,000 Hz.). As expected, these involve devices using voice and music signals that, in turn, are transmitted at much higher carrier frequencies. However, transformers, microphones, speakers and other audio-processing equipment operate at these audio frequencies. Here again, the consumer-market for these products dictate moderately low prices although there are professional recording systems that may require premium components and high quality magnetic materials. The increased frequency range of these devices over line frequencies requires correspondingly improved materials.

Telecommunications Applications
The next group of applications involves the primary electronic operations at much higher frequencies (100 KHz.- 100 MHz.). It includes the areas of telephony, radio and television. Although the latter two are consumer products subject to the economic restraints previously mentioned, the area of telephony is somewhat different as much of the equipment (other than some of the phones) is owned and operated by the telephone companies. Therefore, the importance of quality, efficiency and life expectancy of the devices becomes more important and as a result, higher quality magnetic materials with lower energy losses are used.

Telephony Applications-Aside from some of the permanent magnet materials used in the telephone receiver, there are many soft magnetic materials used in telephony. In addition to the previously described transformer usage, another component is introduced, namely, the inductor. The inductor functions, often in conjunction with a capacitor, to shift the phase of an electrical signal. By combining the two actions, devices called filters can selectively pass certain frequencies while blocking others. Some of the functions of magnetic materials in telephony can be characterized as;

1. Channel filters
2. Wide band transformers
3. Loading Coils
4. Touch-tone generator

Channel Filters-In the first of the telephony applications, the channel filter, the inductor is used in an L-C resonant tuned circuit. Using a series of these tuned circuits, the carrier frequency range can be separated into many different frequency bands, each of which can be used to transmit a separate telephone call. The bandwidth of each call is relatively small representing the voice frequency band. Using

APPLICATIONS-FUNCTIONS-FERROMAGNETIC MATERIALS

this arrangement, many different telephone calls can be made over the same transmission line. The carrier frequency can be several hundred KHz so that the magnetic materials used must be appropriate for those frequencies. An important attribute of the magnetic component for this application is its stability as a function of temperature and time.

Wide Band Transformers-In transmission lines carrying telephone calls, a wide range of frequencies must be passed. The high frequency transformers used in conjunction with these transmission lines must be capable of passing this wide band of frequencies while rejecting those outside the limits. Thus, these transformers are known as wide band transformers.

Loading Coils-When telephone signals are carried over long distances as found in many rural areas, a special problem is encountered. Although the inductance remains the same, due to the insulation of the wire, the capacitance of the circuit increases with the length of the wire. This change would tend to detune the circuit or change the frequency. Since, as we have said earlier, the frequency stability is a prime requisite in telephony and some corrective action is required. The ratio of inductance to capacitance must stay the same. Therefore, the inductance must increase correspondingly. This adjustment is made by passing the wire through special inductance coils at predetermined distances. These inductors are called load coils or Pupin coils.

Touch Tone Generators-The use of Touch-tone dialing is quite familiar to us. Each of the digits in the dialing sequence is represented by a separate audio tone or frequency that is decoded at the central office and routed to the appropriate designee. The tone or frequency for each digit is generated in the Touch-tone headset by a specific L-C resonant circuit. The inductor is usually a wound ferrite with multiple windings so various combinations can produce different frequencies or tones. In this manner, two Touch- tone cores can produce the tones for all the needed digits.

ENTERTAINMENT APPLICATIONS
A very large tonnage of magnetic materials goes into the consumer entertainment market. In the radio and television applications, some of the functions for magnetic cores are as follows;

1. Television picture tube yokes
2. Fyback transformers
3. Power transformers
4. Interstage transformers
5. Pin Cushion transformers
6. Radio and television antennae
7. Tuning slugs

Television Picture Tube Yokes

This application probably uses the highest tonnage of magnetic material for the entertainment segment of the business. The yokes are funnel-shaped rings placed on the neck of the television picture tube. After being wound, their function is to provide the horizontal and vertical deflection of the electron beam that forms the raster on which is superimposed the television signal. The horizontal sweep is produced by a triangular waveform which steadily deflects the beam across the screen and then very rapidly returns (flies back) to the initial point on the next line. This action is repeated until the total screen is covered with the rate of screen change such as the eye sees it as continuous motion (greater than 16 frames per second). The detail per raster line is generated in the horizontal deflection that, in synchrony with the vertical deflection drive, produces the picture per screen

Flyback Transformers

During the flyback period of the horizontal deflection cycle, the large magnetic field stored in the deflection core is rapidly collapsed and the voltage induced is transferred to a single turn primary winding of the flyback transformer. The secondary of this transformer contains thousands of turns producing a very high voltage (about 25,000 V.) which is then placed on the accelerator anodes of the electron gun that, in turn, propels the electron beam. In this scheme, the stored up energy of the horizontal deflection system is recaptured.

Power Transformers

The Power transformers in the system can be in the mains transformer which, since they operate at 50-60 Hz. are of the variety of small transformers described above. Other power transformers may refer to the audio output transformers that operate at the higher frequencies and thus contain the appropriate magnetic materials.

Interstage and Pincushion Transformers

The interstage transformers in both audio and television circuits are used to couple different stages with regard to isolation and impedance matching. The pincushion transformer of the video circuit is used to correct the spherical aberration resulting from the use of a radial or circular sweep on a planar television picture tube. If uncorrected, the linear sweep speed would be much different at the center of the screen than at the ends thus creating a distorted picture.

Antennae for Radio and Television

The wavelengths associated with radio and television are relatively large. To match these wavelengths, the antennae would also be quite large. However, since magnetic materials have the ability to concentrate the received signal or electromagnetic wave by very large factors, antennae made of magnetic material can be quite small. This factor is especially important in small portable radio or television sets.

Tuning Slugs
In the tuner portion of a television set, each channel can be fine-tuned to the proper frequency by adjusting the inductance of a wound coil into which a threaded slug of magnetic material is inserted.

HIGH FREQUENCY POWER SUPPLIES
We have spoken of power supplies for line or mains frequencies and for linear DC power supplies that also operate at line frequencies. In the past twenty years, a new type of power supply has been developed using transistor switching to produce a high frequency square wave. Transformation is done at the high frequency and so, as we shall see later, this reduces the size of the magnetic component drastically and improves the efficiency of the device. These power supplies, known as switched-mode power supplies (SMPS), have been used extensively for providing the DC needed for the bias voltage of semiconductors (transistor and IC) in computers, microprocessors as well as many types of recording devices.

Pulse Transformers
In some cases, the transistors that act as a switch to invert or form the high frequency square wave are free-running and do not require any timing mechanism. However, some transistors must be triggered by pulses that are usually generated by pulse transformers. These usually are small toroidal (or ring) magnetic cores with the ability to perform the rapid voltage rise with controllable voltage waveforms.

SMPS Power Transformers
The power transformers that transform the high frequency input voltage to the usable voltage is at the heart of the SMPS system. It is the device that, because of the efficiency of high frequency transformation, has been able to reduce the size of the large linear 60-Hz. Transformer to a very much smaller one. The material must be able to carry the high power at the high frequency and usually at a higher than ambient temperature.

Switching Regulators
The output of the switching power supply must have very controllable voltage limits or a good degree of regulation. To accomplish this, a device called a switching regulator is used. This involves sampling of the output voltage, comparing it with a known voltage, detecting the difference and feeding back to change the on-off time of the transistor and correct the voltage. A magnetic core is included in the regulator.

Output Chokes

The output DC of the inverter power supply contains a certain degree of unwanted ripple or residual ac., A high frequency output choke is used to remove this ripple. This device is similar to the output choke for mains or linear power supplies. However, the problems with the high frequency ripple do require that special magnetic materials be used.

EMI Applications

In recent years, there has been a proliferation of many advanced systems producing digital electronic signals transmitted either through wires or wireless. To protect these circuits from electromagnetic interference (EMI), special legislation has been introduced in many countries. This legislation would require both the manufacturers of devices producing the EMI as well as those devices sensitive to the EMI to provide appropriate means of minimizing the EMI effect. For many applications such as computer power supplies, the input must be protected against sudden spikes or interference. The presence of these spikes could be very harmful to the information being gathered in the digital circuits employed. To eliminate these input spikes, the power supply is often protected by input or noise filters. There are two types of such devices. The first is the common mode filter in which both of the input or mains wires are encircled by a magnetic core. Normally, the currents are 180 degrees out of phase to each other and pass through unaffected. However, when a sharp spike appears in one wire, the sudden current surge produces a strong magnetic field that is dissipated slowly in the magnetic core. Thus, the core acts as an input protective choke.

In the other type of filter called the in-line filter, there is a magnetic core surrounding each of the input wires. Because each line carries the entire amount of current (as opposed to the difference of the two currents in the common mode type), the type of component and the magnetic material used is expected to be different.

Although the input frequency for the noise filter is line or mains (50-60 Hz.), the frequency or rise time of the spike is consistent with the high frequency range we are discussing here and the materials employed must be appropriate for that frequency. Therefore, this subject is included in the section on high frequency materials.

The previous sections outline the means of protecting the input of the sensitive devices. In addition, the legislation in effect and proposed in many countries mandate that the producer of the noise shall also limit the noise being generated by this device. In most cases, the protective means involves a magnetic core. This is a very fast growing area and will increase more rapidly as new legislation becomes effective.

MICROWAVE APPLICATIONS

At very high frequencies (above 1 GHz.) electrical energy can no longer be transmitted through wires. Instead, it is radiated by means of electromagnetic waves

APPLICATIONS-FUNCTIONS-FERROMAGNETIC MATERIALS 13

similar to light. As such, the common ways of processing electrical energy at lower frequencies are not applicable. Means of controlling the microwave radiation can still be found using magnetic components although the mechanism, technology and materials are quite different. New devices called Faraday rotators, circulators and phase shifters can interact with microwave radiation and perform useful functions. Microwave equipment is used in radar, aircraft and satellite guidance and in space communication systems.

MAGNETIC RECORDING APPLICATIONS

Because of the very large and growing market for all types of magnetic recording and because of the relatively high material cost of the magnetic materials relative to the low frequency power materials for motors and transformers, this application is undoubtedly the highest dollar volume segment. The various components and materials in the broad spectrum of magnetic recording can be classified as follows;

1. Magnetic memory cores
2. Magnetic audio tape
3. Floppy disks
4. Hard disks
5. Video cassette tape
6. Magnetic ink for credit cards
7. Magnetic media
8. Magnetic recording heads
9. Rotary transformers
10. Copier powders
11. Bubble memories
12. Magneto-optic memory disks

Magnetic Memory Cores

The earliest mass memory banks were composed of matrices of small ferrite memory cores with each core representing a logical memory bit. The cores were addressed by selectively sending current pulses through two orthogonal systems of wires threading the cores. This put certain cores into logical "1" conditions while others remained in the "0" state. The cores could be read by sending a reverse pulse through the wires and noting those which experienced a flux change or voltage. This system worked well for many years but is virtually extinct now having been replaced by semiconductor memories, tape systems, cassettes and disks of various types.

Magnetic Recording Media-Tapes, Disks, Drums

As mentioned in the previous section, the original ferrite memory cores have been replaced by new systems. Aside from the semiconductor memories, most of the new systems are magnetic in nature. Most modern magnetic memories are based on small regions or islands of magnetic media that can be magnetized into digital bits for writing and then returned to their initial state for reading the stored bits. Obviously this technology provides more compact memories at low cost and easy access. The media can be particulate (metal or oxide) or in the form of thin films. The same type of media can be used for audio, video or digital computer applications although the associated hardware may be quite different.

Magnetic Recording Heads

Magnetic recording heads are used to write or read magnetic information stored in the media. The overall construction of both types of heads is similar with only the associated electronics varying. Heads can be metallic, ceramic or thin film. Video recording heads present an additional challenge in construction and operation because of the large amount of simultaneous information stored.

Rotary Transformers

As we have said previously, video recording requires special consideration. To read the large amount of data on a video tape, a rotating magnetic head is used. To transfer the data from the head to the ensuing electronic system, a rotary transformer is used. In this case, the part of the core receiving the data moves and the secondary part of the transformer is stationary.

Copier Powders

While a copier is not strictly a magnetic recording device, it is certainly used to store hard copies of the output of computers. This is a case where the only attribute to the carrier powder in xerographic copiers is that it be magnetic. However, in recent years, the carrier powder and the toner powder have been combined into one material. Some magnetic oxide powders fill both functions and therefore can be used as a single powder material

Bubble Memories

These magnetic memories were developed several years ago using thin ceramic magnetic films on which small regions of reverse single domains could be generated, stored and later read by noting their presence or absence in a designated area. While some commercial systems were based on this concept, its application has all but disappeared at present.

MISCELLANEOUS APPLICATIONS

In addition to the applications mentioned in the previous sections, there are some that do not fit into any of the former categories. These applications include;

1. Magnetic transducers-magnetostrictive

APPLICATIONS-FUNCTIONS-FERROMAGNETIC MATERIALS 15

2. Delay lines
3. Reed Switches
4. Electrodes
5. Sensors
6. Ferrofluids
7. Anechoic Chamber Tiles

SUMMARY

We have shown the various applications and functions of magnetic materials in general. Before we can discuss the means of selection of the proper material for each case, we should review the material and component properties insofar as the magnetic and electrical performance is concerned. The next chapter presents the pertinent measured properties of magnetic materials and their associated units.

2 BASICS OF MAGNETISM- SOURCE OF MAGNETIC EFFECT

INTRODUCTION

This chapter introduces the reader to some of the fundamentals of magnetism and the derivation of magnetic units from a physico-mathematical basis. Next, we apply these units to quantify the intrinsic magnetic properties of electrons, atoms and ions. In later chapters, we extend these properties to crystals, and finally, to bulk material. These properties are intrinsic, that is, they depend only upon the chemistry and crystal structure at a particular temperature. Following this examination of intrinsic properties, we will discuss those which in addition to the above, depend upon such characteristics as stress, grain structure and porosity. Finally, we correlate the previously defined units to functional magnetic parameters under dynamic conditions such as those used in electrical devices. At first, the magnetic units are derived primarily from the cgs system that is the more conventional one for basic magnetic properties. When the emphasis is shifted to component and application consideration, both cgs and meter-kilogram-second-ampere (mksa) (SI) units are used.

MAGNETIC FIELDS

A magnetic field is a force field similar to gravitational and electrical fields. That is, surrounding a source of potential, there is a contoured sphere of influence or field. In the case of gravitation, the source of potential is a mass. For electrical fields, the source is a positive or negative electrical charge. Fields (magnetic or otherwise) can be detected only by the use of a probe, which is usually another source of that type of potential. The criterion that is used is the measurement of a force, either repulsive or attractive, that is experienced by the probe under the influence o the field. For gravitation, where the interaction is always attractive, the governing equation is:

$$F = G \times mass(1) \times mass(2) / r^2 \quad [2\text{-}1]$$

where: F = force (in newtons)
G = constant = 6.67×10^{-7} nt-m^2/Kgm^2
$mass(1) \& mass(2)$ = masses (in kg)
r = distance between masses (in meters)

In the case of an electrical field, the corresponding equation is:

$$F = K \times q_1 q_2 / r^2 \quad [2\text{-}2]$$

Where: q_1, q_2 = electric charges (Cs)
K = Electrostatic constant
= 9×10^9 nt-m^2/(C)^2

r = distance between charges (in meters)

The force is repulsive if the two charges are of the same sign and attractive if the signs are different.

Early workers examining magnetic fields found that the origin of the magnetic effect appeared to originate near the ends of the magnets. These sources of magnetic potential are known as magnetic poles. For the magnetic field, there is one main difference compared to the other types of fields. In the gravitational or electrical analogs, the potential producing entities, mass or q, can exist separately. Thus, positive or negative electrical charge can be accumulated separately. In the magnetic case, the two types of magnetic field-producing species appear to be coupled together as a dipole. Thus far, we have not detected isolated magnetic monopoles.

THE CONCEPT OF MAGNETIC POLES

The poles concept was originated a long time ago when the only method of studying magnetic phenomena was based on the interaction of permanent magnets. Although our theories have become much more refined since then, the pole concept is still a useful device in discussions and calculations on ferromagnetism. Poles are fictitious points near the end of a magnet where one might consider all the magnetic forces on the magnet to be concentrated. The strength of a pole is determined by the force exerted on it by another pole. In 1750, John Mitchell measured the forces between magnets and found, for example that the attraction or repulsion decreased in proportion to the squares of the distances between the poles of two magnets. Similar to the gravitational and electrical examples, the force is given by:

$$F = K' m_1 m_2 / r^2 \qquad [2\text{-}3]$$

Where; m_1, m_2 = strengths of the two poles
K' = a constant which has the value of:
= 1 in the cgs system
= 1/4 μ_o in the MKSA system
where $\mu_o = 4 \times 10^{-7}$ henries/m
μ_o = permeability of vacuum

A unit pole (in the cgs system) is defined as one that exerts a force of 1 dyne on a similar unit pole 1 cm away. The force is repulsive if the poles are alike or attractive if they are unlike. Around each pole is a region where it can exert a force on another pole. We call this region the magnetic field. Each point in a magnetic field is described by a field strength or intensity and a field direction which varies with location with respect to the poles. A visualization of the field directions can be made if iron filings are sprinkled on a sheet of paper covering a magnet. The lines indicate the changing directions of the field emanating from the poles. The direction is also that to which a North-seeking end of a compass needle placed at that spot would point. The field strength can be visualized by the density of the lines in any one particular area. The density should fall off according to the inverse square of the distance from the poles as predicted.

BASICS OF MAGNETISM – SOURCE OF MAGNETIC EFFECT

The polarity of the magnet itself must be defined, the assignment being such that the North-seeking pole is the North pole of the magnet. Since opposite poles attract, the north- seeking pole of the magnet is actually the same kind of pole as the South pole of the planet. In other words, the north magnetic pole of the planet is the opposite kind of pole from the North pole of all other physical objects with magnetic properties. The absolute direction of a magnetic field outside of a magnet (represented schematically by arrows in Figure 2.1, for example) is from the north pole to the south pole. Since lines of magnetic field must be continuous, the direction of the field inside the magnet is from south to north poles. The unit of magnetic field intensity called the oersted is defined as that field located 1 cm. from a unit pole. The magnetic field intensity can also be defined in terms of current flowing through a wire loop. In the MKSA system of units, the unit of field strength is the ampere-turn per meter, which then relates the magnetic field to this current flow.

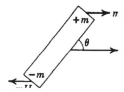

Figure 2.1- Forces acting on a magnet at an angle,θ, to a uniform magnetic field (Chikazumi,1964)

When a magnet of pole strength m, is brought into an external magnetic field (such as that produced by another magnet, the force acting on each pole is given by:

$$F = m H \qquad [2\text{-}4]$$

where m = pole strength, (emu or electromagnetic units)
H = magnetic field strength (oersteds)
  = m_2/kr^2

When a magnetic dipole such as a bar magnet is placed in a uniform magnetic field at an angle, θ , each pole is acted on by forces indicated by Figure 2.1. The result is a couple whose torque is;

$$L = m/H \sin \theta \qquad [2\text{-}5]$$

where: L= Torque
  l = distance between the poles (cm)
  θ = angle between the direction of the magnetic field
    and the axis between the poles (direction of magnetization)

This torque will tend to rotate the magnet clockwise. By measurement of the torque and the angle, θ, we can determine the field strength.

If the axis of a bar magnet is parallel to a uniform field, no force will act on it because the force on one pole will cancel the force on the other. However, a force will result if the field is non-uniform because of the difference in forces experienced by the individual poles. The force is:

$$F_x = ml\, dH/dx \qquad [2\text{-}6]$$

where: F_x = Force in the x direction
dH/dx = Change in the magnetic field per centimeter in the x direction

Figure 2.2 shows this action. The lengths of the arrows represent the field strengths at the two poles and also the difference in forces it creates. In addition to the translational force on the magnet due to its position in a non-uniform field, the magnet will also experience a rotational torque described above if the magnet is at angle to

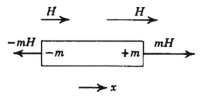

Figure 2.2- Forces acting on a magnet in a non-uniform magnetic field (Chikazumi, 1964)

the external field. Because of the dipolar nature and the combined action of the two poles, any force produced by the magnet in a field is proportional to the term, ml. This is called the magnetic moment that is equivalent to a mechanical moment. In magnetic materials, we are not as much concerned with m or l but with the product, ml, which is a measurable parameter as it was with the magnets. We will call this moment, μ, not to be confused with the permeability, μ (large μ) to be defined later.

To express this property as a material characteristic, we are interested in the magnetic moment per unit volume or the intensity of magnetization. Alternately, this parameter can be called the magnetic polarization or frequently, we shall just refer to it simply as the magnetization, M. The magnetization is given by

$$M = ml/V = \mu/V \qquad [2\text{-}7]$$

where: V = Volume (cm^3)

This definition is important in describing the basic material property that is distinctly separate from the magnetic circuit. When very precise research is conducted, the magnetic moment per unit weight is often used to avoid the problem of density variations with varying temperature or porosity (due to processing condition). In this case, the term is σ, which is the moment per gram. The corresponding M or moment per volume is obtained by multiplying by the density, .

$$M = d\sigma \qquad [2\text{-}8]$$

where ; σ = moment /gm or emu/gm
d = density, gm/cm^3

BASICS OF MAGNETISM – SOURCE OF MAGNETIC EFFECT

It is easy to show that M is also equal to the number of poles per cross sectional area of the magnet

$$M = ml/V = ml/Al \qquad [2\text{-}9]$$
$$M = m/A \qquad [2\text{-}10]$$
where ; A = Cross sectional area (cm^2)

As we shall see later, M can be measured relative to a material (powder, chunk, etc.) or in some cases, electrically relative to a magnetic core. The importance of this alternate definition will become more evident in later chapters when the magnetic circuit is discussed in terms of magnetic flux density of which M is a contributing (often a major) factor.

The magnetization, M, (sometimes called the magnetic polarization) has cgs units called emu/cm^3 or often just electro-magnetic units (emu). The MKSA unit for the magnetization is the Tesla or weber/m^2. There are 796 emu/cm^3 per Tesla or weber/m^2.

ELECTROMAGNETISM

The real beginning of modern magnetism as we know it today began in 1819 when Hans Oersted discovered that a compass needle was deflected perpendicular to a current bearing wire when the two were placed close to one another. It was at this point that electromagnetism was born. Next, Michael Faraday (1791-1867) discovered the opposite effect, namely that an electric voltage can be produced when a conducting wire cut a magnetic field.

ATOMIC MAGNETISM

The work on electromagnetism in the early 1800's clarified the relationship between magnetic forces and electric currents in wires, but did little to explain magnetism in matter, which was the older problem. The theories of that time had assumed that one or more fluids were present in magnetic substances with some separation occurred at the poles when the material was magnetized. In 1845, Faraday discovered that all substances were magnetic to some degree. Paramagnetic substances were weakly attracted, diamagnetic substances were weakly repelled and ferromagnetics were strongly attracted. The French physicist, P. Curie (1895), today best known for his work on of radioactivity, measured the paramagnetism and diamagnetism in a great number of substances and showed how these properties varied with temperature.

Nineteenth-century scientists were still looking for the link between electromagnetism and atomic magnetism. In considering the similarity between magnets and current circuits, Andre Ampere(1775-1836) suggested the existence of small molecular currents which would, in effect make each atom or molecule an individual permanent magnet. These atomic magnets would be pointed in all directions, but would arrange themselves in a line when they were placed in a magnetic field. The expression "Amperian currents" is still used today. The search for a source of these molecular currents ended with the discovery of the electron at the close of the 19th century and reported by J. J.Thompson (1903). By 1905, There was general agreement that the molecular currents responsible for the magnetism in matter were due to electrons circulating in the molecules or atoms.

Bohr Theory of Magnetism

In 1913, Niels Bohr (1885-1962) described the quantum theory of matter to account for many of the effects that physicists of the day could not explain. In this theory, the electrons were said to revolve about the nucleus of an atom in orbits, similar to those of the planets around the sun. The magnetic behavior of an atom was considered to be the result of the orbital motion of the electrons, an effect similar to a current flowing in a wire loop. The motion of the electrons could be described in fundamental units so that the magnetic moment accompanying the orbital moment could also be described. The basic unit of electron magnetism is called the Bohr magneton. Not only a fundamental electric charge but also a magnetic quantity is connected with the electron.

As described by classical in Glasstone (1946), the magnetic moment, μ, resulting from an electron rotating in its orbit can be given by;

$$\mu = ep/2mc \qquad [2-11]$$

Where e = electronic charge of the electron (C)
p = total angular momentum of the electron
m = mass of the electron, g
c = speed of light, cm/s

In the Bohr theory, the orbital angular momentum is quantized in units of $h/2\pi$ (where h is Planck's constant). Therefore, for the Bohr orbit nearest to the nucleus, the orbital angular momentum, p, can be replaced by $h/2\pi$. The resulting magnetic moment can be expressed as:

$$\mu = eh/4\pi mc \qquad [2-12]$$

If we substitute for the known values and constants, we obtain

$$\mu_B = 9.27 \times 10^{-21} \text{ erg/Oersted} \qquad [2-13]$$

This constant, known as the Bohr magneton is the fundamental unit of magnetic moment in the Bohr theory. It is that the result of the orbital motion of one electron in the lowest orbit.

Orbital and Spin Moments and Magnetism

The old Bohr theory was deficient in many aspects and even with the Sommerfeld (1916) variation (the use of elliptical versus circular orbits) could not explain many things. In 1925, Goudsmit and Uhlenbeck (1926) postulated the electron spin. At bout that time, Heisenberg (1926) and Schrodinger(1929) developed wave mechanics which was much more successful in accounting for magnetic phenomena. In

BASICS OF MAGNETISM – SOURCE OF MAGNETIC EFFECT

quantum mechanics, the new source of magnetism is advanced-that of the spin of the electron on its own axis, similar to that of the earth. Since the electron contains electric charge, the spin leads to movement of this charge or electric current that will produce a magnetic moment. Both theoretically and experimentally, it has been found that the magnetic moment associated with the spin moment is almost identically equal to one Bohr magneton. The original equation for the Bohr magneton is changed slightly to include a term, g, known as the spectroscopic splitting factor This factor denotes a ratio between the mechanical angular momentum to magnetic moment. The value of g for pure spin moment is 2 while that for orbital moment is 1. However, the lowest orbital quantum number for orbital momentum is 1 (number of units of $h/2\pi$) whereas the quantum number associated with each electron spin is $\pm 1/2$. The new equation is:

$$\mu = g \times e \times n/2mc \quad (2) \quad [2\text{-}14]$$

where; for orbital moment (lowest state) $g=1$, $n=1$
for spin moment $g=2$, $n=1/2$

We can see why the orbital and spin moment both turn out to be equal to $1\mu_B$. We now have a universal unit of magnetic moment that accommodates both the orbital and spin moments of electrons. The Bohr magneton is that fundamental unit. We have originally defined the magnetic moment in connection with permanent magnets. The electron itself may well be called the smallest permanent magnet.

The net amount of magnetic moment of an atom or ion is the vector sum of the individual spin and orbital moments of the electrons in its outer shells. In gases and liquids, the orbital contribution to magnetism can be important, but in many solids, including those containing the magnetically-important transition metal elements, strong electric fields found in a crystalline structure destroy or quench the effect. Most magnetic materials are crystalline and therefore would be affected by this factor. In the great majority of the magnetic materials we will deal with (those involving the 3d electrons of transition metals), we will not be concerned with the orbital momentum except for small deviations of the g factor from 2. However, when we talk about the magnetic properties of the rare earths, we cannot ignore the orbital contribution. In these cases, the affected 4f electrons are not outermost. Consequently, they are screened from the electric fields by electrons of outer orbitals. This is not the case for the 3d electrons which are in the outermost shell . For the present, however, we will consider the magnetic behavior of most common magnetic materials to be entirely the result of spin moments.

Atomic and Ionic Moments
There are two modes of electron spin. Schematically, we can represent them as either clockwise or counter-clockwise. If the electron is spinning in a horizontal plane and counter-clockwise as viewed from above, the direction of the magnetic moment is directed up. If it is clockwise, the reverse is true. The direction of the moment is comparable to the direction of the magnetization (from S to N poles) of a permanent magnet to which the electron spin is equivalent. It is very common to

schematically represent the two types electron spin as arrows pointed up or down and we shall use this representation in our discussion. A counter-clockwise spin in an atom (arrow up) will cancel a clockwise spin (arrow up) and no net magnetic moment will result. It is only the unpaired spins that will give rise to a net magnetic moment.

In quantum mechanics, the atoms or ions are built up of electrons in orbitals similar to the Bohr orbits. These orbitals are also classified according to the shape of the spacial electronic probability density. This can be visualized as the superimposing of very many photographs of the electron at different times .The shape of the electron cloud that results is the shape of the orbital. For example, for s electrons, this shape is the surface of a sphere. Discrete energy levels are associated with each of these orbitals. As we construct the elements of higher atomic numbers, the higher positive nuclear charge will require more outer electrons. As these are outer electrons are fed in to form the atom, the added electrons go into the lowest unfilled energy levels. Figure 2.3 shows an example of an applicable energy level diagram. The electrons, like balls filling a stepped box would fill from the bottom up. Of interest to us in most magnetic materials, the 3d group of orbitals is especially

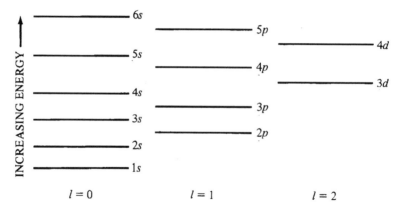

Figure 2.3- Schematic diagram of electronic energy levels

important. Each orbital is further divided into suborbitals each of which can accommodate one electron of each spin direction. The rules of quantum mechanics state when a 3d subshell is being filled, all the electron spins must be in the same direction (unpaired) until half of the subshell is filled at which time they can only enter in the opposite direction or paired. Figure 2.4 illustrates this manner in which the orbitals are filled using a convention previously described. The superscript indicates the number of electrons filling that orbital. The order of addition of subshells is generally from left to right with the exception that the $4s^2$ is added before the $3d^3$. Note that there are four unpaired electrons in the case of the iron atom. To form the Fe^{3+} ion from the iron atom, the two 4s electrons are removed first the one 3d electron giving rise to 5 unpaired electrons. In all these examples, unpaired electrons

BASICS OF MAGNETISM – SOURCE OF MAGNETIC EFFECT

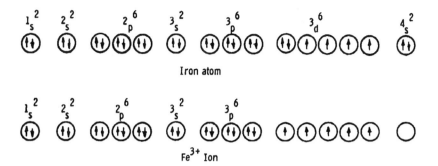

Figure 2.4 - Electronic configuration of atoms and ions

lead to a net magnetic moment. This classifies the atom or ion as paramagnetic, the degree being proportional to the number of unpaired electron spins. Each unpaired spin produced 1 Bohr Magneton as previously mentioned. Table 2.1 shows the number of unpaired directions and thus the number of Bohr magnetons for each element or ion. In compounds, ions and molecules, account must be taken of the electrons used for bonding or transferred in ionization. It is the number of unpaired electrons remaining after these processes occur that gives the net magnetic moment. The spin quantum number, S, has unit multiples if +1/2 or -1/2 depending on orientation. The orbital moment, L, has unit multiples of 1, 2, etc. The vector coupling between L and S is quantized as combined moment, J.

Table 2.1
Numbers of Unpaired Electrons and Bohr Magnetons in Atoms and Ions Involved in Ferro- and Ferrimagnetic Materials

ATOM	NUMBER OF UNPAIRED ELECTRONS (μ_B)
Fe	4
Co	3
Ni	2
ION	
Fe^{++}	4
Fe^{+++}	5
Co^{++}	3
Ni^{++}	2
Mn^{++}	5
Mg^{++}	0
Zn^{++}	0
Li^{+}	0

PARAMAGNETISM AND DIAMAGNETISM

If an atom has a net magnetic moment, (it is paramagnetic), this moment may be partially aligned in the direction of an applied magnetic field. Each atom therefore acts as an individual magnet in a field. The process of rotating these moments against thermal agitation is a difficult one and a large field is necessary to achieve only a small degree of alignment or magnetization.

In many paramagnetic materials such as in hydrated salts, as the temperature is raised, the thermal agitation of the spins reduces even this small amount of alignment. Pierre Curie showed that in these cases, the susceptibility, χ, which is defined as

$$\chi = M/H \qquad [2\text{-}15]$$

where ; χ = susceptibility
M = magnetization or moment, emu/cm^3
H = Magnetic field strength, Oersteds

follows the Curie Law given as;

$$\chi = C/T \qquad [2\text{-}16]$$

Where: C = Curie constant
T = Temperature in Degrees Kelvin

Also; $1/\chi = T/C \qquad [2\text{-}17]$

Figure 2.5 shows the temperature dependence of the inverse of the susceptibility in a paramagnetic. The slope of the line is then $1/C$.

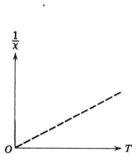

Figure 2.5- Variation of susceptibility of a paramagnetic material with temperature

Diamagnetism is an inherent property of the orbital motion of the individual electron in a field. Since it is even a weaker effect than paramagnetism, it is only observed when the atom does not have a net spin or orbital moment. The orbital mo

BASICS OF MAGNETISM –SOURCE OF MAGNETIC EFFECT

tion even though compensated sets up a field opposite to the applied field in a manner similar to the back emf of Lenz's Law. The effect leads to a negative susceptibity or the actual lowering of the net moment in the material as an external field is applied. Diamagnetism is so weak an effect that a small paramagnetic impurity can offer mask out the effect. Some paramagnetic and diamagnetic susceptibilities are given in Table 2.2. Many ionic materials such as salts (except for aforementioned transition and rare earth metals) are diamagnetic.

Table 2.2
Magnetic Susceptibilities of Some Diamagnetic and Paramagnetic Elements

ELEMENT	MAGNETIC SUSCEPTIBILITY (χ)
Sb	-7.0×10^{-5}
Bi	-1.7×10^{-4}
Cu	$-.94 \times 10^{-5}$
Pb	-1.7×10^{-5}
Ag	-2.6×10^{-5}
Al	$+0.21 \times 10^{-4}$
Nd	3.0×10^{-3}
Pd	8.2×10^{-4}
Pt	2.9×10^{-4}
O_2 (NTP)	17.9×10^{-7}

From: W.T. Scott, Physics of Electricity and Magnetism,
John Wiley, New York, 1959

PARAMAGMETISM OF METALS

The paramagnetic moment we have been discussing is derived from the unpaired electron spins of the free atom or ion. There are many materials for which this treatment for the atoms or ions under the influence of its neighbors (crystal) is valid. These materials include some insulators, rare earth metals and salts, and transition metal salts. These, as expected, would obey the Curie Law of temperature dependence. However, in many metals with unpaired spins such as the alkalis (lithium sodium, potassium) and the transition metals where the localized electron approach cannot be used. These are cases when the moment is derived from unpaired spins of electrons that are the outermost electrons or close to the outermost electrons. In a metal, the outermost electrons of the atom are conduction electrons and as such, do not belong to any atom but form an "electron gas". In the case of the 3d electrons, the electrons with unpaired spins are not outermost but in the shell directly below it. The energy levels of the 3d electrons can overlap with the 4s or outermost electrons and also become partially delocalized from the original atom. This "itinerant electron" model can best be explained by the band theory in which the electrons are distributed in separate bands of the 3d(+), 3d(-), 4s(+) and 4s(-) electrons. The numbers of electrons in each of these states will determine the resultant moment.

The Pauli paramagnets do not obey the Curie Law because the populations in the various states are determined by;

$$M = (3N\mu_B^2 H/2k_B T)(T/T_f) \quad [2.18]$$
$$\chi = M/H = 3N\mu_B^2/2k_B T_f \quad [2.19]$$

where N = number of atoms per unit volume
μ_B = magnetic moment
k_B = Boltzman constant
T_f = Temperature for $E_f = k_B T_f$

where E_f is the energy of the Fermi level which is a constant. Therefore, the paramagnetism of many metals is temperature independent. For the rare earth metals, the 4f electrons that contain the unpaired spins are deep inside the electron shell and are screened from the outermost electrons and thus they behave normally.

In addition to the failure of Pauli paramagnets to obey the Curie Law, they also show reduced paramagnetism from what would be expected by the localized electron model so they would be called "weak paramagnets" as opposed to the other "strong paramagnets". This reduction in moment will be very much more important practically when we examine the ferromagnetism of the 3d metals in subsequent sections.

Paramagnetism of Conduction Electrons

The localized electron theory does not explain the paramagnetic behavior of conduction electrons. In non-conduction electrons, the probability of the electron spin lining up with the magnetic field can be determined by its temperature dependent magnetization. In the electron gas approach there is almost zero probability of the spin alignment. Therefore only the fraction, T/T_f, (Equation 2.18) of the total number of electrons contribute to the susceptibility. This lack of temperature dependence refers to non-magnetic metals and alloys (except for rare earth metals, see below). Figure 2.6 shows the temperature dependencies of the susceptibilities some of the non-magnetic metals. The behavior according to energy bands is shown in Figure 2.7 from Kittel (1971)

Paramagnetism of Transition Metal and Rare Earth Ions

We have said that, in the transition metal elements, the magnetic moment is due almost entirely to electron spin contribution. On the other hand, in the case of the rare earths, there is indeed coupling of the orbital and spin moments to form the vector sum, J. The validity of this assumption is shown in Figure 2.8. The solid line shows the theoretical value based on the J moment and the circles show the experimentally measured values. On the other hand, the plot for the transition metal group in Figure 2.9(dotted line-not b) shows no such correlation with experiment (vertical lines with range). The experimentally measured values come closer to the spin-only, S, theories. Curve a is calculated for strong interaction between orbital moments and surrounding ions while b is calculated for weak interactions. The experimental values lie in between the two cases. In the case of the rare earth ions, the moments are generally larger because of the fact there are 14 electrons in the 4f shell compared to the 10 in the 3d shell and also because the additional orbital moment contribution.

FERROMAGNETISM

Both paramagnetism and diamagnetism are both very important in the study of atomic and molecular structure but these effects are very weak and have no real practical significance. Large scale magnetic effects resulting in commercially important materials occur in atoms (and ions) of only a few metallic elements

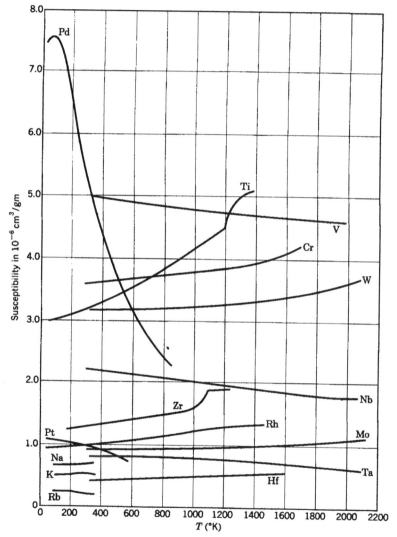

Figure 2.6- T dependencies of the susceptibilities of some non-magnetic metals From Kittel (1971))

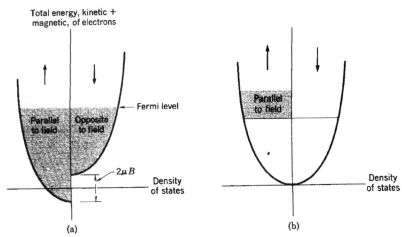

Figure 2.7- Energy bands of a Pauli paramagnet at 0 K ; a)without field b) with magnetic field. From Kittel (1971)

notably Fe, Co, Ni, and some of the rare earths. In alloys or oxides some materials containing these elements and some neighboring ions such as Mn, there is great enhancement of the atomic spin effect. This enhancement comes about from the cooperative interaction of large numbers (10^{13} - 10^{14}) of these atomic spins producing a region where all atomic spins within it are aligned parallel (positive exchange interaction). These materials are called ferromagnetic. An accepted explanation as to why these few elements show this behavior is found in the curve of Bethe (1933) shown in Figure 2.10 from Bozorth (1951). The plot shows the energy of magnetization as a function of the ratio of atomic separation in a crystal to the diameter of the unfilled 3d shell. There is a critical combination of these two factors to permit the positive (ferromagnetic) interaction to occur. Although manganese metal is not ferromagnetic, some alloys of manganese are ferromagnetic because the interatomic distance is changed to allow positive exchange interaction to take place.

The regions of the materials in which the cooperative effect extends are known as magnetic domains. P. Weiss(1907) first proposed the existence of magnetic domains to account for certain magnetic phenomena. He postulated the existence of a "molecular field" which produced the interaction aligning spins of neighboring atoms parallel. W. Heisenberg(1928) attributed this "molecular field" to quantum-mechanical exchange forces. Domains have been confirmed by many techniques and can be made visible by several means.

In ferromagnetic materials (as in paramagnetic materials), the alignment of magnetic moments in a magnetic field at higher temperature is decreased. Since a much greater degree of alignment occurs in ferromagnetics, the effect is even more pronounced. With further temperature increase, the thermal agitation will exceed the exchange forces and at a certain temperature called the Curie Point, ferromagnetism disappears. From complete alignment at 0°K to zero alignments at the Curie point a curve of reduced magnetization, M/M_o,(where M_o = Magnetization at 0°K) plotted against reduced temperature, T/T_c ,(where T_c = Curie point) follows a

BASICS OF MAGNETISM –SOURCE OF MAGNETIC EFFECT 31

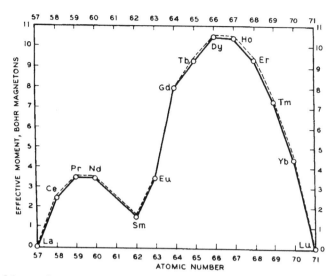

Figure 2.8-Measured moments of the rare earth elements compared with those calculated from spin + orbital moment contributions From Bozoth (1951)

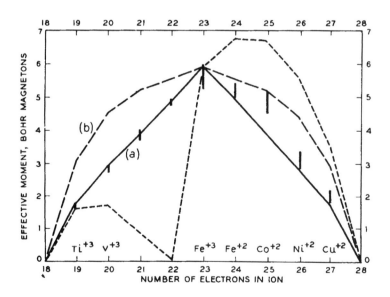

Figure 2.9- Measured moments of the transition metals versus those calculated by spin only and those using spin + orbital moment contributions. From Bozorth (1951)

32 HANDBOOK OF MODERN FERROMAGNETIC MATERIALS

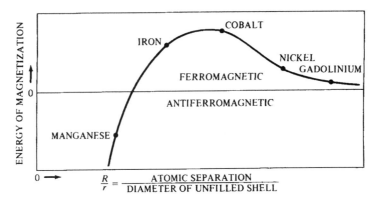

Figure 2.10- Energy of magnetization as a function of ratio of atomic diameter to diameter of unfilled shell. From Bozorth (1951)

similar pattern. Figure 2.11 shows such a universal curve. For the ferromagnetic metals such as Fe, Co, & Ni, the general curve holds fairly well.

Above the Curie point, the ferromagnetic material becomes paramagnetic, the susceptibility of which decreases with temperature. If the reciprocal susceptibility, $1/\chi$, is plotted against T, the curve obeys the Curie Weiss Law;

$$1/\chi = 1/[C\,(T - T_c)] \qquad [2\text{-}20]$$
where: C = Curie Weiss Constant
T_c = Curie Point

Figure 2.12 shows a typical plot.

MAGNETIC MOMENTS IN FERROMAGNETIC METALS AND ALLOYS

In our discussion of weak Pauli paramagnetism in some metals, we found that the paramagnetic moment was reduced from that predicted from the localized electron model due to the contribution of the itinerant electron gas. We excluded the magnetic metals from this group. They are ferromagnetic and their susceptibilities are many orders of magnitude greater than the paramagnetics. However, in considering the atomic moments of the magnetic metals, the band theory is still operative and the moment for each metal is still reduced to the same degree as that anticipated for the paramagnetic metals. In fact, the case for ferromagnetic metals is just an extension of the paramagnetic case. Thus, iron which by the localized electron approach should possess a magnetic moment of 4.0 μ_B, has actually only 2.2 μ_B. Stoner(1933) first proposed the band theory for ferromagnetic metals and it was later expanded by Slater(1936). In it, part of the electrons in the 4s band is transferred to the 3d band because of the overlap of the two bands. However, since the 3d+ band is already full, the excess electrons must go into the 3d- band reducing the net moment. This is shown in the case of Ni in Figure 2.13a, b & f. In a), the con-

BASICS OF MAGNETISM – SOURCE OF MAGNETIC EFFECT

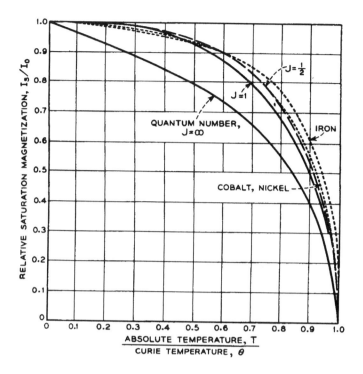

Figure 2.11- Universal magnetization curve showing reduced magnetization, M_s/M_0, or in this terminology, I_s/I_0, versus reduced temperature, T/T_c. Curves for Co, Ni and Fe are shown with some theoretically drawn curves. From Bozorth, Ferromagnetism, D. Van Nostrand, Princeton, 1951

figuration of the free atom is shown. In b) that of the nickel metal is shown with the reduce moment. In f) the tops of the filled levels are at the same height to show the water analogy of filled energy levels. Similar illustrations for other metals are also shown. The case of iron is somewhat different since neither 3d shell is full but the reduction of moment is still there. The distribution of the electrons in the 3d and 4s shells are shown in Figure 2.14a as a function of their distance from the nucleus. Note that with the 3d electrons, the density of states is dense and close to the nucleus. The 4s electrons on the other hand, are spread out rather thinly and extend far from the nucleus where they can overlap orbitals of other atoms. In Figure 2.14b, the levels of the 3d and 4s electrons are shown with the differences between the 3d+ and 3d- shells. Table 2.3 gives the numbers of electrons and holes (vacancies) in the various shells for the metals near iron in the periodic table. If the outer electrons of metallic iron could be localized, then the moment of iron could be increased. An attempt to do just that was made by the addition of nitrogen to make the material ionic and thus have no conduction electrons. Claims of "Giant Magnetic Moments" were made with values of 3.6 μ_B and higher but these have not been thoroughly substantiated.

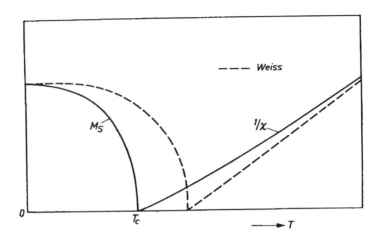

Figure 2.12-Temperature dependence of the saturation magnetization of a ferromagnetic and inverse of the susceptibility above the Curie point. From Smit and Wijn (1959)

Moments of Ferromagnetic Alloys
When a non-magnetic metal (Cu) is alloyed with a ferromagnetic metal (Ni), there is a type of dilution involved For each atom of Cu added per atom of Ni, the effect is the addition of one more electron. In seeking the lowest energy level, it goes into the nickel band rather than the original Cu atom. This would lower the moment since it would go into the 3d- band neutralizing part of the 3d+. Addition of copper linearly reduces the moment until the 3d- band is full. Since there is only 6.uB in the pure Ni, this point occurs at 60% substitution. For zinc with 2 4s electrons, there is twice the effect and have zero moment at 30%. Aluminum with 3 outer electrons has three times the effect and so on. Figure 2.15 show the effect of alloying nickel with metals up to 5 outer electrons with similar results. A theory by Slater (1937)states that in alloys with magnetic metals, the average moment depends only on the number of electrons per atom. When the atomic numbers are only 1 or 2 apart, this theory works well but not as well when they are 4-5 apart. Figure 2.16 shows the atomic moments of alloys of Fe, Ni and Co as a function of their average number of electrons per atom.

ANTIFERROMAGNETISM
In ferromagnetism, the interaction of atomic spin moments was a positive one meaning that the exchange interaction aligned neighboring spins parallel in a magnetic domain. In his study of the paramagnetic susceptibility of certain alloys, Néel (1932) noticed that they did not follow the Curie law at low temperatures but did obey the Curie-Weiss law at high temperatures;

BASICS OF MAGNETISM – SOURCE OF MAGNETIC EFFECT

$$\chi = C/(T+\theta) \quad [2.21]$$

where θ = Experimentally determined constant

Also; $\chi = C/(T - T_N) \quad [2.22]$

where T_N = Neel Temperature

where the extrapolation of the high temperature linear slope of $1/\chi$ vs T resulted in a negative value or a negative Curie point. To accommodate these findings, he postulated a negative exchange interaction aligned the neighboring spins antiparallel. At very low temperatures, the negative exchange force prevented the normal paramagnetic alignment in a field so that the susceptibility was low. As the temperature increased, however, the exchange interaction was weakened. Thus, as the negative exchange diminished, the susceptibility actually increased until a point called the Néel point where the negative interaction disappears. Now the spin system behaves as a paramagnetic with the expected Curie-Weiss law dependence. For a polycrystalline material, the $1/\chi$ versus T curve is shown in Figure 2.17 The negative exchange behavior of material of this type is called anti-ferromagnetism.

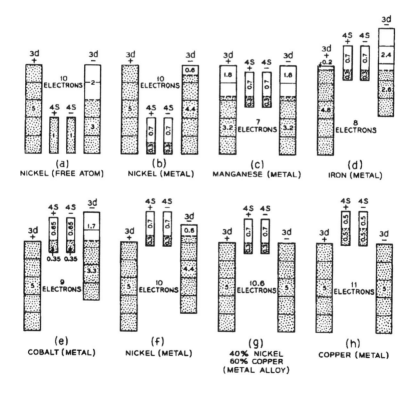

Figure 2.13- The distribution of the electrons in the 3d and 4s orbitals in several metals and an alloy. From Bozorth (1951)

Table 2.3
Numbers of electrons and holes(vacancies) in the various shells for the various metals near iron in the periodic table

Element	Number of Electrons in Following Shells:				Total	Holes in:		Excess Holes in $3d-$ over $3d+$
	$3d+$	$3d-$	$4s+$	$4s-$		$3d+$	$3d-$	
Cr	2.7	2.7	0.3	0.3	6	2.3	2.3	0
Mn	3.2	3.2	.3	.3	7	1.8	1.8	0
Fe	4.8	2.6	.3	.3	8	0.2	2.4	2.22
Co	5	3.3	.35	.35	9	0	1.7	1.71
Ni	5	4.4	.3	.3	10	0	0.6	0.60
Cu	5	5	.5	.5	11	0	0	0

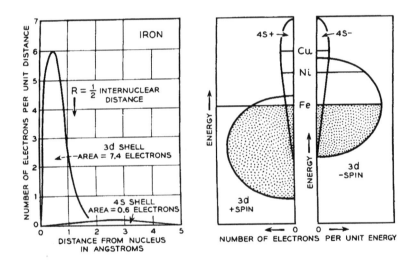

Figure 2.14- a) Distribution of the 3d and 4s electrons as a function of their distance from the nucleus b) Energy levels of the 3d and 4s electrons showing the difference between the 3d+ and 3d- shells. From Bozorth (1951)

Néel (1948) then became concerned with the magnetic behavior of oxides. Now, the magnetic ions in ferrites lie in the interstices of a close packed oxygen lattice. Because the distances between the metal ions are large, direct exchange between the metal ions is very weak. However, Kramers(1934) postulated a mechanism of exchange between metal ions through the intermediary oxygen ions. Néel combined his theory on antiferromagnetism with Kramers ideas on indirect exchange and formulated his new theory for antiferromagnetic oxides and later for ferrites. Later

BASICS OF MAGNETISM –SOURCE OF MAGNETIC EFFECT

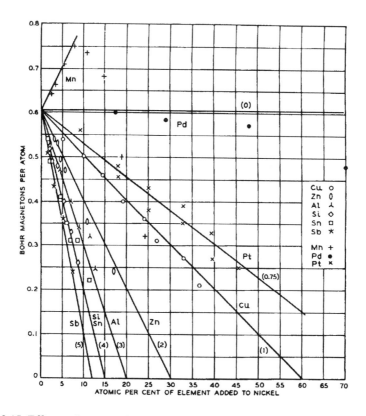

Figure 2.15- Effect on the magnetic moment of nickel on alloying with metals having up to 5 outer electrons. The results are similar in all cases. From Bozorth (1951)

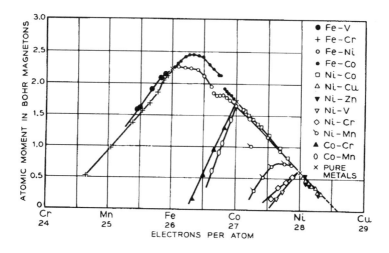

Figure 2.16- The atomic moments of Fe, Ni and Co as a function of their average number of electrons per atom. From Bozorth (1951)

(1950), Anderson put this theory on a mathematical basis and called it superexchange The mechanism assumed that one of the electrons in the oxygen ion could interact with or exchange with the unpaired electrons on one of the metal ions on what we call A sites. To be able to pair with the metal ion spin, the oxygen spin would have to be opposite to that on the metal ion. This would leave the other spin in the oxygen ion orbital free to pair with the unpaired spin of another metal ion preferably located opposite to the original metal ion. Since the

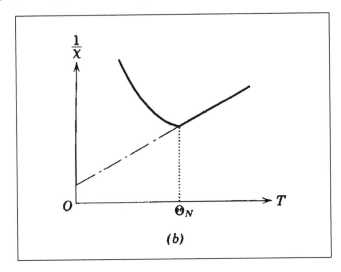

Figure 2.17-Reciprocal of the susceptibility of an antiferromagnetic material showing the discontinuity at the Néel temperature and the extrapolation of the linear portion to the "negative" Curie temperature. From Chikazumi (1959)

second spin of the oxygen ion suborbital is opposite to the first, it can only couple with a spin which is opposite to the original metal ion. This, then, is the reason for the stability of the antiparallel alignment of the two metal ions adjacent to the oxygen ion. Many antiferromagnetic substances are oxides, the classic case being MnO. The theoretical basis of antiferromagnetism was formulated by van Vleck (1941,1951) and Nagamiya (1955) presented an excellent review on the subject.

Zener (1951a,b) has proposed an alternative mechanism to superexchange called double exchange. In this case, the spins of ions of the same element of two different valencies simultaneously exchange electrons through the oxygen ion thereby changing the valences of both. Thus, Fe^{++} O^- Fe^{+++} can change to Fe^{+++} O^- Fe^{++}. Although antiferromagnetic substances have no commercial value and like paramagnetics, are mostly important in theoretical studies, a knowledge of antiferromagnetism is indispensable in the understanding of the magnetic moments in ferrites.

BASICS OF MAGNETISM – SOURCE OF MAGNETIC EFFECT

FERRIMAGNETISM

About the same time that Néel was developing his theory of antiferromagnetism, Snoek (1936,1947) in Holland was obtaining very interesting properties in a new class of oxide materials called ferrites that were very useful at high frequencies. Now, a dilemma had arisen in accounting for the magnetic moment of a ferrite such as magnetite, Fe_3O_4 or $FeO.Fe_2O_3$. The theoretical number of unpaired electrons for that formula was 14, that is, 5 each for each of the Fe^{+++} ions and 4 for the Fe^{++} ion. Theoretically, the moment should be 14 μ_B. Yet the experimental value was only about 4μ_B. Néel then extended his theory to include ferrites. There were still two different lattice sites and the same negative exchange interaction. The difference was that in the case of antiferromagnetics, the moments on the two sites were equal while in the case of the ferrites they were not and so complete cancellation did not occur and a net moment resulted which was the difference in the moments on the two sites. This difference is usually brought about by the difference in the number of magnetic ions on the two types of sites. This phenomenon is called ferrimagnetism or uncompensated antiferromagnetism. Néel(1948) published his theory in a paper called Magnetic Properties of Ferrites; Ferrimagnetism and Antiferromagnetism. In the preceding year, Snoek(1936) in a book entitled New Developments in Ferromagnetic Materials, disclosed the experimental magnetic properties of a large number of useful ferrites.

The interactions of the net moments of the lattice are continuous throughout the rest of the crystal so that ferrimagnetism can be treated as a special case of ferromagnetism and thus domains can form in a similar manner.

Paramagnetism above the Curie Point

Ferrimagnetics also have a Curie Point and one would expect the same type of paramagnetic behavior above the Curie Temperature.(See Fig. 2.12).However, because of the negative interaction such as found in antiferromagnetics, the curve of $1/\chi$ vs T will be concave approaching an asymptotic value which would extrapolate to a negative value which again was found in antiferromagnetics.This type of behavior is strong confirmation of Néel's theory. The $1/\chi$ versus T curve is found in Figure 2.18.

SUMMARY

In this chapter we have discussed the fundamental basis for magnetic behavior in the electronic structure of atoms and ions. The next chapter enlarges our view to the next larger magnetic entity, namely the domain and finally to the bulk material itself. Basically, we will be going from a microscopic view of magnetic behavior to the larger macroscopic picture. We will then be able to define units and thus measure the magnetic performances of the different materials.

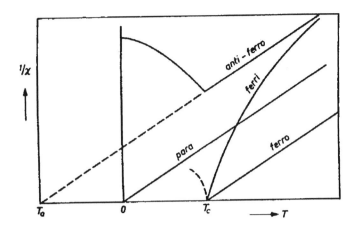

Figure 2.18 - Comparison of the temperature tendencies of the reciprocal susceptibilities of paramagnetic, ferromagnetic and ferrimagnetic materials.

References

Anderson,P.W.(1950) Phys. Rev., 79 ,705
Bethe,H. (1933) Handb. d. physik(5) 24 pt.2,595-8
Bohr,N. (1913) Phil. Mag. 5, 476,857
Bozorth,R.B. (1951) Ferromagnetism ,D.Van Nostrand,New York
Curie,P. (1895) Ann. chim phys.[7] 5, 289
Glasstone,S. (1946) Textbook of Physical Chemistry,D. Van Nostrand, New York
Goudsmit S.(1926) and Uhlenbeck,G.E. Nature, 117,264
Heisenberg,W.(1926) Z. pkysik 38, 411(1928) ibid 49, 619)
Kittel, C.(1971) Introduction to Solid State Physics, J. Wiley and Sons..New York
Kramers,H.A. (1934) Physica, 1, 182
Nagamiya,T.(1955) Yoshida,K. and Kubo,R., Adv. Phy, 4, 1-112,Academic P. N.Y
Neel,L. (1932) Ann. de Phys. 17, 61(1948) ibid. (12) 3, 137
Slater, J.C. (1937)J. Appl. Phys.,8, 441
Snoek,J.L.(1936) Physica (Amsterdam) 3, 463
Snoek,J.L.(1947) New Developments in Ferromagnetic Mat.Elsevier Amsterdam
Sommerfeld,A.(1916) Ann. phys. 51, 1
Thompson,J.J.(1903) Phil. Mag. 5, 346
van Vleck, J.H. (1951) J. Phys Rad, 12, 262
Weiss,P. (1907) J. phys.[4] 6,661-90
Zener, C.(1951a) Phys. Rev. 81, 440
Zener C. (1951b) ibid, 83, 299

3 THE MAGNETIZATION IN DOMAINS AND BULK MATERIALS

INTRODUCTION

Thus far, we have discussed the factors that contribute to the atomic and ionic moments and the effect of their magnetic interactions on the moments of the various crystal lattices. These moments are the maximum values or those measured under saturation conditions, at 0 K., that is, with complete alignment of the net magnetic moments. These values we found were intrinsic properties, that is, they depended only on chemistry and crystal structure (and of course, temperature). We have not discussed the important aspects of domain and bulk material magnetizations. In this chapter, we will expand our scope from the microscopic moment to the larger moment (in domains) and finally to the macroscopic bulk magnetization. Once these are described, we can then turn to the topics of magnetization mechanisms, magnetization reversal, and ultimately to cyclic magnetization, as in alternating current operation. To obtain a clear picture of these topics, the use of domain theory and domain dynamics is indispensable. This chapter will first discuss these subjects and show how they lead to the bulk magnetic properties.

THE NATURE OF DOMAINS

In a ferromagnetic domain, there is parallel alignment of the atomic moments. In a ferrite domain, the net moments of the antiferrimagnetic interactions are spontaneously oriented parallel to each other (even without an applied magnetic field). The term, spontaneous magnetization or polarization is often used to describe this property. Each domain becomes a magnet composed of smaller magnets (ferromagnetic moments). Domains contain about 10^{12} to 10^{15} atoms and their dimensions are on the order of microns (10^{-4} cm.). Their size and geometry are governed by certain considerations. Domains are formed basically to reduce the magnetostatic energy which is the magnetic potential energy contained in the field lines (or flux lines as they are commonly called) connecting north and south poles outside of the material. Figure 3.1 shows the lines of flux in a particle with a single domain. The arrows indicate the direction of the magnetization and consequently the direction of spin alignment in the domain. We can substantially reduce the length of the flux path through the unfavorable air space by spitting that domain into two or more smaller domains. This is shown in Figure 3.2. This splitting process continues to lower the energy of the system until the point that more energy is required to form the domain boundary than is decreased by the magnetostatic energy change. When a large domain is split into n domains, the energy of the new structure is about $1/n$th of the single domain structure. In Figure 3.2, the moments in adjacent domains

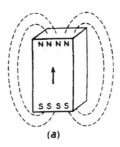

Figure 3.1-Lines of force in a particle of a single domain

are oriented at an angle of 180° to each other. This type of domain structure is common for materials having a preferred direction of magnetization. In other instances, especially where the cubic crystal structure is involved, certain oriented

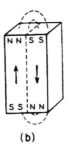

Figure 3.2- Reduction of magnetostatic energy by the formation of domains

domain configurations may occur which lead to lowering of the energy of the system. One of these is shown in Figure 3.3. These triangular domains are

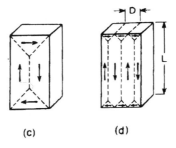

Figure 3.3- Elimination of magnetostatic energy by the formation of closure domains

called closure domains. In this configuration, the magnetic flux path never leaves the boundary of the material. Therefore, the magnetostatic energy is reduced. This type of structure may also be found at the outer surfaces of a material. The size and

THE MAGNETIZATION IN DOMAINS AND BULK MATERIALS

shape of a domain may be determined by the minimization of several types of energies. They are;

1. Magnetostatic Energy
2. Magnetocrystalline Anisotropy Energy
3. Magnetostrictive Energy
4. Domain Wall Energy

In addition, certain microstructural imperfections such as voids, non-magnetic inclusions and grain boundaries may also affect the local variations in domain structure.

Magnetostatic Energy

The magnetostatic energy is the work needed to put magnetic poles in special geometric configurations. It is also the energy of demagnetization. It can be calculated for simple geometric shapes. For an infinite sheet magnetized at right angles to the surface the equation (Bozorth 1951) for the magnetostatic energy per cm^3 is ;

$$E_p = 2 \pi M_s^2 \qquad [3.1]$$

Néel (1944) and Kittel (1946) have calculated the magnetostatic energy of flat strips of thickness, d, magnetized to intensity, M, alternately across the thickness of the planes. The equation is;

$$E_p = 0.85 \, dM^2 \qquad [3.2]$$

The calculations for other shapes come out with the general formula;

$$E_p = (\text{Constant}) \times dM_s^2 \qquad [3.3]$$

Therefore the magnetostatic energy is decreased as the width of the domain decreases. This mathematically confirms our assumption that the splitting of domains into smaller widths decreases the energy from the magnetostatic view. In fact, the energy of the domain structure is one thousandth that of a similar sized single domain.

Magnetocrystalline Anisotropy Energy

Most matter is crystalline in nature; that is, it is composed of repeating units of definite symmetry. Let us take a common geometrical configuration that may form the smallest repeating unit, namely a cube. Atoms or molecules are usually located at corners of the cube and in addition, at either the center of the cube or at the centers of the 6 faces. In most magnetic materials, to varying degree, the domain magnetization tends to align itself along one of the main crystal directions. This direction is called the easy direction of magnetization. Sometimes it is an edge of the cube and at other times, it may be a body diagonal. The difference in energy of a state where the magnetization is aligned along an easy direction and one where it is aligned along a hard direction is called the magnetocrystalline anisotropy energy. This magnetocrystalline anisotropy energy is also that needed to rotate the moment from the easy direction to another direction The energy of the domain can be lowered by this amount by having the spins (ferromagnetics) or moments (ferrimagnetics) align themselves along these directions of easy magnetization. In materials with high uniaxial anisotropy energy the moment of one domain is usually aligned along an easy direction of magnetization. Then, the adjacent domain will have the same tendency to align along the same axis but in the opposite direction. Even in materials with lower anisotropy, the 180° wall is often found. In crystals of cubic symmetry, where many of the major axes are at right angles(such as the cube edges) the 90° domain wall is also a reasonable possibility.

Magnetocrystalline anisotropy is due to the fact that there is not complete quenching of the orbital angular momentum as we postulated originally. With a small orbital moment that is mechanically tied to the lattice, the spin system can couple to it and therefore indirectly affect the lattice or the dimensions of the material.

Magnetostrictive Energy
When a magnetic material is magnetized, a small change in the dimensions occurs. The relative change is on the order of several parts per million and is called magnetostriction. The converse is also true. That is, when a magnetic material is stressed, the direction of magnetization will be aligned parallel to the direction of stress in some materials and at right angles to it in others. The energy of magnetostriction depends on the amount of stress and on a constant characteristic of the material called the magnetostriction constant.

$$E = 3/2 \, \lambda \sigma \qquad [3.4]$$
where; λ = magnetostriction constant
σ = Applied stress

The convention of the sign of the magnetostriction constant is such that if the magnetostriction is positive, the magnetization is increased by tension and also the material expands when the magnetization is increased. On the other hand, if the magnetostriction is negative, the magnetization is decreased by tension and the material contracts when it is magnetized. Magnetostriction as in the case of anisotropy is due to incomplete orbital quenching and the so-called spin-orbit, L-S or Russell Saunders coupling.

Stresses can be introduced by mechanical and thermal operations such as;

1. Metal strip- rolling, slitting, punching
2. Metal powder- pressing, deburring
3. Ferrites- Firing, grinding, tumbling

THE MAGNETIZATION IN DOMAINS AND BULK MATERIALS

These stresses also affect the directions of the moments locally depending on the distribution of the stresses.

Domain Wall Energy
Although Weiss (1907) first came up with the idea of the strong molecular field producing regions of oriented atomic moments or of spontaneous magnetization, it was Bloch(1932) who was the first to present the idea of magnetic domains, with domain walls (sometimes called Bloch walls) or boundaries separating them. In the domain structure of bulk materials, the domain wall or boundary is that region where the magnetization direction in one domain is gradually changed to the direction of the neighboring domain. If δ is the thickness of the domain wall which is proportional to the number of atomic layers through which the magnetization is to change from the initial direction to the final direction, the exchange energy stored in the transition layer due to the spin interaction is;

$$E_e = kT_c/a \qquad [3.5]$$
where kT_c = Thermal energy at the Curie point
a = Distance between atoms

Therefore the exchange energy is reduced by an increase in the width of the wall or with the number of atomic layers in that wall. However, in the presence of an anisotropy energy or preferred direction, rotation of the magnetization from an easy direction increases the energy so the wall energy due to the anisotropy is :

$$E_k = k\delta \qquad [3.6]$$

In this case, the energy is increased as the domain width or number of atomic layers is increased. The two effects oppose each other and the minimum energy of the wall per unit area of wall occurs according to the following equation;

$$E_w = 2(K_a T_c/a)^{1/2} \qquad [3.7]$$
where K_a = Anisotropy constant (described later)

If magnetostriction is a consideration, the equation is modified to;

$$E_w = 2(kT/a)^{1/2} (K_a + 3\lambda_s \sigma/2)^{1/2} \qquad [3.8]$$
here λ_s = magnetostriction constant

Typical values of domain wall energies are on the order of 1-2 ergs/cm^2
The domain wall thickness for the condition of minimum energy is given by the equation;
$$\delta = (\text{Constant}) \times a(E/K)^{1/2}$$

Typical calculated values of δ are about 10^3 Å or about 10^{-5} cm. With some soft magnetic materials the value may be about 10^{-6} cm while in some hard materials,

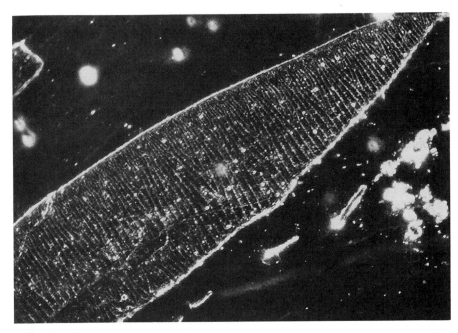

Figure 3.3a- Visualization of magnetic domains by means of the Bitter magnetic particle technique. The white stripes are the domain walls.

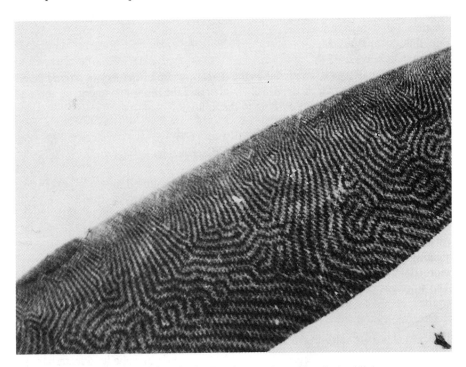

Figure 3.3b- Visualization of domains by Faraday rotation with polarized light.

the value may be on the order of 10^{-4} cm. or about one micron.

The whole array of domains will be arranged in such a way as to minimize the total energy of the system composed mainly of the above four energies.

PROOF OF THE EXISTENCE OF DOMAINS

The earliest experimental indication that domains exist was due presented by Barkhausen (1919) who was able to pick up small voltages due to the discontinuous changes in the magnetizations in these regions. Barkhausen amplified these voltages many times and made them audible on a loudspeaker. Bitter (1931) was first able to visualize domains by spreading over the sample, a suspension of colloidal magnetite. The colloidal particles will be concentrated at the domain boundaries since large field gradients exist there. These arrangements are called Bitter patterns. Figure 3.3a exhibits domain walls using this method. This technique is limited to the static state since the powder prohibits true dynamic observations as well as temperature restrictions. Since light is an electromagnetic wave, it might be expected to interact with magnetic fields and moments. Many so-called magneto-optic effects have been observed. Through this interaction, domains have been made visible microscopically by both reflected and transmitted light. One technique employs a polarized light which has its plane of polarization rotated differently by domains with different magnetization direction. When the rotated light beam is sent through a polarizing medium called the analyzer, the domains will show up because of the contrast in light intensities of the neighboring domains. With reflected light, this phenomenon is known as the Kerr effect. With transmitted light, it is called Faraday rotation. Domain patterns in many magnetic materials have been photographed using this technique. Figure 3.3b is an example of the Faraday technique. Kaczmarek(1992) used the transverse and longitudinal Kerr Effects to observe domains in soft polycrystalline ferrites. Using a laser and fiber optics, he examined hysteresis effects that are in good relationship with bulk measurements

Domain patterns have also been viewed by TEM (Transmission Electron Microscopy). Van der Zaag (1992) studied domain structures in MnZn ferrites using this technique. He found that at a grain size up to 4 microns, the grains were monodomain while above this size, they were polydomain.

THE DYNAMIC BEHAVIOR OF DOMAINS

Two general mechanisms are involved in changing the magnetization in a domain and, therefore, changing the magnetization in a sample. The first mechanism acts by rotating the magnetization towards the direction of the field. Since this may involve rotating the magnetization from an axis of easy magnetization in a crystal to one of more difficult magnetization, a certain amount of anisotropy energy is required. The rotations can be small as indicated in Figure 3.4 or they can be almost the equivalent of a complete 180° reversal or flip if the crystal structure is uniaxial and if the magnetizing field is opposite to the original magnetization direction of the domain. The other mechanism for changing the domain magnetization is one in which the direction of magnetization remains the same, but the volumes occupied by the different domains may change. In this process, the domains whose magnetizations are in a direction closest to the field direction grow larger while those that

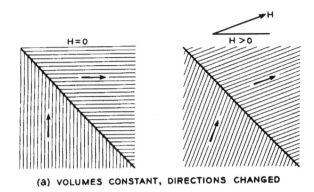
(a) VOLUMES CONSTANT, DIRECTIONS CHANGED

Figure 3.4. Change of domain magnetization by domain rotation.

are more unfavorably oriented shrink in size. Figure 3.5 shows this process which is called domain wall motion. The mechanism for domain wall motion starts in the domain wall. Present in the wall is a force (greatest with the moments in the walls that are at an angle of 90° to the applied field) that will tend to rotate those moments in line with the field. As a result, the center of the domain wall will move towards the domain opposed to the field. Thus, the area of the domain with favorable orientation will grow at the expense of its neighbor.

BULK MATERIAL MAGNETIZATION

We have proceeded through the hierarchy of magnetic structures from the electron through the domain. Although domains are not physical entities such as atoms or crystal lattices and can only be visualized by special means, for the purpose of magnetic structure they are important in explaining the process of magnetization. We now can discuss why a material that has strongly oriented moments in a domain often has no resultant bulk material magnetization. We can also examine why this apparently "non-magnetic" material can be transformed into a strongly magnetic body by domain dynamics discussed above.

The answer, of course, resides in the fact that, if the material has been demagnetized, the domains point in all random directions so that there is complete cancellation and the resultant magnetization is zero (See Figure 3.6). The possible steps to complete orientation of the domains or magnetization of the material are also shown in Figure 3.6.

The Magnetization Curve

We are now ready to look at the bulk magnetic properties of a material. Thus far, the magnetic moment or the magnetization has been given in either atomic units (or Bohr magnetons) or in physical units based on action of magnets. How can we relate these to actual material properties? The Bohr magnetons were based on limiting values at absolute zero and since it was an atomic moment (ferromagnetism) or a resultant or combination of moments (ferrimagnetism), it was in the so called saturated condition. Having said that, there is a zero net moment in unmagnetized bulk

THE MAGNETIZATION IN DOMAINS AND BULK MATERIALS   49

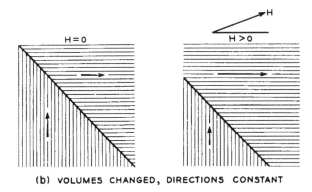

(b) VOLUMES CHANGED, DIRECTIONS CONSTANT

Figure 3.5. Change of domain magnetization by domain wall movement.

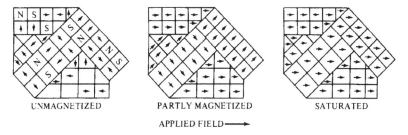

Figure 3.6. Stages in magnetization of a sample containing several crystals.

materials, we can predict that there will be an infinite number of degrees of magnetization between the unmagnetized and saturation conditions. These extreme situations correspond, respectively, to random orientation of domain to complete alignment in one direction with the elimination of domain walls. If we start with a demagnetized specimen and increase the magnetic field, the bulk material will be progressively magnetized by the domain dynamics described previously.

The magnetization of the sample will follow the course shown in Figure 3.7. The slope from the origin to a point on the curve or the ratio M/H has previously been defined as the magnetic susceptibility. This curve is called the magnetization curve. The curve is generally perceived as being made up of three major divisions. The lower section is called the initial susceptibility region in which there are reversible domain wall movements and rotations. Being reversible means that, after changing the magnetization slightly with an increase in field, the original magnetization condition can be returned if the field is reduced to the original value. The second stage of the magnetization curve in which the slope increases greatly is one in which irreversible domain wall motion occurs. The third section of the curve is one of irreversible domain rotations. Here, the slope is very flat indicating the large amount of energy that is required to rotate the remaining domain magnetization in line with the magnetic field.

Units for the Magnetization Curve

We have described the unit of magnetizing field H, from the interaction of magnetic poles. The unit was the oersted, defined as the field experienced at a distance of 1 cm from a unit pole. We have also described the magnetic moment, ml, from the dipole. The pole density in poles per unit cross sectional area is the intensity of magnetization, M, whose units are the same as moment/unit volume = emu/cm^3.

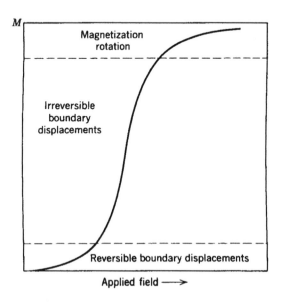

Figure 3.7-Domain dynamics during various parts of the magnetization curve. Source: Kittel, 1956

Conversion between Bohr Magnetons and Saturation

There are times when we have to convert the moment in Bohr magnetons per atom, ion or formula unit in the case of ferrites to units of bulk magnetization, M, in emu/cm^3 or in units of emu/g. The former, M, is more important in magnetic design as part of the magnetic flux. The latter, σ, is important for materials research since with temperature changes, the density must be known accurately at each temperature. The pertinent formula is:

$$M = n \times \mu_B \times N_o \times d/A \qquad [3.9]$$

Where; N_o = Number of atom/mole (6.02 x 10^{23})
       A = Atomic weight
       n = number of unpaired electron spins/atom
       μ_B = value of a Bohr magneton
       d = Density

The value n x μ_B is the moment of the atom or ion in emu.

THE MAGNETIZATION IN DOMAINS AND BULK MATERIALS 51

Flux Lines

Faraday found it convenient to liken magnetic behavior to a flow of endless lines of induction that indicated the direction and intensity of he flow. He called these lines flux lines and the number of lines per unit area the flux density or magnetic induction, B. The flux is composed of H lines and M lines. A schematic representation of the flux is given in Figure 3.8. Note that the lines traverse the sample, leave it at the North pole, and return at the South pole. In cgs units, the induction or flux density, B, is given by;

$$B = H + 4\pi M \qquad [3.10]$$

A unit pole gives rise to a unit field everywhere on the surface of a sphere of unit radius. The area of this sphere is 4π cm^2. The cgs unit of induction is the gauss. The units for the lines of induction or flux are known as maxwells or just plain "lines". Therefore, the units for flux density, B, are maxwells/cm^2. B can also be

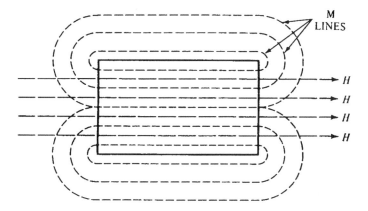

Figure 3.8. Magnetic flux lines composed of *H* (field) lines and *M* (magnetization) lines.

defined by the voltage generated in a wire wound around a core of magnetic material in which there are known variations of flux with time.

Later in this book, as we get more involved with the magnetic circuit and applications, we will concentrate more heavily on the use of induction, B, by itself. The B lines and the H lines are each measured independently when a bulk material is magnetized.

Room Temperature Saturation Inductions of Magnetic Materials

To this point in our discussion, the magnetic moment has usually been measured at 0° K in order to allow the correlation with the number of Bohr magnetons to be made. The magnetization can also be expressed in terms of M_o or the magnetization at 0° K. For practical applications, the room temperature values are much more important. In many cases, the total induction or flux density is measured with the field subtracted out to get the resultant $4\pi M_s$, which is often used interchangeably with B_s for soft magnetic materials. We have detailed the conversion from the Bohr magnetons to the magnetization. In a reverse manner, from value of $4\pi M_s$ (or B_s), the moment or number of Bohr magnetons can be calculated. The saturation induction of several magnetic materials are given at low temperatures and at room temperatures in Table 3.1. Since we are stressing the component properties now, the latter values are of more concern.

MKSA Units

Earlier we stated that, as we got more involved with the circuit aspects of magnetism, it would be useful to introduce the mksa system of units. This may be a convenient time to do so. The mksa unit for magnetic flux is the weber. There are 10^8 maxwells per weber. The unit for flux density is then the weber/m^2 or as it is commonly known, the Tesla, T. There are 10^4 gausses/Tesla. The unit for the

Table 3.1
Saturations of Various Magnetic Materials

Material	Saturation (Gausses)
CoFe (49% Co, 49% Fe 2 V)	22,000
SiFe (3.25% Si)	18,000
NiFe (50% Ni, 50% Fe)	15,000
NiFe (79% Ni, 4% Mo, Balance Fe)	7,500
NiFe Powder (81% Ni, 2% Mo, Balance Fe)	8,000
Fe Powder	8,900
Ferrites	4,000-5,000
Amorphous Metal Alloy(Iron-Based)	15,000
Amorphous Metal Alloy (Co-based)	7,000
Nanocrystalline Materials (Iron-based)	12,000-16,000

magnetic field intensity, H, in the new units is the amp/m. H in oersteds is related to the mksa unit by the equation for the field generated by current through a coil containing N turns.

$$H = .4\pi NI/l \quad [3.11]$$

Where; N = number of turns
I = current in amps
l = length of magnetic path, cm.

Using this conversion, there are 12.57×10^{-3} oersteds/amp/m

HYSTERESIS LOOPS

In magnetic applications, we are interested in how much induction a certain applied field creates. In soft magnetic materials, we want a high induction for a low field. In this case, H is very small compared to 4π M and B is essentially equal to 4πM. In the case of a permanent magnet, the H component can amount to from 50% or more of the total B. If we start with a demagnetized specimen and increase the magnetic field, the induction increases as shown in Figure 3.9.

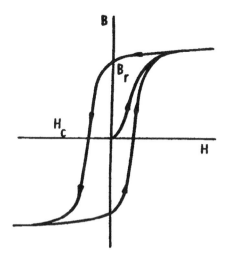

Figure 3.9-Initial magnetization curve and hysteresis loop

At high fields, the induction flattens out at a value called the saturation induction, B_s. If, after the material is saturated, the field is reduced to zero and then reversed in the opposite direction, the original magnetization curve is not reproduced but a loop commonly called a hysteresis loop is obtained. Figure 3.9 shows such a hysteresis loop with the initial magnetization curve included. The arrows show the direction of travel. We notice that there is a lag in the induction with respect to the field. This lag is called hysteresis. As a result, the induction at a given field strength has 2 values and cannot be specified without a knowledge of the previous magnetic history of the sample. The area included in the hysteresis loop is a measure of the magnetic losses incurred in the cyclic magnetization process. The hysteresis losses can also be correlated with the irreversible domain dynamics we had previously mentioned. The value of the induction after saturation when the field is reduced to zero is called the remanent induction or remanence or retentivity, (B_r). The values of the reverse field needed after saturation to reduce the induction to zero is called the coercive force or coercivity, (H_c). The unit for H_c is the oersted and that for B_r is the gauss. Both of these properties are very important and we shall refer to them in almost every magnetic application.

Minor Loops

Thus far we have spoken of the magnetization process when the material is magnetized to saturation. This situation is not always true and loops can be produced when varying degrees of magnetization are produced. When the maximum induction is less than saturation, the loop is called a minor loop. The shape of these minor loops can be vastly different than the saturated loop. When an unmagnetized sample is progressively magnetized it follows the magnetization curve. If we stop part of the way up and then reduce the field to zero and repeat the process to the same value of reverse field the result is a minor loop. A minor loop is shown in Figure 3.10 along with the saturation loop.

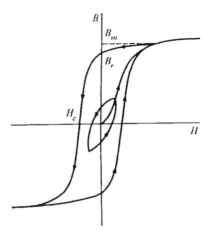

Figure 3.10 A minor hysteresis loop with the saturation loop

PERMEABILITY

We previously defined the susceptibility as the ratio of M to H. For paramagnetics and diamagnetics, this parameter is quite useful. However, in ferromagnetics and ferrimagnetics we are concerned with the total flux density, B, and it is more convenient to define a very important new parameter, μ, the magnetic permeability which is the ratio of induction, B to magnetizing field, H. However, this parameter can be measured under different sets of conditions. For example, if the magnetizing field is very low, approaching zero, the ratio will be called the initial permeability μ_o. Its definition is given by;

$$\mu_o = \lim_{B \to 0} (B/H) \qquad [3.12]$$

THE MAGNETIZATION IN DOMAINS AND BULK MATERIALS

This parameter will be an important in telecommunications applications where very low drive levels are involved. On the other hand, when the magnetizing field is sufficient to bring the B level up to the point of inflection, the highest permeability occurs. This can be seen by visualizing the permeability as the slope of the line from the origin to the end point of the excursion. Since the magnetization curve flattens out after the point, the μ will decrease. Often, it is important to know the position of the max permeability and the course of μ versus B. Such a curve is shown in Figure 3.11.

Factors Affecting the Permeability

Permeability is one of the most important parameters used in evaluating magnetic materials. Not only is it a function of the chemical composition and crystal structure but it is strongly dependent on microstructure, temperature, stress, time after demagnetization and several other factors. We shall discuss these in Chapter 6.

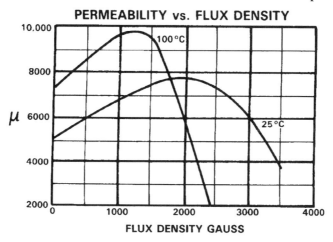

Figure 3.11-Variation of permeability as a function of flux density. Source, Magnetics 1989

In discussing the orientation of domains in a material in this chapter, we have previously briefly described two intrinsic parameters. These two properties were magnetocrystalline anisotropy and magnetostriction. Since they help in determining the equilibrium position of the domains and thus in the movement of these domains, they obviously affect the mechanism of magnetization which also includes the permeability.

MAGNETOCRYSTALLINE ANISOTROPY CONSTANTS

All ferromagnetic and ferromagnetic material possess to a lesser or greater degree a crystal direction or set of directions in which the magnetization prefers to be oriented. To rotate the magnetization from that easy direction requires an energy called the magnetocrystalline anisotropy energy. The energy is expressed in terms of certain anisotropy constants and the direction to which the magnetization is rotated. For the simple case of a uniaxial crystal such as a hexagonal structure, the relevant equation is;

$$E_k = K_1 \sin^2 \mu_B + K_2 \sin^4 \theta + ... \quad [3.13]$$

where K_1 and K_2 = First and second anisotropy constants
θ = angle between the easy axis and magnetization

The conventional units for the anisotropy constants are ergs/cm^3. In hexagonal ferrites, the easy axis is usually the hexagonal or c axis, although certain exceptions were noted earlier. Usually the first anisotropy constant is dominant and is sufficient. For materials with cubic crystalline structure including the spinel ferrites, the anisotropy energy is given in terms of anisotropy constants. In this instance, however, the directions are given in terms of the direction cosines, α_n, which are the ratios of the individual components of the magnetization projected on each axis divided by the length of the magnetization or hypotenuse of the triangle. The appropriate equation for the anisotropy energy is;

$$E_k = K_1 (\alpha_1^2 \alpha_2^2 + \alpha_2^2 \alpha_3^2 + \alpha_3^2 \alpha_1^2) + K_2 (\alpha_1^2 \alpha_2^2 \alpha_3^2) + ... [3.14]...$$

The direction that minimizes the energy will be the most preferred direction. The direction cosines for the principal direction in the cubic structure are;

	α_1 (x)	α_2 (y)	α_3 (z)
Cube edges	1	0	0
Face diagonal	$1/\sqrt{2}$	$1/\sqrt{2}$	0
Body diagonal	$1/\sqrt{3}$	$1/\sqrt{3}$	$1/\sqrt{3}$

When the anisotropy energy for each direction is calculated according to the above equation for cubic materials, the following values for E_k result;

For cube edge (100) $E_k = 0$
For face diagonal (110) $E_k = 1/4 K_1$
For body diagonal (111) $E_k = 1/3(K_1) + (1/27)K_2$

With few exceptions, K_1 will predominate and when K_1 is positive, the easy direction of magnetization will be the cube edge(100) direction and when K_1 is negative, the body diagonal (or so-called (111) direction in Miller indices) will be the preferred direction. In ferrites, with the exception of cobalt ferrites, the value of K_1 is negative, so that the cube diagonal is the easy direction in most ferrites. In soft materials where the domain motion is preferably unrestrained, the anisotropy or K_1 should be quite small in absolute magnitude. In most soft magnetic materials, that is indeed true. To determine the anisotropy constants, the measurements are made on

THE MAGNETIZATION IN DOMAINS AND BULK MATERIALS   57

single crystals that are oriented in the direction in which the anisotropy constant is to be measured. These constants will be correlated later with the properties of the polycrystalline ferrites. Later, in Chapter 11, these constants will be correlated with the properties of the polycrystalline magnetic materials.

In contrast to soft magnetic materials, for those designed to be permanent magnets, we usually want to take advantage of the affinity of the moments and the magnetization for a particular crystallographic direction. In hexagonal materials, as we have seen, this is the c axis. Therefore, hard or permanent magnet materials should have a very high anisotropy.

MAGNETOSTRICTION
Turning now to the magnetostriction constant, the magnetostrictive energy is given by:

$$E = 3/2\, \lambda_s\, \sigma \qquad [3.15]$$

where λ_s = saturation magnetostriction
σ = applied stress

Here again, the energy should be minimized to give the domain freedom of motion. Through magnetostriction, stress, creates high-energy barriers to this motion. The magnetostriction constant or just the magnetostriction is really the sensitivity of the energy to the mechanical stresses. There are stresses produced in soft magnetic materials processing such as those due to thermal and mechanical operations which are difficult to avoid or correct. For good quality soft magnetic materials, a low magnetostriction is highly desirable. On the other hand, in magnetostrictive transducers for such applications as ultrasonic generators, the mechanical motion produced by the magnetic excitation through magnetostriction is used to good advantage. Of course, a high magnetostriction is required in this instance.

Since both anisotropy and magnetostriction are intrinsic properties of a material, by proper choice of the chemistry and crystal structure, we can strongly influence the magnetic parameters for a specific application.

IMORTANT PROPERTIES FOR HARD MAGNETIC MATERIALS
For applications involving cyclic magnetization, the magnetization curves and hysteresis loops are the signatures of the material and contain many of the important parameters. In one major application, the magnetic material is not really cycled. This is the case for the permanent magnet (see Figure 5.1). Here, the core is magnetized to create the magnetic poles. When the magnetizing field is reduced to zero, the B level does not return to zero but follows the hystersis loop to the induction value we have called the remanence. A certain amount of demagnetization occurs on removal of the field depending on the gap length,but with high remanence materials,the net residual can attain a value as high as 10,000 gausses. The coercive force of a uniaxial ferrite such as barium ferrite is given by;

$$H_c = 2K_1/ M_s \qquad [3.16]$$

showing the need for a high crystal anisotropy and low saturation to aid in resisting demagnetization. Because the low M_s conflicts with the need for high remanence, a compromise is usually adopted. The field created from the residual strong poles forms a permanent magnet. Because of a strong uniaxial anisotropy, the coercive forces of permanent magnet materials are quite high ranging from about 500 oersteds for Alnico to 3,000 or higher for other materials. In the case of hard ferrites, the remanence is low because the saturation is so much lower than it is for metals. However, the coercive forces for hard ferrites are much higher than Alnico. In the gap, there are obviously no magnetization, M, lines so that the B lines consist of only H lines resulting from the M lines in the material. An important parameter for a permanent magnet in addition to the coercive force H_c, and the remanence B_r, is the maximum product of B x H occurring in the second quadrant of the hysteresis loop. This unit is frequently known as the energy product and is measured (cgs) in gauss-oersteds or MGO (Mega Gauss-Oersteds) or in kjoules/m^3.

Properties of Recording Materials
In materials used for cores or media for digital recording, we are concerned with creating a material that has two unique states of magnetization. This can easily be realized using the two stable states of magnetic saturation on the hysteresis loop. Such materials are often called square loop materials because of the type of hysteresis loop that is required. Loop properties such as B_s, B_r and H_c are important properties but there is an additional one known as the squareness ratio, B_r/B_s. This ratio was especially important in square loop memory cores, which have all but vanished. However, widely-used recording media such as γ-Fe_2O_3 use H_c as an important specified parameter.

SUMMARY
In this chapter we have examined the importance of domains and how their dynamics can affect the manner in which a material is magnetized. In addition we examined the fundamental units of bulk magnetization and induction in the magnetization curve and cyclically in a hysteresis loop. All of these processes were considered primarily under the influence of a D.C. drive. In the next chapter we shall see the changes that occur when the cyclic traversal is done at a rate such as that produced in the 50-60 Hz. line or mains frequency.

References

Barkhausen,H.(1919) Physik Z. 20, 401
Kaczmarek,R.(1992) Dautain, M. and Barradi-Ishmail, Ferrites, Proc. ICF6, Jap. Soc.,of Powder and Powd. Met, Tokyo, 823
Kittel,C.(1956) Phys. Rev., 70, 965
Neel, L.(1944) J. phys. rad. 5, 241
Van der Zaag, P.J.(1992),Noordermeer, A., Ruigrok, J.J.M., Por, P.T., Rekveldt, M.T.,Donnet, D.M. and Chapman, J.N. Ferrites, Proc. ICF6, Jap. Soc.,of Powder and Powd. Met, Tokyo, 819
Weiss,P.(1907) J. de phys.[4] 6, 661

4 AC PROPERTIES OF MAGNETIC MATERIALS

INTRODUCTION

In the previous chapter we discussed how the hysteresis loop can literally be called the signature of a magnetic material. We have produced this loop by slowly applying a direct current (DC) field in one direction, saturating the material, reducing the field slowly to zero, reversing the field and repeating the same procedure. The loop we get is called the DC hysteresis loop.

DC loops are important for studying the basic properties of a material and are also used in permanent magnet materials design since its operation is equivalent to a DC field. However, the use of magnetic materials is predominantly under alternating current condition, that is, when the magnetizing field is produced by a current varying sinusoidally or according to some other wave-form (square, triangular, sawtooth, etc). This chapter will deal with properties that become important under these conditions.

AC HYSTERESIS LOOPS

As the current goes through one sine-wave cycle, the magnetization will go through one hysteresis loop cycle. This is shown schematically in Figure 4.1. This type of loop is called an ac loop and at low frequencies approximates the D.C. loop. However, certain differences enter as the frequency of the exciting current and therefore the frequencies of traversal of the loop are increased. Losses (that we will describe in the next section) increase the width of the loop. In actual fact, the area contained in the hysteresis loop is indicative of the losses in the material during the cyclic magnetization process.

We have examined the domain dynamics during the traversal of the hysteresis loop. The loss due to irreversible domain changes produces the magnetic hysteresis that is released as heat and is known as hysteresis loss. This type of loss is the same for ac as for D.C. loops. However, as the frequency is increased, internal current loops called eddy currents produce losses that broaden the hysteresis loops. These eddy currents are extremely important in the choice of a magnetic material.

EDDY CURRENT LOSSES

When a material is magnetized cyclically, for example by a sine wave current, there will be induced in the material, a voltage that is in the opposite direction to the voltage producing the magnetizing current and the alternating magnetic field. The induced voltage will set up circular currents in the material which produce magnetic fields opposite to the original magnetic field. If a cylindrical sample is magnetized by a solenoidal winding, there will be eddy currents generated as concentric circles around the central axis of the cylinder. Since each of these current loops produces a

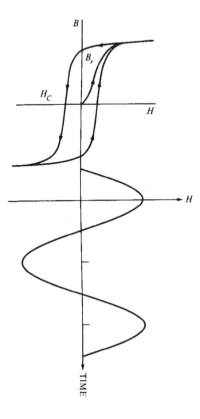

Figure 4.1-Schematic diagram showing the relation of the sine-wave current driving the magnetization through one hysteresis loop per sine-wave cycle.

magnetic field, the eddy current fields will be greatest in the center of the circular cross section where largest number of the field producing current loops surrounding that area are located. The eddy current fields on the outside of the cylinder will be the lowest having the fewest current loops surrounding it.

The induced voltage is a function of dB/dt, or the rate of change of the induction, B, with time. The higher the frequency f, or the 1/t in dB/dt, the greater the induced voltage and the greater the eddy current losses. Eddy current losses will occur in all types of materials but will be greatest in magnetic materials due to their higher permeabilities and therefore the greater change in induction, ΔB. An important example of eddy current losses in non-magnetic material occurs in copper winding of magnetic devices. We will have to consider those losses separately in magnetic components. The eddy current effect is strongly dependent on the resistivity of the material that affects the resistance in the eddy current loop. Since I = E/R, for the same induced voltage, a high resistivity will reduce the eddy currents and

AC PROPERTIES OF MAGNETIC MATERIALS

therefore the opposing field. Since ferrites have high resistivities, eddy currents are not a problem until higher frequencies are encountered.

We have spoken of the depth factor in eddy current losses. At low or medium frequencies, the reduction of the magnetizing field through eddy currents is low if the sample size is not very large. There is a time factor involved in the depth of penetration so that at low frequencies there is sufficient time for the eddy currents to dissipate and for the applied field to reach the center of the sample. At higher frequencies, there is insufficient time for eddy currents to decay and so the reverse fields will become appreciable at distances farther from the center or closer to the surface. Eventually with further increase in frequency, the applied field penetrates to only a small depth. These increased eddy currents shield the inside of the sample from the applied field. In metals where the resistivity is low, to reduce the effect of eddy currents, the strip from which the component is made is rolled thin and insulated from adjacent layers.

The depth of penetration of the applied field can be expressed in terms of the frequency and other parameters. A term frequently used is called the "skin depth" and defined as the point where the applied field is reduced to 1/e th of that on the surface. The equation is;

$$s = 1/2\pi\, \mu f/\rho \qquad [4.1]$$
where s = skin depth(cm)
ρ = resistivity (ohm-cm)
μ = permeability
f = frequency (Hz)

At about three skin depths and below, the applied field is so small as to render the material useless. The energy to generate the eddy currents in this region is used to heat the sample. This is the basis of induction heating and melting. In addition, a phase lag occurs between the flux on the surface and that in the center. It is possible that the magnetization may be in one direction at the surface and the opposing direction inside. The angle of lag, ε, between M at the surface and that at a depth, x, is given by;

$$\tan \varepsilon = 2\pi\, \mu f/\rho\, x \qquad [4.3]$$
where x = depth, cm.

Resistivity

As mentioned earlier, the eddy current losses depend on the resistivity. The resistivity of a material is defined by:

$$R = \rho\, l/A \qquad [4.4]$$
where: R = Resistance in ohms
l = length of specimen, cm
A = area of specimen, cm^2
ρ = resistivity, ohm-cm

Table 4.1 lists the resistivities of several ferrites along with those of some of the metallic ferromagnetic materials

Table 4.1
Resistivities of Ferrites and Metallic Magnetic Materials

Material	Resistivitity, Ω - cm
Zn Ferrite	10^2
Cu Ferrite	10^5
Fe Ferrite	4×10^{-3}
Mn Ferrite	10^4
NiZn Ferrite	10^6
Mg Ferrite	10^7
Co Ferrite	10^7
MnZn Ferrite	10^2-10^3
Yttrium Iron Garnet	10^{10}-10^{12}
Iron	9.6×10^{-6}
Silicon Iron	50×10^{-6}
Nickel Iron	45×10^{-6}

Now, the relationship between Eddy current losses and resistivity is given by;

$$P_e = (\text{constant}) B_m^2 \, f^2 d^2 / \rho \quad [4.5]$$

where (constant) = a factor depending on geometry of the sample
B_m = maximum induction (gausses)
f = frequency (hertz)
d = smallest dimension transverse to flux

Thus, to keep Eddy current losses constant as frequency is increased, the resistivity of the material chosen must rise as the square of the frequency. This relationship reinforces the equation we presented previously on the frequency dependence of the skin depth. The resistivity is an intrinsic property of a material insofar as the crystal lattice or body of a ferrite grain is concerned. However, as we shall see, the influence of the grain boundary can greatly alter the bulk resistivity.

PERMEABILITY
In discussing the D.C. loop we defined a quantity expressing the amount of polarization produced by a unit of magnetizing field. We called it the permeability, μ_{DC}. We related this to the case of movement of the domain walls. Now in A.C. loops, the Eddy current losses which change the hysteresis loop obviously will change the permeability since if the loop is broader, it requires a larger H (magnetizing field) to obtain an equivalent B. We would then expect the ac permeability to be quite frequency dependent and such is indeed the case.

Initial Permeability

A special case of a minor loop-a loop in which the material is not magnetized to saturation-is one in which the H level is extremely small. This is the type of loop encountered in telecommunications such as telephone or radio transmission. Under these conditions, the system is said to be in the Rayleigh region and the permeability under these circumstances is called the initial permeability.

$$\mu_o = \text{limit } (B/H) \quad [4.6]$$
as $B \to 0$ See Figure 4.2

In the mksa system, the term, μ_o, is used as a constant and is defined as the permeability of a vacuum. In that case, the term, μ_i, is often used for initial permeability. We shall use μ_o for initial permeability because of its acceptance in catalogs and much of the literature.

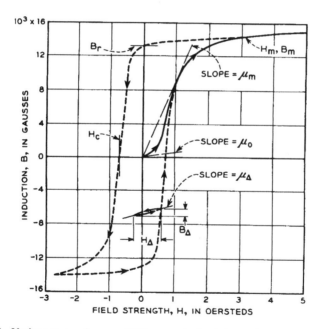

Figure 4.2. Various types of permeabilities are obtained depending on the portion of the loop traversed. Shown are the initial permeability, μ_o, the maximum permeability, μ_{max}, and the incremental permeability, μ. From Bozorth, 1951.

Permeability Spectrum

At high frequencies, the permeability separates into two components μ' and μ''. The first, μ' represents the permeability with the magnetization in phase with the alternating magnetic field and the second μ'' the imaginary permeability with the magnetization that is out of phase with the alternating magnetic field. By the term " in phase", we mean that the maxima and minima of the magnetic field, H, and that of the induction, B, coincide and by the term "out of phase", we mean that the maxima and minima are displaced by 90°.

The combined complex permeability is given in the complex notation by:

$$\mu = \mu' - j\mu'' \qquad [4.7]$$

μ' = real permeability (in phase)
μ'' = imaginary permeability (90° out of phase)
j = unit imaginary vector

The two permeabilities are often plotted on the same graph as a function of frequency. This is known as the permeability dispersion or permeability spectrum. Figure 4.3 from Smit and Wijn (1954) is such a plot. Note that the real component of permeability, μ', is fairly constant with frequency, rises slightly, then falls rather rapidly at a higher frequency. The imaginary component, μ'', on the other hand, first rises slowly and then increases quite abruptly where the real component, μ', is falling sharply. It appears to reach a maximum at about where the real permeability has dropped to about one half of its original value. As the definition of complex permeability implies, these curves are coupled in that the increased losses due to the increase in frequency results in a lowering of the permeability. Earlier, we related this fact to increased eddy current losses.

At the frequencies that we observe the effects in Fig 4.3, there is another type of loss that becomes important and may predominate at certain frequencies. This loss is ascribed to a magnetic phenomenon called ferromagnetic resonance, or, speaking of ferrites, ferrimagnetic resonance. This factor limits the frequency at which a magnetic material can be used. It is also observed that the higher the permeability of the material, the lower the frequency of the onset of ferrimagnetic resonance. Based on the real pioneering theoretical work by Landau and Lifshitz (1935), Snoek (1948) attempted to explain this behavior by assuming that at low fields, domain rotation produces the change in magnetization. Further assuming that the domains were ellipsoidal, he found the following equation applicable;

$$f_r(\mu_o - 1) = 4/3 \gamma M_s \qquad [4.8]$$
where γ = gyromagnetic ratio

This means there is an effective limit to the product of frequency and permeability so that high frequency and high permeability are mutually incompatible. Watson (1980) calculates that the curves in Figure 4.3 can be accounted for by the expression;

$$f_r \mu = 3 \times 10^9 \text{ Hz.} \qquad [4.9]$$

AC PROPERTIES OF MAGNETIC MATERIALS

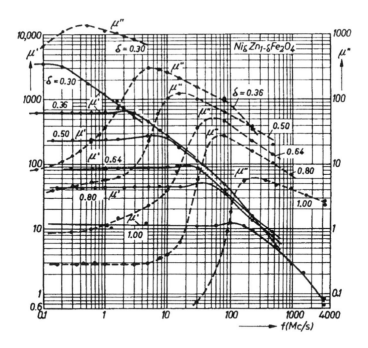

Figure 4.3-Permeability Spectrum Plot of a Nickel Ferrite showing the Frequency Course of the Real and Imaginary Permeabilities. From Smit,J and Wijn, H.P.J., Advances in Electronics and Electron Physics, 6, 105(1954)

Despite Snoek's assumption that the limiting frequency-permeability equation held for rotational processes, his expression is found quite valid under many circumstances involving higher frequencies. Although his effects are found in frequencies approaching microwaves, Rado(1953) found another peak which he attributes to domain wall resonance. Coincidentally, the higher permeability materials always have the lower resistivities. We shall have more to say about ferrimagnetic resonance when we deal with microwave magnetic properties. The ferrimagnetic resonance properties of a material are also considered intrinsic properties.

Loss Tangent and Loss Factor

The ratio of the imaginary part representing the losses in the material to the real part of the permeability is a measure of the inefficiency of the magnetic system. It is called the loss tangent.

$$\tan \delta = \mu'' / \mu' \qquad [4.10]$$

If we normalize the loss tangent per unit permeability, we have a material property describing the loss characteristics per unit of permeability. This property is called the loss factor.

$$LF = \tan \delta / \mu \qquad [4.11]$$

Obviously this parameter should be as low as possible. When we deal with the applications of ferrites in inductors, we shall show how the loss factor fits in with the component requirements.

Loss Coefficients

For low level applications, Legg showed that the losses involved in a magnetic material could be broken into 3 separate categories according to the following equation:

$$R_s/\mu f L = hB_m + ef + a \qquad [4.12]$$

where: R_s = loss resistance
L = inductance
h = hysteresis coefficient
e = eddy current coefficient
a = anomalous loss coefficient

By plotting the quantity on the left-hand side versus the B_m using several different B's, the slope will give the coefficient, h. Plotting the same function against the frequency will give a slope equal to the eddy current coefficient, e. The intercept gives the anomalous loss coefficient, a.

Temperature Factor of Initial Permeability

As we have shown, the permeability is the result of many different effects acting simultaneously. Some are inherent properties depending on chemistry of crystal structure. Some are extrinsic, depending on ceramic microstructure, strains, etc.... With so many parameters that are temperature dependent in themselves, it is not surprising to find wide variation in the shape of the permeability vs temperature curve. A typical curve showing this dependence is given in Figure 4.4.

The peak at the far right drops to zero at the Curie point when ferrimagnetism is lost. The peak to the left is called the secondary maximum of permeability (SMP). This peak most often occurs close to the temperature that the magnetostriction goes through zero. By variation of the chemistry, the peak can be moved to the temperature at which the material will be used. Thus, for low level devices in which the temperature doesn't exceed room temperature, the SMP is usually designed to be around room temperature. In a power material which is meant to be operated at 60° to 100°C, very often the permeability maximum is designed to be in that region.

In the case of Nickel ferrite, since permeability optimization depends on lowering of anisotropy and since K_1 does not cross zero, there is no secondary permeability maximum.

DISACCOMMODATION

A somewhat unique magnetic feature of ferrites is a phenomenon called "disaccommodation". In this type of instability, the permeability decreases with time directly after it is demagnetized. This demagnetization can be accomplisheded by

AC PROPERTIES OF MAGNETIC MATERIALS

heating above the Curie point or it can be done by the application of an alternating current of diminishing amplitude. A factor characteristic of the material called the disaccommodation factor (D.F.) is defined by:

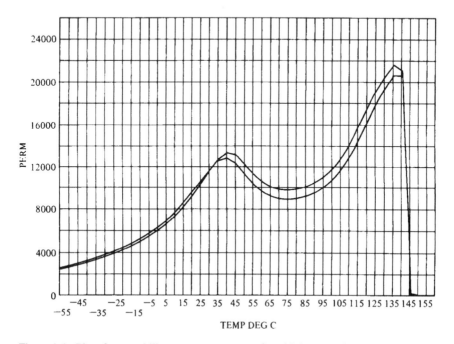

Figure 4.4. Plot of permeability versus temperature for a high-permeability MnZn ferrite.

$$DF = (\mu_1 - \mu_2) / (\mu_1^2 \log t_2/t_1) \quad [4.13]$$

where: μ_1 = permeability shortly after demagnetization (t_1)
μ_2 = permeability after demagnetization later (t_2)

A graph of the change in permeability due to disaccommodation is shown in Figure 4.5. Thus, a material with a permeability of 1,000 measured 100 sec after demagnetization and then remeasured at 990 after 10,000 sec would have a disaccomodation of;

$$DF = (1000-990)/(1000)^2 \log(10000/100)$$
$$= 10^1/(10^6 \times 2) = 5 \times 10^{-6}$$

This factor can then be used in predicting the drop in permeability after different times. As we shall see later, it permits us to calculate the drop in permeability with a gapped core of the same material.

CORE LOSS

Much of our discussion thus far has dealt with the use of magnetic materials in the low level or Rayleigh region. Although this was the major use for ferrites in earlier periods, today a large amount of ferrites are used in power application exemplified by the high frequency switched mode power supplies. Under these conditions of power applications, involving high drive levels, properties such as loss factor and initial permeability are not very useful criteria for power uses. In these cases,

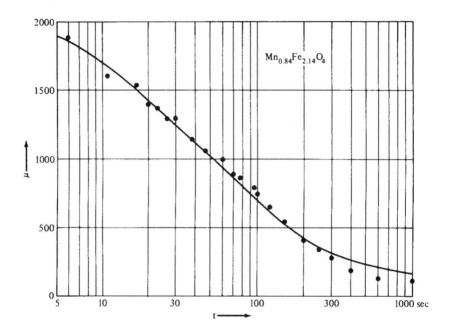

Figure 4.5- Disaccomodation or Time Decrease of Permeability in a Manganese Ferrous Ferrite. From Enz, U., Physica, 24, 622 (1958)

cases, what are needed are low losses in the core at high levels of induction. These losses are called core losses or watt losses. For metals, the units for these losses are watts per pound, (or W/Kg). For ferrites, the units usually used are mW/cm^3, and are often measured at a higher than room temperature (usually at the temperature of intended use). Examples of watt loss curves for some ferrites are given in Figure 4.6.

AC PROPERTIES OF MAGNETIC MATERIALS

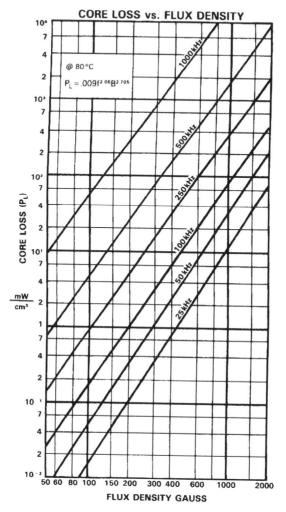

Figure 4.6. Core loss curves of an MnZn power ferrite as a function of frequency and B level. Source: Magnetics, 1989.

MICROWAVE PROPERTIES

The AC uses of ferrites in microwave applications are vastly different from those used at lower frequencies where the magnetic effects produced by ferrites are reflected in the actions of currents & voltages in coils. At microwave frequencies the whole ferrimagnetic exchange interaction breaks down with the elimination of domains. At microwave frequencies, the interaction is between electromagnetic fields and the ferrite materials. Ordinary circuit elements such as switches & transformers are not applicable. The mechanisms of magnetization by domain wall motion and rotation are also not operative. The permeabilities of microwave ferrites are close to 1. How then can ferrites be useful at microwave frequencies? The very high resistivities of some ferrites make them the only magnetic material useful in bulk (thin films of metal have been used). In order to understand the mode of action, it is necessary to return to the concept of electron as spinning top. In the case of a real top, in addition to spinning, the top will precess around an axis in line with the gravitational vector under the influence of gravity. The axis of the spinning top orbits around this axis forming a cone. As the spinning frequency decreases, gravity takes over and the angle of the precession gets larger and finally the top falls flat. In electronic precession, the aligning force is a static D.C. field orienting the unpaired spins of the ferromagnetic atoms or ions (See Figure 4.7). The larger the field, the tighter the spins will be aligned with the field. With lower fields, the angle of precession will increase.

If we liken the precession to the action of a circular pendulum or a type of maypole to form an inverted cone, the original impulse to the ball will create a certain frequency of precession. Again as gravity takes over the angle of precession will decrease. However, if we give the ball another tangential impulse every time it comes around, the energy of the repeating impulse will affect the angle of precession. If the transverse excitation is in phase with the frequency of precession, there is reinforcement of the motion and the angle will get larger. If it is out of phase, there is interference with the motion and the angle will decline. When the frequency of transverse excitation is in phase, we say the system is in resonance. Now the maximum transfer of energy takes place at this frequency. In a microwave

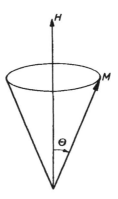

Figure 4.7- Schematic representation of the precession of the magnetization around a static DC field. From Smit and Wijn, 1959

material, the spinning electron is the top. The restoring force is the D.C. field and the transverse excitation is the high frequency electromagnetic field oscillation.

In many microwave applications, the microwave excitation is in the form of a plane-polarized wave (having both magnetic field and electric field components) whose directions are perpendicular to each other. For our purposes, we will concerned only with the magnetic field component, h, which. This plane polarized

AC PROPERTIES OF MAGNETIC MATERIALS

wave is propagated through a metallic wave guide whose dimensions are fixed by the wave-length, which in turn is determined by the microwave frequency in the application. The linearly polarized wave can be considered to be composed of a combination of 2 counter rotating circularly-polarized waves, one designated by a complex permeability, μ_+, rotating clockwise and another designated by a complex permeability, μ_-, rotating counterclockwise. When these two waves travel without any external interaction, the rotational velocities, being equal, cancel and the resultant of the two waves is simply the plane polarized wave with which we started. Both of the circularly polarized waves have real and imaginary components as we mentioned for the lower frequencies (μ_+' and μ_+'' as well as μ_-' and μ_-'')When the circularly polarized waves interact with the ferrite, only the positively rotating h waves, μ_+' and μ_+'', rotating in the same sense as the precession and at the ferrite. For the positively rotating wave, μ_+, there will be produced a typical resonance curve in Figure 4.8. The frequency of the precession is a function of the D.C. field so the resonance can be accomplished by varying the field at constant frequency (Figure 4.8a) or by sweeping the frequency at constant H (Figure 4.8b). The curves for the negatively rotating circularly polarized wave do not show the same effect. This type of phenomena of ferrites can be the basis of a microwave or YIG filter for separating frequencies. To be very discriminative to certain frequencies the absorption peak should be narrow. The width is measured at 1/2 the height of the curve. This width is called the *line width or half line-width*. It is often a figure of merit of the material. The base line off of resonance is called ΔH_{eff}. The total field, $\Delta H = \Delta H_{eff} + \Delta H_{anisotropy} + \Delta H_{porosity}$ according to Schlomann (1956,1958,1971).The main part of the absorption ,μ'' is due to ΔH_{eff} but the sharpness of the separation of the two circularly rotating $\mu'(+)$ and $\mu'(-)$ is due to the other factors of ΔH. After passing through the ferrite medium at resonance, the positively rotating circularly polarized wave is absorbed, decreasing its rotational velocity, while the other (negatively rotating) wave has the same velocity. The resultant of the two waves now is a linearly polarized wave whose plane has been rotated. This phenomenon is called Faraday Rotation and is analogous to the similar effect with light. With the same D.C. field, the rotational sense of the spin precession system will be constant from the point of view of the ferrite. Thus, regardless of the direction of propagation of the original linearly polarized wave, the Faraday rotation will always be in the same sense. In other words, the transmitted and reflected waves in a wave-guide will have different senses of Faraday rotation relative to their propagation directions. This phenomenon is called non-reciprocal and is the basis of several microwave devices. If the initial transmitted wave is rotated 45° by the first interaction with ferrite, the reflected wave will be rotated an additional 45° by the second interaction for a total rotation of 90°. Had the action been reciprocal, the two rotations would have cancelled. By means of this Faraday Rotation, different microwave beams can be isolated or circulated to designated wave-guides. This is the basis of the microwave circulator to be discussed later.

MICROWAVE PRECESSIONAL MODES

We have assumed that in the precession of the spins in the ferrite material around the static D.C. field, the spins are all in phase. When this occurs, we call that mode the uniform precessional mode shown in Figure 4.8. However, there are other modes in which there is a spatial variation of the spins, which variation may be sinusoidal along the propagating direction. The length of one sine-wave alternation is called the spin wave wavelength. The quantized units of energy in

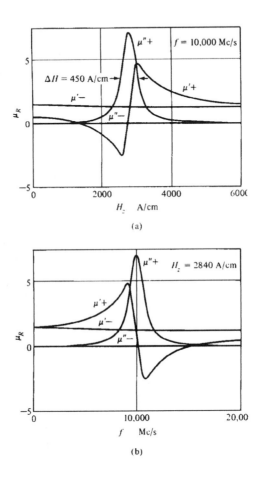

Figure 4.8-Ferromagnetic resonance plots showing the real and imaginary components of the two circularly polarized components of the permeability. In curve (a), the frequency is constant and the DC field is varied while in curve (b), the reverse is true. Source: Brailsford, 1966 p. 251

these spin waves are called *magnons*. The contribution to the ferromagnetic resonance line width due to spin waves is called the *spinwave linewidth*, ΔH_k. When the k modes of this quantity are of the same energies as lattice vibrations (phonons), interchange of energy can take place. There is a thresh-hold of ac current level where the losses become non-linear and increase dramatically. This critical current,

AC PROPERTIES OF MAGNETIC MATERIALS

h_{crit}, level is related to ΔH_k. Therefore, this quantity is an important one under these circumstances

LOGIC AND SWITCHING PROPERTIES OF FERRITES

Although their use in the application has all but disappeared because of semiconductors, the first components of large scale digital memories were ferrite cores. The use was related to the presence of a special type of ferrite called a *square loop ferrite*. The hysteresis loop of such a ferrite is shown in Figure 4.9. This type of ferrite exhibits two metastable states of magnetism, one in which the core (as a toroid) is magnetized in the upper level of saturation and the other the lower. We can then postulate the use of these states as 1's and 0's in digital logic. The core is energized to either state by an appropriate pulse. Then selected cores can be pulsed in the reverse direction to "set" then to 1. The selected cores can be read by resetting them all back to "0". The cores that have been set will have a flux change producing a voltage. In this way, the cores act as "memory cores".

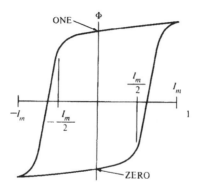

Figure 4.9-A hysteresis loop of a square-loop material. The upper remanence point represents a logical "one" in binary notation while the lower remanence is a "zero". From Albers-Schoenberg, 1954 p.152

PROPERTIES OF RECORDING MEDIA

Instead of using the little toroids just described as memory elements, the equivalent can be achieved with small regions of ferrite deposited on a thin plastic tape. The particles are not toroidal, but by choice of the right material and shape the particles can store logical bits of information. The material must be capable of being magnetized in a temporarily stable magnetic state (0 or 1) so that the material must be somewhat a hard or permanent magnet material, but it should not be so hard that very high fields are needed to demagnetize or read it. Thus the material is "semi-hard". Use is made of materials with intermediate coercive forces of several hundred oersteds. Use is also made of shape anisotropy to prevent demagnetization and to increase coercive force. The longer the particle compared to the cross section, the more the particle approaches a toroid.

SUMMARY

In Chapters 2,3 and 4, we developed the basics of magnetic phenomena including the properties, measurable quantities and units. We are now in a position to review the materials for the various applications mentioned in Chapter 1 in light of the required properties. We earlier said that a useful means of categorizing magnetic materials is according to frequency of operation. We will in general follow this approach in the materials exposition. It is therefore appropriate that we start with the DC applications. One major part of this area concerns the permanent magnet that certainly qualifies as a DC application. Thus, the next chapter will deal with permanent magnet materials.

References

Landau,L.(1935) and Lifshitz, E. Physik Z. Sowjetunion, $\underline{8}$,153
Rado,G.T.(1953) Rev. Mod Phys. $\underline{25}$, 81
Schlomann,E.(1956) Proc. Conf on Magn. and Mag. Mat. $\underline{91}$,200
Schlomann,E.(1958) J. Chem Phys. Sol., $\underline{6}$,242
Schlomann,E (1971) J. Phys., $\underline{32,443}$
Smit,J.(1954) and Wijn,H.P.J. Advances in Electronics and Electron Physics, $\underline{6}$,69
Snoek,J.L.(1948) Physica, $\underline{14}$,207
Watson, J.K. (1980) Applications of Magnetism, John Wiley, New York, 181

5 MATERIALS FOR PERMANENT MAGNET APPLICATIONS

INTRODUCTION

In beginning of our discussion of ferromagnetic materials for electrical applications, the simplest case would seem to be the use as a permanent magnet. We say this because the sole purpose of the magnet is to create a D.C. magnetic field. The permanent magnet is also the popular conception of what a magnetic material is. In some cases, the accessory circuits may be more complex, but the magnetic considerations for the materials are fairly straightforward. The past twenty years have seen great advances in the area of high-energy permanent magnets. Alnico 5 that was the old standby for many years with an energy product of 5 million Gauss-Oersteds has been surpassed with a material whose energy product approaches 50 million Gauss-Oersteds. However, other considerations must be made in the proper choice of a material for a permanent magnet application.

HISTORY OF PERMANENT MAGNETS

Early Permanent Magnet Materials

The earliest known magnet material was, of course, magnetite or lodestone as it was called. The first man-made magnets were prepared by touching or rubbing iron needles on lodestone. These formed the essential part of compasses. Gilbert attached pole pieces to lodestone magnets, making them more powerful. He also found steel bars forged or melted and cooled in the earth's magnetic field were also magnets.

With the development of the electromagnet, interest in permanent magnets waned. It was not until the start of the twentieth century that development of permanent magnet materials started. It is interesting that one of the first of such materials produced in 1901 were the so-called Heusler alloys which strangely enough contain none of the ferromagnetic elements as we know them. Their general composition was Cu_2MnAl

Permanent Magnet Steels

Although iron-carbon alloys were used earlier for permanent magnets, the first improved material was developed in about 1910. The saturation of iron is reduced by carbon additions but the coercive force and energy product are improved greatly in a specimen quenched between 780 and 850° C. Mn added to the carbon steel facilitates rolling without affecting magnetic properties. Tungsten steels were also tried but were supplanted by chrome steels during World War I due to the tungsten shortage. Cobalt steels were invented in 1917 by the Japanese and were called KS or Honda steels. Cobalt increases the saturation of iron and the substituted alloy thus gives them high remanences. There are low, medium and high alloy cobalt steels reflecting the amount of cobalt. Other cobalt steels included the addition of Mo and W. A composition of 12% Co and 17% Mo had commercial use and was known as Remalloy.

AlNi and Alnico Type Permanent Magnet Alloys

The Japanese (Mishima 1932) experimented with alloys containing iron, nickel and aluminum which had higher energy products than Remalloy. The new material had a coercive force, H_c, of 475 Oe. In the U.S., the name for these alloys was Alnico and it was further improved by the addition of Co and later again with the addition of Cu. These additions and appropriate heat treatments gave good properties in many different sizes and shapes of magnets. The standard material for many years after that was Alnico V. An essential part of the heat treatment of this material was the cooling from a high temperature in a strong magnetic field prior to the final hardening treatment. By casting the alloy into a special cooled mold, columnar or directional grain growth could occur which increased the orientation and gave much higher energy products. Values on the order of 7.0 to 7.5 MGO were achieved. Other Alnico materials came later with even greater values up to 9 MGO.

Oxide Permanent Magnet Materials-Hard Ferrites

Although oxidic magnetic materials were investigated earlier, Kato and Takei (Kato 1933) produced the first useful ferrite, cobalt ferrite, by cooling in a magnetic field. The first commercial material was described by Went (1952) and marketed by Philips under the name of Ferroxdure. Hard ferrites are basically barium or strontium ferrites and are available either as non-oriented (isotropic) or oriented (anisotropic) variations. Hard ferrites are made of fairly low cost raw materials and by simple ceramic processes. Their ability to be molded into many complex shapes is an advantage over some of the metallic materials. There are several different variations of the permanent magnet ferrites depending on the properties most desired as well as the cost of the component.

Manganese-Based Magnet Materials

Several manganese-based materials have been tried as permanent magnet materials. The first, MnBi, manganese bismuthide, was studied by Guillaud (1943) and developed by the Naval Ordinance Laboratory under the name of Bismanol. This material is classified as a fine particle magnet since the material must be ground into a fine powder, suspended in a plastic matrix and oriented while the plastic set up. It had a very high theoretical energy product but unfortunately the magnets made from it oxidized so badly that they fell apart after a short time.

Another such material was an alloy of manganese and aluminum. Some of this material was offered as a commercial product by a Japanese company but was

MATERIALS FOR PERMANENT MAGNET APPLICATIONS

Another such material was an alloy of manganese and aluminum. Some of this material was offered as a commercial product by a Japanese company but was probably withdrawn because of quality problems. The coercive force, H_c was listed at 2750 Oe and the energy product 3.5 MGO.

Fine Particle Iron Magnets

Dean and Davis (Dean 1941 produced permanent magnets by pressing fine iron particles made from the electrodeposition of metal on the surface of mercury. The material had an H_c of 15Oe. and a B_r of 9000. Using cobalt or aluminum, even higher values were obtained. The General Electric Company commercialized the material under the name of Lodex. The magnets were categorized as ESD or elongated single domain magnets. The presence of zinc renders the powder non-pyrophoric and inhibits grain growth. The energy product was in excess of 1.6 MGO.

Rare-Earth Cobalt Permanent Magnet Materials

A major advance in the technology of permanent magnet materials came with the development of the rare earth-cobalt materials. Although the theoretical properties of these materials were known for some time, major problems in processing prevented their exploitation. The samarium materials are probably the best known of the family. There are two basic compositions, $SmCo_5$ and Sm_2Co_{17}. The present commercial materials are mostly the 2-17 although the 1-5 materials were developed first in the 1960's. The 2-17 has superior magnetic properties but more of the costly samarium. It is also more difficult to produce consistently. Both materials are made by powder metallurgy and the powder is pressed in a magnetic field to produce an anisotropic material. Because of the high cost of the material, it is not economical to produce the isotropic material.

Neodymium-Iron-Boron Magnet Material

This is the most recent development in permanent magnet materials and represents another significant improvement over previous material, including the samarium-cobalt material discussed in the previous section. The discovery of this material was made simultaneously by the Japanese (Sagawa 1984) and the American General Motors group (Croat 1984). The cobalt availability at the time created problems for the samarium cobalt magnet market. Problems of previous neodymium iron materials were overcome by the addition of boron. The Japanese method involved casting an ingot, followed by breakup into powder and pressing. The General Motors method involved the melt-spinning of the alloy into a fine wire followed by powdering and pressing. Although the properties were not quite as good as those obtained by the Japanese method, the process was much simpler.

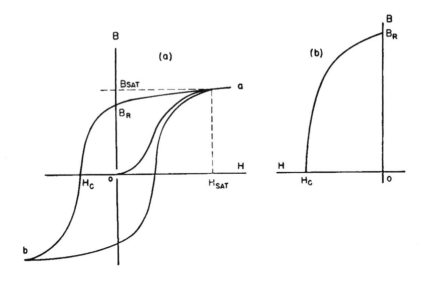

Figure 5.1- Saturation hysteresis loop (a) and demagnetization curve (b) of a permanent magnet material (Underhill 1956)

Miscellaneous Magnet Materials

A novel material with a very special property is an alloy of silver, manganese and aluminum called Silmanol. The energy product is only .085 MGO but it has the ability to withstand severe demagnetizing field. Its use therefore is very limited. Another exotic alloy using very costly raw materials is platinum-cobalt. The energy product is quite high at 5.5 MGO but its cost is so prohibitive that, with the availability of the new high-energy materials to be described next, their production has all but disappeared.

GENERAL PROPERTIES OF PERMANENT MAGNETS

Demagnetization Curves of Permanent Magnets

We have previously mentioned that the most important part of the hysteresis loop (Figure 5.1a) for permanent magnets is the second quadrant. This portion of the loop is known as the demagnetization curve and it is along this part of the curve that the magnet functions (Figure 5.1b). During the process of magnetizing the sample, the magnet is subjected to a field that produces a flux density close to saturation, B_s. When the magnetizing field is reduced to zero, the induction drops back to a value, B_r. If in addition to this drop, the magnetic circuit is opened (a gap is inserted) by removal of soft magnetic flux keepers, the induction drops back to a value less than B_r due to the presence of a demagnetizing field H_d. See Figure 5.2. How much it

MATERIALS FOR PERMANENT MAGNET APPLICATIONS

drops depends on the dimensions of the magnetic circuit and the shape of the demagnetization curve. The ratio of the B level to the H level at the point of operation is called the permeance coefficient and the line from the origin to that point is called the shearing line. Demagnetization curves of commercial permanent magnets usually contain a scale for the permeance coefficient along the top and left side of the plot. The shearing line, when extended to the scale, reads directly in permeance and is equivalent of the permeability of a magnetization curve. (Figure 5.3)

Maximum Energy Products of Magnets

A frequently used criteria of quality of a permanent magnet is the $(BH)_{max}$ product. This is the maximum value that can be obtained by multiplying the corresponding B

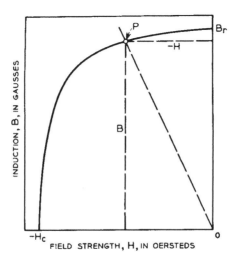

Figure 5.2- Demagnetization curve and shearing line of a permanent magnet material (Bozorth 1951)

and H values at the point of operation on the demagnetization curve. It really represents the maximum amount of energy that can be stored in a given volume of the material. For convenience, contours of constant $(BH)_{max}$ products are superimposed on the demagnetization curve (Figure 5.3). The point of operation can then by correlated with its $(BH)_{max}$ product by noting where it intersects the contours. Since the maximum energy stored occurs at this point, it is recommended that operation of the magnet be designed to be at or near the $(BH)_{max}$ point.

Demagnetization Curves of Permanent Magnet Materials

When a non-oriented (isotropic) permanent magnet material is magnetized and the field removed and the B value drops to (B_r), the moments will be arranged so that

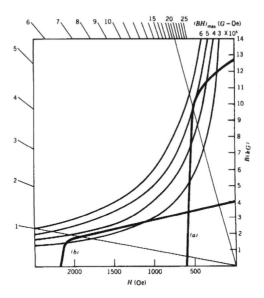

Figure 5.3- Demagnetization curves, $(BH)_{max}$ contours and permeance coefficients for Alnico and b) Ferrite permanent magnet material

their directions will be oriented in the hemisphere closest to the field direction. The average moment will come out to about 1/2 of what one would expect if all the moments were aligned. These magnets have lower B_r than the oriented type but often have somewhat higher H_c. Normally, to effectively saturate the sample requires a magnetizing field of about 4-5 times the coercive force.

If the particles are single-domain and with the magnetization along the easy axis, orienting the particles during the pressing will then align all the moments in one direction. Here, after saturation, there should be practically no degradation of B_r. As in the previous case, one would then expect the B_r to be about twice the B_r of non-oriented sample. However, higher density is sometimes accomplished and more demagnetization effects occur which lowers the coercive force. Figure 5.3 shows the normal demagnetization curves for several hard ferrite materials. Different ferrite magnet materials can be made to optimize H_c, B_r, or the $(BH)_{max}$ product.

Normal and Intrinsic Hysteresis Loops of Permanent Magnet Materials

Thus far, both soft and hard materials have used the same manner of expressing the magnetization and demagnetization process even though there is much difference in the components of the magnetic flux in the two types. It is of interest to examine another way of plotting hysteresis loop and demagnetization curves for permanent magnets. In soft magnetic materials, the magnetization, M, is much larger than H and thus H contributes very little to B. When a material approaches saturation, the

MATERIALS FOR PERMANENT MAGNET APPLICATIONS

magnetization curve will almost flatten out at a constant value (except for small H increase) and the value of B_s is almost equal to $4\pi M_s$. In permanent magnetic materials, since H may be on the same order of magnitude as B (especially with ferrites), we cannot really determine B_s as a material constant from a normal loop. To overcome this deficiency, one can plot 4π M or B-H versus H and then the saturation will be an inherent property of the material alone and the curve will truly flatten out. This type of curve is called the intrinsic magnetization curve for obvious reasons. The saturation induction is B_{si} and the coercive force is H_{ci}. A normal and an intrinsic hysteresis loop are shown in Figure 5.4 We would not expect much difference between the two demagnetization curves for an alnico magnet since H_c is so much lower than B_r. However in some high H_c ceramic magnets, there is much difference (See Figure 5.6). Here, H_{ci} is much larger than H_c and on the order of magnitude of B_r. Note B_r is the same in both systems since H=0. It is impossible for the normal H_c to be as large as B_r, since the demagnetization curve would slope upwards from B_r in the intrinsic curve, a situation that cannot occur.

COMMERCIAL PERMANENT MAGNET MATERIALS - PROPERTIES

Because of the rapidly changing scene in permanent magnet materials development in the past 50 years, the commercial availability of materials is also drastically different. Bozorth (1951) in his section on permanent magnets, lists the steels, the Alnicos, the cobalt alloys such as Vicalloy platinum-cobalt and some nickel alloys such as Misch metal and Magnetoflex. A brief reference at the close of the chapter notes the existence of hard ferrites primarily Vectolite that was a cobalt ferrite. At the present time, the magnetic steels have very little usage in electrical and electronic applications, having been replaced by new generations of oxide and intermetallic materials. This section will list the properties of the presently available materials and look into the possible materials of the future that are now the subjects of research and development.

Commercially Available Permanent Magnet Steels

The MMPA lists the magnetic steels as in Table 5.1 These consist of the chrome steels and the cobalt steels. The ASM handbook lists a tungsten steel but this is mostly of historical interest. Although the B_r's of these materials are quite high, the

Table 5.1-Magnetic Properties of Permanent Magnet Steels

Magnet Material	$(BH)_{max}$ (MGO)	Magnetic Properties	
		B_r (gauss)	H_c/H_{ci} (oersteds)
3½% Cr Steel	.13	10300	60
3% Co Steel	.38	9700	80
17% Co Steel	.69	10700	160
38% Co Steel	.98	10400	230

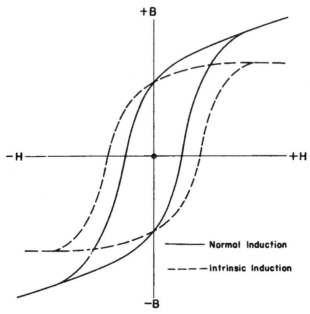

Figure 5.4 Normal and intrinsic hysteresis loops of a permanent magnet material (Parker 1962)

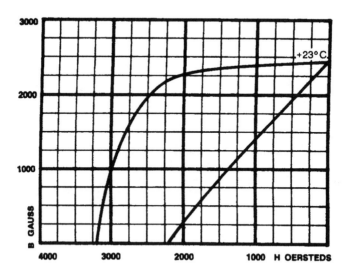

Figure 5.5-Normal and intrinsic demagnetization curves of an anisotropic ferrite material (Fuji 1983)

MATERIALS FOR PERMANENT MAGNET APPLICATIONS 83

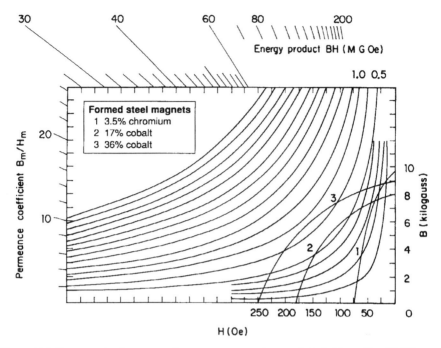

Figure 5.6 - Demagnetization curves for some permanent magnet steels From Jiles (1991)

H_c's are very low and thus are extremely vulnerable to external magnetic fields. Because of the low coercive forces, the maximum energy products are also low. A higher coercive force and maximum energy product can be obtained with cobalt additions but that makes them too costly for the properties obtained. Typical magnetization curves are shown in Figure 5.6

Alnico Magnet Steels
Although Alnico was known at the time of Bozorth's book, the material has been drastically improved in the subsequent years. Bozorth listed the properties of Alnico as;

$$B_r = 12{,}500 \text{ Gausses}$$
$$H_c = 575 \text{ Oersteds}$$
$$(BH)_{MAX} = 4.5 \text{ MGO}$$

HANDBOOK OF MODERN FERROMAGNETIC MATERIALS

The basic Alnico 5 properties have not increased that much but new Alnicos with improved processing techniques have just about doubled the maximum energy products. Alnico magnets come in several different material varieties, starting with Alnico 1 and progressing to Alnico 5. Some such as Alnico 7 are either not available or made by few suppliers. The compositions and properties of various Alnico type alloys are shown in Table 5.2. In addition to the chemistry variations shown, there are also two main different processes used to form the magnet. These are:

1) Cast Alnico
2) Sintered Alnico

Furthermore, there are subdivisions even in these categories, namely the 1) Isotropic or non-oriented 2) Anisotropic or oriented. The oriented materials are made by heat treating the magnets in a magnetic field. Orienting the moments in one hemisphere as opposed to random orientation has the effect of almost doubling the remanence. Thus we will discuss the Alnicos in 4 categories;

1. Isotropic Cast
2. Anisotropic Cast
3. Isotropic Sintered
4. Anisotropic Sintered

Table 5.2- Properties of Alnico-Type Permanent Magnet Materials

MMPA Brief Designation	Original MMPA Class	IEC Code Reference	Chemical Composition*					Magnetic Properties (nominal)			
			Al	Ni	Co	Cu	Ti	Max. Energy Product $(BH)_{max}$ (MGO)	Residual Induction B_r (kilogauss)	Coercive Force H_c (oersteds)	Intrinsic Coercive Force H_{ci} (oersteds)
ISOTROPIC CAST ALNICO											
1.4/0.48	Alnico 1	R1-0-1	12	21	5	3	—	1.4	7.2	470	480
1.7/0.58	Alnico 2	R1-0-4	10	19	13	3	—	1.7	7.5	560	580
1.35/0.50	Alnico 3	R1-0-2	12	25	—	3	—	1.35	7.0	480	500
ANISOTROPIC CAST ALNICO											
5.5/0.64	Alnico 5	R1-1-1	8	14	24	3	—	5.5	12.8	640	640
6.5/0.67	Alnico 5DG	R1-1-2	8	14	24	3	—	6.5	13.3	670	670
7.5/0.74	Alnico 5-7	R1-1-3	8	14	24	3	—	7.5	13.5	740	740
3.9/0.80	Alnico 6	R1-1-4	8	16	24	3	1	3.9	10.5	780	800
5.3/1.9	Alnico 8	R1-1-5	7	15	35	4	5	5.3	8.2	1650	1860
5.0/2.2	Alnico 8HC	R1-1-7	8	14	38	3	8	5.0	7.2	1900	2170
9.0/1.5	Alnico 9	R1-1-6	7	15	35	4	5	9.0	10.6	1500	1500
ISOTROPIC SINTERED ALNICO											
1.5/0.57	Alnico 2	R1-0-4	10	19	13	3	—	1.5	7.1	550	570
ANISOTROPIC SINTERED ALNICO											
3.9/0.63	Alnico 5	R1-1-20	8	14	24	3	—	3.9	10.9	620	630
2.9/0.82	Alnico 6	R1-1-21	8	15	24	3	1	2.9	9.4	790	820
4.0/1.7	Alnico 8	R1-1-22	7	15	35	4	5	4.0	7.4	1500	1690
4.5/2.0	Alnico 8HC	R1-1-23	7	14	38	3	8	4.5	6.7	1800	2020

*Note: Balance iron for all alloys

MATERIALS FOR PERMANENT MAGNET APPLICATIONS

Isotropic Cast Alnico- The isotropic cast materials are found in magnets with lowest cost raw materials and the lowest processing costs and thus have the lowest selling prices. These are known as Alnicos 1,2,3 and sometimes 4. The cobalt content is much lower than in the other Alnicos. Alnico 3 and 4 contain no cobalt. The properties of these materials are shown in Table 5.3 and reflect the lower quality of the magnets.

*Anisotropic Cast Alnico-*These materials are found in Alnicos 5 to 9 and have the higher cobalt contents and thus the higher qualities. Alnico 5 has the highest B_r values, while the succeeding materials are aimed at higher H_c values. Alnicos 8 and 9 have exceptionally high coercive forces. Alnico 8HE (High Coercive Force) has one of the highest of all. In general, when the composition and microstructure favor one of the B_r or H_c values, the other usually is reduced. Alnico 9 is a premium material combining a high coercive force with a relatively high remanence and thus has a high resultant maximum energy product. See Tables 5.4 and 5.5.

Table 5.3-Properties of Alnicos 2,3 and 4-(Isotropic Cast) From Arnold(1994)

	Max. Energy Product Bd x Hd		Residual Induction Br		Peak Magnetic Force		Coersive Force Hc		Recoil Permeability		Permeance Coefficient B/H @ (BdHd) Max.		Induction at Maximum Energy Product	
	MGOe	KJ/m³	G	mT	Oe	KA/m	Oe	KA/m	G/Oe	10⁻⁷Tm/KA	G/Oe	10⁻⁷Tm/KA	G	mT
Alnico 2	1.60	12.7	7200	720	2000	160	560	45	6.2	7.8	12.0	15.0	4400	440
Alnico 3	1.40	11.1	7000	700	2000	160	475	38	5.1	6.4	14.0	17.5	4450	445
Alnico 4	1.35	10.7	5500	550	3000	240	720	57	4.1	5.2	7.0	9.0	3100	310

Arnold Table 5.4--Properties of Alnicos 5, 6 and 800-(Anisotropic Cast) From (1994)

	Max. Energy Product Bd x Hd		Residual Induction Br		Peak Magnetic Force		Coersive Force Hc		Recoil Permeability		Permeance Coefficient B/H @ (BdHd) Max.		Induction at Maximum Energy Product	
	MGOe	KJ/m³	G	mT	Oe	KA/m	Oe	KA/m	G/Oe	10⁻⁷Tm/KA	G/Oe	10⁻⁷Tm/KA	G	mT
Alnico 5	5.50	43.8	12500	1250	3000	240	640	51	3.7	4.6	19.0	24.0	10000	1000
Alnico 5cc	6.50	51.7	13200	1320	3000	240	675	54	2.4	3.0	18.5	23.0	11000	1100
Alnico 6	3.90	31.0	10800	1080	3000	240	750	60	5.6	7.0	14.0	17.5	7400	740
ArKomax® 800	8.10	64.5	13700	1370	3000	240	740	59	2.0	2.5	18.0	22.5	12000	1200
ArKomax® 800 Hi-Hc	8.10	64.5	13200	1320	3000	240	810	64	2.0	2.5	15.5	19.5	11200	1120

Isotropic Sintered Alnico- The sintered Alnicos are used where the ease of producing small or complex shapes with holes or slots by powder metallurgical processes is exploited. In the isotropic sintered, there is only two listings, namely Alnico 2, again a low Co content low cost material and Alnico 8, a high coercive force material. The properties of the isotropic sintered Alnico are given in Table 5.6

Table 5.5--Properties of Alnicos 8, 8HE and 9-(Anisotropic Cast) From Arnold(1994)

	Max. Energy Product Bd x Hd		Residual Induction Br		Peak Magnetic Force		Coersive Force Hc		Recoil Permeability		Permeance Coefficient B/H @ (BdHd) Max.		Induction at Maximum Energy Product	
	MGOe	KJ/m³	G	mT	Oe	KA/m	Oe	KA/m	G/Oe	10⁻³Tm/KA	G/Oe	10⁻³Tm/KA	G	mT
Alnico 8B	5.50	43.8	8300	830	6000	480	1650	131	2.0	2.5	4.5	5.5	5000	500
Alnico 8HE	6.00	47.7	9300	930	6000	480	1550	123	2.0	2.5	5.5	7.0	5750	575
Alnico 8H	5.50	43.8	7400	740	6000	480	1900	151	2.0	2.5	3.5	4.5	4400	440
Alnico 9	10.50	83.6	11200	1120	6000	480	1375	109	1.3	1.6	7.5	9.5	8900	890

Anisotropic Sintered Alnico-Here, there is a variety of materials whose properties are slightly lower than their cast counterparts but which have the advantage in the forming of the magnets. In addition to the Alnico 5 and 6 that have low coercive forces, there are also the Alnico 8 and the 8HC that have the higher coercive forces. The properties of the anisotropic sintered Alnico are given in Table 5.6

OXIDIC OR HARD FERRITE MATERIALS

Just as there are several different classifications of materials in the Alnicos, there are similar variations in the permanent magnet or hard ferrite magnet materials. While the only variation in the major chemistry is the choice of either barium or

Table 5.6 Sintered Alnico 8-Anisotropic and Non-Oriented-Arnold (1994)

	Max. Energy Product Bd x Hd		Residual Induction Br		Peak Magnetic Force		Coersive Force Hc		Recoil Permeability		Permeance Coefficient B/H @ (BdHd) Max.		Induction at Maximum Energy Product	
	MGOe	KJ/m³	G	mT	Oe	KA/m	Oe	KA/m	G/Oe	10⁻³Tm/KA	G/Oe	10⁻³Tm/KA	G	mT
Sint. Alnico 8B	5.00	39.8	8000	800	6000	480	1700	135	1.8	2.3	4.5	5.5	4750	475
Sint. Alnico 8H	5.00	39.8	7000	700	6000	480	1850	147	1.9	2.4	3.4	4.3	4100	410
Unoriented Sint. Alnico 8	2.40	19.1	5500	550	6000	480	1475	117	2.6	3.3	3.8	4.7	3000	300

strontium ferrites, there are other differences. There are the isotropic (unoriented) or the anisotropic (oriented) ferrite magnets. The MMPA designations and some properties of the various types of hard ferrites are shown in Table 5.7.

There are the bulk ceramic magnets as well as the plastic bonded magnets. There are also special process variations to favor H_c, B_r, or $(BH)_{max}$ properties. The original hard ferrites were mainly barium ferrite. However, at present, they are mainly strontium ferrite because of the better properties and the non-toxic property of the strontium.

COMMERCIAL ORIENTED AND NON-ORIENTED HARD FERRITES

Table 5.7 lists the pertinent values of several types of commercially available hard ferrites. Note that there can be many different variations depending on what pa-

MATERIALS FOR PERMANENT MAGNET APPLICATIONS

rameter is important. Thus, with oriented Ba ferrite and Sr ferrite one can get either high B_r or high H_c. Note that the highest $(BH)_{max}$ occurs with the high B_r material. The recoil permeabilities are very similar.

We have said that the highest energy products for ferrite magnets are found in the ones with the highest remanence. Figure 5.7 shows the demagnetization curve and energy product contours for a commercial high remanence magnet that also has a high energy product. The remanence is 4400 Gausses and the energy product is 4.6 MGO. The (B-H) versus H demagnetization curve is also shown along with the energy product contours, The energy product is read at the point closest to the normal demagnetization curve.

Table 5.7- Properties of Permanent Magnet Ferrites

MMPA Brief Designation	Original MMPA Class	IEC Code Reference	Chemical Composition	Magnetic Properties (nominal)			
				Max. Energy Product $(BH)_{max}$ (MGO)	Residual Induction B_r (gauss)	Coercive Force H_C (oersteds)	Intrinsic Coercive Force H_{CI} (oersteds)
1.0/3.3	Ceramic 1	S1-0-1	MO · 6Fe2O3 (M represents	1.05	2300	1860	3250
3.4/2.5	Ceramic 5	S1-1-6	MO · 6Fe2O3 Barium, Strontium or	3.40	3800	2400	2500
2.7/4.0	Ceramic 7	S1-1-2	MO · 6Fe2O3 combination of the two)	2.75	3400	3250	4000
3.5/3.1	Ceramic 8	S1-1-5	MO · 6Fe2O3	3.50	3850	2950	3050
3.4/3.9	–	–	MO · 6Fe2O3	3.40	3800	3400	3900
4.0/2.9	–	–	MO · 6Fe2O3	4.00	4100	2800	2900

NOTE FOR ALL MATERIALS:
Recoil Permeability Range—1.05 to 1.2
Recommended Magnetizing Force—10,000 oersted minimum

For high coercive force materials, some remanence is usually sacrificed. Figure 5.8 shows a demagnetization curves for a high coercive force ferrite magnet. The remanence and energy product are lower than the high remanence magnet.
In addition to the sintered bulk ferrite magnets, there are also plastic bonded magnets. As expected, the properties are inferior to the sintered variety but they offer great manufacturing advantages primarily in unusually-shaped forms. The properties of a plastic-bonded magnet are given in Table 5.8. Figure 5.9 shows the normal and intrinsic demagnetization curves of a ferrite magnet as a function of temperature.

Maximum (BH) of Commercial Hard Ferrite Materials
Since B_r and H_c are almost the same in isotropic ferrite materials and since the curve is a straight line the $(BH)_{max}$ value, or Energy Product as it is commonly known, occurs almost at a point where both B_r and H_c are halved.

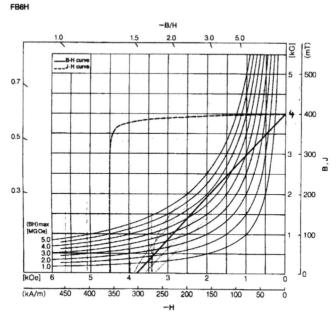

Figure 5.7- Normal and intrinsic demagnetization curves and energy product contours of a high remanence ferrite magnet. From TDK

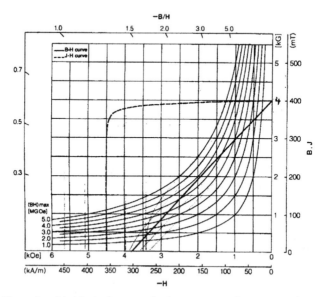

Figure 5.8- Normal and intrinsic demagnetization curves and energy product contours of a high coercive force ferrite magnet From TDK

MATERIALS FOR PERMANENT MAGNET APPLICATIONS

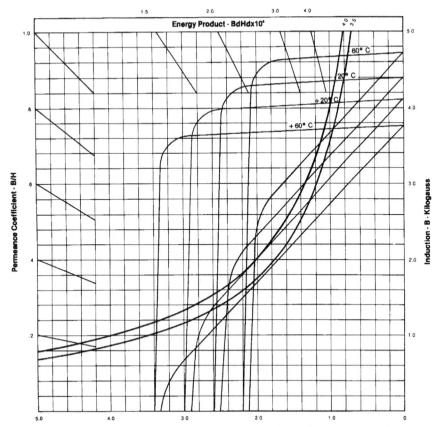

Figure 5.9-Normal and intrinsic demagnetization curves of a ferrite magnet as a function of temperature. Energy product contours and permeance coefficients are shown. From Arnold

$$(BH)_{max} = (B_r/2) \times (H_c/2) = B_r H_c/4 = B_r^2/4 \quad [5.14]$$

For an isotropic material, B_r is about 2000 Gausses and H_c is about 2000 Oersteds (Sometimes designers will use Gausses for both units). According to our approximation $(BH)_{max}$ should be about 1×10^6 and this is indeed true. $(BH)_{max}$ occurs at a B of about 2300 Gausses and an H of about 1700 Oersteds. For a high quality oriented material, B_r is about 4000 while H_c is slightly higher than the isotropic case at about 2500 giving a calculated $(BH)_{max}$ of about 4 MGO which is also close to the actual value. Figure 5.10 shows the various shapes that can be used for the ferrite magnets. One the most important is the curved arc segment for permanent magnet motors.

RARE-EARTH-COBALT PERMANENT MAGNET MATERIALS

Although these materials have been known since the 1950's, technical difficulties prevented their exploitation until much later. The original materials were based on the composition, $RECo_5$ with the main rare earth being samarium ($SmCo_5$). Since then, other compositions primarily the Sm_2Co_{17} have been found to have as good or better properties than the original product. These materials represented a major advance in properties over the then best existing materials mainly the Alnicos. Because of their excellent properties without the instabilities shown by other materials, they are sometimes called the ideal permanent magnet materials. Since the material costs are so high, it is not practical to produce isotropic magnet from them, so all rare earth cobalt magnets are anisotropic.

Properties of Rare Earth-Cobalt Magnets

The remanence values of these materials are variable from about 5000 Gausses to about 11,000 Gausses. The coercive forces can also vary from 4500 to 9000 Oersteds. As in the case of previous materials, the chemistry and processing can be varied to favor H_c, B_r, or BH_{max}. It is important to note that with some of these materials, the intrinsic coercive force can be much higher (almost double in some cases) than the normal coercive force. This property is especially important when the magnet is subject to large demagnetizing fields. The properties of these materials are shown in Table 5.9. The demagnetization curves of these materials are shown in Figure 5.11.

Figure 5.10- Shapes for hard ferrite applications including the arc motor segments

MATERIALS FOR PERMANENT MAGNET APPLICATIONS

Table 5.8 Properties of Plastic Bonded Hard Ferrite Material (Arnold, 1994)

Property	Value* (Units)	
	CGS/U.S.	SI
Intrinsic Parameter[a] (BrHx)	5.2 x 10^6 gauss-oersteds	41.3 kJ/m^3
Residual Induction[1] (Br)	2800 gauss	280 milleteslas
Coercive Force[1] (Hc)	2250 oersteds	179 kA/m
Coercive Force Intrinsic[1] (Hci)	3000 oersteds	238 kA/m
Maximum Energy Product (BH max)	1.9 x 10^6 gauss-oersteds	15.0 kJ/m^3
Thermal Coefficient of Magnetization (reversible)	-0.105% per °F	-0.19% per °C
Thermal Coefficient of Intrinsic Coercive Force (reversible)	0.07% per °F	0.13 % per °C
Peak Magnetizing Force Required	10000 oersteds	800 kA/m

Table 5.9-Properties of Rare-Earth Intermetallic PM Materials

MMPA Brief Designation	IEC Code Reference	Chemical Composition		Magnetic Properties (nominal)**			
		Alloys	Possible Elements	Maximum Energy Product (BH)max MGO	Residual Flux Density Br (gauss)	Coercive Force Hc (oersteds)	Intrinsic Coercive Force Hci (oersteds)
5/16	R5-1	RE Co$_5$	RE = Sm, Nd, MM	5	4700	4500	16000
14/14	R5-1	RE Co$_5$	RE = Sm, MM	14	7500	7000	14000
16/18	R5-1	RE Co$_5$	RE = Sm, Nd	16	8300	7500	18000
18/20	R5-1	RE Co$_5$	RE = Sm, Pr, Nd	18	8700	8000	20000
20/15	R5-1	RE Co$_5$	RE = Sm, Pr, Nd	20	9000	8500	15000
22/15	R5-1	RE Co$_5$	RE = Sm, Pr, Nd	22	9500	9000	15000
22/12	R5-2	RE$_2$TM$_{17}$	RE = Sm, Ce; TM = Fe, Cu, Co, Zr, Hf	22	9600	8400	12000
24/7	R5-2	RE$_2$TM$_{17}$	RE = Sm, Ce; TM = Fe, Cu, Co, Zr, Hf	24	10000	6000	7000
24/18	R5-2	RE$_2$TM$_{17}$	RE = Sm; TM = Fe, Cu, Co, Zr, Hf	24	10200	9200	18000
26/11	R5-2	RE$_2$TM$_{17}$	RE = Sm; TM = Fe, Cu, Co, Zr, Hf	26	10500	9000	11000
28/7	R5-2	RE$_2$TM$_{17}$	RE = Sm; TM = Fe, Cu, Co, Zr, Hf	28	10900	6500	7000
26/20	R7-3	RE$_2$TM$_{14}$B	RE = Nd, Pr, Dy, Tb; TM = Fe, Co	26	10400	9900	20000
27/11	R7-3	RE$_2$TM$_{14}$B	RE = Nd, Pr, Dy, Tb; TM = Fe, Co	27	10800	9300	11000
30/18	R7-3	RE$_2$TM$_{14}$B	RE = Nd, Pr, Dy, Tb; TM = Fe, Co	30	11000	10000	18000
33/11	R7-3	RE$_2$TM$_{14}$B	RE = Nd, Pr, Dy, Tb; TM = Fe, Co	33	11800	10800	11000

* Temperature compensated materials and materials with maximum energy products of 40 MGO are available from various manufacturers.
** To achieve the properties shown in this table, care must be taken to magnetize to technical saturation.

NEODYMIUM-IRON-BORON PERMANENT MAGNET MATERIALS

These materials have advanced the properties of permanent magnet materials yet another step past the above named samarium cobalt materials. In some respect, they really are an extension of the former work with the use of other rare earths and the addition of the boron. Although the first materials contained the neodymium and iron other rare earths have been substituted for some of the neodymium and cobalt has been substituted for part of the iron. Here again, the anisotropic materials are the only ones being produced. However, there is a large-scale use of the plastic-bonded material

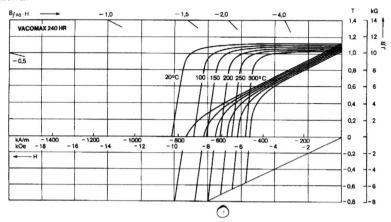

Figure 5.11- Demagnetization Curves of a RE-Co PM material. From Vacuumschmelze

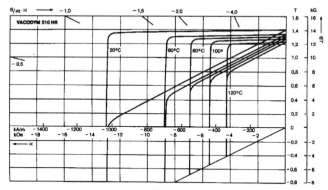

Figure 5.12-Demagnetization Curves of NdFeB Material From Vacuumschmelze (1997)

Properties of Neodymium-Iron-Boron Magnets

The properties of the neodymium-iron boron magnet materials are shown in Table 5.9. The progression shows increases in both the remanences and coercive forces. The demagnetization curves for the NdFeB material is shown in Figure 5.12.

Some less expensive versions of these materials made by melt spinning are also available. The raw materials are less costly than the samarium cobalt materials, but-as will be pointed out later, means to prevent corrosion and other instabilities may

MATERIALS FOR PERMANENT MAGNET APPLICATIONS

increase the overall cost of this material. The properties for the neodymium plastic and rubber bonded materials are given in Table 5.10 and the demagnetization curves in Figure 5.13. The demagnetization curves of the NdFeB materials are compared to those of Alnico and a hard ferrite in Figure 5.14.

Table 5.10-Properties of Plastic Bonded NdFeB Material From Arnold (1994)

Property	Units*			
	2002A		2002B	
	CGS/U.S.	SI	CGS/U.S.	SI
Intrinsic Parameter[a] (BrHx)	50 x 10^6 gauss oersteds	400 kJ/m^3	20 x 10^6 gauss oersteds	160 kJ/m^3
Residual Induction[1] (Br)	4900 gauss	490 milleteslas	5100 gauss	510 milleteslas
Coercive Force[1] (Hc)	4100 oersteds	330 kA/m	3800 oersteds	300 kA/m
Coercive Force Intrinsic[1] (Hci)	15000 oersteds	1200 kA/m	7500 oersteds	600 kA/m
Maximum Energy Product (BH max)	5 x 10^6 gauss oersteds	40 kJ/m^3	5 x 10^6 gauss oersteds	40 kJ/m^3
Thermal Coefficient of Magnetization (reversible)	– 0.08% per °F	– 0.15% per °C	– 0.07% per °F	– 0.13% per °C
Peak Magnetizing Force Required	35000 oersteds	2800 kA/m	30000 oersteds	2400 kA/m

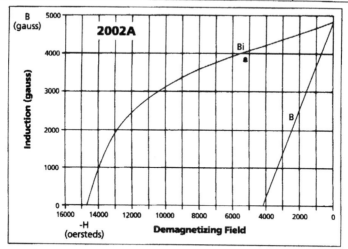

Figure 5.13 - Demagnetization curves for NdFeB plastic-bonded magnet material. From Arnold

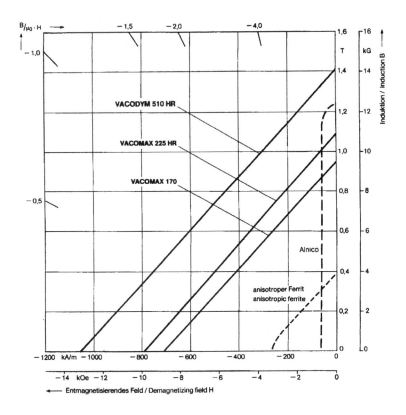

Figure 5.14-Demagnetization curves of NdFeB, Alnico and hard Ferrite.(Vacuumschmelze)

Table 5.11- Properties of FeCrCo Permanent Magnet Materials

Brief Designation	MMPA Class*	IEC Code Reference	Magnetic Properties		(Nominal) BH$_{max}$ (MGO)
			B$_r$ (kilogauss)	H$_c$ (oersteds)	
ISOTROPIC					
1.6/0.46	Fe Cr Co 1		8800	460	1.6
1.6/0.35	Fe Cr Co 2	R6	9900	350	1.6
1.0/0.20	Fe Cr Co		10500	200	1.0
ANISOTROPIC					
5.2/0.61	Fe Cr Co 5	R6	13500	600	5.25
2.0/0.25	Fe Cr Co 250	R6	14000	250	2.0

*Composition is 15 to 35 weight percent chromium, 5 to 20 cobalt, balance iron with minor amounts of other elements present.

From MMPA (1990)

MATERIALS FOR PERMANENT MAGNET APPLICATIONS 95

IRON-CHROMIUM-COBALT MAGNET MATERIALS
These materials are produced in either the isotropic or anisotropic varieties. It is necessary to heat treat them to achieve desired properties. However, the material can be rolled, drawn or otherwise worked before the final heat treatment. It is this property that is of value in their use. Properties for this material are in Table 5.11.

DUCTILE PERMANENT MAGNET ALLOYS
These materials are important when the magnet must be formed into sheet, wire or otherwise worked. As a rule, magnetically hard materials are also mechanically hard. These materials are different in this respect.

Ductile Permanent Magnets-Cobalt Alloys
The characteristics of Remalloy-type material are shown in Figure 5.12. Replacing some of the cobalt with chromium gave a slightly higher energy product and lowered the cost. Remalloy was the first magnet material not using carbon for precipitation hardening. Remalloy could also be hot rolled and machined before final hardeningAnother series of ductile permanent magnet alloys are those containing cobalt and vanadium. The vanadium lowers the magnetic properties but is necessary for fabrication and mechanical considerations. These alloys go under the names of Vicalloy (I and II) and P6 alloy. The P6 alloy is particularly useful as a hysteresis alloy as it has a large area of hysteresis at low drive level.

Ductile Permanent Magnets-Copper Alloys
Many permanent magnet materials are hard mechanically as well as magnetically. At times, there are applications requiring a ductile permanent magnet material. Two such materials are ones containing copper. These two are called Cunife (Copper-nickel-iron) and Cunico (Copper-nickel-cobalt).

MISCELLANEOUS PERMANENT MAGNET MATERIALS
Primarily for historical significance, certain miscellaneous permanent magnet materials are listed. The platinum-cobalt material has been known for some time and has good properties but its cost is prohibitive. The ESD magnets developed by GE had some interesting applications but is rarely used. MnBi (Bismanol) also has good properties, but corrosion problems preclude its use. The same might be said for the MnAlC material. The properties of these materials and some others are listed in Table 5.12

CRITERIA FOR CHOOSING A PERMANENT MAGNET MATERIAL
With the wide variation of properties available in permanent magnet materials, one must use some guidelines or criteria to specify the optimum material to be used in a specific application. These guidelines are meant to correlate the restraints imposed on the material to be used and can be categorized according to the following conditions;

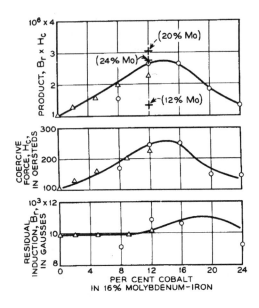

Figure 5.15-Properties of Remalloy permanent magnet materials. From Bozorth (1951)

CRITERIA FOR CHOOSING A PERMANENT MAGNET MATERIAL
With the wide variation of properties available in permanent magnet materials, one must use some guidelines or criteria to specify the optimum material to be used in a specific application. These guidelines are meant to correlate the restraints imposed on the material to be used and can be categorized according to the following conditions;
From MMPA (1990)
  1. Application-Magnetic Field Requirement
  2. Physical or Mechanical- Space Factor, Weight
  3. Stability Requirements
  4. Ductility Requirements
  7. Cost

Each of these factors will be examined separately. As is the case in many material selection problems, one can not get the best of all requirements in one material. Therefore, the choice is made as to which are the most critical and somehow a compromise is reached defining the most appropriate material. We must add that very often the last parameter namely, cost, may narrow the choice considerably.

MATERIALS FOR PERMANENT MAGNET APPLICATIONS

Table 5.12-Properties of Several Permanent Magnet Materials

Magnet Material	$(BH)_{max}$ (MGO)	Magnetic Properties	
		B_r (gauss)	H_c/H_{ci} (oersteds)
3½% Cr Steel	.13	10300	60
3% Co Steel	.38	9700	80
17% Co Steel	.69	10700	160
38% Co Steel	.98	10400	230
Ceramic 2	1.8	2900	2400/3000
Ceramic 6	2.45	3200	2820/3300
Bonded Ceramic	1.5	2500	2100/2200
Alnico 4	1.35	5600	720
PtCo	9.0	6450	4000
Vicalloy 1	.80	7500	250
Remalloy	1.0	9700	250
Cunife 1	1.4	5500	530
ESD 31	2.3	5000	1000
ESD 32	3.0	6800	960
ESD 41	1.1	3600	970
ESD 42	1.25	4800	830
MnA1C	5.0	5450	2550/3150
Bismanol	5.3	4800	3650
Vectolite	.5	1600	900

Magnetic Field Requirement

If the highest magnetic field available is the prime requirement without regard to other factors such as gap size or magnet length, the choice must lie with the materials with the highest B_r. This is so because the field in the gap is equal to the flux density in the material. The choice in this case would be with the Alnico type materials that have remanences on the order of 12,000-13,000 Gausses. The reader must remember that this analysis is for field intensity only. To achieve the high field may mean very small gap and long length of magnet.

Space and Weight Requirements

We will show in the section on design that the highest field that can be obtained for a given volume of material in a specific gap is in a material having the highest maximum energy product. Thus if minimum space for a given field is the main objective, materials such as rare-earth cobalt or neodymium- iron-boron are the most advantageous. If a lower cost and a high-energy-product material are needed, Alnico 9 would be the next nearest competitor. Insofar as weight goes, the ferrites

Stability of Permanent Magnet Materials

One of the important factors affecting the performance of a permanent magnet is its stability. This includes changes in properties due to the following conditions;

1. Temperature
2. Corrosion
3. Mechanical shock or vibration
4. Tensile and Flexural Strength
5. External demagnetizing fields

Temperature Stability of Permanent Magnets

Among the thermal properties of concern in Permanent magnets are the following items;
1. Curie Temperature
2. Maximum Service Temperature
3. Reversible Temperature coefficient of Remanence and Intrinsic Coercive Force.

The importance of the Curie Temperature and Maximum Service Temperature is obvious. The third factor is to be considered in the choice of the material with regard to constancy of field produced. The values for these three properties of the materials discussed are given in Table 5.13 for the RE materials and for the Alnicos in Table 5.14. Figure 5.9 shows the change of magnetization curves of a ferrite magnet as a function of temperature. Note that B_r is reduced with temperature, as expected and H_c is increased. Most materials show an irreversible decrease on heating initially followed by a reversible decrease that is a function of sample dimensions namely the l/d ratio.

Corrosion in Permanent Magnet Materials

This is a serious consideration in some magnetic materials. Some materials such as manganese bismuth were completely eliminated from consideration because they were subject to corrosion even in air at room temperature. On the other hand, the ferrite magnets because of their oxidic nature have no such problem. In materials such as the neodymium-iron-boron materials, it is necessary to protect the material from corrosion by a plastic coating. The coating used to protect the NdFeB permanent magnet materials are listed in Table 5.15 .

Shock and Vibration Stability

The shock stability is related to the brittleness of the material. Because ferrites are ceramics, they are quite brittle and thus cannot take severe shocks without breakage. On the other hand, the ductile materials are able to handle high shocks without breakage.

MATERIALS FOR PERMANENT MAGNET APPLICATIONS

Tensile and Flexural Strength

Depending on the conditions of the application, this factor can assume some importance on the choice of a material. When high tensile forces are encountered, ferrites with their low tensile and flexural strength would not be a good choice. This is the reason that ferrites are used as stators in motors rather than as rotors. The physical characteristics for ferrites are given in Tables 5.16 and 22.2

Table 5.13- Physical and Thermal Properties of the RE Intermetallic Materials

	1-5 Alloys	2-17 Alloys	Nd-Fe-B
Mechanical Properties:			
Modulus elasticity	23×10^6 psi	17×10^6 psi	22×10^6 psi
Ultimate tensile strength	6×10^3 psi	5×10^3 psi	12×10^3 psi
Physical Properties:			
Density	8.2 g/cc	8.4 g/cc	7.4 g/cc
Coefficient of thermal expansion			
Perpendicular to orientation	$13 \times 10^{-6}/°C$	$11 \times 10^{-6}/°C$	$-4.8 \times 10^{-6}/°C$
Parallel to orientation	$6 \times 10^{-6}/°C$	$8 \times 10^{-6}/°C$	$3.4 \times 10^{-6}/°C$
Electrical resistivity	53μ ohm cm.	86μ ohm cm.	160μ ohm cm.
Magnetic Properties:			
Curie temperature	750°C	825°C	310°C
Reversible temperature coefficient			
of residual induction (-100°C to +100°C)	-.043%/°C	-.03%/°C	-.09 to -.13%/°C
Recoil permeability	1.05	1.05	1.05
Max. service temperature ∗	250°C	300°C	150°C

Table 5.14- Thermal Properties if the Alnicos From MMPA (1990)

Brief Designation	Original MMPA Class	IEC Code Reference	Reversible Temperature Coefficient % Change per °C			Curie Temperature		Max. Service Temperature	
			Near B_r	Near Max. Energy Prod.	Near H_c	°C	°F	°C	°F
1.5/0.57	Alnico 2	R1-0-4	-0.03	-0.02	-0.02	810	1490	450	840
5.5/0.64	Alnico 5	R1-1-1	-0.02	-0.015	+0.01	860	1580	525	975
3.9/0.80	Alnico 6	R1-1-4	-0.02	-0.015	+0.03	860	1580	525	975
5.3/1.9	Alnico 8	R1-1-5	-0.025	-0.01	+0.01	860	1580	550	1020
5.0/2.2	Alnico 8HC	R1-1-7	-0.025	-0.01	+0.01	860	1580	550	1020
9.0/1.5	Alnico 9	R1-1-6	-0.025	-0.01	+0.01	860	1580	550	1020

NOTE: The above data is a composite of information available from industry and research sources.

External Demagnetizing Fields

The coercive Force, primarily the intrinsic coercive force, H_{ci}, is the property that is used to determine resistance to external magnetic fields. Inspection of the demagnetization curve will show that as the coercive is increased, the B value is affected less when external fields increase the H value. Design of the circuit in later sections will show this same connection.

Table 5.15-Coatings Used for Protecting NdFeB Materials from Corrosion From Vacuumschmelze (1997)

Tabelle 4 Oberflächenbeschichtungen
Table 4 Surface Coatings

Oberfläche / Surface	Verfahren / Method	Mindestschichtdicke für Korrosionsschutz / Min. layer thickness for corrosion protection	Farbe / Colour	Härte / Hardness	Lösemittelbeständigkeit / Resistance to solvents	Temperaturbereich / Temperature range	Besonderheiten / Comments
Zinn / tin	galvanisch / galvanic	>15 μm	silber halbglänzend / silver semibright	HV 20[1]	sehr gut / very good	<160°C	reinraumtauglich; sehr gute Beständigkeit im Feuchteklima / suitable for clean-room, excellent resistance to humid atmosphere
Nickel / nickel	galvanisch / galvanic	>10 μm	silber halbglänzend / silver semibright	HV 350[1]	sehr gut / very good	<200°C	reinraumtauglich; sehr gute Beständigkeit im Feuchteklima / suitable for clean-room, excellent resistance to humid atmosphere
Elektrotauchlacke / Electrocoating	KTL Gestell rack / Durchlauf bulk	>15 μm / >6 μm	schwarz / black	4H[2] / 4H[2]	sehr gut / very good	<130°C / <150°C	sehr gute Klima + Salzsprühbeständigkeit; excellent climatic + salt spray resistance;
Ni + Sn / Ni + Sn	galvanisch / galvanic	Ni >5 μm, Sn >10 μm	silber halbglänzend / silver semibright	HV 20[1]	sehr gut / very good	<160°C	reinraumtauglich; hervorragende Beständigkeit im Feuchteklima / suitable for clean-room, superior resistance to humid atmosphere
Aluminium gelb chromatiert	IVD	>10 μm	gelb halbglänzend / yellow semibright	HV 20[1]	sehr gut / very good	<500°C	sehr gute Klima + Salzsprühbeständigkeit / excellent climatic and salt spray resistance

1) Vickershärte (Richtwerte) / Vickers hardness (nominal values)
2) Bleistifthärte / Pencil hardness

If one looks at the recoil lines of ferrite magnet materials, one sees that they coincide very nearly to the demagnetization curves. In fact, there is very little change in the recoil line since the curve is so linear itself. Therefore, ferrite magnets especially isotropic ones are quite resistant to change in properties as the gap is changed or as external demagnetization or magnetization occurs.

STABILIZATION OF PERMANENT MAGNETS
One method of stabilization of ferrite magnets against external fields is by subjecting them to a temporary demagnetizing field after magnetization. This method increases stability by lowering the induction. An ac field is often used.

COST CONSIDERATIONS IN PERMANENT MAGNET MATERIALS
Just as there are great varieties of magnetic materials and applications, so there are large variations of prices of permanent magnets. Much of the differences are due to the spreads in material costs but in analyzing the size and shape of the magnet, there also is a spread within one material due to the size and complexity involved. Small cores, in general, are more costly per pound of magnet material because of the added cost in manufacture. This difference is more exaggerated in the lower cost materials such as in permanent magnet ferrites.

MATERIALS FOR PERMANENT MAGNET APPLICATIONS

Raw Material Costs

Some of the low-cost low-quality magnet steels used for inexpensive applications may carry the lowest prices for permanent magnets. However, for the bulk of magnets used for the bulk of electrical and electronic applications, the lowest price permanent magnet material is the hard or permanent magnet ferrite. Barium ferrites appeared first and were followed by strontium ferrites. At first, strontium ferrites were more costly than barium ferrites (about 10% more), but as strontium ferrites came into greater usage, the price for equivalent grades is about the same for both. Ferrite powder before forming sells for from 20 to 30 cents per pound as of this writing. There is a great price differential as we go to the next material namely, Alnico. For Alnico, the starting alloys material costs are about $5 per pound but this price is quite sensitive. The cost of cobalt has remained stable for many years now but in the 1970's, it had risen dramatically due to World supply problems. Not surprisingly, it was during this period that the usage of ferrite magnets really took off.

The next magnet material, cost-wise, is the most recently developed Neodymium-Iron-Boron Magnets. The cost of the raw materials depends on the method of manufacture.

The cost of the basic alloy by vacuum melting or reduction-diffusion is between 12-20 dollars per pound. As a product for resale, one supplier of this material sells the powder for plastic bonded magnet manufacture for 15-20 dollars per pound. This material is made by rapid quenching of the molten alloy or by melt spinning, producing a semi-amorphous powder. The next most expensive permanent magnet material would be the rare-earth cobalt material. Present practice favors the production of the 2-17 material rather than the 1-5 material. The 2-17 material is superior in properties and uses samarium but is more difficult to process. The cost of vacuum melted, cast and hydrogen-decrepitated material is about $80 per pound. If the material is made by calcium reduction and diffusion, the cost is about $60 per pound. Thus, from the ferrite powder at 20-30 cents per pound to the rare earth-cobalt material at 60 to 80 dollars per pound, one can see the wide variation in raw material cost mentioned earlier.

COST OF FINISHED MAGNETS

Hard Ferrites

According to Abraham(1994), typical prices for anisotropic hard ferrite magnets are about $1.20 per pound for flat shapes (rings and blocks). Prices for motor arcs are in the range of $2.20 per pound. The largest tonnage is in wet process anisotropic magnets.

Alnico Magnets

Typical prices for Alnico magnets are in the $15-20 range.

Nd-Fe-B Magnets.
The cost of manufacturing a Nd-Fe-B magnet from powder is about $40 per pound giving a finished price for the magnet of not less than $50 per pound. In addition to this cost, to protect the magnet from oxidation, a coating must be applied after preparation. Even before coating, the surface must be pacified. These steps increase the cost of the magnet greatly. In a disk drive actuator magnet, a 50cent magnet will need about a 25cent coating.

Rare-Earth-Cobalt Magnets
The rare-earth cobalt magnet that started with the highest raw material cost has about the same manufacturing cost as the Nd-Fe-B magnets. Rare-earth-cobalt magnets sell for from 120 to 150 dollars per pound.

CALCULATIONS AND DESIGN OF MAGNETS

Practical Magnetic Circuits-The Air Gap
We can create a toroidal permanent magnet with no gap. However, the usefulness of a permanent magnet in all practical circuits involves the use of a gap, fixed or variable. To the lay person, the shape of a permanent magnet is generally either a horseshoe or a bar. The permanent magnet circuit, in addition to requiring the piece of permanent magnet material that provides the source of the field, usually consists of an air gap in which the useful field is located and connecting sections of a soft magnetic material to direct the flux to the desired location. In this manner, the field can be made more uniform and avoid flux leakage.

Dimensions and Units in the Magnetic Circuit
The concept of flux, ϕ, (in maxwells (cgs), webers (MKSA) or the just line/cm^2 or lines/in^2 (English)) being the product of the flux density, B, and the cross sectional area, A, has been dealt with previously. This will be especially useful in permanent magnet design. In closed magnetic circuits, flux lines are continuous from magnet through the gap (excluding leakage).

$$\phi = B_m A_m = B_g A_g \quad [5.1]$$

where m subscripts = material properties
g subscripts = gap properties

Another important concept not discussed earlier is that of magnetomotive force which is the product of the magnetic field strength, H, and the length of the section of the magnetic circuit involved. The definition of the field, H, is given by

$$H = F/m \quad \text{(Oersteds)} \quad [5.2]$$

MATERIALS FOR PERMANENT MAGNET APPLICATIONS

where F = Force experienced 1 cm. away
m = pole strength

The energy of a unit pole moved the length, l, in a magnet circuit is then;

$$E = H \times l \times m \qquad [5.3]$$

Since the unit pole has an m =1

$$E = H \times l \; (1) \qquad [5.4]$$

The magnetomotive force, Hl, is the magnetic potential. In a closed magnetic circuit, there cannot be any change in energy or potential so that the sum of all the Hl's must be 0. Thus;

$$H_m l_m = H_g l_g \qquad [5.5]$$

Since there is no source of magnetization in the gap, $H_g = B_g$. For constancy of flux through the circuit and if $A_m = A_g$, then;

$$H_m l_m = B_m l_g \quad \text{and} \qquad [5.6]$$

$$B_m/H_m = l_m/l_g \qquad [5.7]$$

This ratio is the permeance coefficient which we spoke of earlier and is somewhat equivalent to the permeability of a material in the magnetization curve. Therefore, from the dimensions, we should be able to predict the operating point on the demagnetization curve (excluding leakage) and therefore the field we can expect in the gap. If A_m is not equal to A_g, the equation becomes

$$B_m/H_m = A_g l_m / A_m l_g \qquad [5.8]$$

If we are attempting to design the dimensions for most optimum use of the magnet per volume of magnetic material used we can examine this situation mathematically. By multiplying the two equations relating flux and magnetomotive force and rearranging, we obtain;

$$l_m A_m = B_g^2 l_g A_g / B_m H_m \qquad [5.9]$$

and;
$$V_m = B_g^2 V_g/(BH)_m \quad [5.10]$$

Getting the smallest volume of material for a given field in a given gap volume or conversely, getting the highest field in a set gap for a given volume of material, we should pick the point of operation as the one which gives the highest product of B_m and H_m. This $(BH)_{max}$ can be found from the contours of the particular material. The permeance coefficient at that point will give the ratios of the dimensions and other requirements of cross sectional area and length of gap will optimize the design. The permeance of the circuit may be considered the equivalent of the magnetic conductance with the reciprocal the equivalent of the resistance. This reciprocal is called the reluctance, R, it being then l_g/A_g (Actually μ_g/A_g with μ for air essentially equal to 1). In fact, this reluctance can now be used in an equation equivalent to Ohm's Law of Magnetism. It may be stated as;

$$F = \phi \times R \quad [5.11]$$
Where F = Magnetomotive Force
ϕ = Flux,
R = Reluctance

In most cases, it is the permeance which is used in design calculations but the reluctance idea is important because like resistances in series, they can be additive.

Leakage and Leakage Factors
In our equations thus far, we have not considered the effect of leakage or for flux that exists outside of the magnetic circuit (magnet, soft flux paths and gap. Parker (1962) describes two types of leakage. One is near the gap where some lines bow away from the gap and therefore can not be used. The other is that which leaves the body of the magnet proper. The first leakage will reduce the field in the gap and therefore the length of the magnet must be increased to have the same magnetomotive force in the gap.

$$H_m l_m = f H_g l_g \quad [5.12]$$

The second will reduce the flux in the core and therefore the flux in the gap so additional flux or larger magnet area is required to compensate for the leakage flux. The design equation is;

$$B_m A_m = F B_g A_g \quad [5.13]$$

where f and F are the components of the total leakage. Leakage factors can be calculated rigorously with involved mathematics. There are graphical and empirical methods of estimating leakage coefficients. However, the reader can refer to vendors tables or to books on the subject. Recently, designs of magnet gap factors have

MATERIALS FOR PERMANENT MAGNET APPLICATIONS 105

been studied using finite element analysis. In considering the leakage constants, the magnetomotive forces of the soft magnetic material connectors are also factors.

Table 5.16-Physical Properties of Ceramic Magnets

Property	Typical Value*
Density	4.9 g/cm^3
Coefficient of thermal expansion (25°C to 450°C) —Perpendicular to orientation	10 x 10^{-6} cm/cm/°C
—Parallel to orientation	14 x 10^{-6} cm/cm/°C
Thermal conductivity	.007 cal/cm-sec°C
Electrical resistivity	10^6 ohm-cm
Porosity	5%
Modulus of elasticity	2.6 x 10^7 psi
Poisson ratio	0.28
Compressive strength	130,000 psi
Tensile strength	5000 psi
Flexural strength	9000 psi
Hardness	7 (Mohs)

*NOTE: The above data is a composite of information available from industry and research sources.

OPTIMUM SHAPES OF FERRITE AND METAL MAGNETS

Figure 5.5 shows the demagnetization curves for 2 types of permanent magnetic materials, one for an alnico permanent magnet and the other for a typical ceramic barium ferrite magnet. The BH$_{max}$ product is slightly higher for the alnico than for the barium ferrite but the most prevalent difference between the two curves is the shape. As we would expect the B_r of alnico is much higher than the ferrite but the H_c behaves conversely. This is graphically illustrated by comparing the permeance coeffients at the $(BH)_{max}$ point. For alnico, this value is 20 and for the ferrite it is 1. Analysis of permeances shows that for the alnico, the large B_m needs only a small A_m for a required flux but the small H_m needs a large l_m for the same magnetomotive force. For a hard ferrite, the reverse is true. Thus, optimum shape for alnico is long and thin (prolate), and for ferrite, short with a large cross section (oblate).

RECOIL LINES-OPERATING LOAD LINES

After a magnet is magnetized and operating on its open circuit load line, narrowing the gap will lower the demagnetizing field but the operating point does not move up on the magnetization curve, but, instead, traces another minor loop. See Figure 5.4. The system will equilibrate at a different point and a new operating load line will be established. Now the slope of all these minor loops average out to be a common value known as the recoil permeability. This value is normally given in commercial catalogs and can be used in magnet design.

SUMMARY

The permanent magnet materials of this chapter are among the few magnetic materials that do not operate in an electromagnetic manner. Thus, no windings are used to create the magnetic field in the material. Most of the rest of the magnetic materials in this book do operate electromagnetically. Sometimes, the current used is DC, but mainly it is ac of various frequencies. As a result, the next chapter will deal with the electromagnetic usage of magnetic materials at DC and low frequencies for power applications.

References

Arnold (1992) Plastiform Magnet Material

Arnold (1993) Alnico Permanent Magnets

Bozorth, R.M.(1951) Ferromagnetism, Van Nostrand Reinhold Co., New York

Dean, R.S. (1941) and Davis, S.W. U.S. Patent 2,239,144 App. 7/11/38

Guillaud, C. (1943) Thesis Strasbourg, 1-129

Kato, Y.(1933)and Takei, T. J. Inst. Elect Eng. Jap.53, 408

MMPA (1988) Permanent Magnet Guidelines PMG-88

MMPA (1990) Standard Specifications for Permanent Magnet Materials-0100-90

Mishima, T. (1932) Ohm, 19, 353

Parker, R.J. (1962) and Studders, R.J.,Permanent Magnets and their Application, John Wiley and Sons, New York.

TDK (1998) TDK Ferrite Magnets-(BME-018A)

Underhill, E.M.(1956)Permanent Magnet Design, Crucible Steel Co., Pittsburgh,Pa.

Vacuumschmelze (1997)Rare Earth Permanent Magnets –PD-002

Went, R.S. (1952) Rathenau, G.W., Gorter, E.W. and van Oosterhout, Philips Tech. Rev. 13, 194

6 DC AND LOW FREQUENCY APPLICATIONS

INTRODUCTION
Whereas the preceding chapter dealt with magnetically hard materials, the remainder of the book will deal mainly with soft magnetic materials. The difference between the two types can be surmised from their respective hysteresis loops. As indicated by their coercive forces, a much greater reverse field is needed in the case of the permanent magnet material than that for the soft magnetic material. This feature is consistent with the function of the hard material, that is, to create a magnetic field employing its permanent magnetization or polarization. On the other hand, soft magnetic materials prior to usage are mostly in the demagnetized state and have little or no permanent magnetization. In usage, they are mostly acted on by a electrical current. In some cases, the current is a direct current (D.C.) but in a vast majority of the applications, it is an alternating (ac) current whose frequency can extend from 50-60 Hz. to the GHz. (1,000,000,000 Hz) or microwave) region. For non-microwave applications, the current is generally applied through a wire coil surrounding the magnetic core. The discussions on soft magnetic materials are arranged to start at the lowest frequency and extend to mid range and finally, to the highest frequencies. Most of the low frequency ac applications are in the 50-60 Hz. range, but we have included 400 Hz. applications as well since they are the power equivalents in the aircraft sector of usage. The metallic strip materials are mostly suited for the lower frequencies, so they will be discussed first, followed by the higher frequency metallic powder materials.

MATERIAL REQUIREMENTS FOR DC AND LOW FREQUENCIES

Functions of a Magnetic Material at D.C. or Low Frequency.
Applications at D.C. and low frequencies are considered to be power applications i.e., they operate at high induction levels. As a result, the prime function of a the magnetic material under these conditions is to store large quantities of electrical energy which can then be transformed into;

    1) Mechanical energy (motors)
    2) The reverse (generators)
    3) Other voltages or currents (transformers).

According to Finke (1975), the energy stored in a magnetic field is

$$E = HB/2 \qquad [6.1]$$

Where H= Magnetizing Field in A/m

If H= 500 A/m and B= 20 KG;
$E = 5 \times 10^{-2}$ Ws/cm3,
For a capacitor at 10 KV/cm
$E = 5 \times 10^{-6}$ Ws/cm3

Therefore, a magnetic field can store about 10,000 times the amount of energy than the electric field of a capacitor. In a device, the power is given by

$$P = HBfV_C/2$$
Where V_C = Core volume in cm3

For a specific volume and magnetizing current, to obtain the maximum power we can either increase B or f. Increasing frequency leads to higher Eddy current losses especially in metals. Reducing the thickness of the strip can reduce the effect of these increased losses but this option can be quite costly. Since most of the metallic materials have high saturations and acceptable losses at the low frequencies, the metallic magnetic material is the ideal choice at these applications. The need for low losses is especially critical since the heat generated by these losses can reduce the saturation to an inefficient or even catastrophic level. In terms of efficiency the lowered B means lower power delivered. The catastrophic aspect is derived from a situation where the operating induction approaches the lowered saturation level of the material. Under these circumstances, the current in the winding can become extremely high leading to severe damage to the core or surrounding components. To prevent such occurrences, the designer will de-rate the operating induction to a safer level. However, since the de-rated level is a percentage of the original level, the material with the highest saturation consistent with acceptable losses at that frequency is chosen. Since many of the low frequency applications are consumer items such as small motors, relays, or appliance transformers, the cost of the device or component must be comparatively low. Since the raw material cost is a significant part of the overall component cost, the material must be readily available and easy to purify. Mechanically, the material must be ductile enough to be formed or rolled easily. The rolling requirement is necessary since the most important form of the material is as fairly thin strip. The component is formed by either stacking laminations of the thin strip or by winding toroidal tape cores. Through this method, the Eddy current losses are lowered especially for ac applications. Since iron has a very high saturation and low cost, all the DC and low frequency material will consist off iron and iron-based alloys. We can now discuss the materials available for this particular application.

METALLIC MATERIALS FOR DC OR LOW FREQUENCIES

Having determined the requirements for a magnetic material for use at D,.C. or low frequency ac applications, we can now enumerate the appropriate materials. They are listed in Table 6.1.

Table 6.1- Metallic Materials for Low-Frequency Power Applications

1. Iron
2. Low-carbon Steel
3. Non-Oriented Silicon Steel
4. Oriented Silicon Steel
5. 2-V Permendur and Supermendur-400 Hz. Applications
6. Iron-based Amorphous Metal
7. Iron-based Nanocrystalline material
8. Sintered powdered iron

These materials are listed chronologically as they were developed. They also are listed pretty much according to their cost except for the last material. They all have high saturations but their losses at the low frequencies discussed here vary greatly due primarily to their differing resistivities and also their purity. Other distinguishing features affecting the losses will be discussed later. The first five materials are ductile enough to be rolled into thin gage strip. An interesting situation regarding these materials is the fact that some of them, depending on the processing and strip thickness, can have dual uses such as low and higher frequency versions. This feature can be found in materials such as iron, silicon-iron, and amorphous metal alloys.

History of Magnetic Materials for Low Frequencies

As expected, iron was the first material used for magnetic applications. The high carbon content caused many quality and processing problems. The next material developed was low-carbon steel and produced a significant improvement such that the material is still in great use albeit with many new technical changes. The real big change in magnetic steel quality came in 1882 when Hadfield (1889) accidentally produced an alloy of 1.5% silicon. After intermediate studies and improvements, the first commercial production was made in Germany in 1903 (1912). The first commercial production in the U.S. was made in the same year. The improvement over previous produced iron was fourfold.

1. Permeability was increased
2. Hysteresis Loss decreased
3. Eddy current loss decreased
4. There was no deterioration with aging

Gumlich (1912) in Germany and Yensen (1915,1924) in the U.S. improved the material by control of impurities. The work on the oriented silicon steel proceeded mainly in the 3.5% Silicon alloy. We should point out that alloys between 1-3% silicon were also produced and are still being produced with properties that may not be as good as the 3.5% but with lower costs for lower quality applications. The next area of improvement was in the cold-rolling step. This led to the discovery and a patent by Goss in 1934 for grain oriented silicon steel which was the second milestone in magnetic steels. By a combination of chemistry, hot and cold rolling steps area of improvement was in the cold-rolling step. This led to the discovery and a patent by Goss in 1934 for grain oriented silicon steel which was the second milestone in magnetic steels. By a combination of chemistry, hot and cold rolling steps

Table 6.2-Chemistry of Pig Iron

Element	Per Cent	Element	Per Cent
Fe	93	Si	1.0
C	4	P	0.2
Mn	1.5	S	0.03

From Bozorth (1951)

Table 6.3-Chemistry of Open Hearth Iron

Element	Niagara Per Cent	Westinghouse Per Cent	National Radiator Per Cent
C	0.006	0.02	0.010
Mn	.000	.01	.001
Si	.005	.01	.004
P	.005	.05	.003
S	.004	.003	.003
Al		.01	.009
Cu	.015		.001

From Bozorth (1951)

and an important transformation anneal, properties of this material improved significantly. Through the orientation of the grains, the magnetic flux could be oriented in the most preferred plane and direction which is arranged to be the rolling plane and direction. Further improvements in the grain oriented silicon steels were made by Littman and others. The next advances were made by Japanese workers who developed an improved oriented silicon iron called Hi-B or Orientcore, both names having secondary meanings. High B refers to increased operating induction due to the second term, namely better orientation. This was accomplished by a new insulation additive and improved processing. The next improvement involved laser or mechanical scribing to improve orientation. The latest developments in soft magnetic materials for low frequency lie in the new amorphous materials first introduced by the Allied Corporation and by the Japanese with the nanocrystalline materials.. The techniques is quite revolutionary requiring no rolling as the material is formed in the final gage and needs only a low temperature anneal. Another recent development described later is the increased use of sintered powdered iron.

SOFT IRON AS A MAGNETIC MATERIAL

Pure iron is quite an attractive material because of its high saturation and low cost. It enjoys the last advantage because of its large commercial use in many other nonmagnetic applications and its ready availability. As shown in Table 6.2, pig iron contains about 1% silicon, 1.5% manganese, and 4% carbon. However, purification in the open hearth process (Table 6.3) will reduce the carbon to .02% and the silicon

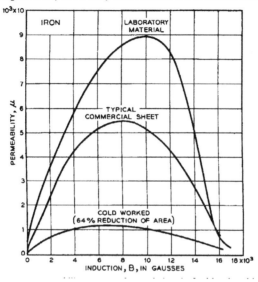

Figure 6.1-Properties of annealed and cold-reduced iron. From Bozorth (1951)

to .002% with the iron content at 93-99.7%. Iron of even higher purity can be made electrolytically or by the iron carbonyl process.

Magnetic Properties of Soft Iron

The magnetic properties of pure iron can vary in maximum permeability from 5,000 in the commercial product to 350,000 for the highest laboratory value. The lowest coercive force is about .01 Oersteds. The wide range in permeability is shown in Figure 6.1. The reported maximum permeabilities of iron are shown in Figure 6.2 over a period of years. The magnetic properties are affected strongly by the the heat treatment according to the temperature, time and rate of cool especially in the 800-900° C. range as we shall see later. Impurities such as carbon can affect the aging of the material. Coercive force and hysteresis loss of some specimens will increase 100% or more after holding at 100 °C. for 200 hours and will change appreciably even at 25° C. This phenomenon is caused by precipitation of C or N if their solubilities are exceeded. The solubilities of some impurities in iron are affected by the presence of others. Manganese reduces the solubility of sulfur and carbon reduces the solubility of silicon. Other elements such as Ti and V counteract the effect of aging.

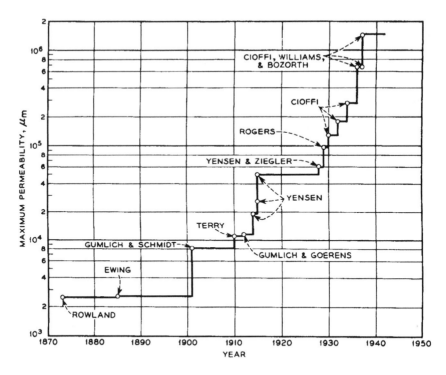

Figure 6.2 Improvement in reported maximum permeability of iron. From Bozorth (1951)

Applications of Soft iron

Although soft iron can be purified and made into an attractive material with high permeability and low coercive force, the cost of processing would be so high that it is only used in certain special applications. One area in which soft iron with moderate amount of impurities is sometimes applied is in electromagnets. The ultra high purity is not necessary in this and other applications. Areas where lower purity soft iron may be used include DC relays, plungers, pole pieces, solenoids and brakes for intermittent use where the magnetic requirements are low and the cost must be low. In most cases, the iron is used in the bulk form (sintered or wrought)since Eddy currents are not operative at DC Soft pure iron is not really appropriate for AC applications because of its low resistivity. Properties of some magnetic irons are given in Table 6.4. As substitutes for soft iron, the next materials with improved properties at low frequencies are the low carbon steels

LOW CARBON STEELS

We have noted that the magnetic applications for soft iron have been limited to mostly DC devices. Modern steel-making practice has produced a series of soft magnetic materials with higher manganese content, silicon levels of less than .6% and controlled amounts of carbon, sulfur, nitrogen and phosphorus. These materials have intermediate properties in some low frequency magnetic applications and

DC AND LOW FREQUENCY APPLICATIONS

Table 6.4-Magnetic Properties of Iron

Curie temperature, °C	θ		770
Saturation magnetic moment/g at 20°C [36W2]	σ_s		217.75
Saturation magnetic moment/g at 0°K	σ_0		221.89
Saturation intensity of magnetization at 20°C	I_s		1 714
Saturation induction at 20°C [37S4]	B_s		21 580
Saturation induction at 25°C [41S1]	B_s		21 580
Saturation intensity of magnetization at 0°K	I_0		1 735.2
Saturation induction at 0°K	$4\pi I_0$		21 805
Bohr magnetons per atom	n_0		2.218
Maximum permeability, commercial product	μ_m		5 000
Highest maximum permeability (polycrystalline)	μ_m		350 000
Coercive force, commercial product	H_c		0.9
Lowest coercive force	H_c		0.01

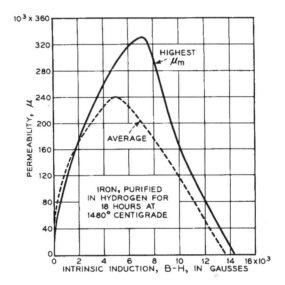

Figure 6.3- Improvement in permeability of iron by treating in hydrogen at 1480° C. From Bozorth (1951)

prices that can be quite attractive. This steel can be the same material of that car bodies are made and thus is a mass-produced item. The tonnages for these materials are actually greater than for the more efficient and costlier silicon steels. In the U.S. alone, over one million tons of these materials are produced annually representing more than 57% of the electrical sheet produced there. In recent years, particularly those after 1980, further improvements in these materials have created considerable interest and made them even more desirable than before.

Chemistry of Low Carbon Steels
A representative chemistry for the impurity content for low carbon steels is given in Table 6.5

Table 6.5- Typical Chemistry of Low Carbon Steels

C	.04-.06%
P	.05-.15%
Mn	.35-.8 %
S	.006-.0255
Si	.05-.25%

The higher manganese content is effective in reducing the free sulfur and oxygen in the steel. It also permits easier recrystallization and increases resistivity. The higher silicon content also increases resistivity over soft iron but since it is lower than silicon steels, the sheet is softer mechanically and easier to punch laminations. This saves die wear, an important economic advantage.

Processing of Low Carbon Steels
Improved steel-making practices has led to lower impurities in the products and continuous casting has provided a more uniform hot band at a low cost. The hot band is pickled to remove surface oxide, cold rolling to an intermediate gage and annealing. Both continuous and batch annealing can be used. The low carbon steels can be supplied in several forms;

1. Full hard or cold rolled- Type 1
2. Semi-processed or cold rolled, annealed and temper rolled -Type 2
3. Fully processed or annealed- Type 2S
4. Hot rolled for D.C. applications

Decarburization of full hard and semi-processed material is necessary to remove carbon and grow grains. The anneal is usually done at about 800 °C degrees in a nitrogen-hydrogen atmosphere having a dew point of 13°C. The 800 °C temperature is chosen to avoid the α-γ transformation (about 910 °C.) and the subsequent need for a very slow cool through the transition point. The fully processed material is coated for inter-laminar resistance. The semi-processed material has a Si+Al content of below .25%, a C content of less than .015%, a Mn content of about .8 % and a S content of less than .01%. The surface roughness should be between 5-12 microns for good stacking factors. Rastogi (1976) has obtained improved magnetic results by developing (100) and (110)textures by special processing

Magnetic Properties of Low Carbon Magnetic Steels
Semi-processed low carbon steels have permeabilities of about 2000 and Watt losses of 5.5 W/kg. The permeability is usually measured at 15kG(1.5 Tesla)

DC AND LOW FREQUENCY APPLICATIONS

Magnetic Properties of Low Carbon Magnetic Steels
Semi-processed low carbon steels have permeabilities of about 2000 and Watt losses of 5.5 W/kg. The permeability is usually measured at 15kG(1.5 Tesla)

Applications of Low Carbon Steels
Low carbon steels are used in ac applications in which the operating conditions are not severe and the performance is not that critical. This would include small motors for intermittent use, generators, pole pieces, laminations for relays and other electromechanical equipment.

Table 6.6

Typical properties of 0.46 mm thick low-carbon steels at 1.5 T and 60 Hz

Steel	Condition	Core loss (W kg^{-1})	Relative permeability
Standard product (0.05% C, 0.40% Mn, 0.12% P, 0.025% S)	full-hard	8.50	2000
Con-Core[a]	fully processed	9.90	
Locore N[b] (silicon steel)	fully processed	9.90	
Vacuum-degassed extra-low-carbon steel (0.006% C)	fully processed	8.40	
Locore M-47[b] (silicon steel)	fully processed	8.40	
Standard product	semiprocessed	6.60	2200
Improved product (0.04% C, ≤0.60% Mn, 0.12% P, 0.025% S)	semiprocessed	6.10	2300
New steel[c]	semiprocessed	5.50	2300

a US Steel trade name b Republic Steel trade names c Inland product

Reprinted from Evetts, Concise Encyclopedia of Magnetic and Superconducting Materials, © 1992, Pergamon Press, p.196, with permission from Elsevier Science.

SILICON STEELS
One method of lowering the core losses in magnetic sheet materials is to raise the resistivity. This is simply done by increasing the silicon content. The low carbon steels have a Si content of about .25% and are distinguished from the silicon steels which content over 1% Si. We shall see that the most widely used Si steels are those which contain about 3.25%Si. Although higher silicon contents increase the resistivity and lower the Watt losses, the material becomes too brittle for normal processing. There is, however, a material which contains about 6.5% Si which has very good magnetic properties but must be specially processed because of its brittleness. We shall also see that there is another classification distinguishing the various types

(small motors and generators. The oriented materials are used in transformers and larger continuous use equipment

Chemistry of Silicon Steels

Silicon increases the resistivity of these magnetic steels (See Figure 6.4) thus lowering Eddy current losses. It also increases permeability by lowering the magnetocrystalline anisotropy and magnetostriction. The effect of Si on saturation is to lower it as shown in Figure 6.4. The effect is explained by the diluent action at low silicon contents (See Figure 6.5) and somewhat greater at higher Si contents. The Curie point is also reduced. The lower silicon materials also have higher permeabilities. The silicon content of these steels varies from 1-3.75% with the greatest use at 3.25%. At this level the chemistry is a compromise between the mechanical properties (ductility) and magnetic properties. Another factor affecting the choice of chemistry is related to the processing and is shown on the Si-Fe phase diagram (Figure 6.6). The region in which the α-γ transition takes place extends to a Si content of about 2.5%. Below that value, if the temperature of the heat treatment is above the lower temperature boundary, the cool rate needed to avoid the high temperature phase is uneconomically slow. However, if the Si content is higher than the 2.5% level, the α-γ transformation does not take place avoiding the problem. The critical value of silicon is also dependent on the C content as shown in Figure 6.7. If the C content is about .07%, the Si content must be about 7% to avoid the transition. Manganese is an important additive to electric sheet. One function is to increase the ductility without affecting the electrical properties. In recent years it has also become useful to add aluminum since its action is quite similar to the silicon. The levels of the interstitial impurities such as C, S, N, P and O must be kept very low.

Processing of Silicon Steels

The silicon steels are either melted in a basic oxygen furnace or continuously cast, the latter method being the more economically attractive one. They are then hot rolled and then acid-pickled to remove the oxide scale. It is then cold rolled with some possible intermediate anneals. After reducing to final gage, the fully- processed material must be given a final stress-relief anneal. The non-oriented materials are then coated but the oriented materials must be subjected to further processing to develop the necessary texture. The transformation process will be described in the section on oriented silicon steels.

Non Oriented Silicon Steels

These are silicon steels whose grains are randomly oriented relative to the rolling plane and direction. The principal attraction for this material is primarily in cost at the expense of some of the magnetic properties including permeability and losses. They may be sold in either the semi-processed or fully processed condition similar to the situation in low carbon steels. In the case of the semi-processed material, the user must provide the stress-relief anneal. The best properties are obtained from the semi-processed material because the stress-relief anneal eliminates the damage done in the punching or other mechanical forming process. They are sold with either the

DC AND LOW FREQUENCY APPLICATIONS

organic (C-3) or inorganic (C-5) coating which provides a good mechanical punchability and the insulating interlaminar resistance. The non-oriented materials are usually sold in either 14, 19 or 26 mil gages. Annealing non-oriented silicon

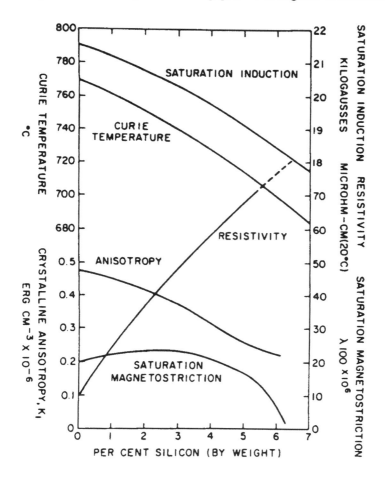

Figure 6.4 Variation of properties of FeSi as a function of Si content. From Littman (1970)

Magnetic Properties of Non-Oriented Silicon Steels

Since the saturation induction, B_s, is an intrinsic magnetic property depending only on the chemistry, the orientation will not have any effect and is given in Figure 6.5. It is important to note that when specifying the magnetic properties of these materials that the measuring conditions are given. The induction B and the permeability is often given as B_{80} or B_{800} referring to the H value in A/m. This corresponds roughly of the strip. The permeabilities of the non-oriented silicon steels are in the range of to 1 and 10 Oersteds respectively. Another factor that must be specified is the thickness about 1500 at B_{800} induction level. The value of B_{800} for non-oriented silicon

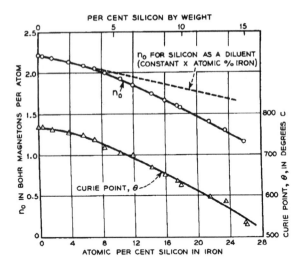

Figure 6.5-Magnetic moment and Curie point of FeSi versus Si content showing the reduction due to dilution effect. From Bozorth (1951)

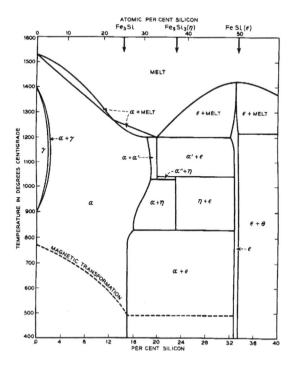

Figure 6.6 Si-Fe phase diagram. From Bozorth (1951)

DC AND LOW FREQUENCY APPLICATIONS

steels is about 1.4 T (14,000 Gausses). The B_{max} level is usually at 1.5 T (15,000 Gausses) although newer materials may be measured at higher B levels. The comparable losses are given in Table 6.7. ASTM Document A677-77 lists the values for non-oriented silicon steels while A683-77 gives similar values for oriented grades. steels at higher temperatures to remove C and S also leads to larger grain size that increases permeability and reduces hysteresis losses.

Oriented Silicon Steels

Goss (1934) developed the grain-oriented silicon steels. Interestingly enough, his name can be an acronym for the product namely Grain-Oriented-Silicon-Steel. The grain orientation produces a so-called texture that is the superposition of the orientation of all the grains in the surface of the sheet. In addition to the orientation of the planes with the surface or rolling plane, there is also the alignment of the crystallographic directions with the rolling direction. There is a decided preference for the

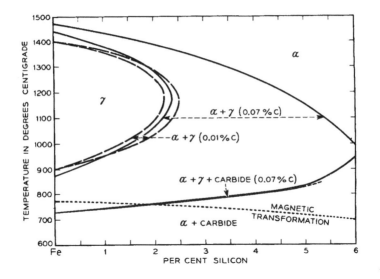

Figure 6.7- Change in FeSi phase diagram with carbon content. From Bozorth (1951)

flux to flow more easily in certain planes and directions and this varies according to the material. Figure 6.8 shows this preference of direction for several materials. In the case of silicon iron, the 110 plane and the 100 direction are the most favored so there is an incentive to produce these orientations. This texture is called the cube-on edge (COE) orientation. Figure 6.9 shows the (110) plane in the rolling plane and the <100> direction in the rolling direction. Figure 6.10 shows how the losses for the cube-on edge texture varies with the permeability as compared to the simple cube-on-face (001) <100> orientation. The losses are lower for the COE for similar permeabilities. Perfection of the COE orientation leads to a higher permeability as shown in Figure 6.11. For the perfectly oriented sample, the B(800) or the B at about 10 Oersteds is 2.03 Tesla. For misorientations of 3°, 6° and 22.5°, the B_{800}

would be 1.8 and 1.4 Teslas respectively. The orientation is produced in a transformation anneal.

Effect of Grain Size and Thickness

In both non-oriented and oriented materials the effect of increasing grain size is to increase permeability (Littman (1971). For non-oriented material, it is the high tem-

Table 6.7-Core Losses for Various Grades of Silicon Steel

AISI designation	Type[a]	Thickness (mm)	Maximum core loss at 60 Hz, 1.5 T[b] (W kg^{-1})	Saturation induction (T)	Nominal Si + Al content (wt%)
M47	NO	0.64	10.8	2.11	1.05
		0.47	10.14	2.11	
M45	NO	0.64	7.94	2.07	1.85
		0.47	6.72	2.07	
M43	NO	0.64	5.95	2.04	2.35
		0.47	5.07	2.04	
M36	NO	0.46	5.29	2.02	2.65
		0.47	4.52	2.02	
		0.36	4.19	2.02	
M27	NO	0.64	4.96	2.02	2.80
		0.47	4.19	2.02	
		0.36	3.97	2.02	
M22	NO	0.64	4.81	2.00	3.20
		0.47	4.08	2.00	
		0.36	3.70	2.00	
M19	NO	0.64	4.59	1.99	3.30
		0.47	3.84	1.99	
		0.36	3.48	1.99	
M15	NO	0.47	3.70	1.98	3.50
		0.36	3.20	1.98	
M6	CGO	0.35	1.46 (2.07)	2.00	3.15
M5	CGO	0.30	1.28 (1.83)	2.00	3.15
M4[c]	CGO	0.35	(1.79)	2.00	3.15
		0.28	(1.68)	2.00	
M3	CGO	0.28	1.09 (1.57)	2.00	3.15
		0.23	0.96 (1.44)	2.00	
M2	CGO	0.18	0.89 (1.40)	2.00	3.15
M2H	HP	0.30	(1.55)	2.00	2.90–3.15
		0.28	(1.46)	2.00	
M1H	HP	0.30	1.05 (1.41)	2.00	2.90–3.15
		0.28	1.02 (1.38)	2.00	
M0H	HP	0.30	1.03 (1.36)	2.00	2.90–3.15
		0.28	0.92 (1.30)	2.00	
		0.23	0.88 (1.21)	2.00	3.15

a NO = nonoriented products, $B_{800} \simeq 1.4$ T; CGO = conventional grain-oriented products, $B_{800} \simeq 1.8$ T; HP = high-permeability products, $B_{800} > 1.88$ T   b Values in parentheses are at 1.7 T   c Below M4, these are industry-accepted values and have not yet been established by AISI or ASTM

perature anneal that also helps remove harmful impurities and lowers losses. In the oriented material, large grains improve orientation. For the COE orientation, the losses reach a minimum value at about .5mm as shown in Figure 6.12 .The lower losses following the minimum are due to improved orientation.

As predicted from the Eddy Current equation (Equation 4.9), the major effect of reduction of sheet thickness would be to lower losses. However, as shown in Figure 6.13, there is an effect of increased losses with thinner sheet possibly due to surface domain wall pinning effects. See Littman (1967).

Newly Improved Oriented Silicon Iron Materials

Developments in Japan by Nippon Steel workers in 1968 were able to improve greatly the quality of oriented silicon steel materials. The material was known as HiB material or Orientcore. The misorientation of the (110) was reduced to about 3%, much lower than previous materials. This factor allowed the material to be used at higher inductions than previously used (17 KG or 1.7 T). The core losses of this material is shown in Table 6.8. Several manufacturing developments were the major reason for the improvements. One was the use of a second grain boundary

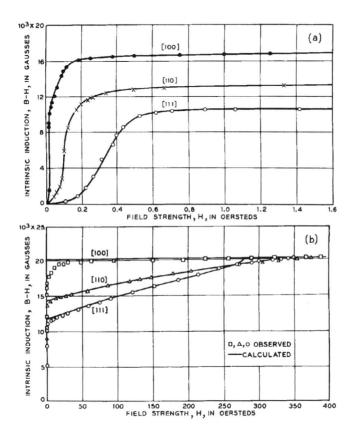

Figure 6.8-Intrinsic induction versus field strength for the principal crystallographic directions of SiFe. From Bozorth (1951)

inhibitor. The one previously used was MnS. The new one was AlN or aluminum nitride. Another feature was a one-step large-reduction rolling step. Further modifications, such as the use of antimony also helped. In addition to the above changes, a new inorganic insulation was used which increased the tensile stress on the material. This also improved the magnetic properties. See Table 6.8.

The magnetic losses of the Hi-B material were 7-20% lower than previous materials for transformer usage. Further improvements were made in the material by the use of laser or mechanical scribing. Finally, there were reductions in the thickness of the material that also lowered the losses. See Figure 6.14.

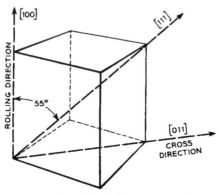

Figure 6.9- A cubic cell showing the different crystallographic directions. The <100> planes are the cube faces. From Bozorth (1951)

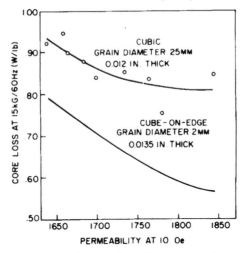

Figure 6.10-Core losses for cube-on-edge (COE) and cube-on face orientations of SiFe of the same permeability. From Littman 1967

IRON-BASED AMORPHOUS METAL MATERIALS.

For many years, it had been predicted by many magneticians (including the author) that amorphous magnetic materials could only be paramagnetic. During the 1960's, small laboratory specimens of weakly ferromagnetic amorphous materials were made culminating in 1969 with the resulting continuous casting of the Metglas^{TM} alloys by Allied in the U.S. Over the next 20 years or so, a large number of amorphous magnetic alloys have been developed by Allied in production equipment and in fairly wide widths. The general composition of these alloys is $M_{80}B_{20}$ in

DC AND LOW FREQUENCY APPLICATIONS

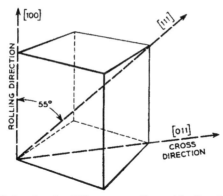

Figure 6.9- A cubic cell showing the different crystallographic directions. The <100> planes are the cube faces. From Bozorth (1951)

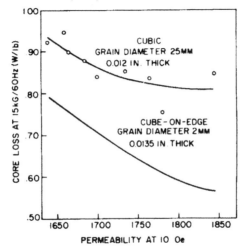

Figure 6.10-Core losses for cube-on-edge (COE) and cube-on face orientations of SiFe of the same permeability. From Littman 1967

IRON-BASED AMORPHOUS METAL MATERIALS.

For many years, it had been predicted by many magneticians (including the author) that amorphous magnetic materials could only be paramagnetic. During the 1960's, small laboratory specimens of weakly ferromagnetic amorphous materials were made culminating in 1969 with the resulting continuous casting of the Metglas®* alloys by Allied in the U.S. Over the next 20 years or so, a large number of amorphous magnetic alloys have been developed by Allied in production equipment and in fairly wide widths. The general composition of these alloys is $M_{80}B_{20}$ in

---

* Metglas® is a registered trademark of Allied-Signal Inc.

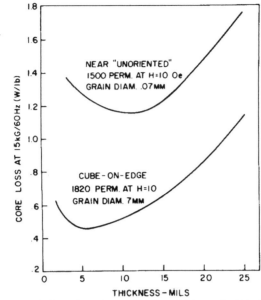

Figure 6.13- Variation of core loss with sheet thickness in SiFe. From Littman 1967

which M is a magnetic metal such as Fe, Ni or Co and B is a non-metal such as B or a combination of non metals including B, Si, C, and P. The composition of the alloys can be tailored to accentuate one or more of the magnetic parameters such as permeability, saturation, squareness and low magnetostriction. Even in a single composition, the material can also be processed for high or low frequency applications.

Since, in this chapter, we are dealing with low frequency power materials, we will concentrate exclusively with the alloys that are best suited for that application. These are the iron based materials in which the composition is close to 80% iron and 20% boron + silicon +carbon. At present, there are about 4 such materials manufactured by Allied under the name of Metglas^{TM} 2605 series. The 26 stands for iron whose atomic number is 26 and the 05 for boron whose atomic number is 5.

These alloys are listed in Table 6.9 with their major magnetic properties. The best combination of characteristics for applications such as 60 Hz. distribution transformers, power transformers and motors would be the 2605TCA because of its low losses at 60 Hz, its high D.C. permeability and its moderately high saturation. These figures are measured on samples heat treated in a longitudinal magnetic field (See later under processing) but as such have only 30% of the losses if an M-2 oriented silicon steel. For airborne applications at 400 cycles where weight and volume are critical a higher saturation 2605CO containing cobalt might be better suited. Its characteristics are shown in Catalog page 20. The comparison of two 2605 alloys versus 4 mil (very thin)3% SiFe is shown in Figure 6.15. The core losses of Metglas

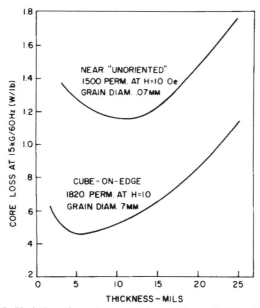

Figure 6.13- Variation of core loss with sheet thickness in SiFe. From Littman 1967

which M is a magnetic metal such as Fe, Ni or Co and B is a non-metal such as B or a combination of non metals including B, Si, C, and P. The composition of the alloys can be tailored to accentuate one or more of the magnetic parameters such as permeability, saturation, squareness and low magnetostriction. Even in a single composition, the material can also be processed for high or low frequency applications.

Since, in this chapter, we are dealing with low frequency power materials, we will concentrate exclusively with the alloys that are best suited for that application. These are the iron based materials in which the composition is close to 80% iron and 20% boron + silicon +carbon. At present, there are about 4 such materials manufactured by Allied under the name of Metglas® 2605 series. The 26 stands for iron whose atomic number is 26 and the 05 for boron whose atomic number is 5.

These alloys are listed in Table 6.9 with their major magnetic properties. The best combination of characteristics for applications such as 60 Hz. distribution transformers, power transformers and motors would be the 2605TCA because of its low losses at 60 Hz, its high D.C. permeability and its moderately high saturation. These figures are measured on samples heat treated in a longitudinal magnetic field (See later under processing) but as such have only 30% of the losses if an M-2 oriented silicon steel. For airborne applications at 400 cycles where weight and volume considerations are critical, a higher saturation alloy, 2605CO containing cobalt might be better suited. The comparison of two 2605 alloys versus 4 mil (very thin) 3% silicon-iron is shown in Figure 6.15. The core losses of Metglas®

Table 6.9
Listing of Various Amorphous Metal Magnetic Materials including the Allied Metglas™ Materials From Jiles (1991)

Alloy	Shape	As cast			Annealed		
		H_c (A/m)	M_r/M_s	μ_{max} (10^3)	H_c (A/m)	M_r/M_s	μ_{max} (10^3)
Metglas #2605 $Fe_{80}B_{20}$	Toroid	6.4	0.51	100	3.2	0.77	300
Metglas #2826 $Fe_{40}Ni_{40}P_{14}B_6$	Toroid	4.8	0.45	58	1.6	0.71	275
Metglas #2826 $Fe_{29}Ni_{44}P_{14}B_6Si_2$	Toroid	4.6	0.54	46	0.88	0.70	310
$Fe_{4.7}Co_{70.3}Si_{15}B_{10}$	Strip	1.04	0.36	190	0.48	0.63	700
$(Fe_{.8}Ni_{.2})_{78}Si_8B_{14}$	Strip	1.44	0.41	300	0.48	0.95	2000
Metglas #2615 $Fe_{80}P_{16}C_3B$	Toroid	4.96	0.4	96	4.0	0.42	130

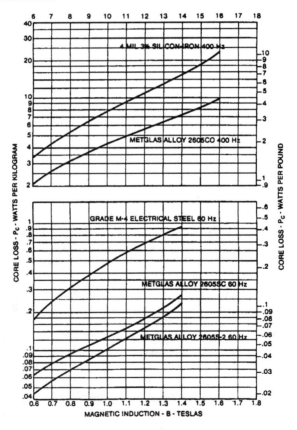

Figure 6.15- Losses in several Metglas™ alloys compared to a 4 mil SiFe alloy. From Jiles (1991)

DC AND LOW FREQUENCY APPLICATIONS

Table 6.9
Listing of Various Amorphous Metal Magnetic Materials including the Allied Metglas® Materials From Jiles (1991)

Alloy	Shape	As cast			Annealed		
		H_c (A/m)	M_r/M_s	μ_{max} (10^3)	H_c (A/m)	M_r/M_s	μ_{max} (10^3)
Metglas #2605 $Fe_{80}B_{20}$	Toroid	6.4	0.51	100	3.2	0.77	300
Metglas #2826 $Fe_{40}Ni_{40}P_{14}B_6$	Toroid	4.8	0.45	58	1.6	0.71	275
Metglas #2826 $Fe_{29}Ni_{44}P_{14}B_6Si_2$	Toroid	4.6	0.54	46	0.88	0.70	310
$Fe_{4.7}Co_{70.3}Si_{15}B_{10}$	Strip	1.04	0.36	190	0.48	0.63	700
$(Fe_{.8}Ni_{.2})_{78}Si_8B_{14}$	Strip	1.44	0.41	300	0.48	0.95	2000
Metglas #2615 $Fe_{80}P_{16}C_3B$	Toroid	4.96	0.4	96	4.0	0.42	130

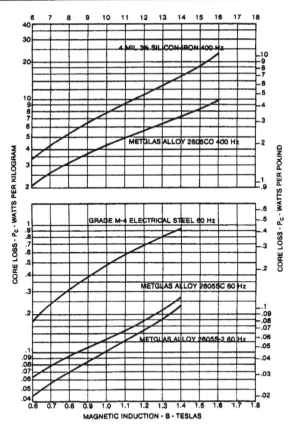

Figure 6.15- Losses in several Metglas® alloys compared to a 4 mil SiFe alloy. From Jiles (1991)

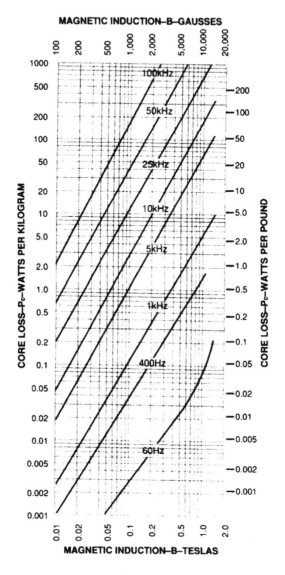

Figure 6.16- Core losses of Allied Metglas® 2605TCA material. From Allied (1994)

Si level above that percentage resulted in difficulty in rolling and other mechanical working because of brittleness. Actually, zero magnetostriction for FeSi is obtained at about 6.5% Si. This alloy has not been used very much because of the difficulty in processing. Schemes such as warm rolling of the alloy have been used but that makes the material costly and the results are not always consistent. Recently, though, this material has been offered a s a commercial product. NKK of Japan says

DC AND LOW FREQUENCY APPLICATIONS

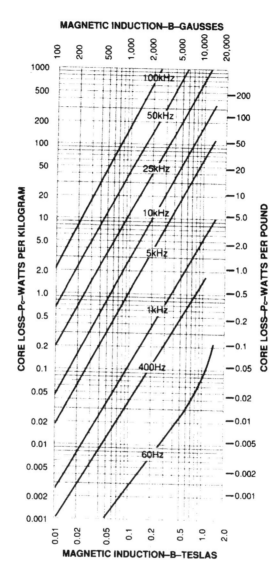

Figure 6.16- Core losses of Allied Metglas® 2605TCA material. From Allied (1994)

Si level above that percentage resulted in difficulty in rolling and other mechanical working because of brittleness. Actually, zero magnetostriction for FeSi is obtained at about 6.5% Si. This alloy has not been used very much because of the difficulty in processing. Schemes such as warm rolling of the alloy have been used but that makes the material costly and the results are not always consistent. Recently, though, this material has been offered a s a commercial product. NKK of Japan says rial, they say the mechanical properties are better. At any rate, the 6.5% Si has lower losses than the grain oriented 3.5% material.

Table 6.8
Magnetic Properties of 6.5% SiFe vs Other Metallic and Ferrite Materials

(1 lb = 0.453 kg)

Material	Thickness(mm)	Induction(B$_s$ (T))	Core losses(W/kg)					Maximum mag Permeabiity
			W10/50	W10/400	W5/1k	W2/5k	W1/10k	
NK Super E-Core	0.05	1.25	0.85	7.0	5.2	8.0	5.1	16,000
	0.10	1.25	0.72	7.3	6.2	11.8	9.7	18,000
	0.20	1.27	0.55	8.1	8.4	19.0	16.7	19,000
	0.30	1.30	0.53	10.0	11.0	25.5	24.5	25,000
Oriented 3%Si steel sheet	0.10	1.85	0.72	7.2	7.6	19.5	18.0	24,000
	0.23	1.92	0.29	7.8	10.4	33.0	30.0	92,000
	0.35	1.93	0.40	12.3	15.2	49.0	46.0	94,000
Non-oriented 3%Si steel sheet	0.10	1.47	0.82	8.6	8.0	16.5	13.3	12,500
	0.20	1.51	0.74	10.4	11.0	26.0	24.0	15,000
	0.35	1.50	0.70	14.4	15.0	38.0	33.0	18,000
Amorphous Fe-Si-B system	0.03	1.38	0.13	1.5	2.2	4.0	4.0	300,000
Ferrite	Bulk	0.37				3.0	3.0	12,000

NANOCRYSTALLINE MATERIALS

The most recent of the soft magnetic metal materials are the nanocrystalline materials. They do have the relatively high saturation required for low-frequency power applications. However, the main application for these materials appears to be in the higher frequency, higher permeability area. First, the brittleness of the material (as it is in the amorphous metal material) dictates that toroids are the only practical shapes. Second, it is doubtful that the cost of the material can compete with the other power magnetic materials mentioned earlier in this chapter. The nanocrystalline materials will be dealt with in later chapters for other applications.

SINTERED SOFT IRON FOR MAGNETIC APPLICATIONS

There has been a significant increase recently in the use of sintered soft iron for magnetic parts for both DC and low to medium frequency ac applications. The used of pressed and sintered compacts by powder metallurgical techniques has been substituted for magnetic lamination sheet materials. By the use of insulated particles, the eddy currents have been lowered permitting their use at higher frequencies than conventional gage SiFe or lamination steels. They are recommended to frequencies up to about 800 Hz. Several suppliers including Hoeganas of both Sweden and the U.S and Magnetics International now offer the new iron powders. The MI powder is marketed under the name of Accucore while Hoeganas's trade name is Somaloy. An obvious advantage to the use of powder metallurgy is the possibility of pressing the parts to near finished dimensions so stacking laminations with the size uncertainty and additional processing is eliminated. Complex shapes such as multiple-

DC AND LOW FREQUENCY APPLICATIONS 131

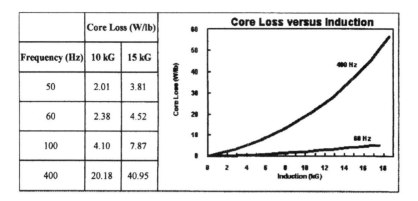

Figure 6.18-Core losses of Accucore sintered iron cores at several frequencies. From Magnetics International

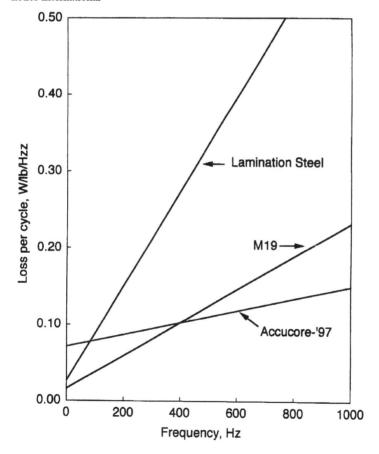

Figure 6.19- Core losses versus frequency curves of Accore sintered iron core compared to lamination steel and SiFe. From Magnetics International

pole motor stators can be formed easily. Also the shapes are isotropic compared to the directional flux in laminations. Figure 6.18 gives the core losses versus of induction for Acccore. Figure 6.19 gives the core losses of the MI Acucore material versus frequency. The breakdown of losses in the lamination material and the sintered iron material is shown in Table 6.11.

Table 6.11
Breakdown of Hysteresis and Eddy Current Losses in Lamination and Sintered Iron Materials

Lamination Steel $P_h \approx 50\%$ $P_e \approx 50\%$

Pressed Core $P_h \approx 95\%$ $P_e \approx 5\%$

The densities and thus the maximum inductions and permeabilities are lower for the sintered material explaining the difference in hysteresis loss percentages. However, the insulated particles give higher bulk resistivities for the sintered materials explaining the difference in eddy current loss percentages.

Hoeganas America uses a fluidized bed process to coat the powders. This equpment is shown in Figure 6.20.

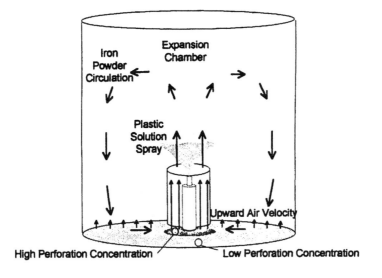

Figure 6.20- Fluidized bed process for coating iron powder for sintered magnetic iron powder cores From Hoeganaes America

DC AND LOW FREQUENCY APPLICATIONS 133

Figure 6.21 shows the permeability versus frequency curves for three coated sintered iron powder materials. Material 1 is a coarse iron powder coated with 0.6 wt. % plastic. Material 2 is a fine iron powder coated with 0.75 wt. % plastic and

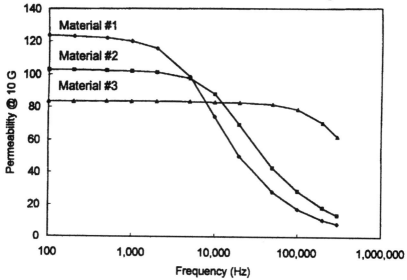

Figure 6.21- Permeability versus frequency curves of 3 different iron powders for sintered magnetic part applications. From Hoeganas America

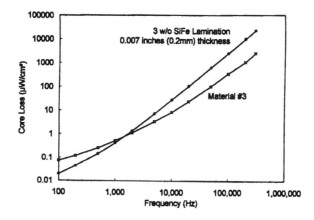

Figure 6.22- Core loss versus frequency curves for a sintered insulated metal powder material versus a 7mil SiFe lamination material. From Hoeganas America

material C used a fine distribution of iron powder that was chemically treated to produce a coating similar to the one used in powder (dust) core applications. See

Chapter 10. As expected, the roll-off frequency is inversely proportional to the permeability. The core losses of material 3 are compared to that of 7 mil SiFe (thin

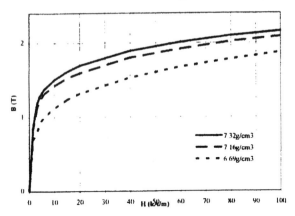

Figure 6.23- Magnetization curves for Somaloy iron powder with Kenolube lubricant as a function of density. From Hoeganas Sweden

gage) in Figure 6.22. At somewhat higher frequencies the sintered iron material has lower losses. We should point out, however, that the measurements are made at extremely low flux levels and do not represent the properties as a power material.

For the case of the Somaloy material from Hoeganas of Sweden, the powder is coated with a lubricant, Kenolube, compacted at 600-800 MPa and heat treated at 500° C. in air or 470-580 ° C. in steam. Alternately it is coated with a binder-lubricant LB1 presses at the same pressures as before and heat treated at 275 ° C. in air. The inductions as function of densities for the Kenolube 0.5% coated material are given in Figure 6.23 and the losses of the coated powder material is compared with a lamination steel lamination (Lam 1018) and a SiFe lamination material in Figure 6.24.

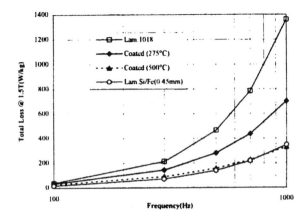

Figure 6.24- Core losses versus frequency curves of two coated iron powder materials compared to two metal lamination materials. From Hoeganas Sweden

SUMMARY

In this chapter, the soft magnetic materials most suitable for low frequency power applications were listed. As required, the all had high saturations, relatively low losses at 50-60 Hz., higher resistivity than iron. Losses generally decreased with increasing silicon content. They all had reasonably low cost and except for the amorphous metal alloys, good workability. For consumer-oriented applications, such as distribution transformers and appliances, the tonnages are very large and lower costs are needed. For industrial equipment, such as in large generators, longer life expectancy requires higher quality at the expense of somewhat costlier materials. The silicon(+ Al) content is generally limited to about 3.25 percent. As we continue to increase the frequency of operation, in the next chapters, we will have to rely on somewhat different materials

References

Allied Corp, Technical Bulletins Metglas 2605TCA
Bozorth, R.M. (1951) Ferromagnetism, D.Van Nostrand Co. New York
Cioffi, P.P.(1932)Phys. Rev. 39, 363
Cioffi, P.P.(1934) ibid, 45, 742
Cioffi, P.P.(1934), Williams, H.J. and Bozorth, R.M.,ibid, 51, 1009
Goss, N.P.(19374) U.S. Patent No. 1,956,959, (App. 8/7/33)
Goss, N.P.(1935) Trans. Am.Soc.Metals,23, 515
Gumlich, E. (1912) and Goerens, P. Ferrum, 10,33
Hadfield, R.A. (1889),J. Iron and Steel Inst.(London)222
Jiles, D. (1991) Magnetism and Magnetic Materials, Chapman and Hall, London
Littman, M.F. (1967) J. Appl. Phys., 38 1104
Littman, M.F. (1971) IEEE Trans. Mag. **MAG-7**, 48
Rastogi, P.K.(1976) A.I.P.Conf. Proc. 34, 61
Yensen, T.D., (1915)Trans. A.I.E.E., 34, 2601

7 SOFT COBALT-IRON ALLOYS

INTRODUCTION
The last chapter discussed the magnetic materials that were most useful at line or mains frequencies. They all possessed relatively high saturation values and low losses at 50-60 Hz. This chapter involves the use of materials at somewhat higher frequencies, 400-1600 Hz. and especially those in which a premium is placed on weight and volume limitations, such as in airborne electrical generators and similar specialty applications. The 400 Hertz operation reduces the size of the component and in addition, materials with the highest saturation are desired. These are the cobalt-iron based alloys. In this chapter, we will review the chemistry, processing and magnetic properties of these alloys. Since the cost of cobalt raises the price of these materials, they again would only be used in the special applications noted above.

HISTORY OF IRON-COBALT ALLOYS
The high saturation of iron-cobalt alloys was first noted by Preuss (1912) and also by Weiss (1912) on the alloy Fe_2Co. The 50% Co alloy with a higher permeability and only a slightly lower saturation was reported by Ellis (1927) but had been patented by Elman(1929 a year earlier. The name Permendur was applied to this alloy because the permeability endured even at high saturations. In 1932, White and Wahl (1932) patented a modification of this alloy which added vanadium to it to make it easier to cold work. Later, Gould and Wenny (1957) refined the process by using high purity materials and special processing. They called the material Supermendur.

CHEMISTRY AND STRUCTURE OF COBALT-IRON ALLOYS
The room-temperature maximum saturation of the binary cobalt-iron alloys occurs at a composition containing 35% Co. This is shown in Figure 7.1. Because of the high H field needed to saturate (17,000 Gauss)for the measurement, the appropriate saturation is given as the intrinsic saturation or B-H instead of the usual B value. This composition corresponds closely to the Fe_2Co found in the earliest studies. With further refinements, Williams(1915) was able to reach a value of 25,800 Gauss.The large drop-off at 95% Co is due to a transition from the Gamma to

138 HANDBOOK OF MODERN FERROMAGNETIC MATERIALS

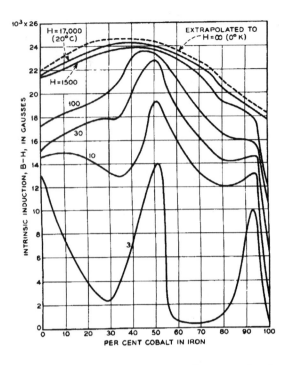

Figure 7.1-Intrinsic induction (B-H) of binary CoFe Alloys as a function of Co content. From Bozorth (1951)

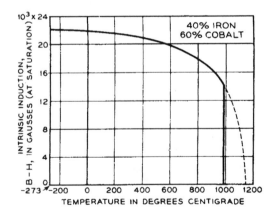

Figure 7.2-Intrinsic induction, (B-H) of 60%Co alloy with temperature. From Bozorth (1951)

epsilon phases. After the magnetizing field drops to about 100 Oersteds or less, the saturation maximum appears at about 50% Co. Temperature-wise, the 60% alloy shows a peculiar drop in the saturation at about 980°C. due to an alpha-gamma transformation point. Figure 7.2 shows this drop along with the extrapolation to the virtual Curie point. In the 50% alloy, another peculiar effect of temperature

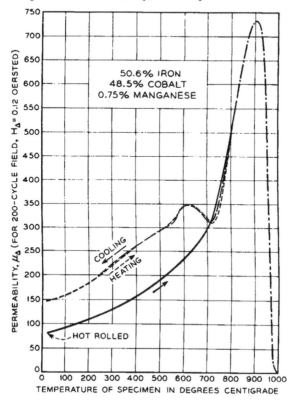

Figure 7.3- Permeability of 48.5% Co alloy after various heating and cooling programs. From Bozorth (1951)

was noted by Forrer (1931) related to an order-disorder transformation, This was further studied by Ashworth. Here, in the 48.5% Co alloy, the sample was first hot rolled, then reheated to 830°C., cooled to R.T. and again reheated to 830°C. The course of the permeability is shown in Figure7.3. The hot rolling produced disorder maintained at R.T. After re-heating and slow cooling, ordering was accomplished at 720°C. and the permeability rose on further cooling. During the reheat and cool, the same curve was traced.

Insofar as permeability is concerned, Elmen (1929 measured the initial permeability of the 50% cobalt alloy and found it higher than iron. Figure 7.4 shows the μ_o and μ_{max} for samples annealed at 850oC. And 1000oC. The peak at about 50% Co is evident for both permeabilities. The value of μ_{max} was found to be higher for the sample annealed at the lower temperature. (850°C.) Figure 7.5 shows the permeabilities of iron and cobalt-iron as a function of induction level. Note that the permeabilities of the cobalt-iron are greater at the higher inductions than the iron. By using material melted in hydrogen, real high permeabilities were accomplished by Cioffi(1935). This is shown in Figure 7.6. It is interesting to note that the Curie point for the binary alloy is higher than either of its components. For

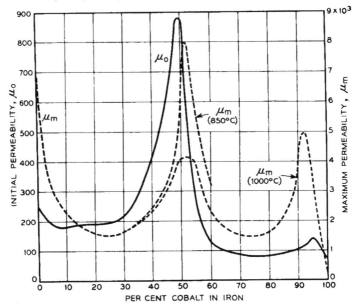

Figure 7.4-Initial and maximum permeabilities of CoFe alloys as a function of Co content for two different annealing temperatures. From Bozorth (1951)

an examination of the make-up of the permeability, the first anisotropy constant, K_1, for iron is positive and decreases with cobalt addition. It is zero near the 50% Co ordered phase. Not surprisingly, the permeability also peaks at this point. The magnetostriction in the (111) direction is zero near 25% Co and is about 25×10^{-6} at 50% Co. The high magnetostriction in the (100) direction qualifies it as a magnetostrictive transducer material (except for the cost).

The high saturation of the 35% alloy led Stanley and Yensen to look at the use of chromium additions to enable the material to be rolled into thin sheets. It was possible to produce these sheets with a saturation of 2.4 Teslas, a coercive force of 50 A/m, a remanence of 1.15 T, an initial permeability of 650, a max perm of 10,000 and a resistivity of 20 μ-ohm-cm. A commercial product of this type having 27% Co and .6% Cr is called Hiperco 27. Another variation has 35% Co but is only be used in castings or machined parts.See Table 1. The most useful cobalt-iron alloy contains 49% Co, 49% Fe and 2%V (Vanadium). It is sold commercially as 2V Permendur or Supermendur. The properties for these materials are given in Table 3. Major(1975) added 4.5% Ni which enhanced the high temperature-ductility without significant loss of properties. This permitted its use as rotor material in high-speed generators. The high temperature ductility was increased by a factor of four as well as increasing the resistivity. In 1989 he proposed the use of .25% Ta (tantalum) or

Nb (niobium) in place of the vanadium. It improved the ductility as well as increasing the saturation.

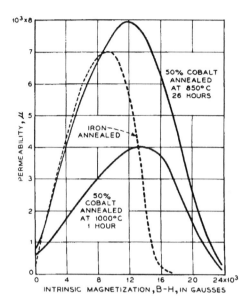

Figure 7.5-Intrinsic induction of 50% Co alloy at high B levels. From Bozorth (1951)

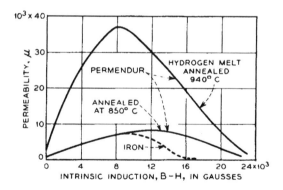

Figure 7.6 Permeability vs intrinsic induction of the 50% Co alloy (Permendur) after hydrogen or conventional annealing and compared to that of iron. From Bozorth (1951)

Processing of Soft Cobalt-iron alloys

The processing of the cobalt-iron alloys requires a great deal of care because of its limited ductility. The S should be lower than .05%, the C below .04%, the Si below .08% and the Mn between .3-.4%. For the 2V alloy, after melting and casting, the ingot is broken down at 1000-1200° C. and kept at 1000° C.during the hot rolling (

to .1 inch) operation. It is then rapidly quenched in an ice-salt-brine mixture. It is then cold rolled to gage and annealed for one hour at 850oC. The 2V composition is a compromise between the magnetic properties and ease of production.

For processing the Hiperco, about .1%C is added to improve forgeability. It is deoxidized with Si and Ti. It is hot rolled, quenched from 910° C. and cold rolled to .025 inch. In the lab, strip as thin as .001" has been made. The strip is then annealed at between 815-925° C. (below the 950° C. phase transformation point).

Magnetic Properties of Soft Co-Fe alloys

As mentioned previously, the most widely used of these alloys are the 2V Permendur and the Supermendur alloys. The properties of these materials are given in Table 7.2. The high saturations are common to all of these alloys but some show lower coercive forces and others(Supermendur) can be magnetically annealed to produce a square loop material. Still others have the addition of niobium and carbon for grain refinement and higher yield strength for motor rotor applications. Another important feature of the Co-Fe alloys is the very high Curie point that permits its use at very high temperatures.

COMMERCIAL AVAILABILITY OF SOFT COBALT-IRON MATERIALS

The 2V Permendur (49% Co, 49%Fe 2%V) and Supermendur are available as strip in gages from .0005" to .004" for tape cores, cut cores and laminations. Hiperco 27 is available in thin sheet form but for the Hiperco 35 material, it is only available in bulk form for machining into useful parts. A comparison of the magnetization curves for Permendur, Supermendur and other metallic strip materials is given in Figure 7.2. The high inductions of the cobalt-iron alloys are very evident.

Table 7.1
Physical and Magnetic Properties of Hiperco 27 and 35.

Property	Hiperco "27"	Hyperco "35"
Tensile Strength(MPa)	517.5	
Yield Strength(Mpa)	262.20	
Elongation(%)	10	
Reduction of Area(%)	50	
Rockwell Hardness	B85/92	
Coercive Force(A m^{-1})	198.95	198.95
Permeability (max)	2800	
Flux density at Permeability (T)	1.6	
Saturation Induction (T)	2.36	2.4
Resistivity ($\mu\Omega$ cm)	19	40

Reprinted from Evetts, Concise Encyclopedia of Magnetic and Superconducting Materials ,© 1992, Pergamon Press, p.201, with permission from Elsevier Science.

APPLICATIONS OF SOFT COBALT-IRON MATERIALS

We have said that the major uses for these materials are for the airborne and special applications where weight and volume limitations require high saturation.

Table 7.2
Properties of Permendur and Supermendur (High Purity CoFe-2% V)

Property	Permendur	Supermendur
Electrical Resistivity	25×10^{-6}	25×10^{-6}
Saturation Induction (T)	2.4	2.4
Magnetic Permeability	4000-8000	92500
Coercive Force (A m^{-1})	3.98×10^2	15.92
Magnetostriction Coefficient[a]	60×10^{-6}	60×10^{-6}

a-At 7.96×10^3 A m^{-1}

Reprinted from Evetts, Concise Encyclopedia of Magnetic and Superconducting Materials, © 1992, Pergamon Press, p.196, with permission from Elsevier Science.

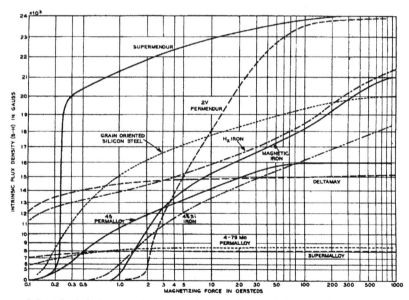

Figure 7.7-Intrinsic induction of various soft magnetic materials From Gould (1957)

materials. These uses would include aircraft generators and motors, compact pole pieces and other flux concentrators, high speed motors generators and alternators, magnetic bearings, high temperature magnetic components, magnetostrictive transducers and specialty transformers.

MAGNETIC COMPONENTS USING COBALT-IRON ALLOYS.
The principal components of soft cobalt iron alloys are tape-wound cores, motor laminations, transformer laminations and cut cores.

SUMMARY
The last two chapters have dealt with magnetic materials for power applications at low frequencies. There are other applications in which the frequencies can be DC or somewhat higher where the magnetic material is used to shield. The power levels are considerably lower and materials with much higher permeabilities are needed. The next chapter will deal with these materials for shielding applications

References

Bozorth, R.M. (1951) Ferromagnetism, D.Van Nostrand Co. New York
Ellis, W.C. (1927) R.P.I. Bull. Eng. Sci. Serv. $\underline{16}$, 1
Elmen, G.W. (1929U.S.Patent1,739,752)Appl.12/18/27
Gould, H.L.B.(1957) and Wenny, D.H., AIEE Spec. Publ. **T-97,** 675
Major, R.V. (1975) Martin, M.C., and Branson, M.W.,2nd EPS Conf. on Soft Mag. Mat. Wolfson Cent. Cardiff ,103
Preuss, A. (1912),Dissertation, Zurich, 190
Stanley, J. K. (1947) and YensenT.D., AIEE, $\underline{66}$, 714
Weiss, P.(1912) and Preuss, A.,Trans. Faraday Soc. $\underline{8}$, 160
White, J.H. (1932),and Wahl, C.V.,U.S. Patent 1,862,559 Appl. 8/14/31
Wohlfarth, E.P. (1980), Editor, Ferromagnetic Materials, North-Holland Publ. Co. Amsterdam,

8 METALLIC ALLOYS FOR MAGNETIC SHIELDING APPLICATIONS

INTRODUCTION

The last three chapters have dealt with magnetic materials that were useful for from D.C. to relatively low frequency. In the same frequency range (D.C. to about 1200 Hz.), there is another requirement which must be satisfied by a soft magnetic material. Surprisingly, the need is caused by interference brought about by magnetic devices. In these low frequency, low impedance circuits, stray magnetic fields can cause severe problems to sensitive electronic devices such as cathode ray tubes, transformers, magnetic tape data containers and photo-multiplier tubes which may be in the vicinity of the field- producing material. A magnetic shield acts to greatly reduce the disturbing field reaching the sensitive device. Sometimes, the shield is placed around the emitting device such as a motor or power supply. It is strange that one main solution to the reduction of stray magnetic field would be a strongly magnetic material. However, if you can think of the high permeability material surrounding the sensitive device as a shunt or flux diverting mechanism, the effect is not difficult to understand. This is shown in Figure 8.1. First, in Figure 8.1a, the magnetic field lines are shown schematically. This field induces a strong magnetization in the shield (shown without field) in Figure 8.1b. The superposition if the 2 fields is shown in Figure 8.1c. The concentration of field lines would be greatest for materials having the highest magnetic permeabilities. These materials turn out to be special nickel-iron alloys that we shall introduce in this chapter. In the next chapter we shall show more applications for these materials for telecommunications applications, but we mention them here because they fit in the frequency range order we have been developing.

SHIELDING FACTOR AND ATTENUATION

If we attempt to define the efficiency of a shield, a good place to start is with the ratio of the field outside of the shield to that inside he Shield. This ratio is called the shielding factor, S. It given by;

$$S = H_o/H_i \quad [8.1]$$
where H_o = Field in Oe. Outside of the shield
And H_i = Field inside the shield

Another often-used figure of merit is the attenuation that is given by;

$$A = 20 \log S \quad \text{in dB} \quad [8.2]$$

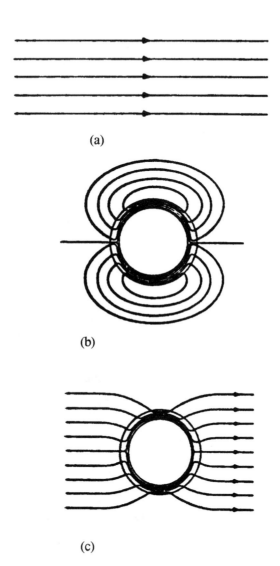

Figure 8.1-a) External magnetic field; b) Induced magnetic flux in the shielding material; c) Superposition of both fluxes, a) and b)

For a cylindrical shield where the height is at least four times the outer diameter, the shield factor can be approximated by;

METALLIC ALLOYS FOR MAGNETIC SHIELDING

$$S = \mu t/2\rho \quad [8.3]$$

Where μ = Permeability of the shield material
And t = thickness of the material in inches or cm.
ρ = Outer radius of the shield in inches or cm.

Doubling the thickness of the shield would only double the shielding factor but if a second shield of the same thickness very closely surrounded the first, the combination could produce attenuation about 40 times the first. For multiple layer shields, the second multiplies the attenuation by roughly;

$$Sij = \mu t^2/\rho^2 \quad [8.4]$$

Thus a useful technique in shielding design is the use of multiple shields
For shapes other than a cylinder, the shielding factor of a sphere is given by;

$$S = (4/3 \; \mu d/D) + 1 \quad [8.5]$$

Where D = Diameter
d = thickness

and for a cube ;

$$S = (4/5 \; \mu d/a) + 1 \quad [8.6]$$

Where a = edge of cube

The flux density in a cylindrical shield is approximately:

$$B(Gauss) = 2.54 \; d \; H_o/2T \quad [8.7]$$

Where d = shield diameter in inches
B = Flux density (Gauss)
H_o = External fields (Gauss)
T = Shield thickness in inches

Thus, a 2"- diameter .025" thick shield and an external field of 1 Oersted will operate at a flux density of 100 Gauss.

MATERIALS FOR MAGNETIC SHIELDING

We can see from Equation 8.1 that the shielding factor is proportional to the permeability of the shield material. The highest permeabilities of the magnetic alloys are developed in the nickel iron alloys in the neighborhood of 80% nickel. Other nickel iron alloys in the 50% nickel range have lower permeabilities but about double the saturation which may be effective in very high stray fields where the flux density in the shield may exceed the saturation of the 80% alloy. At even higher field, higher saturation materials such as silicon iron may be needed. Still another available material is the cobalt based amorphous alloy in the form of thin foil. Tables 8.1a and 8.1b lists the most widely used shielding materials. The highest initial permeabilities (at very low flux level are the 80% NiFe alloys. They also have very high permeabilities at a flux density of 200 Gauss which is typical of the induction produced by low stray magnetic fields. The induction at μ_{max} that produces the most efficient shielding for these materials is 3000 Gauss. The coercive force is low at .004-.1 Oersteds. The saturation occurs at 7500 Gauss but the permeability there is zero.

Table 8.1a- Various Materials for Magnetic Shielding From Amuneal(1990)

Material	Saturation (Gauss)	Permeability μMax	Permeability μ40	Maximum Shielding Efficiency H_o/H_{in}	Maximum Shielding Efficiency dB
Amumetal (80% Ni)	8,000	400,000	60,000	100×10^3	100
Amunickel (48% Ni)	15,000	150,000	12,000	19×10^3	85
Silicon Steel (3% Si-Fe)	20,000	5,000	3,000	1.3×10^3	62
1010 Carbon Steel	22,000	3,000	1,000	$.75 \times 10^3$	58

Table 8.1b- Various Materials for Magnetic Shielding ;From Vacuumschmelze (1989)

Table 1 Materials for magnetic shielding (nominal values)

Material	Permeability μ_4	Maximum permeability μ_{max}	Flux density B_{opt} T	Flux density B_m T	Coercivity H_c A/cm	Saturation flux density B_s T	Curie temperature °C	Electrical resistivity $\Omega \cdot mm^2/m$	Density g/cm³
VACOPERM 70	30000	70000	0.3	0.65	0.03	0.8	400	0.55	8.7
VACOPERM 100	50000	130000	0.2	0.55	0.015	0.78	400	0.6	8.7
CRYOPERM 10	at 4.2 K or 70 K similar properties to VACOPERM 70 at 20°C						430	0.35	8.7
PERMENORM 5000 H2/H3	4000	30000	0.5	1.35	0.10	1.55	440	0.45	8.25
PERMENORM 3601 K2/K3	3000	15000	0.4	1.0	0.20	1.3	250	0.75	8.15
VITROVAC 6025 X	25000	100000	0.15	0.45	<0.01	0.55	250	1.35	7.7

Static values
Other materials like RECOVAC, PERMENORM 5000 S2, VACOFLUX, VACOFER S2 etc. on request.

In fact, the permeability drops from μ_{max} to saturation. Therefore for fields which will produce more than 3000 Gauss, the thickness of the shielding can be reduced to lower the flux density or another higher saturation material should be chosen. The alloys containing about 48% Ni are the medium permeability materials. Their initial permeabilities are about 12,000, their μ_{200} values are only slightly higher and their μ_{max} values are only half that of the 80% alloys but still appreciable. Most important, the μ_{max} occurs at about 5000 Gauss (2000 more that the 80% materials). The saturation induction is also higher at about 15,000 Gauss. The coercive force ranges from .01-.1 Oersted. For operation at higher fields, the lower permeability higher saturation materials are chosen. The silicon irons have initial permeabilities of 300, and μ_{max} values about 8,000 but this occurs at an induction of 8,000 Gauss or 5,000 Gauss more than the 80% Ni alloys. The saturation values are about 20,000 Gauss and the coercive force ranges from .7-.9 Oersted. Still higher fields can be shielded with a higher saturation low perm material namely iron. Here, the μ_o is 50, μ_{200} is

150 and μ_{max} is 5000 . However, μ_{max} occurs at a value of 12,000 Gauss, The saturation is 21,5000 Gauss. Still another interesting material that can be used as a thin foil to shield wire and other unusual shapes is the Co-based amorphous alloy. We have mentioned the iron based amorphous alloys in connection with the low frequency power materials. A useful property of this Co-based alloy is that is a zero-magnetostriction material and its properties are not reduced by mechanical work such a wrapping the foil or fitting it around the shaped object. This means that it does not have to be annealed after wrapping. The initial permeability is 15,000, the μ_{max} about 100,000 and the induction at μ_{max} is 1500 Gauss. The saturation induction is 5500 Gauss. The material is only available in a thickness of about 30 microns or about .001".

PROCESSING OF SHIELD MATERIALS

The production of the Ni-based strip material will be discussed in Chapter 6. The shield are mostly custom-designed by the shield fabricator who cuts, punches, draws, spins or otherwise forms the shield. Openings and discontinuities are kept to minimum for continuity of flux. Welding may also be needed. All these mechanical operations severely reduce the magnetic quality of the original material because of the residual stresses. The formed shield must therefore be subject to a stress relief anneal in an inert or preferably reducing atmosphere (usually hydrogen). Especially for the high permeability NiFe alloys, the conditions for the anneal especially during the cool are quite critical. The 80% alloys are annealed for 2 hrs at about 1100°C in dry hydrogen and cooled to about 300°C. at a cool rate of about 200°C per hour. The same type of ordering takes place as mentioned for the cobalt-iron alloys. This also will be expanded in cool rate is not critical. The silicon iron materials can be annealed at about 900°C. For one and cooling as rapidly as possible. It is important that after annealing, especially the highest permeability shields that the shield not be mechanically disturbed in any manner, that is squeezing, bending or shocking. It is also advisable to demagnetize (primarily for D.C. fields) the shield before use to assure that the material is at zero flux rather than at some remanence value. This demagnetization is accomplished by subjecting it to a 60Hz. field 100 times the coercive force and gradually decreasing it to zero.

MEASUREMENT OF SHIELDING MATERIAL

Shielding efficiency of the magnetic material is usually accomplished in a device called a Helmholtz Coil Assembly. The equipment is shown in Figure 8.2 and described in ASTM Test 698 Procedure. Two parallel large usually 4 feet diameter) coil are wound with 48 turns of #14 coated wire to provide a uniform magnetic field. A 60 Hertz a.c. drive is used. A search coil wound with 100,000 turns of #40 coated wire is placed in the center of the large coils with its axis parallel to the exciting field direction. The current is turned up to yield 1 Volt in the search coil. A standard cylindrical shield of about 2 inch diameter and about 6 inches high is the placed around the search coil. The same current that produced the 1 volt is applied

150   HANDBOOK OF MODERN FERROMAGNETIC MATERIALS

with the shield placed over the coil. The induced voltage is then noted. Assume that the induced voltage with the shield is 1 mV. The shield factor is then 1000:1. For measurements on custom fabricated shields, the vendor and user will usually devise their own similar test.

Figure 8.2- A Typical Helmholtz Coil Apparatus for Measuring Effectiveness of Shielding Material. Courtesy of Magnetic Division of Spang & Co.

METALLIC ALLOYS FOR MAGNETIC SHIELDING

COMMERCIAL FORMS OF SUPPLY

The magnetic strip for shielding is made in thickness of .014-.062". For foil, the thickness are .002-.010" for high permeability foil and .004" for the 50% NiFe foil. The NiFe strip material is usually available in thickness up to about 10 inches. Often, the supplier of the material may provide the material in cut to order sheets for the fabricators convenience. Figure 8.3 shows the variation of Permeability in a Permalloy 80 material with strip gage.

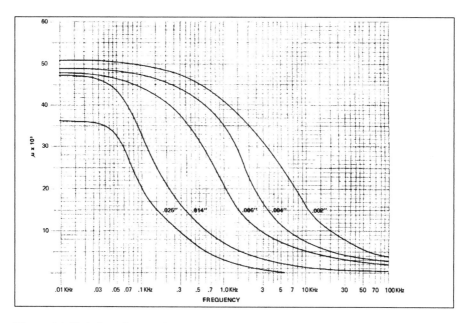

Figure 8.3 Variation of permeability with frequency for different gage strip of Permalloy 80. From Amuneal(1990)

SHIELDING FACTOR AS A FUNCTION OF FREQUENCY

The properties for the shielding materials and the equations for shielding factor are those obtained at D.C. The 50 Hz. ac shielding factors for the 80% and 50% alloys for different thickness and flux levels are given in Table 8.4. Because the magnetic permeability drops with frequency, the values at 50-60 Hz. are somewhat lower than the D.C. values. Shielding at D.C. and low frequency a.c. occurs because of the high permeability path through which the flux can flow. For higher frequencies, especially in low resistivity metals, Eddy currents are progressively established which provide additional shielding in addition to the magnetic effect. As the frequency increases, the usefulness or efficiency of the magnetic shielding drops to decrease of flux penetration. However, the lower resistivity (higher conductivity) metals such as copper or aluminum provide even more Eddy current shielding. At some frequency, the shielding factor for the copper or aluminum will be higher than the magnetic

152 HANDBOOK OF MODERN FERROMAGNETIC MATERIALS

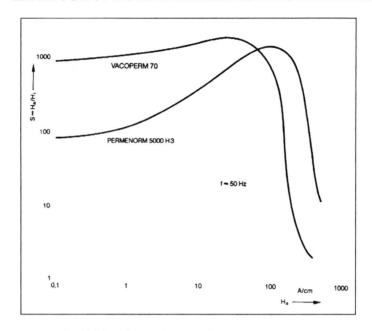

Figure 8.4- Magnetic shielding factors for Permalloy 80 (Vacoperm 70) and Permalloy 50 (Permanorm 5000) as a function of frequency. From Vacuumschmelze (1989)

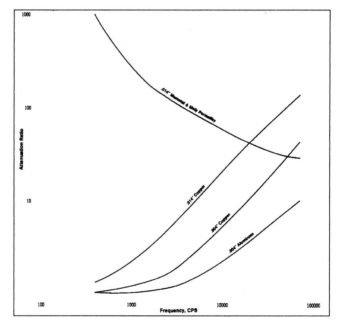

Figure 8.5- Attenuation ratio for various Materials subjected to a 1 Oersted field as a function of frequency. From Allegheny (1974)

METALLIC ALLOYS FOR MAGNETIC SHIELDING 153

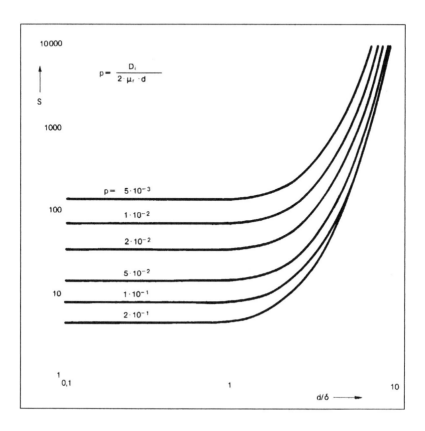

Figure 8.6-Shield factors versus frequency for hollow magnetic spheres as a function of $D/2\mu d$, where D = sphere diameter and d= thickness of metal. From Vacuumschmelze (1989)

alloys. This is shown in Figure 8.5. We have spoken about flux penetration or skin depth of magnetic sheet. The skin depth of the strip is related to the frequency by

$$\delta = \sqrt{2.54 \times 10^5 \rho / \mu f}$$

where δ = skin depth in mm
ρ = resistivity in μohm-m
f = frequency Hz.
μ = Permeability

Figure 8.6 shows how the shield factor, S, increases when the ratio of d/δ exceeds 2.

SUMMARY

The materials listed in this chapter are used mostly at frequencies from DC to about 800 Hz. Iron and SiFe are used for the lower frequency range (50-60 Hz.) and higher field strengths whereas the high permeability NiFe alloys are useful at higher efficiencies and lower fields. The 50 percent NiFe alloys have properties intermediate to the two mentioned materials. The next chapter will deal with the NiFe alloys for high permeability low-level applications other than shielding. These include the telecommunications and instrument applications. The frequency range for these materials will extend to the upper kilohertz region.

References

Amuneal (1990) Complete Guide to Magnetic Shielding
Vacuumschmelze (1989) Magnetic Shielding, FS-M9
Allegheny-Ludlum Steel Co.(1974) Magnetic Shielding Electrical Materials

9 HIGH-PERMEABILITY HIGH-FREQUENCY METAL STRIP

INTRODUCTION

In chapters 6 and 7, the use of high magnetic saturation materials for high-level power applications was discussed. We can say that these applications were saturation-driven. Their core losses were relatively low at those frequencies and for the iron and silicon-iron materials, the cost was also relatively low as would be expected for consumer-related products. The cobalt-iron was used primarily for airborne, high-temperature or specialty applications. In the last chapter (Chapter 8), we introduced the nickel iron alloys for magnetic shielding purposes. The frequencies were still quite low (D.C. to 800 Hz.). In this chapter, we will expand the use of these alloys to the non-shielding applications at higher frequencies. These will include the low-level electronic transformer and inductor applications. While earlier materials were mostly the so-called "non-linear" materials (near the μ_{max} region), the initial permeability region (Rayleigh region) is where materials in this chapter operate. The electronic signals of telecommunications are quite small. These materials would then be described as linear magnetic materials. The emphasis here, as it was in the shielding case, is on the high initial permeability of the materials. The highest permeability is found in the nickel-iron alloys. There is also great versatility in choosing compositions and heat treatments to enhance selected properties in combination with the high permeability. This is possible because of the extremely wide range of ductility in the nickel-iron alloys. This feature permits the choice of compositions where the fundamental intrinsic parameters such as anisotropy, magnetostriction, and saturation can be optimized. This was a feature not possible in the iron, silicon iron or cobalt-iron strip materials. In addition, the nickel-iron alloys are sensitive to such process variations as atomic ordering, and magnetic annealing which adds to the ability to change the shape of the hysteresis loop and also increase the permeability. This means that the electronic designer can choose the material that best fits his needs. Mechanically, the increased ductility of the nickel iron alloys allows the rolling down to extremely thin gages (.00025 " or 1 micron) which improves high frequency operation. Along with the nickel iron alloys, we will also deal with the new amorphous alloys and the newest nanocrystalline alloys, both of which may be used for the same applications as the nickel iron alloys.

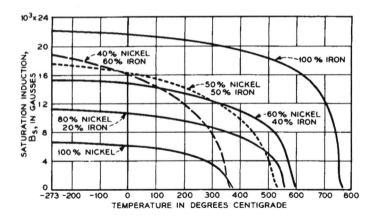

Figure 9.1-Saturation induction of some NiFe alloys. From Bozorth (1951)

HISTORY OF THE NICKEL IRON ALLOYS

Nickel iron alloys were used for other applications before they were discovered for magnetic applications. The low expansion alloy, Invar, and the glass sealing alloys are examples. Hopkinson (1890) was the first to study the magnetism of these alloys. He examined them in the range of 1-73% nickel. Barrett (02) only looked at the alloys with 31% or less nickel. In 1910, several investigators measured the properties of these alloys at relatively high drive levels. They found the highest saturation at about 47% nickel. They also determined the Curie points for the alloys. None of these workers looked at the properties at low fields. There were few applications requiring these properties at the time. In 1913, Elmen (1913) was the first to examine these materials in weak or medium fields. He was searching for a better material than silicon iron for telephone apparatus operating at a few hundred Gauss. He found that the permeabilities of the 30-90% Ni alloys were higher than any known material. He also discovered the special heat treatment known as the "Permalloy" treatment that further increased the permeability of the alloys in the 50-90% Ni region. The increase was greatest for the alloy at 78% Ni. The maximum permeabilities increased as well. Practical use in relays was made of Elmen's work as early as 1921. His patent on Permalloy was issued in 1917 (Elmen 1917) and his patent for the spcial heat treatment was issued in 1926(Elmen 1926). The first scientific paper on Permalloy was published in 1923 (Arnold 1923). In addition to the 78 Permalloy, an alloy containing 45% Ni found use in telephone apparatus. It had high saturation and high resistivity. In 1924, Yensen had a patent issued for a 50% Ni alloy (Hipernik) which was heat treated in hydrogen and also had the same permeability obtained by Elmen in 78 Permalloy. It also had the high saturation and resistivity of 45 Permalloy. Yensen's use of vacuum melting provided improved quality but the hydrogen annealing is probably his most important contribution. In 1921 (Elmen 1921, 1926) added chromium to the high permeability alloys to increase resistivity and in the same year (Elmen 1928), he found that the addition of large amounts of cobalt produced Perminvar, a wasp-waist hysteresis loop material

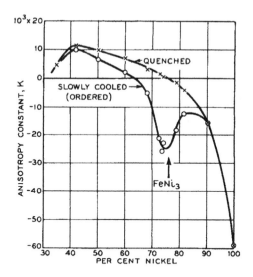

Figure 9.2- Magnetocrystalline anisotropy constants of NiFe alloys as a function of Ni content. From Bozorth (1951)

in which the permeability was almost independent of drive up to 1-2 Oersteds. Smith and Garnett invented Mumetal in 1924 when they added several percent copper. Later molybdenum was added to increase the resistivity and thus the Eddy current losses. The alloy with 4% Mo and 79% Ni is widely used today. In addition to the high permeability alloys, the 29-33% Ni alloys have the lowest Curie points (around room temperature) and as such, can be used as temperature compensating alloys for permanent magnets and instruments. These materials will be discussed in the chapter on miscellaneous applications.

The first magnetic amorphous alloys were pioneered by Allied in the early 1970's under the trade name of Metglas®. These alloys were the Ni or Co based materials. These are closest in applications to the ones we are discussing in this chapter. The nanocrystalline alloys are the newest materials for these applications and some commercial products are already in use.

HIGH PERMEABILITY NICKEL IRON ALLOYS

The groups of commercially available nickel iron alloys are listed with the most outstanding property in Table 9.1. Several of these alloys are available with combinations of either round, square or flat hysteresis loops. Figure 9.1 shows the dependence on composition of some of the intrinsic parameters that affect the properties listed above. The maxima for the saturation and Curie point are shown in Fig. 9.1. Those for anisotropy and magnetostriction are given in Fig. 9.2 and 9.3. The permeability is maximum for the case when K_1 is close to zero or at about 80% Ni. The effect of composition on the magnetostriction for the binary alloys is very large. In fact, it is not possible to have zero magnetostriction at the same composition. However, by additions of Mo and other additives, it is possible to achieve this

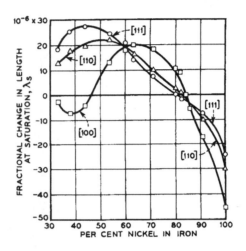

Figure 9.3-Magnetostrictions of some NiFe alloys versus percent Ni. From Bozortth (1951)

Table 9.1
Major Electrical and Magnetic Properties of NiFe Alloys vs Ni Content

Alloy (%Ni)	Major Property
36	Highest Resistivity
45	
48	Highest Saturation
50-55	
65	Highest Curie Point
76-82	Highest Permeability

combination. We will first deal with the binary alloys in the 80% range and then with the substituted case

BINARY NICKEL-IRON ALLOYS IN THE 80% NI RANGE

The highest permeability in the binary alloys occurs at 78.5% Ni. The processing started with the melting of Armco iron and high purity nickel in a high-frequency induction furnace with a covering of borax and about .3% Mn for desulfurization. Aluminum or magnesium was added for deoxidation. The ingots were rolled to .006". Commercial high permeability NiFe alloys are usually melted in a vacuum

HIGH PERMEABILITY-HIGH FREQUENCY METAL STRIP 159

Figure 9.3a- A laboratory vacuum induction melting furnace. From Magnetics

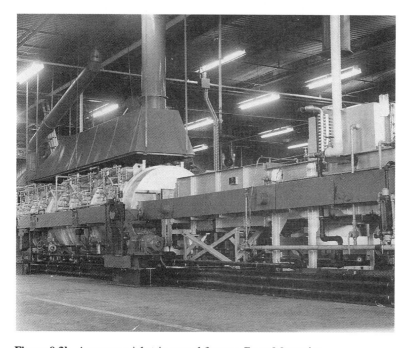

Figure 9.3b- A commercial strip anneal furnace. From Magnetics

induction furnace, hot rolled to about .1" and cold rolled to gage. They are then strip-annealed and slit to width. A laboratory vacuum melting furnace is shown in Figure 9.3a and a commercial strip anneal furnace is shown in Figure 9.3b For the heat treatment, Elmen used 3 variations. The first consisted of heating for one hour at 900°C.-950°C. and cooling at a maximum rate of 100°C. per hour. The second was his "Permalloy" treatment or double treatment. This consisted of heating and cooling as before then reheating to 600°C. and cooling in open air by placing it on a copper plate. Later, the first cooling to room temperature was omitted. The third method consisted of heating to 900-950°C. in a closed pot and then held at 450°C. for about 20 hours. The results of this experiment are

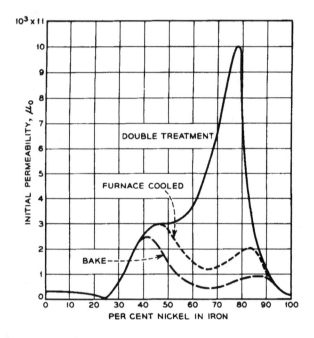

Figure 9.4-Initial permeabilities versus Ni content of NiFe alloys heat-treated by several different methods. From Bozorth (1951)

shown in Figure 9.4 and 9.5 for the initial and max permeabilities respectively. The coercive force results are shown in Figure 9.6. The effect of cool rate on the two permeabilities are shown in Figure 9.7. The cool rates peak at 20 °C. per minute for the μ_o and 80 degrees per minute for μ_{max}. Kelsall (1934) found that the maximum permeabilities of the 78 alloy could be increased many fold if heat treated above 400°C. in a magnetic field. The initial permeability actually decreased. This behavior is shown in Figure 9.8. The magnitude of the field necessary to do this magnetic annealing is shown in Figure 9.9. The hysteresis loops for 65% Ni alloy with and without the magnetic field during the furnace cool is given in Figure 9.10. For laboratory and commercial materials, the μ vs B curve is given in Figure 9.11. There is

HIGH PERMEABILITY-HIGH FREQUENCY METAL STRIP

not much commercial production for the binary alloys today. Cartech does mention a High Permeability 78.5 alloy.

Nickel-Iron Alloys (80% Ni) containing Copper

Mumetal, the first NiFe alloy containing copper was invented by Smith and Garnett (1924). It contained 5% Cu and 78% Ni. The initial permeabilities of this alloy annealed by the two methods described is shown in Fig.9.12. The permeability vs induction for several workers is shown in Fig.9.13. The alloy as now produced commercially contains 5% Cu and 2% Cr. The nickel content is between 72-76%. It has relatively high permeability and resistivity and low loss in low fields.

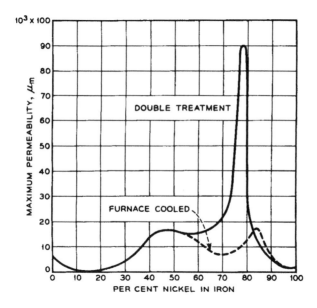

Figure 9.5-Maximum permeabilities vs Ni content of NiFe alloys heat-treated by several different methods. From Bozorth (1951)

Bozorth(1951 gives the highest initial permeabilities as 30,000 and that of the average commercial product as 15,000-20,000. However, recent commercial catalogs give values from 30,00-60,000 and μ_{max} values to 150,000. The saturation like that of the other 80% Ni alloys is about 8,000 Gauss. Annealing in hydrogen is recommended in the final anneal which should be at 1100-1200 °C.

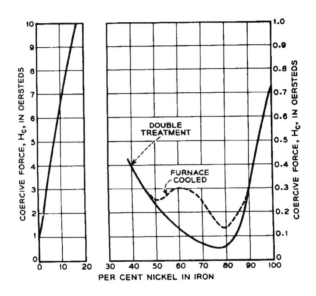

Figure 9.6-Coercive forces versus Ni content of NiFe Alloys heat-treated by several different methods. From Bozorth (1951)

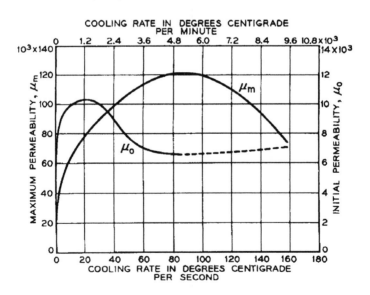

Figure 9.7- Effect of cool rate on initial and maximum permeabilities of 78 Permalloy From Bozorth (1951)

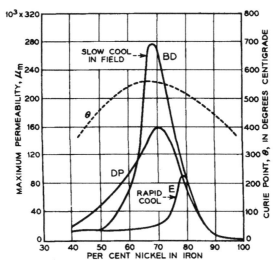

Figure 9.8- Effect of cooling 78 Permalloy in a magnetic Field versus other heat treatments without field. From Bozorth (1951)

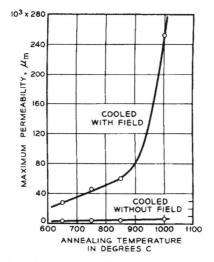

Figure 9.9- Variation of maximum permeability of Permalloy 78 with strength of magnetic field during cool. From Bozorth (1951)

Nickel-Iron Alloys with Molybdenum- 4-79 Permalloy

We have said that it is not possible to have simultaneous zero anisotropy and magnetostriction in the binary alloys, but that with addition of molybdenum (and some

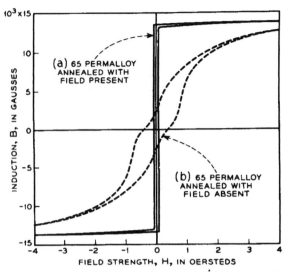

Figure 9.10- Hysteresis loops of 65 Permalloy annealed with and without a magnetic field. From Bozorth (1951)

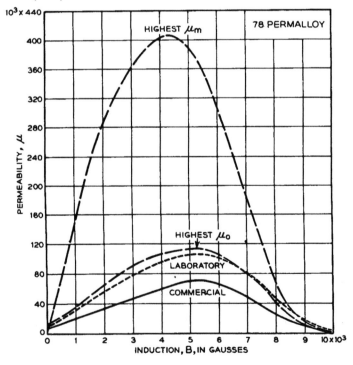

Figure 9.11- Permeabilities of laboratory and commercial grades of Permalloy 78. From Bozorth (1951)

HIGH PERMEABILITY-HIGH FREQUENCY METAL STRIP

times Cu), this indeed is possible. A portion of the triangular composition diagram(Fe-Ni-Mo) is shown in Figure 9.14 with the lines for zero anisotropy (K1) lines and zero magnetostriction 111 and λ_s (polycrystalline) are shown for the 480 final annealing temperature and for that at 550° C. Superimposed on this diagram is the values for 0, 5 and 10% copper addition. Several zero crossings for each of these conditions are found with the changes in the other three elements necessary to achieve this. High initial permeability is obtained at crossing of zero K_1 and that of λ_s. When the 560°C. heat treatment is used, the cooling must be rapid or the K_1 line will be shifted to higher Mo contents. The 80% Ni-5%Mo and the 77%Ni-4%Mo-5% Cu are both close to the zero K_1-zero λ_s lines. The 80% Ni-5% Mo alloy is a premium material known as Supermalloy when high purity optimized processing is used, It has the highest permeability of any known material. The 77-4-5 material is of course Mumetal. Material with 79% Ni and 4% is often referred to as 4-79 Permalloy and is a common high permeability material Mo whose properties are usually somewhere between Mumetal and Supermalloy. The optimum cool rate for the 4-79 Permalloy is very much dependent on the composition.

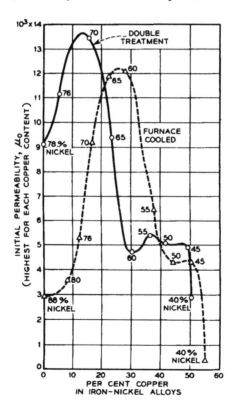

Figure 9.12-Initial permeabilities of various NiFe alloys with Cu additions with several heat treatments From Bozorth (1951)

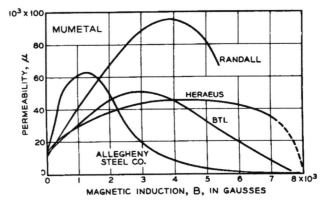

Figure 9.13 - Permeabilities of Mumetal reported by several different manufacturers. From Bozorth (1951)

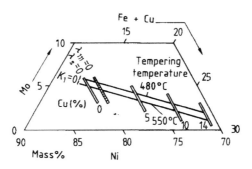

Figure 9.14 - Portion of FeNiMo phase diagram with Cu substitutions for the Fe. Areas of zero anisotropy and magnetostrictions are shown. From Pfeiffer (1980)

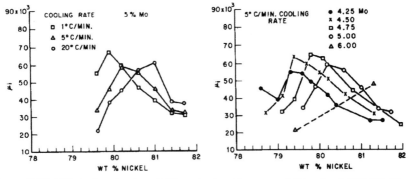

Figure 9.15 - Effect of cool rate and Ni content on the permeability of a 4.75% Mo-Permalloy. From Cohen (1967)

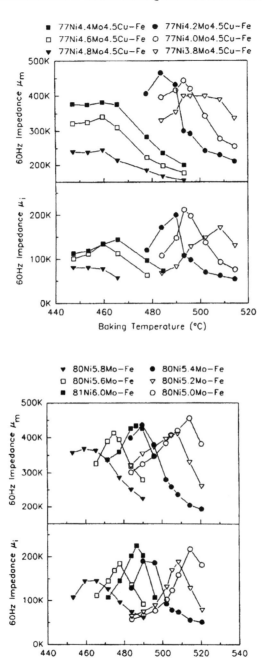

Figure 9.16- Initial and maximum impedance permeabilities in a 5% Mo Permalloy versus bake temperature. Top curve is for samples containing Cu and the bottom one is for samples without Cu. From Masteller (1997)

For the variation of cool rate with nickel content for the 4.75% Mo alloy, Figure 9.15 (Cohen 1967) gives the series of curves. In Figure 9.16 (Masteller, 1997) the initial and maximum permeabilities are correlated with the Ni content and Bake Temperatures for a Cu containing and Cu-free 5% Mo (Supermalloy) material. The cool rate was kept at 165° C./h. 9.17 shows the results of a commercial heat of the 80-5 alloy at various take-out temperatures(Hattendorf 1991). The material was heated in an electric arc furnace, vacuum-oxygen decarburized and cast into ingots. These were then hot and cold rolled to .07 mm. And wound into tape cores. The final anneal was at 1200°C. And cooled at the rate of .9°C. Per minute. The cores were taken out of the furnace at the indicated temperature. The initial perm μ_4 (measured at 4mA/cm (.005 Oersted) and the max perm increase when the temperature reaches 480-490oC. At the highest μ_o, K_1 is zero. At lower take-out temperatures or slower cooling, $K_1<0$ but the λ_{111} (easy direction) is close to zero. This gives rise to the square loop as shown. At higher than permeability maximum take-out temperatures, $K_1>0$ and λ_{100} (easy direction is not zero. Stresses are induced and the permeability drops sharply and the loop is skewed or flat. If a flat μ versus T curve is needed at a certain temperature, Figure 9.18 (Hattendorf (1991) shows differences in take-out temperatures that permit a wide range of of temperatures of use.

For pulse applications, the use of transverse magnetic annealing of a high permeability NiFe alloy produces a flat hysteresis loop which provides a large ΔB and a high pulse permeability. For applications of the Ni-Fe alloys as recording head materials, another requirement, that of good wear resistance is also needed. With small additions of either Ti and Nb or the ceramic material, aluminum oxide, this can be achieved and commercial materials such as Hitachi's Tufperm are available.

NICKEL-IRON ALLOYS WITH 65% NICKEL
The alloys with 65% Nickel are a combination of the high permeability possessed by the 80% alloys and the high saturation of the 50% alloys. They are mostly used in tape cores and can be processed to yield round, square or flat hysteresis loops.
The square or flat loops can be obtained by magnetic annealing with the field in the longitudinal or transverse directions respectively. The hysteresis loops of a 65% alloy with and without a longitudinal magnetic anneal is shown in Figure 9.10. When the 65 Permalloy is annealed at 1400° C. in hydrogen and magnetically annealed the permeability is much higher as shown in Figure 9.29.The emphasis on either the high saturation or high permeability can be controlled by adjustment of the chemistry. The 65 Permalloy has the highest Curie point of the nickel iron alloys. This is not a very widely used material.

NICKEL-IRON ALLOYS IN THE 50% NICKEL RANGE
These alloys ranging from 45-50% nickel alloys are characterized by having the highest saturations. They also have high resistivities. They may contain up to .5% Mn and .35% Si. The most common alloy contains 48% Ni(highest saturation) and is known as 48 alloy. There are two grades available. Rotor grade with μ_{40}(40

Gauss) of 5-6000 and transformer grade with μ_{40} of 10,000-12,000 and max perms of about 90,000. The latter is used in audio and instrument transformers and instrument relays. The rotor grade is not oriented and useful for rotors and stators. The transformer grade is cube-oriented with the easy directions parallel and perpendicular to the rolling direction. This is achieved by a 95% final cold reduction so that recrystallization occurs during the following anneal. This particular alloy lends itself well to a powder metallurgical method of preparation. A material with a flat hysteresis loop is possible using a transverse magnetic anneal. Using a magnetic anneal can increase the initial and maximum permeabilities in these materials by a factor of 3-4 fold after the high temperature anneal. This increase is shown in Figure 9.20. When material with the same composition but without the cube texture was magnetically annealed, the effect was very much smaller. Aside from permeability considerations for the NiFe alloys, another important consideration is the core loss that we know is dependent on the frequency and thickness of the strip. Core loss curves for some of the thinner gages of several of these alloys are given in Figures 9.21-24. The curves in Figure 9.21 are of Alloy 48 alloy; in Figure 9.22 the alloy is Orthonol(Orthonol is a cube oriented 50-50 alloy). The Square Permalloy of Figure

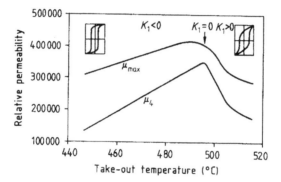

Figure 9.17-Variation of permeability with take-out temperature for a 5% Mo-80%Ni material. From Hattendorf (1991)

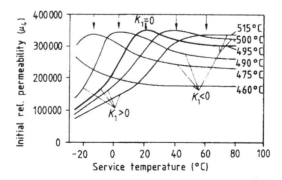

Figure 9.18- Variation of permeability with service temperature for a 5% Mo-80%Ni material with varying take-out temperatures. From Hattendorf (1991)

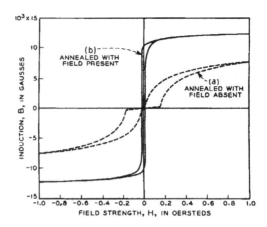

Figure 9.19- The effect of magnetic anneal on 65 Permalloy which was annealed at 1400°C. in hydrogen From Bozorth (1951)

9.23 is of the 4-79 square loop material and the Supermalloy of Figure 9.24 is of the 5% Mo composition.

NICKEL-IRON ALLOYS IN THE 36% NICKEL RANGE

The 36% Ni alloys are characterized by their high resistivity and high saturation. Permeability is relatively low 4,000 for μ_o and 50,000 for μ_{max}. With a fine-grained structure and high resistivity, the ac properties are good. They are used in transformers, pulse transformers and relay. The 40% alloy also has high resistivity and a higher saturation (1.48 T) and a higher permeability (9,000).

As previously noted, the 30% nickel alloys have Curie points near room temperature and as such can be used as temperature compensating alloys. The saturation is strongly dependent on the temperature. The Curie points can range from 30° C.- 120°C. Powder metallurgy- produced parts assure the accurate control of composition needed for this application. These are available commercially. There must be a linear dependence of flux density with temperature. See Figure 9.25.

AMORPHOUS MATERIALS- HIGH PERM-HIGH FREQUENCY

We have spoken of the iron-based amorphous Nickel-iron alloys for low-frequency, high power applications in Chapter 6 and the Ni-based amorphous alloys in Chapter 8. In this section, we will deal with the nickel and Co based amorphous alloys for much the same applications of the crystalline nickel-iron alloys discussed in this chapter. These are the applications for high permeability and high frequency operation. The first commercial amorphous alloy offered for sale was Allied Metglas® 2826. As in the case of the iron based 2605 alloys, the number indicates the major elements in the composition. Thus 28 stands for nickel(atomic number 28) and the 26 stands for iron (atomic number 26). The composition was $Fe_{40}Ni_{40}P_{14}B_6$. It has since been replaced with a better alloy, Metglas® 2826MB ($Fe_{40}Ni_{38}Mo_4B_{18}$). It

HIGH PERMEABILITY-HIGH FREQUENCY METAL STRIP 171

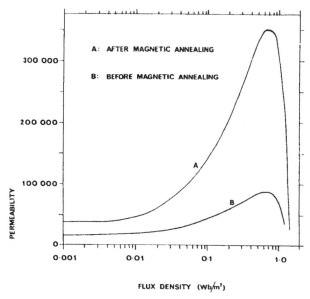

Figure 9.20-Permeability versus flux density for a 53% cube-oriented NiFe alloy before and after magnetic annealing. From Kang (1967)

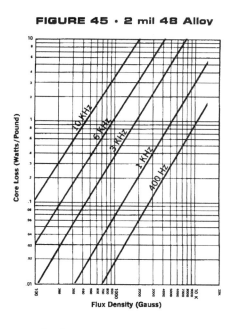

Figure 9.21-Core loss curves for .002" Alloy 48. From Magnetics (1995)

172 HANDBOOK OF MODERN FERROMAGNETIC MATERIALS

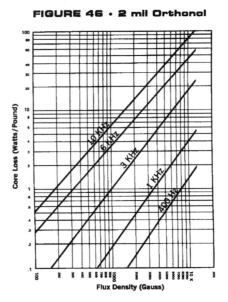

Figure 9.22-Core loss curves for .002" Orthonol. From Magnetics (1995).

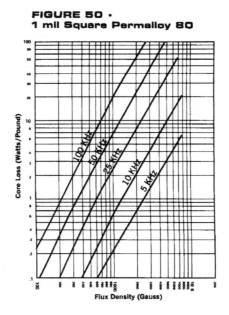

Figure 9.23-Core loss curves for .001" square Permalloy 80. From Magnetics (1995).

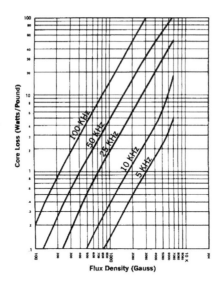

Figure 9.24-Core loss curves for .001" Supermalloy .From Magnetics (1995)

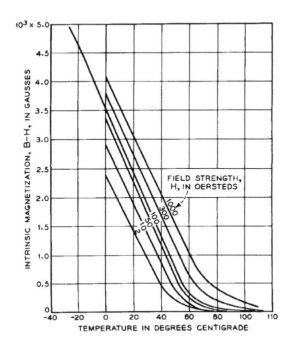

Figure 9.25- Intrinsic flux density (B-H) for a 30% NiFe alloy From Bozorth (1951)

most closely resembles the crystalline NiFe alloys. It has a μ_{max} of 800,000 after annealing. It has an improved saturation magnetostriction of 12 ppm. The resistivity is higher than the crystalline alloys at 138 $\mu\Omega$-cm. And the Curie temperature is 353°C. It can be used in high frequency transformers and field sensors and shielding. It can be magnetically annealed for either square or flat hysteresis loops. There are two cobalt-based materials produced by Allied, Metglas® 2705M and Metglas® 2714A. Both have essentially zero magnetostriction. The iron-based amorphous alloys had high magnetostrictions. The 2705 material has a relatively high μ_{max} at

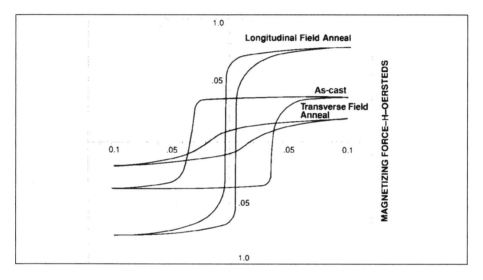

Figure 9.26a - Hysteresis loops with various anneals for Metglas ® 2826MB
From Allied (1995)

600,000 but it has another important property related to that. The permeability as cast (no anneal) is still a moderately high 290,000. This is a great saving for shield fabricators. This alloy is useful in high frequency cores and for magnetic sensors. It has a saturation of 7700 Gauss. The 2714A alloy has the highest μ_{max} of any amorphous alloys at 1,000,000. The saturation magnetostriction is much less than 1 ppm. Its resistivity is 142 $\mu\Omega$-cm but its Curie point is only 225°C. The saturation is 5700 Gauss. It has extremely low core loss and excellent corrosion resistance. It can be used in high frequency transformers, high-sensitivity matching transformers, ultra-sensitive current transformers and in sensor applications. With all these excellent properties, it is not surprising that this is the most expensive amorphous alloy. As in the other amorphous alloys, the thickness is set at 20-30 microns (about .001") dictated by the strip casting process. It has additional properties for recording head material because of its reat hardness. However, its high brittleness must also be

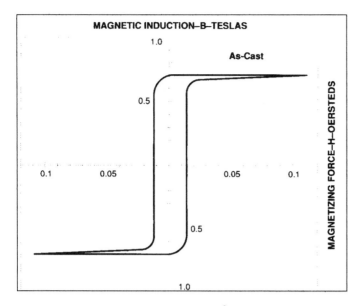

Figure 9.26b- Hysteresis loop for Metglas ® 2705 From Allied (1995)

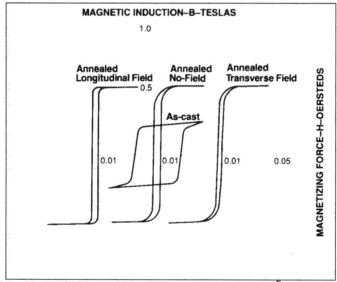

Figure 9.26c- Hysteresis loops with various anneals for Metglas ® 2714A. From Allied (1995)

considered. The hysteresis loops for the 3 materials mentioned are found in Figures 26a,b and c. The impedance permeability and core loss curves for the 2714A material are given in Figures 27 and 28. Vacuumschmelze also produces several Ni and Co based amorphous alloys. Vitrovac 4040 is a FeNi based alloy that has a μ_{max} at

50 Hz. of 250,000, an H_c of 4 mA/cm (.005 Oersted) a saturation of 8700 Gauss, and a saturation magnetostriction of 1.35 ppm. The two Co based alloys are essentially zero magnetostriction. Vitrovac 6025 has a high μ_{max}, a saturation of 5500 Gauss, a resistivity of 135 $\mu\Omega$-cm and a saturation magnetostriction of less than .3ppm. Vitovac 6030 has a μ_{max} of 300,000, $\lambda_s = .3$ but B_s is 8000 Gauss. The resistivity is 135 $\mu\Omega$-cm. This alloy may be used in low loss inductive components especially at high frequency and also for magnetic sensors(anti-theft) and for recording heads.

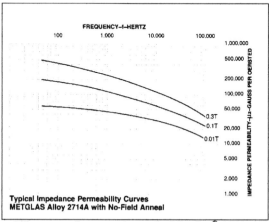

Figure 9.27- Permeability versus frequency losses for Metglas® 2714A. From Allied (1995)

NANOCRYSTALLINE MATERIALS - HIGH FREQUENCY OPERATION

The iron-based nanocrystalline materials are the newest ones and have the potential of being the next generation of magnetic alloys. They are the characterized by a very small grain size which lends itself to high frequency operation. The saturation and permeability for low magnetostriction materials are given in Figure 9.29 The first reported material with excellent magnetic properties was by Yoshizawa (1988). It came as an outgrowth of the amorphous metals studies. While earlier workers had tried to reduce the high magnetostriction of the partially crystallized iron based amorphous alloys by the use Nb, Mo or Cr, the properties were not good. Crystallization of the amorphous materials by heat treatment above the crystallization temperature, produced precipitates such as iron borides. The important contribution of Yoshizawa was the use of copper that prevented this precipitation effect. This permitted the use of the Nb to lower the magneostriction without harmful effects. The formula for his material that was called Finemet^{TM} by Hitachi Metals was $Fe_{73.5}Cu_1Nb_3 Si_{13.5}B_9$. The amorphous metals were made by the single roller method

HIGH PERMEABILITY-HIGH FREQUENCY METAL STRIP 177

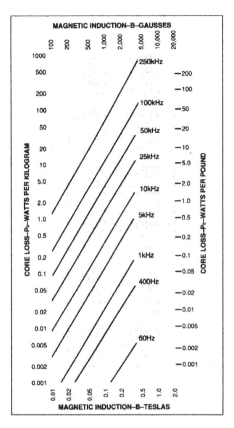

Figure 9.28- Core losses for Metglas ™ 2714A. No field anneal. From Allied (1995)

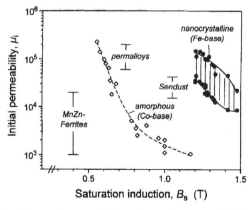

Figure 9.29 Saturation and permeability of some low magnetostriction materials. Reprinted from Herzer, Handbook of Magnetic Materials Vol. 10, ©1997), p. 454, with permission from Elsevier Science.

in widths of 5mm. and thickness of 15-20μm. There were made into cores with an OD of 19 mm, an ID of 15 mm. and annealed at 380-650°C. The magnetic properties as a function of Cu content are given in Figure 9.30. The saturation is high at 1.2-1.3 T. The permeability at 1 Khz. Is 100,000, the H_c was .5A/m. The Cu markedly improved the properties of the FeNbSiB alloy with respect to all magnetic properties but the saturation. The crystallization temperature was 488°C. For the 1%- Cu alloy, the copper has the effect of nucleating the bcc-Fe solid solution and preventing the precipitation of the borides. In the alloy, there are Fe rich regions and Cu-Nb rich regions. The iron rich region crystallizes because of the lower crystallization temperature while the CuNb region remains amorphous (higher crystallization temperature). The iron -rich region cannot grow into the Cu-Nb region so the grain size remains small and the texture is random. Therefore the saturation magnetostriction is low (2 x 10-6)compared to the iron based amorphous alloys. In a later paper (Yoshizawa 1989) was able to magnetically anneal these materials and obtain square or flat loops by longitudinal or transverse anneal respectively. The square loop material was inferior to other materials but the flat loop (transverse) material showed good properties to 150 KHz. and the permeability was 30,000. Results are shown in Figure 9.31. Hitachi offers standard tape-wound toroids of the Finemet material. Late Suzuki (1990) developed another nanocrystalline alloy system based on Fe-Zr-Nb-B. This was later expanded and reported by Makino(1997). Permeabilities up to 160,000 were obtained and in other alloys, saturations to 1.7 T (17,000 Gauss). The material is called Nanoperm . Most of the materials described here are experimental and laboratory results. Aside from the Hitachi tape cores of

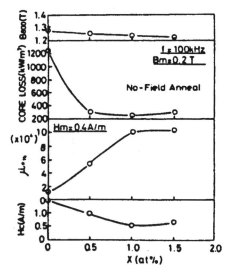

Figure 9.30-Magnetic properties of nanocrystalline material as a function of Cu content. From Yoshizawa (1988)

HIGH PERMEABILITY-HIGH FREQUENCY METAL STRIP 179

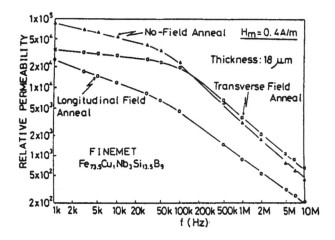

Figure 9.31- Permeability versus Frequency for Finemet material with different anneals From Yoshizawa (1989)

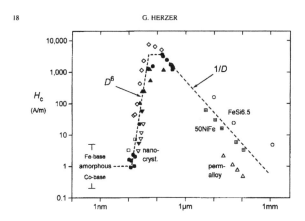

Figure 9.32-Variation of coercive forces of various materials with grain size Reprinted from Herzer, Handbook of Magnetic Materials Vol. 10, ©1997), p418, with permission from Elsevier Science.

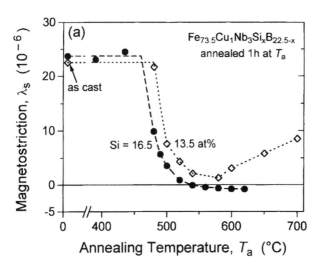

Figure 9.33-Dependence of magnetostriction on annealing temperature for a nanocrystalline material with varying Si content From Herzer (1997)

the Finemet material, Vacuumschmelze offers a commercial product called Vitroperm which is a low magnetostriction alloy with high permeability and exceptionally low core losses. The saturation is 12,000 Gauss. Herzer (1989) has pointed out the anomaly of the nancrystalline materials with respect the influence of grain size on the coercive force, H_c, In conventional crystalline materials H_c varies inversely with the grain diameter (1/D). Thus, the larger the grain size, the lower the coercive force. In the nanocrystalline materials, the coercive force varies as the sixth power of the grain size (D^6) Therefore, the coercive force decreases strongly as the grain size decreases. This is shown in Figure 9.32. Herzer explains this by noting that the small size of the grains are smaller than the ferromagnetic exchange length and thus, the magnetization cannot follow the easy axis of magnetization. The texture is then random and the H_c is low and the permeability is high. In addition to the anisotropy, the magnetostriction is also low. The change in magnetostriction with annealing temperature is shown in Figure 9.33. The magnetostriction is also a function of Si content.. The permeability and loss factor for the nanocrystalline material as compared to the MnZn ferrite or Co based amorphous material is shown in Figure 9.34. It shows improved properties over the ferrite and similar properties to the amorphous material. In addition, the saturation of the nanocrystalline material is higher and also has better thermal stability than the other two. Hilzinger points out that in addition to the switched mode power application (to be discussed in a later chapter), there are low level uses as in signal transformers, common mode chokes, pulse transformers, current transformers and ground fault interrupter cores.

This new generation of magnetic materials is still in its infancy and it is hard to predict the extent that it will improve our use of magnetic materials.

HIGH PERMEABILITY-HIGH FREQUENCY METAL STRIP

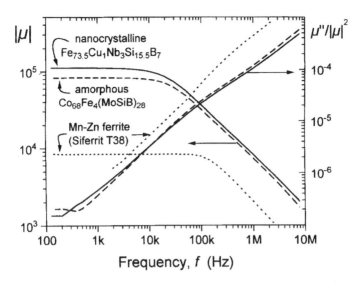

Figure 9.34-Permeability versus frequency for various materials. Reprinted from Herzer, Handbook of Magnetic Materials Vol. 10, ©1997, p455, with permission from Elsevier Science.

SUMMARY

While this chapter has dealt with materials with high permeability and high frequency operation, The next chapter will also deal with high frequency low level uses but the permeabilities will be much lower. The reason for its use lies in the ability to resist temperature, DC bias and other changes, prevent saturation in the magnetic material and act as inductors in medium frequency L-C resonant circuits.

References

Allied Corp. Metglas® Technical Bulletin, Metglas® Products, 6 Eastman Rd. Parsippany NJ 07054
Arnold, H.D. (1923) and Elmen, G.W. J. Franklin Inst. **195**,621
Barrett, W.F. (1902) Brown, W. and Hadfield, R.A.,J. IEE, **31**, 674
Bozorth, R.M. (1951) Ferromagnetism, Van Nostrand New York
Cartech (1988) Carpenter Soft Magnetic Alloys, Carpenter Technology Corp., Reading, PA 19612
Cohen, P. (1967) J. Appl. Phys. **38**, 1174
Elmen, G.W. (1917) Can. Patent, 180,539 Appl. 10/4/16.

Elmen, G.W. (1926) U.S. Patent 1,586,884 Appl. 5/3/21
Elmen, G.W. (1928) J. Franklin Inst. **206**, 317
Hattendorf, H. (1991)-(unpublished)Krupp, VDM Werdohl, Germany in Huebner, U., Material Science and Technology,VCH Publishing Co.Weinheim, 348
Herzer, G. (1997) Handbook of Magnetic Materials, Vol. 10, Elsevier Science B.V. Amsterdam, 418,444,454,455
Hitachi(1998) Finemet ® Technical Data Sheet -Internet
Hopkinson, J. (1889),Proc. Roy. Soc.(London) **47**, 23
Hopkinson, J. (1890),ibid **48**, 1-13
Kang, I.K., (1967), Scholefield, H.H.,and Martin, A.P., J. Appl. Phys. **38**, 1178
Kelsall, G.A. (1934),Physic, **5**, 169
Magnetics (1995) Tape-Wound Cores Design Manual, TWC-400, Magnetics, Division of Spang and Co., Butler, PA 16001
Makino, A. (1997),Hatanai, T., Naito, Y. Bitoh, T.,Inoue, A., and Masumoto, T., IEEE Trans. Mag. **MAG33**, 3793
Masteller, M.S. (1997) Trans. Mag **MAG33**, 3769
Pfeifer, F. (1980) in Concise Encycl.of Magnetic and Superconducting Materials Ed. by J. Evetts, Pergamon Press,Oxford, 349
Smith, W.S.(1924) and Garnett, H.J., Br. Patent 224,972 Appl. 8/25/23
Suzuki, K. (1990), Kataoka, N., Inoue, A., Makino, A. and Masumoto, T.,Mater. Trans. JIM **31**,743
Vacuumschmelze (1996) Soft Magnetic Materials, PHT 001, Vacuumschmelze GMBH, P.O. B. 2253. D-63412, Hanau, Germany
Yensen, T.D.(1924) Trans. AIEE, **43**,145
Yoshizawa, Y (1988) Oguma, S. and Yamaguchi, K.,J. Appl. Phys.,**64**, 6044
Yoshizawa, Y (1989) and Yamaguchi, K., IEEE Trans. Mag. **MAG25**, 3324

10 METAL POWDER CORES FOR TELECOMMUNICATIONS APPLICATIONS

INTRODUCTION

In Chapter 9, metallic magnetic materials with high permeability and ability to operate at high frequency were reviewed. These were meant for low level applications such as those in telephony and radio where the portion of the hysteresis loop traversed was the Rayleigh or initial permeability region. In the same region, there is also a need for magnetic components that are very stable under environmental or operating conditions. These applications are also in the telecommunications industry. For example, outdoor telephones may have to operate at temperatures from about – 40°C. (also –40°F) to about +40° C(104°F). Magnetic components in overhead telephone transmission lines must also be stable under these temperature extremes. In addition the lines may be struck by lightning which subjects the line to a high D.C. current shock which may also disturb the magnetic components. Aside from the environmental effects, telephone lines must also carry D.C. for bell ringing purposes. Aside from stability aspects, the telephone application may also need very sensitive frequency control devices called filters which pass or reject certain frequency bands. The frequencies can be fairly low for audio applications or higher for R.F. carrier frequencies in the range from hundreds of KHz. to the Megahertz range. The high frequency and stability aspects can be met by using some of the same magnetic materials previously discussed in compressed powder form. Reducing the smallest dimensions of a magnetic material reduces Eddy current losses thus allowing higher frequency operation. In the course of reducing dimensions, the three dimensional bulk can be reduced to strip (two dimensions) then to wire (one dimension) and ultimately to powder. All four shapes have at one time been used for magnetic purposes.

Linearity of permeability with drive level is also desirable as it was with the strip materials. However, in some of the telephony applications the linearity or constancy of slope with a D.C. bias is needed. This extends the range of initial permeability to regions where the permeability would change in the strip material even though the ac drive was still low. Moreover, the strip materials would saturate leading to catastrophic results. The protection against saturation is another important property of powder cores. Powder cores may stay linear even with the superposition of fields as high as hundreds of oersteds. Some of the same stability effects can be obtained by placing a discreet gap in the magnetic circuit. The powder core is effectively a gapped core but the gap is distributed to the spaces between the powder particles. This "distributed" gap may be more useful than the discreet gap in that it has no fringing flux at the gap and also the small particles allow for high frequency operation. A useful equation for separation of the losses in low level magnetic cores was developed be Legg (1936). By a series of measurements on cores at different frequencies and flux levels the so-called Legg coefficients can be determined in his equation. The Legg equation is;

$$R_m/\mu f L_m = hB + ef + c \qquad [10.1]$$

Where R_m = Resistance due to material
L_m = Inductance
μ = Permeability
f = frequency
h = Hysteresis loss coefficient
e = Eddy Current loss coefficient
c = Anomalous loss coefficient

There are two main types of materials used for the linear applications in this chapter. They are the iron powder cores and the 2-81 moly-perm (2% molybdenum-81% Ni balance Fe) powder cores (MPP). There are many different varieties within these two broad categories for different properties and price ranges. The iron powder cores will be dealt with first followed by the MPP cores. The use of the above powder cores or variations of them will also be treated later in the section on high frequency high power applications and also in the section on EMI materials.

HISTORY OF POWDER CORES FOR TELECOMMUNICATIONS
Heaviside (1887) was the first to study to look at iron powders impregnated in wax. The first practical use for powder cores was as load coils in telephone transmission lines. The load coils (or Pupin coils) were components used to compensate for the increased capacitance of long transmission lines. This capacitance increase would change the frequencies of the signals being carried. Before the First World War, load coil cores made of iron wire were placed around the cable to provide the inductance to restore the proper frequencies in the cable. During the war, American telephone companies could not get the required wire from Europe so they looked at other materials including powdered iron. The toroidal powdered iron cores were used in the U.S. as early as 1918 and were found to be superior to the steel wire cores (Speed and Elmen 1921). Soon, Europe followed the same path. The original cores only had permeabilities of about 30 but were subsequently improved to about 120. The iron powders for the cores were made by electrolytic deposition. Processes for making mechanically disintegrated powders were complicated and also caused explosions. In 1930 I.G. Farben of Germany devised a chemical process for making the fine iron powder. The iron was reacted with carbon monoxide to form the liquid iron carbonyl. This was then distilled and decomposed to form extremely fine and pure spherical iron powders. In 1935, Siemens made powdered cores of the car-

bonyl iron. Carbonyl iron is still used extensively today for high quality powdered iron cores. For use in radio coils, the first article describing the application was by Crossley (1933). Meanwhile, in 1923, the permalloys were discovered and in 1927, powdered permalloy was used in powder cores (Shackelton 1928). In 1940, the moly permalloy powder core was invented by Bell Laboratories (Given 1940). This became the standard material for higher quality powder cores and the same material is widely used today. In the U.S. and Canada where telephone transmission lines in rural areas are still mostly above ground requiring the temperature stability of the MPP core. They are still used as load coil cores. However, in Europe and Japan, where the transmission lines are below ground and do not need the temperature or lightning stability, the shift has been to ferrite cores for the same purpose mostly for economy reasons. With the change to fiber optics and underground transmission, the use of MPP cores as load coils in the U.S. and Canada is diminishing. The use in electronic circuitry is stable and in some areas such as in noise filters (EMI) and output chokes in switched mode power supplies (SMPS), the use is growing.

IRON POWDER CORES FOR AUDIO AND RF APPLCATIONS

The properties of iron powder cores depend greatly on the chemical purity, particle size and shape of the powder used. High purity larger high-density particles give higher permeabilities at low and medium frequencies. Smaller or layered particles (such as those prepared by the carbonyl process) give lower permeabilities at low frequencies but permit permit operation with low losses at high frequencies. Higher density cores using lower insulation percentages and higher forming pressures have higher permeabilities. The highest permeability is about 100 and is effective to about 100 Khz. On the other hand, carbonyl iron cores with relative permeabilities of about 5-10 are usable up to several hundred Megahertz. For low permeability cores, the temperature coefficient is about 10-50 ppm per degree C. For higher permeability powder cores, the T.C. is on the order of several hundred ppm per degree C. Similarly with D.C. bias, an iron powder core with a permeability of 20 retains that permeability with a superimposed D.C. field of 1600 A/m (20 Oersteds) while one with a relative permeability of 60 decreases 15% under the same conditions.

A general rule for choice of type of iron used states that up to permeabilities up to 35, carbonyl iron is used. For permeabilities of 35 and over, hydrogen-reduced iron or atomized iron is generally used. One consideration for the use of carbonyl iron cores is the cost. The carbonyl iron material is expensive at about $10,000 per metric ton compared to $2,000-4,000 per metric ton for the hydrogen-reduced material.

Two of the carbonyl iron powders, the E-type and the C-type can represent two variations of these powders. The E-type is the powder prepared by the carbonyl process. The structure of the particle is made up of concentric shells with a space in between the shells. This is often called an "onion-skin" structure. Because of the internal spaces, the effective particle size should be much smaller than those measured by seiving since that only measures the external dimension. The C-type powder is made from the E-type by long heat treatment in hydrogen at 400°C. giving powders than were carbon free. Samples of these two powders along with an electrolytic iron powder were separated by or sedimentation sifting into various particle size fractions. Then cores were made of each powder There is a direct correlation between the particle size and the permeability. The smaller the particle size, the lower the permeability as we have stated before. The losses also decrease as the particle size is decreased. All three of the Legg coefficients in the Legg equation (given earlier in this chapter) are lowered as the permeability decreases. The lowering of the Eddy current loss makes it possible to operate at higher frequencies. The relation between particle size and the hysteresis and Eddy current loss coefficients is shown in Table 10.1.

PROCESSING OF IRON POWDER CORES

The processing scheme to prepare the iron powder has been discussed previously. The production of the cores will follow two separate processes depending on whether carbonyl or hydrogen reduced iron powder is used. Both will be described although for the R.F. applications of this chapter, the carbonyl method(except for the 40 perm core) is generally used. The two methods may be called the high pressure method and the low pressure method. The carbonyl irons use the low pressure scheme. Because of the fineness of the powder and the less dense core needed for the low permeabilities, consolidation without the use of an organic polymer binder would be impossible. Therefore the powder is oxidized, and phosphated to give it a thin inorganic insulation It is then immersed in a polymer solution in a solvent such as acetone. The solvent is evaporated while the iron slurry is stirred. The result is a powder that is coated with the polymer . The core is then die-pressed and heated to cure the polymer. A theoretical curve of permeability vs binder content was formulated by Kornetzsky (1936) and is shown in Figure 10.1. The equation for this relation is;

$$\mu_i = \mu_k(1+a/3\mu_k) = 3/a \quad [10.2]$$
where: μ_k = initial permeability
a = binder content

For low perm cores (10-20) the binder content is high volume-wise but since the density of the polymer is about 1/7 that of the metal, the wt % is about 2%. The pressing is done at low pressure hence the name of the process. In the case of the hydrogen reduced powder, dense compacts are needed without the dilution effect of a polymer binder to lower the permeability. A thin coating of inorganic insulation must be used and high compaction pressures up to 50 tsi (tons/sq) are necessary to plastically deform the iron particles. The core then must be heat treated at high temperatures to relieve the stresses induced in the pressing operation. Without this anneal, the magnetic properties would be very poor. In the case of the MPP cores discussed later, the high pressure method is always used. A serious consideration in the manufacture of powder cores relates to problems that may be encountered by the

Table 10.1- Particle Sizes, Permeabilities and Legg Loss coefficients for various types of Iron Powder Core Materials From Richards (1953)

Material	Particle size (microns)	Permeability	Magnetic loss coefficients		
			$h \cdot 10^6$	$c \cdot 10^6$	$e \cdot 10^6$
(a) Electrolytic (sifted)	82	31	360	1400	41
	66	32	300	1400	26
	22	29	260	1300	14
	12	23	230	1100	4·5
(b) Electrolytic (pebble mill)	18	12	239	1800	9·5
	15	11	232	2100	8·3
	16	10	190	2300	6·2
	14	10	195	2300	4·1
	12	10	173	2400	2·9
(c) C-type powder	7·0	17	90	1100	3·9
	4·6	12	44	700	0·8
	3·0	10	32	800	0·3
(d) E-type powder	4·0	8·4	11	200	0·2
	3·5	8·1	7·4		
	3·1	7·5	6·7		
	2·8	6·9	5·8	not measured	
	2·2	7·0	4		
	1·7	6·8	4		

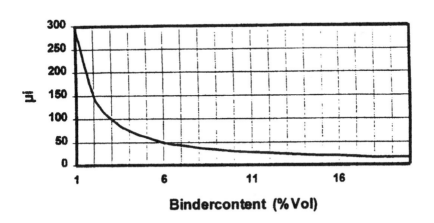

Figure10.1-Permeability versus binder content in iron powder cores From Grohs (1996)

winding house using powder core toroids. This relates to the so-called "break-strength" of the core. The winding process may put very strains on the core and possible crack it. The core after heat treatment must be coated with a durable and moisture resistant paint. This paint does increase the break strength of the core The finish must also be able to withstand a voltage breakdown test of up to 500-1000 V. Parylene coating is another option.

Magnetic Properties of Iron Powder Cores

The effect of the type of iron powder used in cores for R.F applications on the permeability, temperature stability and resonant frequency range is given in Table 10.2. Most of them are of the carbonyl type except for the 40 perm core. The material with a permeability of 1(same as air) is a phenolic which just serves as a coil former. It has the highest frequency. The carbonyl iron(Mix 17) would probably be the finest powder and extends to the highest frequency of the irons (250 MHz.). It also has the least change of permeability with temperature. The hydrogen-reduced material that has the largest particle size has the highest permeability, the lowest frequency of operation and the highest temperature coefficient. The variations of permeability with D.C. bias and with ac flux density are given in Figures 10.2 and 10.3. Again the same pattern is repeated with D.C. bias. The lowest perm material drops the least and the highest perm the most. With flux density, the changes in perm behave the same way with regard to the iron materials. Core loss while not as important in the lower flux conditions must still be considered. If losses are high enough, the heat generated may change the saturation of the core.

Table 10.2
Permeabilities, Temperature Stabilities and Resonant Frequency Ranges for Various Iron Powder Core Materials for Telecommunications Applications From Micrometals (1993)

		GENERAL MATERIAL PROPERTIES FOR RF CORES		
Mix #	Basic Iron Powder	Material Permeability (μ_0)	Temperature[1] Stability (+) (ppm/°C)	Resonant Circuit Frequency Range (MHz)
1	Carbonyl C	20	280	.15-3
2	Carbonyl E	10	95	.25-10
3	Carbonyl HP	35	370	.02-1
4	Carbonyl J	9.0	280	3-40
6	Carbonyl SF	8.5	35	3-40
7	Carbonyl TH	9.0	30	1-25
8	Carbonyl GQ4	35	255	.02-1
10	Carbonyl W	6.0	150	15-100
12**	Synthetic Oxide	4.0	170*	30-250
15	Carbonyl GS6	25	190	.15-3
17	Carbonyl	4.0	50	20-200
42	Hydrogen Reduced	40	550	.03-.80
0	Phenolic	1	0	50-350

[1]Temperature stability values listed are for closed magnetic structures.

The quality factor of an inductor in a resonant frequency application is the so-called Q of the core. The Q for a specific core and winding has a curve as shown in Figure

METAL POWDER CORES FOR TELECOMMUNICATIONS

10.4. If Q is an important requisite, the core is chosen with the Q peaking at the particular frequency of operation. The percent frequency variation is given by

$$\Delta f/f = 1/Q \qquad [10.3]$$

Therefore, the higher the Q, the less the frequency change and the greater the frequency specificity. Of course the inductance required and the temperature stability must also be considered. Permeability of the iron powder cores. For broadband applications, the use of iron powder cores is limited to cases of moderate bandwidth, low loss and good stability. Ferrites are more suitable when the bandwidth is large. In broadband applications, Q is not an important factor so frequency of operation is wider. It is determined by the impedance characteristics as a function of frequency with the ranges of high impedance attenuating those frequencies and those of low impedance passing the particular frequency range. A typical impedance versus frequency curve is given in Figure 10.5

Shapes of Iron Powder Cores

The greatest usage of iron powder is in the form of toroidal cores. It is a closed magnetic circuit that optimizes the permeability and losses, is more compact than E-cores and only has one part. However, the toroid configuration requires more complex winding equipment than a bobbin wound core. Another configuration for RF cores is the balun shown in Figure 10.6. This shape is best for broadband applications since it provides more impedance per unit length of wire than a toroid

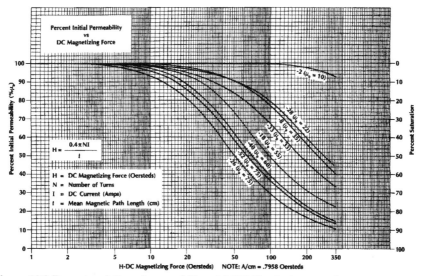

Figure 10.2-Percent variation of permeability with DC bias for iron powder cores of different permeabilities From Micrometals (1993)

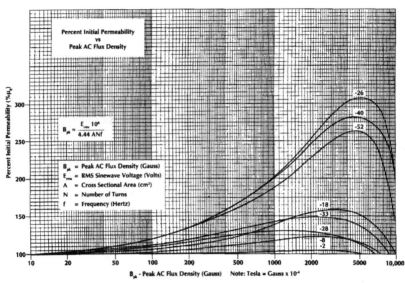

Figure 10.3-Percent variation of permeability with flux density for iron powder cores of different permeabilities From Micrometals(1993)

and so will operate over a wider frequency range. Impedance versus frequency curves for baluns are shown in Figure 10.7. They are however, more difficult to wind. Iron powder cores are also available in plain cores (cylinders), hollow cores (tubes) sleeves, threaded cores, E-cores, U-cores, cups, and bobbins with bobbin sleeves. Many of these such as the plain cores, bobbins, E-cores, U-cores and cups have the economical advantage of simple winding (spool-type). Threaded cores are often used in adjustable slug tuners in prepared coil forms. The shielded assembly is available commercially. Still another variation is the cylinder molded around a wire with leads.

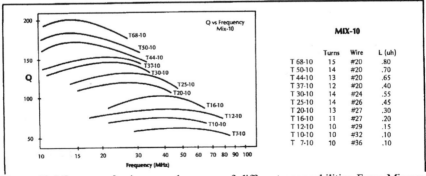

Figure 10.4-Q curves for iron powder cores of different permeabilities From Micrometals(1993)

MOLY-PERM POWDER CORES FOR TELECOMMUNICATIONS

During the time period between the introduction of powdered iron cores and that of the invention of 2-81 moly-permalloy powder cores, there were attempts to use the straight 80-20 NiFe permalloy powder cores (Shackleton, 1928). However, once the 4-79 moly perm material was discovered, the research indicated that the best equivalent material for the powder version would be the 2-81 (2% molybdenum-81% nickel) composition. It has become the standard moly-permalloy composition and is still widely in use today. The improved permeability, lower losses better high frequency operation and great stability have kept it as a superior material for many applications. In addition to its use in loading coils, other applications as high frequency filter transformers, inductors and chokes are examples especially where added stability is needed. The frequency of operation extends to about 150 MHz. that is slightly under the highest frequency at which the lowest permeability powdered iron cores can function.

Processing of Moly-Permalloy Powder Cores (MPP)

About 200 pound ingots of 2% Molybdenum, 81% Nickel and 17% Iron with slight additions of sulfur and a small amount of manganese are melted in air in an electric induction furnace. The ingots are hot rolled to about ½ inch thick and then cold rolled to about 1/4 inch thick. The proper sulfur content will permit the hot rolling and also embrittle the material during the cold rolling. The proper grain size is established during the hot rolling operation. With the use of hammer and attrition mills, the powder is broken down to an appropriate particle size. A typical distribution is shown in Figure 10.8. The hot rolling operation for preparing brittle for Moly-perm powder cores is shown in Figure 10.9. The powder is sieved to remove the coarse particles and the annealed at 850°C. The powder is then broken up again and a insulated with few percent of an inorganic mixture. The problem is to coat the particles with a minimum thickness that will not break away during the pressing operation and will not fuse and flux the magnetic particles during the heat treatment. The separation between the particles is on the order of the wavelength of light. The thickness of the insulating film is approximately $rt/300p$ where r is the percent insulating material by volume, p is the packing factor of the magnetic material and t is the rms particle diameter(spherical particles). The relation between packing factor, percent dilution and permeability is given in Figure 10.10. In addition to the other requirements, insulating film is approximately $rt/300p$ where r is the percent insulating material by volume, p is the packing factor of the magnetic material and t is the rms particle diameter(spherical particles). The relation between packing factor, percent dilution and permeability is given in Figure 10.10. In addition to the other requirements, the ceramic coating must not react with the hydrogen during the anneal. The actual process is a carefully held secret by the manufacturers. However, the literature de scribes the insulation as one containing sodium silicate, magnesium hydroxide and talc. The powder is then pressed into rings at a pressure of 100-150 psi. The toroids are then annealed at 600-650°C.to remove the strains and develop the magnetic properties. The core is then coated with a insulating paint. This paint serves to keep the moisture out which would change the properties of the core and also to prevent

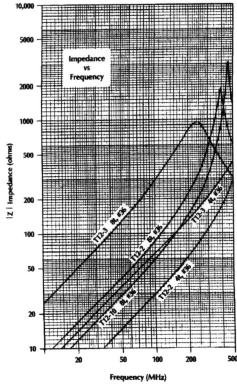

Figure 10.5 Impedance versus frequency for iron powder core toroids From Micrometals (1993)

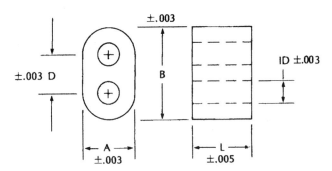

Figure 10.6- A balun configuration for iron powder cores for RF applications From Micrometals (1993)

METAL POWDER CORES FOR TELECOMMUNICATIONS

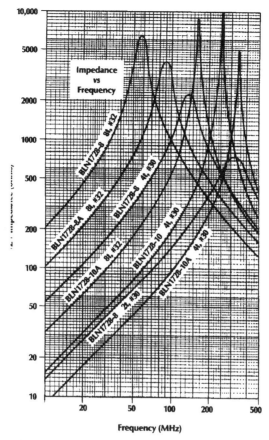

Figure 10.7- Impedance versus frequency curves for a iron powder core balun for RF applications From Micrometals (1993)

the scraping off of the wire insulation when the core is wound. The core is tested for permeability and other pertinent magnetic properties. A flow chart of the processing is shown in Figure 10.11. The dependence of the permeability on the density of the core is shown in Figure 10.12. At 2% insulation, the density is 7.7 or about 90% of the bulk density of 8.65. The tensile strength is 300-400 psi. The relation between pressing pressures and both density and strength is given in Figure 10.13

For temperature compensation where needed, an alloy of 12% molybdenum-80% nickel is added in small amounts (tenths of a percent is added to the 2-81 alloy. The 12-80 material has a Curie point about 70oC. For lower permeabilities, the particle size distribution. Pressing pressure and insulation percentage are adjusted accordingly. Several theories on the relation of core permeabilities to the packing fraction of the core give the following equation

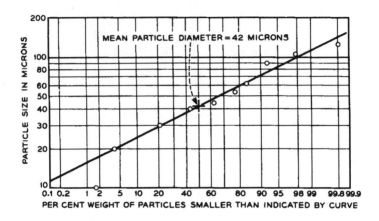

Figure 10.8-Particle size distribution of powder to be used in a MPP core. Fom Bozorth (1953)

$$\mu = \mu_i P \log \mu = p \log \mu_i \quad [10.4]$$
Where μ = permeability of the core
P = packing fraction
μ_i = permeability of the material

Owens (1955) has determined the discreet air gap that is the equivalent of the combined inter-particle gaps. The equation appropriate for this calculation is

$$\mu_c = \mu/[r(\mu-1)+1] \quad [10.5]$$
Where $r = l_a/l_c$
l_a = length of air gap
l_c = length of magnetic material
μ_c = core permeability
μ = permeability of the magnetic particles

Figure 10.14 shows the variation of core permeability with different air gaps for a range of permeabilities. For 125 perm MPP cores, the insulation comprises .8% of the flux path length. Since the average particle size is 40 microns, the mean

METAL POWDER CORES FOR TELECOMMUNICATIONS 195

Figure 10.9- A hot rolling operation for preparing brittle for moly-perm powder cores, Picture courtesy of Magnetics

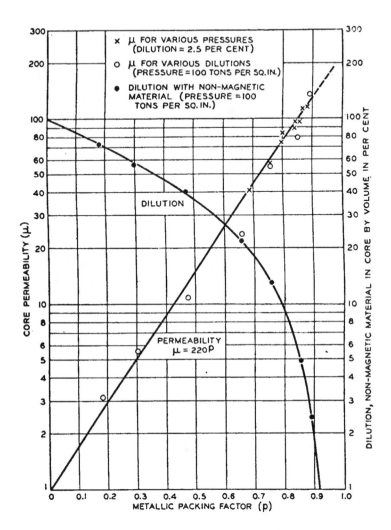

Figure 10.10- Relationship between metallic packing factor, permeability and percent dilution. From Legg (1940)

thickness of the insulation between particles would be about .32 microns. The highest perm offered commercially is 550 perm. This material is made by converting the powder into flake before pressing and is known as a flake core.

Properties of Moly Perm Powder Cores (MPP)

The commercially offered moly-perm powder cores comes in about 10 permeabilities. The permeabilities along with the frequency ranges where they are normally used is given in Table 10.15. For load coil applications, the 125 permeability core

METAL POWDER CORES FOR TELECOMMUNICATIONS

has been the one of choice. For high frequency transformers or inductors, the 14 and 26 perms are frequently used. Many of the higher perm (200 and over) are used in power applications. The other permeabilities are used in LC resonant circuits, and

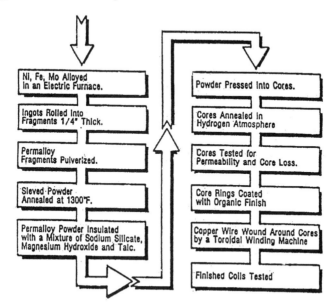

Figure 10.11 Flow chart of process for making MPP cores

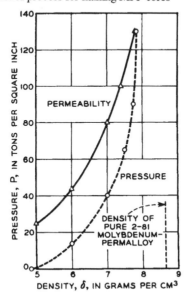

Figure 10.12-Relation between pressure, density and permeability in the making of MPP cores Fom Bozorth (1951)

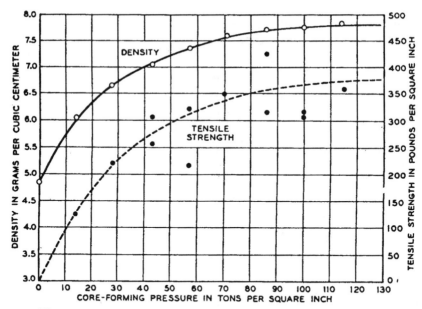

Figure 10.13. Pressing pressures, density and tensile strength in the forming of MPP cores. From Legg (1940)

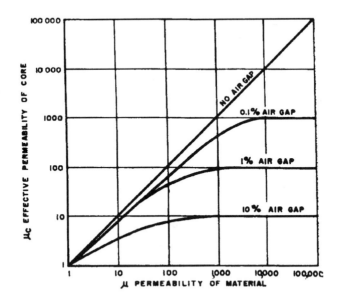

Figure 10.14-Material permeability, effective permeability with various air gaps in MPP cores From Owens(1956)

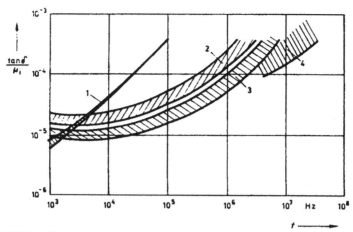

Figure 10.15 Loss factor versus frequency for several types of powdered metal cores 1-Sendust (μ = 80) and MPP core (μ = 125); 2-Carbonyl Iron C (μ = 50);3- Carbonyl Iron HF(μ = 14);4- Carbonyl Iron HFF(μ =5-10)From Heister in Heck (1974)

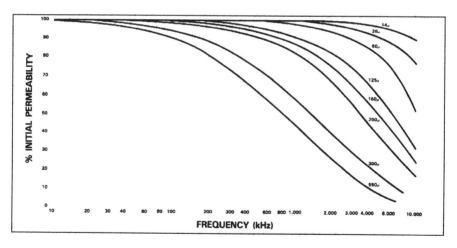

Figure 10.16-Permeability variation with frequency for MPP cores of different permeabilities. From Magnetics (1997)

RFI filters.The variation of permeability with frequency is shown in Figure 10.16. Within the permeability ranges, there are varying degrees of temperature stability of permeability. First, there are cores that are unstabilized. Then, there are 3 or 4 types of flat temperature stability over different temperature ranges. A typical comparison of permeability variations in stabilized and unstabilized cores is given in Figure 10.17. Sets of curves describing this stability of unstabilized and stabilized cores are given in Figures 10.18a and 10.18b. Last, there is a linear temperature stabilized core that is meant to compensate for the negative temperature characteristics of a polystyrene capacitor. The stability can be affected by external conditions such moisture, winding stresses and potting compounds. The effects of moisture on the weight of insulation and core permeability are shown for 125 perm cores in Figure 10.19. Procedures for user stabilization of the cores can be obtained from the vendors' catalogs.

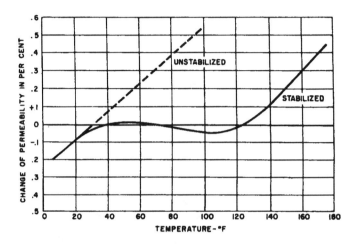

Figure 10.17- Changes of permeability with temperature for unstabilized and stabilized MPP cores From Owens (1956)

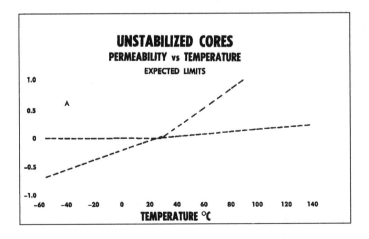

Figure 10.18a- Expected limits in permeability variation with temperature for unstabilized MPP cores From Magnetics (1997)

METAL POWDER CORES FOR TELECOMMUNICATIONS

For resonant circuit applications, the Q is a very important property to be considered, This depends on core permeability, core size, winding and frequency. Powder core vendors display the Q curves of the different cores with various windings. A sample of such a Q curve is given in Figure 10.20.

Another important item of stability is the variation of permeability with drive level for both DC and ac. Figure 10.21 shows the decrease in permeabilty with DC bias for several permeabilities. A critical test for the performance of a core under high D.C. bias and residual effects is the so-called "butterfly" curve. In this case, the core is first subjected to a positively increasing D.C. bias to over 100 Oersteds. Then, the bias is lowered and reversed and finally returned to the starting point. Figure 10.22 shows such curves for a 125 perm core. The change in permeability after such cycling is about .3% attesting to the great stability of the cores. The variation of permeability with ac flux level is shown in Figure 10.23. While the 4-79 Moly-Permalloy strip has a maximum permeability at about 300-4000 Gauss, for the MPP cores it occurs at only about 1000 Gauss. In many typical applications the flux level for MPP cores is under 100 Gauss.

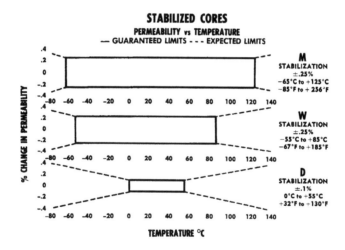

Figure 10.18b-Guaranteed and expected limits of permeability variation With temperature for stabilized MPP cores From Magnetics (1977)

Another important property of the MPP cores is the core loss. The Eddy current core loss limits the usage to about 300 KHz. The total core loss is composed of the three factors listed earlier under the Legg coefficients. The total core loss limits expresses in ohm/Henry x μ as well as the Legg coefficients are given in Table 10.3. These are suitable below 200 Gauss. The core losses in Watts/lb. for various frequencies are usually plotted against flux density. Such a plot is shown in Figure

202 HANDBOOK OF MODERN FERROMAGNETIC MATERIALS

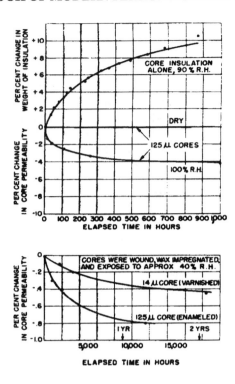

Figure 10.19- Effect of moisture on the permeability of uncoated and coated MPP cores. From Owens (1956)

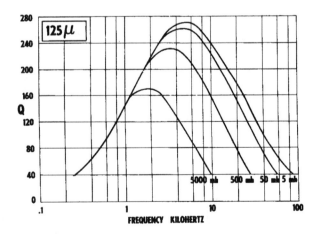

Figure 10.20- Q Curves for 125 perm MPP cores. From Magnetics (1977)

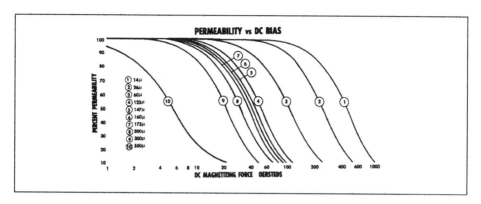

Figure 10.21- Variation of permeability with DC bias for various permeability MPP Cores. From Magnetics (1977)

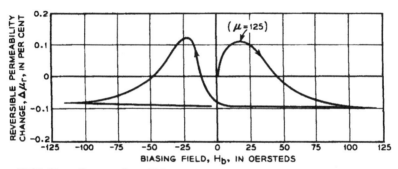

Figure 10.22- Butterfly curve for a 125 perm MPP core. From Bozorth (1951)

10.24. A Steinmetz type equation is also listed with the appropriate coefficients for the material. Lastly, and connected to the core losses is the temperature rise. The rise in temperature (which includes core losses and copper losses) can be given as a function of the core losses by the following equation.

Rise (°C) = Total power dissipated(mW)/Surface Area(cm^2)   [10.6]

The copper losses can be calculated from the winding data and the core losses can be obtained from the core loss curves. The surface areas for the cores are given by the vendor in the dimensional tables.

HANDBOOK OF MODERN FERROMAGNETIC MATERIALS

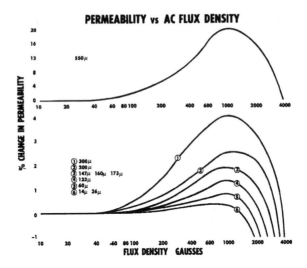

Figure 10.23- Variation of permeability with flux Density for various permeability MPP cores. From Magnetics (1977)

Table 10.2
Core Loss Table for Various Permeability MPP Cores From Magnetics (1977)

CORE LOSS TABLE

Permeability μ	MAXIMUM CORE LOSSES			TYPICAL LOSS COEFFICIENTS			Maximum Perm. Shift After DC Magnetization*
	$\frac{R_{AC}}{\mu L}$ OHMS / HENRY x μ	TEST CONDITIONS		EDDY CURRENT e	HYSTERESIS a	RESIDUAL c	
		FLUX DENSITY GAUSS	FREQUENCY KHZ				
14	60.0	4	75.0	6.5×10^{-9}	7.0×10^{-6}	143×10^{-6}	±0.1%
26	70.0	4	75.0	7.0×10^{-9}	4.0×10^{-6}	96×10^{-6}	±0.2%
60	1.5	10	8.0	7.5×10^{-9}	1.5×10^{-6}	50×10^{-6}	±0.3%
125	0.20	20	1.8	15×10^{-9}	0.9×10^{-6}	32×10^{-6}	±0.5%
147	0.20	20	1.8	16×10^{-9}	0.9×10^{-6}	28×10^{-6}	±0.5%
160	0.20	20	1.8	17×10^{-9}	0.9×10^{-6}	25×10^{-6}	±0.5%
173	0.20	20	1.8	20×10^{-9}	0.8×10^{-6}	23×10^{-6}	±0.5%
200	0.25	20	1.8	25×10^{-9}	0.7×10^{-6}	21×10^{-6}	±0.5%
300	0.25	20	1.8	30×10^{-9}	0.8×10^{-6}	30×10^{-6}	±0.5%
550	0.40	20	1.8	27×10^{-9}	1.5×10^{-6}	88×10^{-6}	±3.5%

SUMMARY

Chapters 6-10 have dealt with the metallic magnetic materials both in strip and powder core forms. Although we started at the low frequencies, we gradually advanced to some of the intermediate and higher frequencies. The next part of the book is concerned with the ceramic magnetic materials, generically classified as ferrites. Because of their high resistivities, they will be much more effective at the very high frequencies and thus constitute a very important addition to our inventory of magnetic materials. The next chapters discuss in the following order, the crystal structure, chemistry, microstructure and processing of the magnetic ferrites

METAL POWDER CORES FOR TELECOMMUNICATIONS

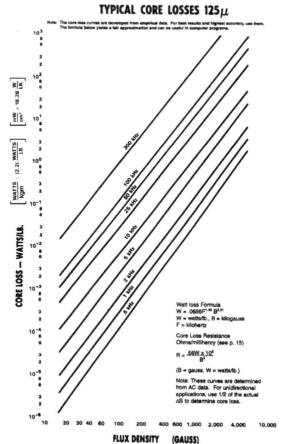

Figure 10.24-Core losses for various permeabilities of MPP cores From Magnetics (1994)

References

Bozorth, R.M. (1951) Ferromagnetism, Van Nostrand Reinhold, New York
Grohs, P. (1996) Gorham-Intertech Conf., Feb. 26-28, 1996, San Francisco, CA
Kornetski, M. (1936) and Weis, H., Wiss. a.d. Siemens Werken **XV**,95
Legg, V.E.(1940) and Given, F.J., Bell Syst. Tel. J., **19**,385
Legg, V.E.(1936) Bell Syst. Tel. J. **16**,39
Magnetics (1997) Molypermalloy Powder Cores –MPP400- 6E
Micrometals (1990),Iron Powder Cores for RF applications-Catalog 3D,Micrometals, 11190 N. Hawk Circle, Anaheim CA, 92807
Owens, C.D. (1956)Conference Paper-Presented at Conference on Magnetism and Magnetic Materials,June 14-16, 1955, Pittsburgh, PA
Richards,C.E.(1953)in Soft Magnetic Materials for Telecommunications. Interscience Press, New York, 233
Shackelton, W.J.(1928) and Barber, I.G., Trans. AIEE, **47**,429
Speed, B. (1921) and Elmen, G.W., Trans. AIEE, **40**, 1312

11 CRYSTAL STRUCTURE OF FERRITES

INTRODUCTION

In Chapter 1, we built up a series of magnetic structures of increasing complexity starting with the electron, progressing to the atom (or ion) and finally, focusing on the domain. Although the domain is important in explaining cooperative magnetic phenomena, the next larger physical magnetic entity after the ion is the ferrite unit cell or the crystal structure. The crystal structure of a ferrite can be regarded as an interlocking network of positively-charged metal ions (Fe^{+++}, M^{2+}) and negatively-charged divalent oxygen ions ($O^=$). Hereafter in the sections on ferrites, we will be dealing with ions rather than atoms and specifically in magnetic oxide ceramics. Since the crystal contains a network of ionic bonds, we can think of the crystal as a giant molecule. The arrangement of the ions or the crystal structure of the ferrite will play a most important role in determining the magnetic interactions and therefore, the magnetic properties.

CLASSES OF CRYSTAL STRUCTURES IN FERRITES

In the magnetic ceramics, the various crystal structures start with the arrangement of the oxygen ions. Let us consider a layer of these oxygen ions closely packed so that the lines connecting their centers form a network of equilateral triangles. The next layer of oxygen ions is also closely packed so that their centers lie directly over the centers of the equilateral triangles of the first layer. Now, the third layer can be arranged in two different ways. First, it can repeat the positions of the first layer in which case we call it a hexagonal close-packed structure in a type of ababab arrangement. This leads to a structure that has a unique crystal axis that we find in some ferrites. Second, the oxygen ions can be so placed that their centers lie directly over the centers of the equilateral triangles adjacent to the ones used for the hexagonal close-packed structure. The fourth layer would then repeat the first so that the pattern would be abcabcabc. This gives rise to a crystal structure called the cubic close-packed. The spinel ferrite is an example of this class. The type of crystal structure preferred is determined by the size and charge of the metal ions that will balance the charge of the oxygen ions and the relative amounts of these ions. Some oxides such as yttrium oxide (Y_2O_3) may form more than one class depending on the ratio of Y_2O_3 to Fe_2O_3. Thus, $Y_2O_3.Fe_2O_3$ and $3Y_2O_3.5Fe_2O_3$ have different crystal structures.

The crystal structure is frequently related to the ultimate application. For example, BaO combines with Fe_2O_3 to form a hexagonal structure with a unique crystal axis predisposing to a permanent magnet application. On the other hand, the cubic crystal structure has many equivalent crystal directions and so will be useful when it is advantageous to avoid a preferred direction.

Spinel Chemistry

The spinel is by far the most widely used ferrite, so much so that the term is almost synonymous with the word, "ferrite". The spinel structure is derived from the mineral, spinel, ($MgAl_2O_4$ or $MgO \cdot Al_2O_3$) whose structure was elucidated by Bragg (1915).

Analogous to the mineral spinel, magnetic spinels have the general formula $MO\,Fe_2O_3$ or MFe_2O_4 where M is the divalent metal ion. The trivalent Al is usually replaced by Fe^{+++} or by Fe^{+++} in combination with other trivalent ions. Although the majority of ferrites contain iron oxide as the name might imply, there are some "ferrites" based on Cr, Mn, and other elements. Although Mn and Cr are not ferromagnetic elements, in combination with other elements such as oxygen and different metal ions, they can behave as magnetic ions. Thus, chromites and manganites are possible but not important commercially. In the magnetic spinels, the divalent Mg^{++} can be replaced by Mn^{++}, Ni^{++}, Cu^{++}, Co^{++}, Fe^{++}, (Li^+) Zn^{++}, or more often, combinations of these. The presence of Fe^{+++}, Fe^{++}, Ni^{++}, Co^{++} and Mn^{++} can be used to provide the unpaired electron spins and therefore part of the magnetic moment of a spinel. Other divalent ions such as Mg^{2+} or Zn^{2+} (or monovalent Li) are not paramagnetic but are used to disproportionate the Fe^{+++} ions on the crystal lattice sites to provide or increase the magnetic moment.

Magnetic Moments of the Individual Ions in Spinels

Chapter 2 touched on Néel's treatment of the moments of ferrites according to the interactions between the spins of magnetic ions on two different sublattices. Before we discuss the interactions between the various ions in the spinel lattice, it may be helpful to review the number of unpaired electron spins for each ion involved in spinels so that we may determine the net moment after the magnetic interactions. These were listed in Table 2.1 with the theoretical number of Bohr magnetons they produce.

Ionic Charge Balance and Crystal Structure

The spinel lattice is composed of a close-packed oxygen arrangement in which 32 oxygen ions form a unit cell that is the smallest repeating unit in the crystal network. Between the layers of oxygen ions, if we simply visualize them as spheres, there are interstices that may accommodate the metal ions. Now, the interstices are not all the same; some which we will call A sites are surrounded by or coordinated with 4 nearest neighboring oxygen ions whose lines connecting their centers form a tetrahedron. Thus, A sites are called tetrahedral sites. The other type of site (B sites) is coordinated by 6 nearest neighbor oxygen ions whose center connecting lines describe an octahedron. The B sites are called octahedral sites. In the unit cell of 32 oxygen ions there are 64 tetrahedral sites and 32 octahedral sites. If all of these were filled with metal ions, of either +2 or +3 valence, the positive charge would be very much greater than the negative charge and so the structure would not be electrically neutral. It turns out that of the 64 tetrahedral sites, only 8 are occupied and out of 32 octahedral sites, only 16 are occupied. If, as in the mineral, spinel, the tetrahedral

CRYSTAL STRUCTURE OF FERRITES

sites are occupied by divalent ions and the octahedral sites are occupied by the trivalent ions, the total positive charge would be $8 \times (+2) = +16$ plus the $16 \times (+3) = +48$ or a total of $+64$ which is needed to balance the $32 \times (-2) = -64$ for the oxygen ions. There would then be eight formula units of $MO \cdot Fe_2O_3$ or MFe_2O_4 in a unit cell. A spinel unit cell contains two types of subcells (See Figure 11.1). The two types of subcells alternate in a three-dimensional array so that each fully repeating unit cell requires eight subcells.

As we said in Chapter 1, the mechanism of ferromagnetism involves the negative exchange interaction of atomic moments of ions on two different lattice sites. Many of the properties of useful ferrites can be predicted by an understanding of these interactions and the site preferences of the metallic ions.

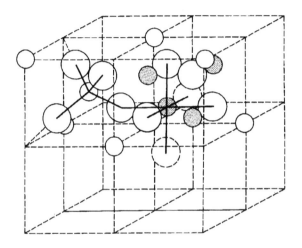

Figure 11.1- Two subcells of a unit cell of the spinel structure. From Smit, J and Wijn, H.P.J., Advances in Electronics and Electron Physics, , 72 (1954)

Site Preferences of the Ions

The preference of the individual ions for the two types of lattice sites is determined by;

1. The ionic radii of the specific ions
2. The size of the interstices
3. Temperature
4. The orbital preference for specific coordination

The most important consideration would appear to be the relative size of the ion compared to the size of the lattice site. The divalent ions are generally larger than the trivalent (because the larger charge produces greater electrostatic attraction and so pulls the outer orbits inward). Table 11.1 lists several of the applicable ionic radii. The octahedral sites are also larger than the tetrahedral (See Table 11.2). Therefore, it would be reasonable that the trivalent ions such as Fe^{+++} would go into the tetrahedral sites and the divalent ions would go into the octahedral. Two exceptions are found in Zn^{++} and Cd^{++} which prefer tetrahedral sites because the electronic configuration is favorable for tetrahedral bonding to the oxygen ions. Thus Zn takes preference for tetrahedral sites over the Fe^{+++} ions. Zn^{2+} and Co^{2+} have the same ionic radius but Zn prefers tetrahedral sites and Co^{2+} prefers octahedral sites because of the configurational exception. Ni^{2+} and Cr^{3+} have strong preferences for octahedral sites, while other ions have weaker preferences.

Table 11.1
Radii of Metal Ions involved in Spinel Ferrites

Metal	Ionic Radius (Angstrom Units, A°)
Mg	.78
Mn^{+++}	.70
Mn^{++}	.91
Fe^{++}	.83
Fe^{+++}	.67
Co^{++}	.82
Ni^{++}	.78
Cu^{++}	.70
Zn^{++}	.82
Cd^{++}	1.03
Al^{+++}	.57
Cr^{+++}	.64

From: Handbook of Chemistry and Physics, Chemical Rubber Publishing Co., Cleveland, OH, 1955

Table 11.2

Radii of Tetrahedral and Octahedral Sites in Some Ferrites

Ferrite	Tetrahedral Site Radius	Octahedral Site radius
$MnFe_2O_4$	.67 A.	.72 A.
$ZnFe_2O_4$	.65 A.	.70 A.
$FeFe_2O_4$	.55 A.	.75 A.
$MgFe_2O_4$	.58 A.	.78 A.

Unit Cell Dimensions

The dimensions of the unit cell are given in Angstrom Units which are equivalent to 10^{-8} cm. Table 2.3 lists the lengths, a_o, of some spinel unit cells. If we assume that the ions are perfect spheres and we pack them into a unit cell of meas-

CRYSTAL STRUCTURE OF FERRITES

ured (X-ray diffraction) dimensions we find certain discrepancies that show that the packing is not ideal. The positions of the ions in the spinel lattice are not perfectly regular (as the packing of hard spheres) and some distortion does occur. The tetrahedral sites are often too small for the metal ions so that the oxygen ions move slightly to accommodate them. The oxygen ions connected with the octahedral sites move in such a way as to shrink the size of the octahedral cell by the same amount as the tetrahedral site expands. The movement of the tetrahedral oxygen is reflected in a quantity called the oxygen parameter which is the distance between the oxygen ion and the face of the cube edge along the cube diagonal of the spinel subcell. This distance is theoretically equal to $3/8a_o$.

Table 11.3
Unit Cell Lengths of Some Simple Ferrites

Ferrite	Unit Cell Length (Å)
Zinc Ferrite	8.44
Manganese Ferrite	8.51
Ferrous Ferrite	8.39
Cobalt Ferrite	8.38
Nickel Ferrite	8.34
Magnesium Ferrite	8.36

Interaction Between Magnetic Moments on Lattice Sites

With regard to the strength of interactions between moments on the various sites, the negative interaction or exchange force between the moments of two metal ions on different sites depends on the distances between these ions and the oxygen ion that links them and also on the angle between the three ions. The interaction is greatest for an angle of 180° and also where the interatomic distances are the shortest. Figure 2.2 shows the interatomic distances and the angles between the ions for the different types of interactions. In the A-A and B-B cases, the angles are too small or the distances between the metal ions and the oxygen ions are too large. The best combinations of distances and angles are found in the A-B interactions.

For an undistorted spinel, the A-O-B angles are about 125° and 154°. The B-O-B angles are 90° and 125° but in the latter, one of the B-O distances is large. In the A-A case the angle is about 80°. Therefore, the interaction between moments on the A and B sites is strongest. The BB interaction is much weaker and the most unfavorable situation occurs in the AA interaction. By examining the interactions involving the major contributor, or the A-B interaction which orients the unpaired spins of these ions antiparallel, Néel(1948) was able to explain the ferrimagnetism of ferrites The interaction between the tetrahedral and octahedral sites is shown in Fig. 2.3. An individual A site is interacted with a single B site, but each A site is linked to four such units and each B site is linked to six such units. Thus, to be consistent throughout the crystal, all A sites and all B sites act as unified blocks and are coupled antiparallel as blocks

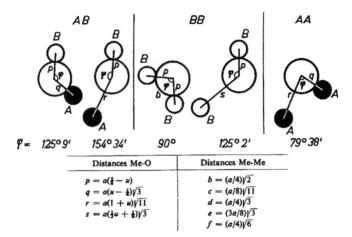

Distances Me-O	Distances Me-Me
$p = a(\frac{3}{8} - u)$	$b = (a/4)\sqrt{2}$
$q = a(u - \frac{1}{4})\sqrt{3}$	$c = (a/8)\sqrt{11}$
$r = a(1 + u)\sqrt{11}$	$d = (a/4)\sqrt{3}$
$s = a(\frac{1}{2}u + \frac{1}{8})\sqrt{3}$	$e = (3a/8)\sqrt{3}$
	$f = (a/4)\sqrt{6}$

Figure 11.2- Interionic distances and angles in the spinel structure for the different type of lattice site interactions. From Smit,J and Wijn, H.P.J., Ferrites, John Wiley, New York, 1959, p.149

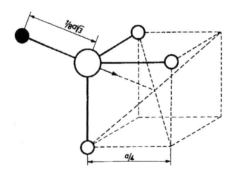

Figure 11.3- Nearest neighbors to an oxygen ion showing A-B interaction through the oxygen ion. From Smit,J and Wijn, H.P.J., Ferrites, John Wiley, New York, 1959 p. 139

Normal Spinels

In a unit cell of spinel lattice, eight tetrahedral and sixteen octahedral sites are occupied by metal ions or by 1 tetrahedral and 2 octahedral for each formula unit. In the case of zinc ferrite, the tetrahedral sites are occupied by zinc ions, which being non-paramagnetic (having no unpaired electronic spins) produce no anti-ferromagnetic orientation of the ions on the octahedral sites that are occupied by Fe^{+++} ions. The Fe^{+++} (B-B) interactions are so weak as to be unimportant. Therefore, zinc ferrite is not ferrimagnetic. This type of arrangement is called a normal spinel structure. (Figure 2.4)

CRYSTAL STRUCTURE OF FERRITES

TYPE OF FERRITE	METAL IONS ON LATTICE SITES				RESULTANT MOMENT
	A (TETRAHEDRAL SITES)		B (OCTAHEDRAL SITES)		
	IONS	MOMENTS	IONS	MOMENTS	
Zinc ferrite— $ZnFe_2O_4$ (Normal spinel)	Zn^{++}	—	Fe^{++}	↑↓	0
Nickel ferrite— $NiFe_2O_4$ (Inverse spinel)	Fe^{+++}	↓	Fe^{+++} Ni^{++}	↑ ↑	↑
Nickel-zinc $Ni_{.5}Zn_{.5}Fe_2O_4$	Fe^{+++} Zn^{++}	↓	Fe^{+++} Ni^{++}	↑ ↓	↑

Figure 11.4- Metal ion distribution in ferrites

Inverse Spinels

The mineral, spinel, was found to be a case of a normal spinel. However, Barth and Posnak (1915) found many cases in which the trivalent ions preferred the tetrahedral or A sites and filled these first. They were able to use X-ray diffraction to distinguish between the ions on the various sites when the scattering power of the two ions was quite different. Spinels showing this type of structure are known as inverse spinels.

There are analogs of inverse spinels in magnetic ferrites. Let us consider the case of nickel ferrite in which eight units of $NiFe_2O_4$ go into a unit cell of the spinel structure. The ferric ions preferentially fill the tetrahedral sites, but there is room for only half of them (eight). The remaining eight go on the octahedral sites as do the eight Ni^{++} ions. The antiferromagnetic interaction orients these eight Fe^{+++} moments and eight nickel moments antiparallel to the eight Fe^{+++} moments on the tetrahedral sites. The Fe^{+++} ion moments will just cancel, but the moments on the nickel ions give rise to an uncompensated moment or magnetization. This type of ferrite is called an inverse ferrite.(Figure 2.4) . Many of the commercially important ferrites are inverse spinels.

Magnetic Moments of Inverse Spinels

In the previous case, the net magnetic moment in the nickel ferrite was the result of the moments of the eight Ni^{2+} ions on the octahedral sites. We have previously assigned to the Ni^{2+} ion the value of $2\mu_B$ per ion or $16\mu_B$ for a unit cell containing eight formula units. We can predict the magnetic moments of the other inverse spinels in a similar manner. These predicted values are listed in Table 2.4 along with the measured values. Because the effect of thermal agitation on the magnetic moments will lower the magnetic moment, the correlation of the moment to Bohr magnetons is always referred to the value at absolute zero or 0° K. This is usually done by extrapolation of the values at very low temperatures. The deviations from the theoretical values can be attributed to several factors, namely:

1. The ion distribution on the various sites may not be as perfect as predicted.
2. The orbital magnetic contribution may not be zero as assumed.
3. The directions of the spins may not be antiparallel in the interactions. In other words, they may be canted.

Table 2.4
Magnetic Moments of Some Simple Ferrites

Ferrite	Magnetic Moment (μ_B)	
	Measured	Calculated
$MnFe_2O_4$	4.6	5
$FeFe_2O_4$	4.1	4
$CoFe_2O_4$	3.7	3
$NiFe_2O_4$	2.3	2
$CuFe_2O_4$(Quenched)	2.3	1
$MgFe_2O_4$	1.1	0
$Li_{.5}Fe_{2.5}O_4$	2.6	2.5
γ-Fe_2O_3	2.3	2.5

Normal versus Inverse Spinels

Although some spinels are either normal or inverse, it is possible to get different mixtures of the two. Often, the ratio of the two will depend on the method of preparation. Some of the first ferrites studied by Néel (1948) were ones that contained Mg and Cu which by thermal treatment reduced the Fe^{+++} on the tetrahedral A sites of the inverse spinel. As a result, there was an imbalance of the Fe^{+++} ions on the two sites and thus a magnetic moment. Even Zn ferrite with a higher than 50 mole percentage of Fe_2O_3 and a special firing can have a small moment. It is customary to represent the spinel formula by placing the tetrahedrally-situated ions before the brackets and the octahedrally-located ions between the brackets. Thus, the formula for zinc ferrite would be written as $Zn[Fe_2]O_4$.

Ferrous Ferrite

Ferrous ferrite or magnetite ($FeO.Fe_2O_3$ or Fe_3O_4) is a completely inverse spinel (Shull, 1951) with a moment of about $4\mu_B$ (theoretical moment =4) totally due to the Fe^{++} ions in the octahedral sites. Below 119° K, it transforms to an orthorhombic structure (Bickford, 1953).

Zinc and Cadmium Ferrites
Zinc ferrite is, of course, a normal spinel under standard preparation conditions (Hastings and Corliss,(1953). The same is true for cadmium ferrite.

Manganese Ferrite
Manganese ferrite, $MnO.Fe_2O_3$ of $MnFe_2O_4$, was originally thought to be inverse but was later found to be about 80% normal (Hastings and Corliss,1956). It has a moment at $0°$ K of about $4.6\mu_B$ compared to a theoretical 5.

Cobalt Ferrite
Cobalt ferrite, $CoO.Fe_2O_3$ or $CoFe_2O_4$ is shown by neutron diffraction to be completely inverse (Prince,1956) The measured moment is $4\mu_B$ even though the theoretical value is $3\mu_B$.

Nickel Ferrite
According to Corliss and Hastings (1953), nickel ferrite, $NiO.Fe_2O_3$ or $NiFe_2O_4$,is 80% inverse and has a moment of $2.3\mu_B$ compared with a theoretical value of $4\mu_B$

Copper Ferrite
Copper ferrite, $CuO.Fe_2O_3$ or $CuFe_2O_4$, is partially inverse at high temperatures. This structure can be quenched to maintain the partial inverse structure at room temperature. This has been shown by X-ray (Bertaut,1951) and by magnetic measurements by Neel (1950) and Pauthenet (1950).

Magnesium Ferrite
Magnesium ferrite, $MgO.Fe_2O_3$ or $MgFe_2O_4$, can be made partially normal at high temperatures and this normal structure can also be quenched in a manner similar to copper ferrite. In fact, the same workers listed above for copper ferrite studied the magnesium ferrite at the same time. Slow cooling gives inverse structure and rapid cooling gives normal structure.

Lithium Ferrite
Because of the monovalent nature of Li, there has to be an excess of Fe^{+++} ions to maintain charge neutrality. Thus, the formula is $Li_{.5}Fe_{2.5}O_4$. The additional .5 Fe^{+++} is used for charge balance. Every fourth Fe^{+++} in the octahedral sites is replaced by a Li^+. The magnetic moment is $2.6\mu_B$ compared with a predicted $2.5\mu_B$.

Gamma Ferric Oxide
An interesting material with the spinel structure is γ-Fe_2O_3. This magnetic material (discussed later in Magnetic Recording in Chapter 14, has no divalent ons. How-

ever, two thirds of the octahedral sites normally occupied by the divalent ions are occupied by Fe^{+++} ions while the other third is vacant This arrangement produces the equivalent charge of the replaced divalent ions but also leads to an imbalance in the numbers of Fe^{+++} ions on the two different sites and therefore to a magnetic moment. The measured moment is $2.3\mu_B$ versus a calculated $2.5\mu_B$.

Neutron Diffraction
Barth and Posnak (1915) used X-ray diffraction to distinguish the various metal ions on the different lattice sites and thus distinguish normal from inverse spinel structures. The situation is somewhat more difficult in the magnetic spinels. The X-ray scattering power of the Fe^{+++} ions is almost the same as that of the other metal ions involved. Thus, no definitive structure can be deduced. However, the interaction of the magnetic moment of the neutron with the spinel structure can make this distinction. The use of neutron diffraction has confirmed the structure of normal and inverse spinels (Shull and Koehler, 1951) and (Hastings and Corliss, 1953). A good review of the use of neutron diffraction to determine magnetic structure is found in Bacon (1955).

Mixed Zn Ferrites
The preference of Zn ions for tetrahedral sites is used to good advantages in mixed Zn ferrites where Zn replaces some of the magnetic divalent ion with the same stoichiometric amount of Fe^{+++} present. Let us assume that 50% of the divalent magnetic ion (eg. Ni^{++}, Mn^{++}) is replaced with Zn^{++}. The Zn^{++} goes on to half the tetrahedral (A) sites leaving room on the other half of the A sites for Fe^{+++} ions. The remaining Fe^{+++} ions go on the octahedral sites. The Fe^{+++} moments on the tetrahedral sites orient all the octahedral site moments antiparallel to them so that the Fe^{+++} moments on the tetrahedral sites neutralize only one third of the octahedral Fe^{+++} ions leaving a large percentage (the other two thirds) oriented, but uncompensated giving a net magnetic moment.

Additional magnetic moment comes from the magnetic M^{++} ions that have also been oriented antiparallel. The total uncompensated, oriented ions consists of one-half of all the Fe^{+++} ions originally present plus all the M^{++} ions (except Zn^{++}) giving a large magnetic moment (Figure2.4). The non-magnetic Zn ions cannot be substituted for the magnetic M^{++} ions without limit as the A-B interaction weakens because of great dilution. The effect of adding ZnO to an inverse spinel can be predicted for each ferrite for varying Zn additions. Figures 2.5 and 2.6 show the actual measured moments from two different investigators- (Guillaud [1949, 1950a, 1950b, 1951a, 1951b] and Gorter[1950]) for some Zn-substituted inverse ferrites. The increase in moment with Zn content is originally quite linear as expected. However, at a value of about 50% replacement, the curve heads downward. This is due to the dilution of the spin moments which weakens the A-B interaction. That is, the average distance between interacting spins gets larger. Note that if the linear portions of the curves were extrapolated to 100% replacement, the value would be 10

CRYSTAL STRUCTURE OF FERRITES

μ_B. Of course this result is absurd since all that would be left would be Zn ferrite which we have shown is non-magnetic.

The site preference can be dependent on temperature. Some sites may particularly favor certain ions over others at high temperatures, but reverse the order of preference at low temperatures. Such a ferrite is Mg ferrite which because Mg^{++} is a non-magnetic ion, should not give any net moment if the Fe^{+++} ions are equally split between the tetrahedral and octahedral. At high temperatures, some Mg^{++} goes

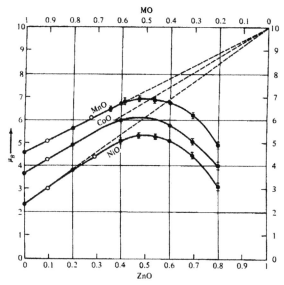

Figure 11.5- Effect of zinc substitution on the magnetic moments of some ferrites. From Guillaud,C.,J. Phys Rad., 12,239 (1951)

into the tetrahedral site leaving the remaining tetrahedral (A) sites for only part of the Fe^{+++} ions. The remainder (more than one-half) of the original Fe^{+++} goes to octahedral sites. This gives rise to a disproportionate occurrence of Fe^{+++} moments similar to the mixed Zn ferrites. If the high temperature site distribution can be maintained at room temperature by rapidly quenching the Mg ferrite from a high temperature, an inverse structure (Bertaut, 1951) with respectable magnetic moment (Neel, 1948) can be obtained. A similar situation exists for Cu ferrite.

Sublattice Magnetizations

In developing his theory of ferrimagnetism, Néel (1948) postulated two separate sublattice magnetizations corresponding to the two sublattices. In other words, each sublattice could be treated as possessing its own magnetization and the resulting ferrite being the superposition of the two magnetizations. This can be shown in Figure 2.7a in which each sublattice magnetization as well as the resultant is shown as a function of temperature. In the simple ferrites, the resultant curves are all quite

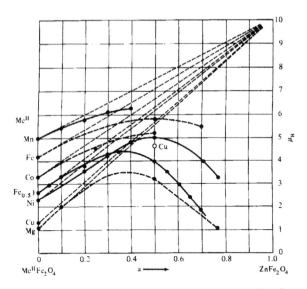

Figure 11.6- Variation of magnetic moment with increasing zinc substitution. From Gorter, E.W. , Philips Research Reports, 9, 321 (1954)

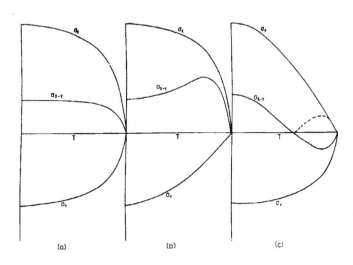

Figure 11.7-Superposition of various combinations of two opposing sublattice magnetizations producing differing resultants including one with a compensation point (right-hand plot).

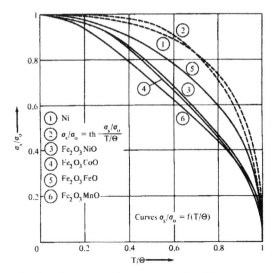

Figure 11.8- Plot of reduced magnetization versus reduced temperature for several ferrites. From Pauthenet,R., Comptes Rendus, 230,1842,(1950)

similar to each other. In fact, the universal curves showing the reduced magnetic moment (σ_T/σ_o) vs the reduced temperature (T/T_c) for a number of different ferrites appear quite similar to the ones found for ferromagnetics (see Figure 11.8) Neel (1948) predicted a large variety of different possible M vs T curves which are shown in Figure11.9. The type that we have seen thus far has been the one for Figure 11.9Q. Yafet and Kittel (1952) have pointed out that the ones shown in Figures 11.9M, 11.9V and 11.9R are not possible because of the non-zero slope at 0° K. An interesting variation is shown in Figure 11.9N, in which the magnetization falls to zero as the temperature is raised and then appears to rise again and finally

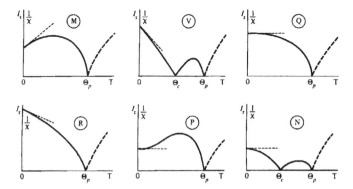

Figure 11.9- Various possible theoretical temperature dependencies on the saturation moments of ferrites. From Néel ,L. Ann. Phys., 3,13,(1948)

drops to zero at the normal Curie point. The breakdown into sublattice magnetizations for this case shows a somewhat different situation as shown in Figure 11.7c. At a point below the Curie point, the two sublattice magnetizations are equal and thus appear to have no moment. This temperature is called the compensation temperature. Below this temperature one sublattice magnetization is larger and provides the net moment. Above this temperature the other magnetization does dominates and the net magnetization reverses direction. By simply measuring the magnetization, this reversal cannot be detected but neutron diffraction can observe the change. There are a number of materials, for example, some rare earth garnets (which we will discuss later in this chapter) display this type of behavior. Gorter(1953) reported such a curve in a LiFeCr ferrite. The M vs T (or in this case, σ vs T) curve is shown in Figure 11.10. This work is one of the most convincing pieces of evidence supporting Néel's theory.

HEXAGONAL FERRITES

This class of magnetic oxide (Went,1952) is called a "magnetoplumbite" structure from the mineral of the same name. Whereas the symmetry of the spinel crystal structure is cubic, that for the magnetoplumbite structure is hexagonal. Thus, it has a major preferred axis called the c axis and a minor axis called the a axis. The preferred direction is used to good advantage as a permanent magnet material.

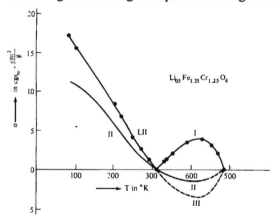

Figure 11.10-Temperature dependence of the magnetic moment in $Li_{.5}Fe_{1.25}Cr_{1.25}O_4$ showing a compensation point.From Gorter, E.W. and Schulkes,J.A., Phys. Rev., 90,487 (1953)

The oxygen ions are closely packed as they are in the spinel structure but there are oxygen layers which now include the Ba^{++}, Sr^{++} or Pb^{++} ions which have about the same ionic radii as the oxygen ions and therefore can replace them in the lattice.

The magnetoplumbite unit cell shown in Figure 11.11 contains a total of ten layers, two of which contain the Ba^{++}: four layers of four oxygen ions each; followed by a layer of three oxygen ions and one Ba^{++} ion; again followed by four layers of four oxygen as having ions each; and another layer containing three oxygens and one Ba^{++} ion but situated diametrically opposite to the Ba^{++} ion in the previous layer containing Ba^{++}. The Fe^{+++} ions are located in the interstices of these ten layers. There are octahedral and tetrahedral sites plus one more type not found in the spinel structure in which the metal ion is surrounded by 5 oxygen ions forming a trigonal bi- pyramid in the same layer as the Ba^{++} ion.

The Magnetoplumbite formula is $MFe_{12}O_{19}$ or $MO \cdot 6Fe_2O_3$, where M can be Ba, Sr, or Pb. There are two formula units per unit cell. The moments of the 12 Fe^{+++} ions are arranged with the spins of 12 in the up direction and 8 in the down direction per formula unit, giving a predicted net moment of 4 Fe^{+++} ions per formula unit times $5\mu_B$ per ion or a total of 20 μ_B per formula unit. The measured value is found to be close to this figure. All the magnetic moments in this structure are oriented along the c axis including the net moment listed above.

Workers at the Philips Research Laboratories where the hexagonal ferrites were discovered found a series of other compounds possessing the hexagonal structure in addition to the magnetoplumbite type just discussed. These compounds were made by combining the magnetoplumbite composition with various spinel ferrite compositions in differing ratios. Thus, layers of spinel sandwiched between layers of magnetoplumbite. The magnetoplumbite material was designated the "M" material. The various combinations containing different ratios of the M material with spinel were given other letter designations. Some of the structures in the series and their formulae are given in Table 11.5. Several of these materials such as all of the Y compounds and the compounds of the W and Z series in which the divalent spinel ion is Co^{++} have an interesting property. The c axis that, for the M series was the preferred axis for the moment or magnetization to be oriented, now becomes the difficult or hard direction of magnetization. Thus, the residual moment now possesses a preferred plane of magnetization. This gave rise to the term "Ferroxplana " (Jonker,1956). These materials are closer to the spinel in application rather than the permanent magnet application. They are used at very high frequencies. The Fe_2O_3 in the magnetoplumbite and other hexagonal structures can be partially replaced with Al, Ga, Cr, or Mn.

Table 11.5
Designation and Composition of Several Hexagonal Ferrites

Ferrite Designation	Chemical Composition
M	$BaO \cdot 6Fe_2O_3$
W	$BaO \cdot 2MeO \cdot 8Fe_2O_3$
S(spinel)	$MeO \cdot Fe_2O_3$
Z	$3BaO \cdot 2MeO \cdot 12Fe_2O_3$
Y	$2BaO \cdot 2MeO \cdot 6Fe_2O_3$

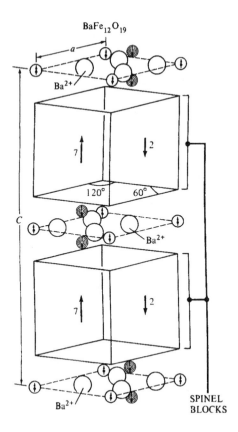

Figure 11.11-Unit Cell of barium ferrite, $BaFe_{12}O_{19}$ From Went,J.J.et al, Philips Tech. Rev., 13, 194 (1952)

MAGNETIC RARE EARTH GARNETS

Magnetic garnets crystallize in the dodecahedral or 12-sided structure related to the mineral garnet. The general formula is $3M_2O_3.5Fe_2O_3$ or $M_3Fe_5O_{12}$. Note that in this case, the metal ions are all trivalent in contrast to the other two classes. In the magnetic garnets of importance, M is usually yttrium (Y) or one of the rare earth ions. Even though yttrium is not a rare earth, it behaves as one and therefore is included in the designation "rare earth" garnets. The ions, La^{+++}, Ce^{+++}, Pr^{+++}, and Nd^{+++} are too large to form simple garnets but may form solid solutions with other rare earth garnets.

Magnetic Garnets were discovered by Bertaut and Forrat(1956) and independently at about the same time by Geller and Gilleo(1957a). Their crystal structure elucidated by Geller & Gilleo.(1957b) (Figure 11.12) In garnets, there are three different types of sites.These are tetrahedral(a) octahedral (b), and dodecahedral(c) sites. The unsubstituted garnets having only trivalent ions are very stoichiometric so that preparation problems is simplified compared to the spinels. The rare earth

CRYSTAL STRUCTURE OF FERRITES

ions are large so that they occupy the large dodecahedral sites. There are 16 octahedral, 24 tetrahedral and 16 dodecahedral sites in a unit cell containing 8 formula units. One formula unit, $3M_2O_3 5Fe_2O_3$ is distributed as follows:

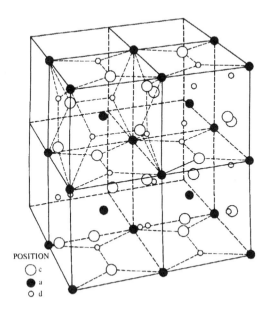

Figure 11.12-Unit cell of a rare earth garnet

$3M_2O_3$ - dodecahedral (c)
$3Fe_2O_3$ - tetrahedral (a)
$2Fe_2O_3$ - octahedral (b)

The moments of the Fe^{+++} ions on the octahedral sites are antiferromagnetically coupled to the moments of the Fe^{+++} ions on the tetrahedral sites. The moments of the M^{+++} ions on the dodecahedral sites are also coupled to the tetrahedral sites similarly and as previously mentioned may contribute to the magnetization of that sublattice. In the absence of the rare earth ion contribution as in the most important of the series, $3Y_2O_3.Fe_2O_3$ all the moments are due to the Fe^{+++} ions. For this formula unit, the net resultant moment is due to the $2Fe^{+++}$ on the tetrahedral. This gives 2 x 5 μ_B for Fe^{+++} = $10\mu_B$.

As noted previously, only spin moments are important for spinels and hexagonal ferrite because the orbital angular momentum was quenched by the strong crystal field of the lattice. In the case of 3d magnetic ions, they were in the ions outer shell. In the case of the rare earth elements, there are electron shells, (namely, the 5s, p, and d) which are surrounded the 4f electron. This helps shield the 4f electrons from the crystalline field and allows for some orbital contribution. If the rare earth ion is also paramagnetic, its moment will be opposite to the $10u_B$ of the Fe^{+++} ions. The moment of each electron of the rare earth ions is the sum of the orbital contribution designated by the orbital quantum number, L, and the spin contribution designated by 2S (since the spin quantum number per electron is either + or - 1/2. The total of these when added vectorially is know as J and is equal to L+2S.

Since there are six M^{+++} ions per formula unit given above, the rare earth moment is 6(L+2S) and combining this with the Fe^{+++} net moment described above the calculated total moment per formula unit will be 6(L+2S)-10. If the orbital contribution is not included (L=O), the corresponding moment would be 12S-10. Table 11.6 (Bertaut and Pauthenet 1957) shows the measured value of the moments (at 0°K.).Figure 2.8 also shows the measured moments for the rare earth ions along with the values calculated based on each of the above assumptions (Including or excluding the orbital contribution). For L=0(as in Y,Gd and Lu), the agreement with the 6(L+2S)-10 value is good but in the other cases, the real value is somewhere between the calculated value for spin plus orbital moments and that for spin only moments. A reasonable explanation in this case is that the orbital interaction is partially quenched similar to the case with the 3d electrons. Y^{+++} and Lu^{+++} have no unpaired electrons (not paramagnetic) and therefore will not contribute to the magnetic moment, hence the moment is due completely to the Fe^{+++} ions($10\mu_B$).

As we have observed in the section on sublattice magnetization, some of the rare earth garnets possess compensation temperatures. This can be shown in Figure 11.13 from Bertaut and Pauthenet(1957) which plots the μ_B vs T for some of the rare earth garnets. As expected, Y and Lu garnets do not have compensation temperatures and therefore follow Neel's type Q course. The garnets that have very high moments at 0° K (Gd, Tb,etc) appear to be of type V which was postulated to be impossible. For Yb, there does not appear to be a compensation temperature and thus resembles the M of Néel's scheme.

Table 11.6
Magnetic Moments of Some Rare Earth Garnets

Rare Earth Garnet	Magnetic Moment, 0°K (u_B)
$Y_3Fe_5O_{12}$	5.01
$Sm_3Fe_5O_{12}$	5.43
$Eu_3Fe_5O_{12}$	2.78
$Gd_3Fe_5O_{12}$	16.0
$Tb_3Fe_5O_{12}$	18.2
$Dy_3Fe_5O_{12}$	15.2
$Ho_3Fe_5O_{12}$	10.2
$Tm_3Fe_5O_{12}$	1.2
$Yb_3Fe_5O_{12}$	0
$Lu_3Fe_5O_{12}$	5.07

CRYSTAL STRUCTURE OF FERRITES

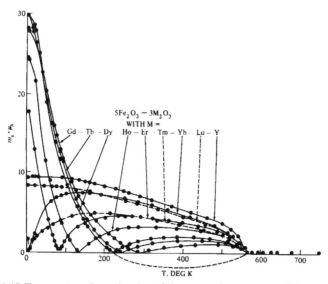

Figure 11.13-Temperature dependencies of the magnetic moments of the rare earth garnets. Note the occurrence of a compensation point in several of the curves. From Bertaut F., and Pauthenet,R., Proc. IEE, 104, Suppl.#5, 261,(1957)

Substituted Garnets

Often, the moment of a garnet such as YIG has to be lowered. The substitution of Al^{+++} for Fe^{+++}. results in the smaller Al^{+++} going into the tetrahedral site. Because the tetrahedral sites had the surplus Fe^{+++} spins, this substitution decreases the moment on the tetrahedral sites and therefore the total moment of the garnet. Large ions such as In^{+++} or Sc^{+++} occupy the octahedral sites which will increase the difference and therefore the total moment. In other substitutions for the Fe^{+++}, various combinations of ions of other valencies can be used. For example, equal numbers of divalent and tetravalent average out to a +3 valence and can be substituted. Some such examples are;

Divalent-- Fe^{++}, Ni^{++}, Co^{++}, Mn^{++}
Tetravalent --Si^{++++}, Ge^{++++}

An interesting case is that of V^{+++++} coupled with Ca^{++} (Nicholas,1980). Two Ca^{++} will combined with 1 V^{+++++} to have the equivalent number of ions and same total positive charge as 3 trivalent rare earth ions. By assuming that the substituted ions fit into the lattice, a garnet can be prepared with a total or large percentage of this substitution. Such a garnet which also contains trivalent bismuth is a "rare earth" garnet without the costly rare earth. For this reason, it has been the subject of much research into garnet materials (sometimes called CalVanBIG) is not reproduced consistently.

In bubble memory applications, specially substituted rare earth garnets are used to form the cylindrical bubble domains. A discussion of bubble memories will be found in the section on magnetic memories (See Chapter 21)

Miscellaneous Structures

Orthoferrites

Aside from the spinels, hexagonal ferrites and garnets, the next most important structure is the orthoferrite or perovskite structure. The formula is $MFeO_3$ where M is usually yttrium or a rare earth. The structure is orthorhombic rather than cubic. The weak ferromagnetism is due to the canting or non-parallel alignment of the antiferromagnetically coupled ions.

The orthoferrites are of the perovskite structure which is shown in Fig. 11.14. Jonker and van Santen (1953) found ferromagnetic perovskites of the type La-CaMnO$_3$ or LaSrCoO$_3$. Bertaut and Forrat(1956) determined the structure of the rare earth orthoferrites or perovskites and it was during that period that the rare earth garnets were accidentally discovered.

The practical importance of the orthoferrites rested in their application in the original bubble memory structures. They have since been replaced by anisotropic rare earth garnets.

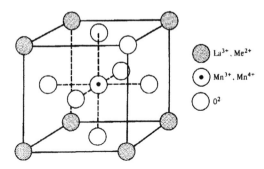

Figure 11.14-Unit cell of a perovskite. From Jonker,G.H., and van Santen,J.H., Physica, 9, 120 (1950)

Chalcogenides

Another magnetic material related to the ferrimagnetic spinels is the chalcogenide group of the type $CdCr_2S_4$, $CdCr_2Se_4$, or $HgCr_2Se_4$ (Baltzer,1965 and Menyuk, 1966). In this case, the oxygen of the ferrites is replaced by S or Se. These substances are interesting because in addition to being ferromagnetic, they are also semiconducting. It cannot compete as a semiconductor or in a purely magnetic application but in cases in which both properties are needed, there are possibilities of usage.

References

Bacon,G.E.(1955) Neutron Diffraction, Clarendon Press, London
Baltzer,P.K.(1965),Lehman,M.H. and Robbins,M.,Phys. Rev. Lett. 15, 493
Barth, T.W.F.(1915) Z. Krist. 30, 305
Bertaut,F.(1951) J. Phys Rad. 12,252
Bertaut,F.(1956) Comptes Rendus, 242,382
Bertaut,F.(1957) and Pauthenet,R.Proc. IEE, 104, Suppl.#5 ,261
Bickford,L.R.(1953) Rev. Mod.Phys, 25,75
Bragg,W.H.(1915) Nature, 95,561
Geller,S.(1957a) and Gilleo,M.A.,J.Chem.Phys.Solids, 3,30
Geller,S.(1957b) and Gilleo,M.A.,Acta Cryst. 10,239
Gorter,E.W.(1950),Comptes Rendus 230,192
Gorter,E.W.(1953) and Schulkes,J.A.,Phys. Rev., 90,487
Gorter,E.W.(1954) Philips Res. Rep., 9,321
Gorter,E.W.(1955) Proc. IEE, 104,Suppl. #5,1945
Guillaud,C.(1949) and Roux,M. Comptes Rendus, 229,1133
Guillaud,C.(1950a) and Crevaux,H.,ibid, 230,1256
Guillaud,C.(1950b) and Crevaux,H.,ibid, 230,1458
Guillaud,C.(1951a) and Sage,M.,ibid, 232,944
Guillaud,C.(1951b) J. Phys. Rad. 12,239
Hastings,J.M.(1953) and Corliss,L.M.,Rev.Mod.Phys. 102,1460
Hastings,J.M.(1956) and Corliss,L.M.,Phys. Rev., 104,328
Jonker,G.H.(1950) and van Santen,J.H. Physica, 19,120
Jonker,G.H.(1956), Wijn,H.P.J. and Braun,P.B.,Philips Tech. Rev., 18,145
Menyuk,N.(1966),Dwight,K.,Arnott,R.J.,and Wold,A.,J.Appl. Phys., 37,1387
Neel,L.(1948) Ann. de Phys. 3,137
Neel, L.(1950) Comptes Rendus, 230,190
Okamura,T.(1952) and Kojima,Y.,Phys. Rev. 86,1040
Pauthenet,R.(1950) Comtes Rendus, 230,1842
Prince,E.(1956) Phys.Rev. 102,674
Shull,G.C.(1951),Wollan,E.O.,and Kohler,W.C., Phys. Rev., 84,912
Shull,G.C.(1959) J.Phys. Rad. 20,109
Verwey,E.J.W.(1947) and Heilmann,E.L.Jr., J. Chem. Phys.15,174
Went,J.J.(1952),Rathenau,G.W., Gorter,E.W.and van Oosterhout, Philips Tech.Rev. 13,194
Yafet,Y.(1952) and Kittel,C.,Phys.Rev. 87,290

12 CHEMICAL ASPECTS OF FERRITES

Introduction

Thus far, we have discussed the important electrical and magnetic properties useful in describing ferrites. To obtain the desired properties, it is a difficult, if not hopeless, task for the designer of ferrite materials to depend strictly on random magnetic testing for these properties in the search for the optimum material. It is much more fruitful to correlate these properties with the relevant chemical and ceramic characteristics and by controlling these characteristics, to have a better method of reproducing the material. The next two sections will focus on the examination of these correlations and in the following chapter, we will try to apply what we have learned to the processing of these materials. Through our increased knowledge of these correlation factors and because of the availability of more sophisticated means of measuring these properties, improved and more reproducible materials have been developed.

Intrinsic and Extrinsic Properties of Ferrites

The chemical aspects of ferrites have already been considered in discussing the magnetic moments and thus the saturation magnetizations. In addition, the temperature dependencies of these properties and the Curie points have been examined. Other intrinsic properties of ferrites such as magnetostriction, anisotropy and ferrimagnetic resonance have been mentioned. Many magnetic characteristics such as permeability and losses depend partially on these factors and partially on other structure-dependent properties such as those relating to microstructure. We will examine this combination in this section. A clean separation of these sets of properties is not possible because some chemistry considerations indirectly affect the extrinsic properties, which, in turn, depend on the processing or past history of the material. This section deals with the chemistry of ferrites showing how the major elements affect the intrinsic properties and then points out the elements (usually minor elements) that may affect the microstructure. The next section will deal with the microstructural characteristics of ferrites and their effects on the extrinsic properties.

Magnetic Properties under Consideration

For soft magnetic ferrites, the properties of concern to the user or designer are:

1. Saturation Induction or Flux density
2. Temperature dependence of Saturation
3. Permeability

4. Temperature stability of permeability
5. Time Stability of Permeability, Disaccomodation
6. Low level losses, tan δ, Loss Factor
7. Power Level losses, core losses
7. Temperature Dependence of Power Losses
8. Flux Density Dependence of Power Losses
9. Coercive Force, H_c and Remanence, B_r

For hard ferrites, in addition to the coercive force, H_c, and the remanence, B_r, the maximum energy product, $(BH)_{max}$ is a very important property. For microwave ferrites, such parameters as the ferrimagnetic resonance line-widths are pertinent.

Mixed Ferrites for Property Optimization

One of the most important attribute or advantage of ferrites is their very high degree of compositional variability. Most of the original fundamental intrinsic properties on ferrites are made on the simple ferrites such as $MnFe_2O_4$, $CoFe_2O_4$ and $NiFe_2O_4$. However, most commercially important ferrites are of the mixed variety and actually consist of solid solutions of the various simple ferrites with an infinite number of combinations possible. The solid solutions are more thermodynamically stable than the separated mixtures of the individual spinels. The effect of each simple ferrite can be used to balance or otherwise optimize a specific property of the final mixed ferrite. Depending on the requirements of the ultimate application, various combinations of different properties can be obtained by blending a judicious choice of the simple ferrites in a proper ratio. The great variety of possible combinations was proposed by Gorter(1954, 1955) and has been the subject of much subsequent research in ferrites.

Saturation Induction of Zn-substituted Ferrites

In Chapter 11, we noted that Zn substitution increases the moment of Mn and Ni ferrites at $0°$ K. We also stated that Zn lowers the Curie point of Zn-substituted ferrites. Since the lowering of the Curie point would also lower the saturation values at temperatures near the Curie point relative to the unsubstituted case, we would then expect to see a crossover in the temperature-dependence curves of the magnetization. This is indeed the case. Figure 12.1 shows the temperature course of B_s for some Zn-substituted Mn ferrites while Figure 12.2 gives the same for Nickel-Zinc ferrites.

Permeability Dependence on Chemistry

The magnetic permeability may be considered as a measure of the efficiency of a magnetic material. That is, it tells you how much you get out in terms of polarization or magnetization our input of magnetizing force will give us. In metallic magnetic materials which have much higher saturation magnetizations than ferrites, the highest permeabilities are not usually present in the materials with the

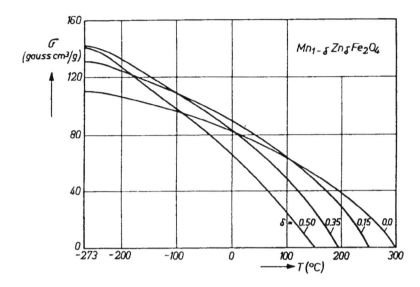

Figure 12.1 Temperature dependencies of the magnetizations of Zn-substituted manganese ferrites. From Smit,J. and Wijn,H.P.J., Ferrites, John Wiley, New York,(1959) p.158

highest saturations. The reason why is the dominance of other intrinsic factors such as low magnetostrictions and anisotropies coupled with good extrinsic factors such as very clean grain boundaries and low residual stresses after annealing. In ferrites, on the other hand, the saturations are lower, the microstructure is not as clean (thicker grain boundaries) and many firing stresses remain. For the DC or low frequency permeability, advantage must be taken of materials with higher saturations. For reversible rotational processes, the permeability is given by Chikazumi (1964) as;

$$(\mu-1) = (\text{constant}) M_s^2 \sin^2 \theta / K_1 \quad [5.1]$$
where θ = angle between M and H

For reversible wall processes, the permeability is;
$$(\mu-1) = (\text{constant}) M_s^2 S/\alpha \quad [5.2]$$
where S= wall surface area
α = second derivative of the wall energy with respect to wall displacement

We can see that both the permeability due to domain rotation and that due to domain wall movement are in one manner or another related to $(M_s)^2$. It is not surprising to find the higher permeabilities in ferrites in the higher saturation MnZn ferrites compared to the others. Of course, the low anisotropy and magnetostriction

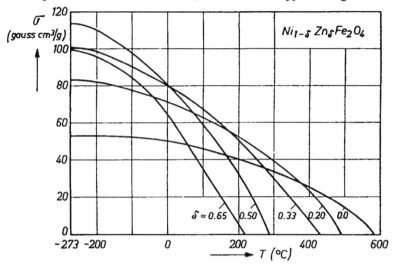

Figure 12.2 Temperature dependencies of the magnetizations of Zn-substituted nickel ferrites. From Smit,J. and Wijn, H.P.J.,Ferrites, John Wiley,New York, (1959)

also help. In fact, high saturation is certainly not the only requirement for high permeability as magnetite has a high saturation but very low permeability even at low a.c. frequencies.

The above discussion of permeability has not included the effects of higher frequencies. Manganese zinc ferrites certainly have higher permeabilities at low or medium frequencies. When the frequencies involved are in the upper megahertz range, the permeability drops off due to the great increase in losses as shown in the previous chapter. Under these conditions, NiZn, MgMn or other higher resistivity ferrites are the materials of choice.

Effect of Iron Content on Permeability

The metal ion present in the largest concentration molewise in ferrites is Fe^{3+} and since it has a high ionic moment (5 u_B) it has the highest potential for controlling the magnetic characteristics. In the completely inverse ferrites such as $NiFe_2O_4$, the large moments of the two Fe^{3+} sublattices cancel each other and no advantage is taken of the potential Fe^{3+} moment. It is in the zinc-substituted mixed ferrites in which the Fe^{3+} ions on the two sublattices are disproportionated that the large moment of Fe^{3+} is used. Of course, these effects are not chemical (i.e., they are not related to the Fe_2O_3 content) but crystallographic (i.e., they are related to lattice site distribution). Actually, with the great variety of possible chemistries of spinel ferrites, the Fe_2O_3 content of the finished ferrite is varied least of all of the metal ions since it is pegged at 50 mole percent by the spinel formula ($MO.Fe_2O_3$). In most commercially important MnZn ferrite materials, the starting mix may contain slightly more than 50 mole percent Fe_2O_3. The purpose of the extra iron is to improve the magnetic properties by the formation of Fe^{2+} ions. One such basic property is the magnetostriction that we defined as the change in the length of a material when it is subjected to a magnetic field.

The magnetostrictions of several ferrites are shown in Table 12.1. Note that the only ferrite that has a positive magnetostriction is ferrous ferrite. Therefore, it would seem useful to minimize the net magnetostriction by compensating the negative values of the other ferrites through the incorporation of ferrous ferrite as part of the solid solution. Ferrous ferrite, $FeO \cdot Fe_2O_3$ or Fe_3O_4 is actually magnetite, a naturally occurring, but technically unimportant ferrite. The additional iron for the divalent Fe^{++} is usually added in the original mix as Fe_2O_3, but then is reduced to FeO or Fe_3O_4 in the sintering or firing process to maintain the 50 percent Fe2O3 requirement. The molar excess of iron oxide may be as high as 5 percent.

Guillaud(1957) found that for a fixed Mn concentration, increasing the Fe content in the initial mix made the μ vs T slope less steep at room temperature. Many manganese ferrites containing ferrous ions have a secondary permeability maximum (SPM), the primary maximum occurring just before the Curie Point. Figure 4.4 illustrates the two maxima. As the iron content is increased, the secondary maximum moves to lower temperatures. Figure 12.3(Guillaud, 1957 and Figure 12.4(Lescroel and Pierrot,1960) show this effect. If the highest permeability is desired at room temperature, the iron content is usually chosen so the secondary permeability maximum occurs at room temperature. This change in the temperature dependence of the permeability is due to the movement to lower temperatures of the point where the magnetostriction goes through zero. The zero crossing of the magnetostriction will closely coincide with the permeability maximum. This correlation is shown for MnZn ferrite in Figure 12.5,a,b for MnZn ferrites whose MnO contents were varied between 32-39 mole % with all samples prepared in the same manner (Guillaud, 1957). However, in another series of samples in which all had the same composition (53% Fe2O3, 30% MnO and 17% ZnO) but fired differently, a minimum of permeability was found at the minimum of magnetostriction as shown in Figure 12.5c (Guillaud, 1957). While the latter case is an exception, it shows the caution that must be exercised with this correlation. In the case of Nickel Ferrite, the effect of the Fe content is shown in Figure 12.6 (Guillaud, 1960). Witrh regard to the anisotropy, the work of Ohta (1963) shows the relation between anisotropy and permeability in MnZn ferrites (Figure 12.8 and 12.9). The variation of composition is given for Mn ferrite in Figure 12.9 (Miyata,1961) and for Ni ferrite in Figure 12.10 (Usami 1961). In instances where the permeability maximum does not occur at zero anisotropy, the deviation may be due to the presence of increased porosity that lowers the permeability prematurely. (Stuijts 1964).

In some cases, the substitution of other trivalent ions such Al^{+++}, Ga^{+++}, or Cr^{+++} for Fe^{+++} is made for special magnetic or electrical functions. These may include the reduction of the saturation magnetiztion, increasing the temperature stability or increasing the resistivity. In the spinel ferrites, the substitution of non-magnetic ions such as Al^{+3} on octahedral sites in an inverse spinel such as nickel ferrite will reduce the saturation magnetization since it has the net uncompensated moment. At a certain value of x in $NiAl_xFe_{2-x}O_4$, the two sublattices are equal and the material is antiferromagnetic. Further substitution leads to reversal of the magnetization. This is another example of a compensation point. Trivalent scandium or chromium The variation of anisotropy with concentration is given for Mn ferrite in Fig. 12.9.

Table 12.1
Saturation Magnetostrictions of Some Ferrites

Ferrite	Saturation Magnetostriction λ_s
$MnFe_2O_4$	-5×10^{-6}
$FeFe_2O_4$	$+41 \times 10^{-6}$
$CoFe_2O_4$	-110×10^{-6}
$Ni_{.8}Fe_{2.2}O_4$	-17×10^{-6}
$Li_{.5}Fe_{2.5}O_4$	-8×10^{-6}
$CuFe_2O_4$	-6×10^{-6}
$Mn_{.6}Zn_{.1}Fe_{2.1}O_4$	$+3 \times 10^{-6}$

From: Bozorth,R.M., Tilden,E.F., and Williams,A.J. ,Phys. Rev. 99, 1788 (1955)

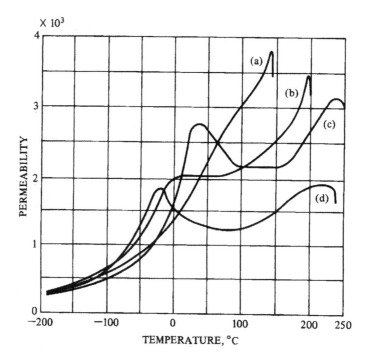

Figure 12.3-Variation of temperature dependencies of the permeability of some Mn ferrites with composition From Guillaud (1957)

substituted for ferric ion on octahedral sites acts in a similar fashion. Substitution of Al^{3+} for Fe^{3+} in magnesium ferrite also lowers the moment. There are also concurrent increases in the permeability and the resistivity. In nickel ferrite, substitution of equimolar amounts of additional Ni^{2+} and Ti^{4+} can also replace ferric ions on octahedral sites and reduce the moment.

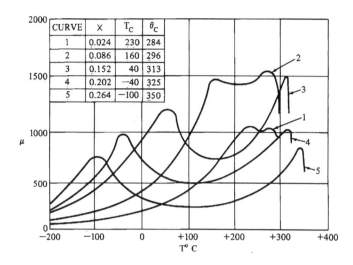

Figure 12.4-Initial permeabilities of some Mn ferrites [Fe2O3,x(FeO),(1-x)MnO] as a function of x. From Lescroel and Pierrot,(1960)

Gilleo(1958) investigated the substitution of many ions for the ferric ions in garnets. His main interest was to reduce the saturation for microwave resonance properties and to increase the temperature stability. In the garnets, Al^{3+} is often used to reduce the magnetization. However, for garnets, the substitution occurs on the tetrahedral sites which are the ones that have the uncompensated moments. Ga^{3+} accomplishes the same purpose. On the other hand the larger Sc^{3+} and In^{3+} ions occupy the octahedral sites and therefore increase the magnetization. Cr^{3+}, though smaller than Fe^{3+}, acts anomalously and also occupies the octahedral sites, leading to an increased moment. As always happens when non-magnetic ions are substituted for magnetic ions, the Curie temperature is lowered regardless of the site of the substitution. Divalent ions of metals such as Mn, Fe, Ni, and Co can be substituted for Fe^{3+} if equal amounts of quadrivalent Si or Ge are also added to maintain the equivalent 3+ charge of two ferric ions by one divalent and one trivalent. Sattar (1997) used trivalent rare earth ions for iron substitution in CuZn ferrites. The rare earths were La, Nd, Sm, Gd and Dy. The resistivity was increased as well as the thermoelectric power. Their results are explained by an electron hopping mechanism.operation and because of its higher T_c, can function at higher temperatures. In commercial practice, cost may be a consideration. Mg ferrite may be used in lower requirement television yokes and flyback transformers because of the lower cost of Mg versus Mn and because its higher resistivity eliminates the need for taped insulation between yoke and winding. Mg ferrite may be used in conjunction with other divalent ions (Mn) in higher frequency applications for resistivity reasons. The main divalent ions

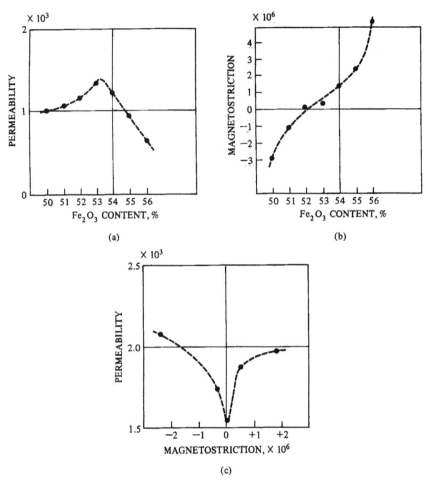

Figure 12.5- (a) Permeability of Mn ferrites as a function of Fe_2O_3 content. (b) Variation of magnetostriction of the same Mn ferrites as a function of Fe_2O_3 content. Figure (c)- Variation of permeability of Mn ferrites with magnetostriction. From Guillaud (1957)

used for compensation in the ferrite lattice are Fe^{++}, Zn^{++}, and Co^{++}. The use of Fe^{++} has already been discussed.

Dormann(1989) in a very detailed theoretical study reviewed the magnetic structures of substituted ferrites. This review included ferromagnetic order, canted spin states and phase diagrams.

Effect of Divalent Ion Variation

As mentioned earlier, in spinel ferrites, the divalent ion can be Mn^{++}, Ni^{++}, Co^{++}, Mg^{++}, Cu^{++}, Zn^{++}, and Fe^{++}. The choice is determined by the specific application. or

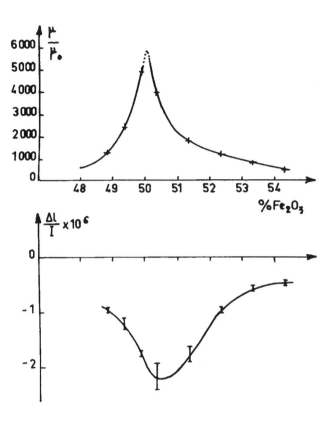

Figure 12.6 - Variation of permeability and magnetostriction as a function of Fe_2O_3 content. From Guillaud (1960)

materials where large magnetic moments are needed, such as in power applications, the magnetic metals ions with the most unpaired spins are chosen.. As seen from Table 2.1, this is one reason manganese ($5u_B$) ferrite and ones that contain uncompensated Fe^{3+} ($5u_B$) ions are useful. Although Ni^{2+} has a lower moment ($2u_B$) than Mn^{2+}, $NiFe_2O_4$ has higher resistivity for high frequency operation and, because of its higher T_c, can function at higher tempratures. In commercial practice, Mg ferrite may be used in lower requirement television yokes and flyback transformers because of the lower cost of Mg over Mn and because its higher resistivity eliminates the need for taped insulation between yoke and winding. Mg may be used in conjunction with other divalent ions (Mn) in higher frequency applications for resistivity reasons. The main divalent ions used for compensation in the ferrite lattice are Fe^{++}, Zn^{++}, and Co^{++}. The use of Fe^{++} has already been discussed

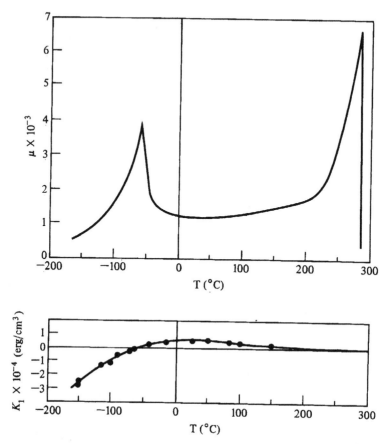

Figure 12.7 Variation of permeability and magnetostriction as a function of Fe_2O_3 content. From Guillaud (1960)

Permeability Dependence on Zinc

Most commercially important low frequency ferrites contain zinc. As noted earlier in this chapter, zinc ion substitution for other divalent ions can increase the effective magnetic moment. It also contributes to an increase in magnetic permeability. Very often, it is the ratio of ZnO to the other divalent oxides as well as the degree of divalent Fe substitution that gives ferrite material developers the greatest latitude in optimizing the properties of a specific ferrite.

In the sintered ferrite, the Zn content will depend on the amount put in originally minus that lost in the sintering process. Since zinc is a rather volatile ion, incorrect firing will cause its loss that will lead to a gradient in zinc content across the thickness of the ferrite. In addition to losing the chemistry influence of the Zn, strains will be introduced further deteriorating the ferrite.

Zn not only increases the moment but it also lowers magnetostriction and anisotropy. The anisotropy of $MnFe_2O_4$ is -28×10^3 ergs/cm^3 (Bozorth 1955) and that of $Mn_{.45}Zn_{.55}Fe_2O_4$ is -3.8×10^3 ergs/cm3(Galt, 1951). In Ni Ferrite the λ_s is reduced from -26×10^{-6} (Smit & Wijn, 1954) to -5×10^{-6} for $Ni_{.36} Zn_{.64} Fe_2O_4$ (Enz, 1955). MnZn Ferrites also have so much Fe^{++} ion that λ_s is usually lower than 1×10^{-6}.

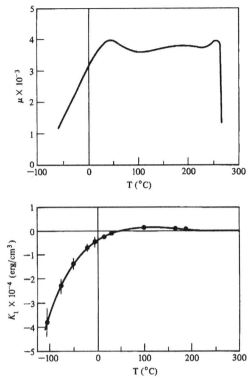

Figure 12.8- Variation of permeability and crystal anisotropy constant, K_1, with temperature in a MnZn ferrite containing $17MnO-21ZnO-62Fe_2O_3$. From Ohta (1963)

Zinc ferrous ferrites were prepared by Brabers (1997) with rather high saturation magnetizations. In fact, they were the highest recorded for spinel ferrites. For applications, however, the permeability must be increased without great reduction in the magnetization NiZn ferrites for anisotropy compensation

Effect of Cobalt on Permeability

The effect of cobalt is evidenced primarily on the anisotropy. As seen in Table 12.2, Co ferrite or mixed ferrites with Co are the ones that have positive anisotropies. Therefore, the anisotropy of cobalt ferrite can be used to compensate the negative anisotropies of other ferrites. Cobalt ferrite is quite frequently used in Ni or NiZn ferrites for anisotropy compensation.

Both Akashi (1971,1972) and Roess (1977) have used cobalt to lower losses and for temperature compensation. Another advantage of cobalt is its ability to make the ferrite susceptible to magnetic annealing. In this process, the sample is subjected to a D.C. magnetic field while cooling through a certain temperature range which is usually in the vicinity of the Curie point. (Kornetski 1955,1958)

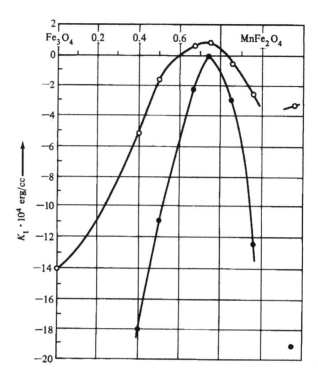

Figure 12.9- Variation of crystal anisotropy constant, K1, with composition in Mn ferrites. From Miyata (1961)

Various types of hysteresis loops can be obtained using cobalt combined with magnetic annealing. We can get an induced isoperm loop with Co-doping with Fe excess plus a magnetic anneal in the direction of the measuring field (Kornetski & Brockman, 1958).

After the firing soak, iron rich ferrites, should be cooled slowly from the Curie Point to room temperature to improve properties. Excess Fe creates cation vacancies which improve Co^{++} diffusion (Iida, 1960). Vacancies are lattice sites in either oxygen or metal ion positions giving rise to a defect structure.

High frequencies losses are also lowered by the addition of Co^{++}, permitting their use in the high megahertz range. (Stijntjes 1971) used Co stabilization was also by Stijntjes in combination) with TiO_2 in low loss, low level applications whereas Buthker (1982) are used Co in high power applications.

Oxygen Stoichiometry

With the exception of the change in zinc content through volatilization during the high temperature sintering process, the elements that were added in the initial raw materials mixture will be present in the final ferrite composition. There is, however, one element whose content may be varied in the sintering step by its equilibration

Table 12.2
Magnetocrystalline Anisotropy Constants of Some Ferrites

Ferrite	Anisotropy Constant, K1
Fe Fe$_2$O$_4$	-1.1 x 10^3
Co$_{.8}$Fe$_{2.2}$O$_4$	+3.9 x 10^6
Co$_{1.1}$Fe$_{1.9}$O4	+1.8 x 10^6
Co$_{.3}$Zn$_{.2}$Fe$_{2.2}$O$_4$	+1.5 x 10^6
Co$_{.3}$Mn$_{.4}$Fe$_2$O$_4$	+1.1 x 10^6
Mn$_{.45}$Zn$_{.55}$Fe$_2$O$_4$	-3.8 x 10^3
Mn Fe$_2$O$_4$	-28 x 10^3
Ni$_{.8}$Fe$_{2.2}$O$_4$	-39 x 10^3
Ni Fe$_2$O$_4$	-63 x 1010^3
Ni$_{.7}$Co$_{.004}$Fe$_{2.2}$O$_4$	-10 x 10^3

From: Bozorth,R.M.,Tilden,E.F.,and Williams,E.T.,Phys. Rev., 99,1788 (1955)

with the firing atmosphere. That element is oxygen and small changes in its content may drastically alter the properties of the ferrite. Part of the need for oxygen balance is brought about by the iron content. In the ferrites in which more than 50 mole percent of Fe$_2$O$_3$ is used in the initial mix(especially in MnZn ferrites), the excess iron is normally converted to FeO or Fe$_3$O$_4$ if the spinel stoichiometry is to be preserved. In some situations non-equilibrium atmospheres for the stoichiometric spinel may create conditions for slight oxygen surpluses or deficiencies without the apparent presence of a second phase. These situations occur by the formation of either cation or anion vacancies in the spinel lattice. As mentioned in Chapter 2, γ-Fe$_2$O$_3$ exemplifies a structure in which the divalent ion charge has been replaced by a combination of Fe^{3+} ions and vacancies. In effect, the presence of vacancies in a ferrite like MnZn ferrite may be considered a solid solution of γ-Fe$_2$O$_3$ and the stoichiometric spinel. In addition to altering the properties through the chemical or crystallographic effect, vacancies, as will be shown in the next section, can exert an influence on the microstructural aspects by influencing the diffusion rates of the cations.

Tanaka(1975a,b,c) studied the effects of oxygen non-stoichiometry as well as composition on several magnetic and mechanical properties of MnZn ferrites. In a ater paper, (Tanaka,1978) he looked at how the initial permeability, the frequency and temperature responses of the initial permeability, and the domain structures were influenced by the same oxygen non-stoichiometry. The oxygen content was controlled by the firing atmosphere as will be shown in the next chapter. The oxygen non-stoichiometry was denoted by a term he calls the"oxygen parameter" (δ) not to be confused with the oxygen parameter mentioned in Chapter 2).Tanaka's oxygen parameter, (δ), is given by;

(MnZnFe)O$_{4+\delta}$)

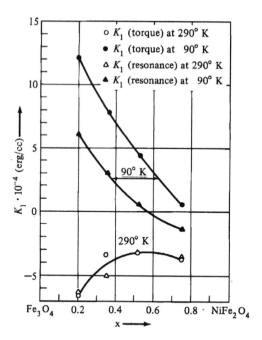

Figure 12.10-Dependence of crystal anisotropy constant, K_1, with composition in Ni ferrites

The oxygen parameter was determined by the partial pressure of the oxygen in the firing atmosphere and the correlation of this with the oxygen content of the ferrite, which was determined independently. Tanaka found that the domain structure of a single crystal MnZn ferrite with a negative or zero oxygen parameter was dependent on the crystal anisotropy and the preferred directions for K_1. At high oxygen levels, the domain structure was more complex and random being determined by the local internal stresses and the distortions of the lattice. In polycrystalline ferrites fired to give low oxygen parameters. The domains are continuous across the grain boundaries while with high oxygen parameters, the patterns are random and unclear.

The initial permeability of several ferrites with varying Fe content and fired to different oxygen parameters are given in Table 12.3 and plotted in Figures 12.11,a,b.) as a function of frequency. At lower frequencies, the highest permeabilities are found in the ferrites with low oxygen parameters, maximizing at x= .02. The permeability decreased with oxygen content up to x= .03 and maximized at zero oxygen parameter for x= .04. At higher frequencies (3 MHz.), the highest permeability occurs at a smaller value of x (.01) and high oxygen parameters. At the same frequency, the permeability increased with increasing oxygen up to x=.03. The permeability at the highest frequency was not dependent on the permeabilities at low frequencies. As will be seen in the section on microstructure, the

Table 12.3
Permeability, Loss Factor and Resistivity of MnZn Ferrites $(MnO)_{.3-x}ZnO.2(Fe_2O_3)_{.5-x}$ with varying Oxygen Parameter, γ

x	γ	10 kHz	100 kHz	3 MHz	10 kHz (10^{-6})	100 kHz (10^{-6})	3 MHz (10^{-3})	ρ (Ω cm)
0.00	0	2030	1890	663	19	39	1.9	
	−0.0025	2700	2580	739	16	33	2.0	
0.01	0.01	1250	1240	831	5.2	14	1.1	17
	0	1740	1700	767	5.9	14	1.6	7.1
	−0.0025	3460	3320	503	7.9	20	4.6	2.7
0.02	0.01	1700	1690	685	2.3	11	2.1	2.2
	0	2530	2540	430	2.7	14	4.7	1.5
	−0.0025	5910	5730	350	4.9	24	8.3	1.0
0.03	0.01	1600	1600	656	2.1	12	2.1	
	0	2470	2470	603	2.3	14	2.8	
	−0.0025	3350	3360	503	6.2	23	3.7	
0.04	0.01	2530	2540	492	1.4	14	4.5	0.93
	0	4140	3940	415	2.4	18	5.5	0.78
	−0.0025	1850	1830	319	7.1	36	5.8	0.39

Source: T. Tanaka, Jap. J. Appl. Phys. 17, 349 (1978).

permeability is affected by the grain size and the porosity, but in this case, these effects were minimized by keeping these factors constant.

The loss factors in ferrites fired with excess oxygen decreased with oxygen content at both high and low frequencies. With regard to the temperature dependence of the initial permeability, for $x = .02$, the μ vs T curve gets flatter with increasing oxygen parameter. For $x = .04$, the typical movement of the secondary permeability maximum to higher temperatures with increasing oxygen parameter is observed.

The first anisotropy constant, K_1, for several single crystal ferrites with similar compositions were used to calculate the permeability based on domain rotation processes, assuming that the polycrystalline ferrites had the same constants. The calculated permeability was about 400 while the measured low frequency values were between 1250 and 5910. Therefore domain wall displacement must be the prime mechanism of magnetization at these frequencies. At higher frequencies, the permeability is determined by the changes in domain structure and reduction of the losses.

The non-stoichiometry and phase stability of magnesium-ferrous $(Mg_xFe_{1-x})_{3-\delta}O_4$ ferrite at 1000°C. as a function of Mg content was studied by Kang (1997) by coulometric titration. The non-stoichiometry, δ, was a function of the log p_{O2} and strongly dependant on the Mg content, x. As x increases from 0 to 0.29, the titration curves shift to higher oxygen potential, indicating that x increases the concentration of cation vacancies. He (Kang,1992a) also studied the phase stability

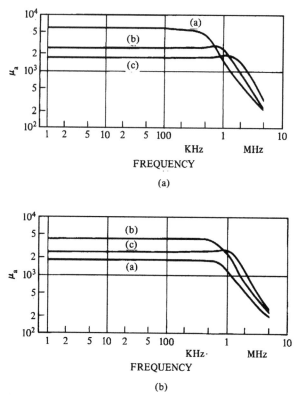

Figure 12.11 - Frequency dependencies of initial permeability for MnZn ferrites

and magnetic properties of Mg ferrite against temperature, cationic composition and oxygen partial pressure using conductivity and thermoelectricity. With these results he constructed phase diagrams of p_{O2} vs T for fixed compositions and also composition vs p_{O2} at fixed temperatures.

Effect of Purity on Permeability

With respect to removing harmful impurities, there is a great deal of difference between the processing of metals and that of ceramics. In metals processing, many interstitial impurities such as C,O,S and N can be removed either by outgassing or by deoxidants which are later removed in the slag. In addition, many foreign metallic elements present in the raw materials can be removed in the melting process. In conventional ferrite processing, the impurities that are present in the raw materials will be present in the finished ferrite. This lack of an economical purification scheme puts a considerable burden on the ferrite producer to use as pure raw materials as is economically reasonable. Ironically, the addition of some minor elements in low concentrations can sometimes increase the quality of the ferrite to a large degree. We will discuss the useful minor elements and note those which are detrimental.

Effect of Foreign Ions on Permeability

Guillaud (1957) studied the effect of alkali (Na,K) and alkaline earth (Mg,Ca,Ba) impurities on the permeability of Mn-Zn and Ni-Zn ferrites. The impurities were added either by coprecipitating with the original composition or by impregnating the mixture with a salt of the impurity and then decomposing the salt to the oxide. Grain sizes were kept constant. At low impurity concentrations, there is an initial increase in permeability. The general trend of increase a function of impurity concentration is shown in Figure 12.12 for potassium. The steep part of the curve is attributed to the impurity that dissolves in the lattice whereas the flat portion is the result of saturation and deposition at the grain boundary. The increase in permeability was greater the larger the ionic radius of the impurity, the effect being greater for alkaline earths (2+ valence) than for alkalies (1+ valence). The effect of ionic radius is shown in Figure 12.13 for both groups. At higher impurity concentrations, the effect will reverse and eventually the permeability will decrease due to microstructural effects such as grain size and porosity that will be discussed later in this chapter. The general trend of degree of reduction of permeability as a function of concentration is shown in Figure 12.14. Permeability decrease may originate from the mismatch of ionic size of the impurity and the subsequent distortion and strain of the lattice. The only exception in the materials studied was in the case of CaO, which shows improved overall electrical properties even at high concentrations. It should be noted that the beneficial effect of CaO extends to a concentration of about .2 weight percent after which the properties slowly degrade. An impurity that is controlled very carefully in advanced ferrite technology is SiO_2. This oxide appears to act together with CaO possibly through the formation of $CaSiO_3$. Akashi (1961, 1963,1966) showed that its presence improved the properties of MnZn ferrites at low concentrations, reached a maximum effectiveness and then dropped off at higher levels. At still higher concentrations, impurities such as SiO_2 create a duplex structure of giant grains within a fine grain matrix that is very detrimental to the permeability. This will be discussed later. Many metallic elements such as Mo, V, Cu, Cd, and Al show similar behavior with increases in permeability of up to 50% at very low concentration (Lescroel and Pierrot, 1960).These materials may increase the permeability initially by acting as a sintering aid to increase the density and the grain size. Mizushima (1992) found that some rare earth oxides (M_2O_3) where M is Ce, Pr , Eu, Dy , Ho or Er improved the values of σ_s (saturation magnetization in emu/gm) and μ_r when added to a single crystal MnZn ferrite. Especially Er shows a max value of 10.56 Wb/Kg when 0.05% of erbium oxide is added. This effect was also found in the polycrystalline material. The effect is related to the dissolution of the erbium oxide into the ferrite. Its use is expected in magnetic recording heads. Rezlescu (1997) studied the effect of rare earth doping in NiZn ferrites. By introducing small amounts of the rare earths, important modifications in structure and properties can be obtained. They tend to flatten the μ-T curve, shift the Curie point to lower temperatures and increase the resistivity. The judicial addition or control of these minor constituents is generally considered to be used by all of the major ferrite producers for power materials.

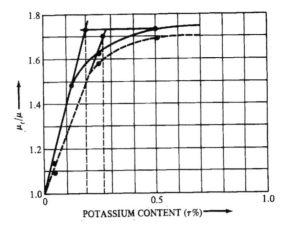

Figure 12.12 Variation of permeability of a MnZn ferrite with potassium impurity content. μt is the permeability with the K impurity present and μ is the permeability without the addition. From Guillaud (1957)

Kimura (1989) warns, however, that in a $(MnO)_{.36}(ZnO)_{.10}(Fe_2O_3)_{.54}$ ferrite, exaggerated grain growth can occur with additions of SiO_2 over 200 ppm even when there is no prolonged milling of the calcined ferrite powder. The addition of V_2O_5 was found to suppress the exaggerated grain growth caused by the silica additions even in the case of large additions. Increased grinding time decreased grain size but increased number of exaggerated grains. Postopulski (1989) presented a new method of describing the granular structure of ferrites.

Roelofsma(1989) measured the solubilities of CaO and SiO_2 in the presence of each other. He found that the presence of one reduced the solubility of the other inside of the grain (lattice). At 1350°C., for low SiO_2 levels, the solubility of CaO was >.49% and at low CaO levels, the solubility of SiO_2 was .07%. In the presence of a nominal .288 % CaO content, the SiO_2 content in the grain was only .02% while the nominal SiO_2 (from the addition) was .156%. Conversely, in the presence of a nominal .156% SiO_2 addition the CaO was only .193% as opposed to a nominal .288% from the CaO addition. There is no maximum solid solubility found for the molar ratio $CaO/SiO_2 = 2$ as might be expected. Obviously, the implication is that some of the excess SiO_2 and CaO is deposited at the grain boundary which may increase the grain boundary resistance and lower eddy current losses. During the cool some of the calcium and silica may remain in the lattice, which will affect the magnetic properties.

Roess (1986) has indicated that the addition of CaO and proper processing will substantially improve the high frequency response(See Figure 12.15) Apparently little information has been about optimum CaO and silica additions for optimum properties. The amount obviously depends on the application, raw materials and processing used but this question remains a critical factor in Mn Zn ferrites.Mochizuki et al (1985 reports the use of CaO And SiO_2 to increase the grain

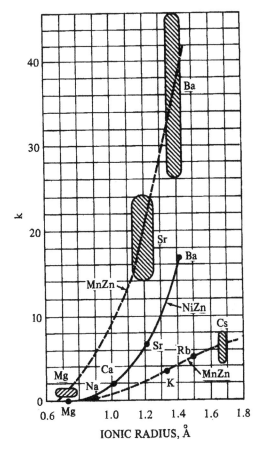

Figure 12.13- Effect of ionic radius of impurities on permeability in a MnZn ferrite. $k=[(\mu_t/\mu)-1]$, where μ_t is the permeability of the ferrite with the impurity present and τ is the molecular percentage of the impurity. From Guillaud (1957)

boundary resistivity. Stijntjes (1989) cites references that show CaO content to be higher at the grain boundaries than in the bulk. The CaO present in the lattice can cause microstresses. Based on on Roelofsma's work, the judicious combination of CaO and SiO_2 may be used to control the lattice concentrations of the other.

Temperature Dependence of Initial Permeability

Earlier in this chapter, we showed that increasing the iron content will move the secondary maximum of permeability to lower temperatures. By choosing the right chemistry we could arrange to have the highest permeability at the operating temperature (usually room temperature). There is yet another reason for adjusting them secondary maximum, namely to provide the material and eventually the the magnetic component with the required temperature stability over a specified

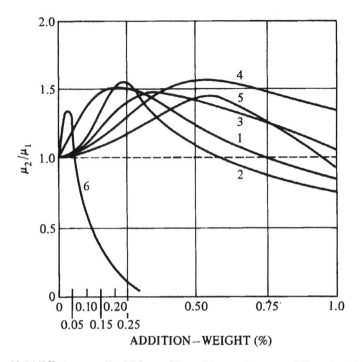

Figure 12.14 Effect on small additions of impurities on the permeability of a MnZn ferrite. μ_2 is the Permeability with the Impurity and μ_1 is the Permeability without it. The impurities are;(1) MoO (2) CuO (3) CdO (4) C (5) Al_2O_3 (6) La_2O_3. From Lescroel and Pierrot (1960)

temperature range. The variation in the permeability with temperature is often a very important consideration in a magnetic component. If we plot the permeability as a function of temperature, the slope over a specific temperature range of operation can be expressed as a material parameter called the temperature factor which is defined as:

$$T.F = \Delta\mu/\mu^2 \, \Delta T$$

where $\Delta\mu$ = Difference in permeability (μ_2-μ_1) at T_2 and T_1 respectively
ΔT = Change in temperature (T_2-T_1)

The temperature factor can be used to predict the variation in magnetic properties of a magnetic component. to those reported By moving the secondary maximum of permeability through variation of the Fe_2O_3 content, the room temperature permeability can be maximized as previously shown. The T.F. of the ferrite can be also be altered dramatically by the same method including a change in sign. This can be shown in Figure 12.16 where the Fe content is increased progressively in .15 mole percent increments. This device will be used in regulating the temperature coefficient of a component. Johnson (1978) has reported a strong effect of iron content on the temperature factor of NiZn ferrites in addition for MnZn ferrites.

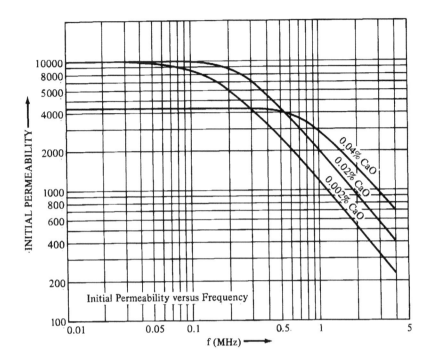

Figure12.15-Effect of CaO additions on the high frequency properties of the permeability. The permeability, though lower, with the Ca extends to a much higher frequency before rolloff. From Roess, Proc. 3rd Int. Conf. on Phys. of Mag. Mat.,Szczyrk- Bita, Poland, Sept.9-14,1986.

In addition to the adjustment of the magnetostriction through the iron content, several other mechanisms of controlling the T.F. have been proposed. Most of these involve anisotropy compensation as opposed to magnetostriction compensation based on iron control. For anisotropy control, the natural choice, as previously seen, is the use of Co ferrite. This technique has been used by several workers with good results on MnZn ferrites and Ni ferrites. In addition, Stijntjes (1971) reported on the use of TiO_2 which is effective in creating two zero crossings of the anisotropy constant, K_1 (one at room temperature and another at a lower temperature). The result is a flattening out of the permeability curve. Jain (1980) has reported the use of SnO_2 for the same purpose in MnZn ferrites. Another case by Park (1997) in which additives were used to alter the temperature coefficient of the permeability was one in which nickel oxide, NiO, or chromium oxide, Cr_2O_3, reduced the temperature coefficient of a MgCuFe ferrite from 15.5×10^{-6} to 10.3×10^{-6} and 5.5×10^{-6} for the Ni and Cr respectively. The reduction in slope is attributed to the change in the magnetocrystalline anisotropy. With NiO the grain size decreased at low levels but produced abnormal grain growth at high levels.

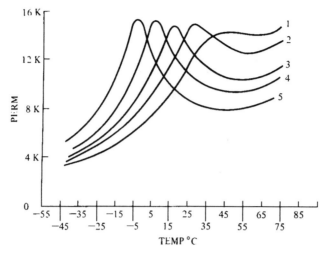

Figure 12.16- Change of temperature dependence of the initial permeability of a high permeability MnZn ferrite with Fe_2O_3 content. In going from 1 through 5, the Fe_2O_3 content is increased in .15 mole percent increments.

Time Dependence of Permeability (Disaccomodation)

Disaccomodation is an important stability factor in many electronic circuits where changes in properties over time can be very detrimental. Again, the component operation can be predicted from the disaccommodation factor defined in the last chapter. It is generally accepted that disaccomdation occurs by diffusion of certain ions through the ferrite lattice at room temperature. The ions thought to be responsible for this phenomenon are Fe^{2+} ions and the mechanism for diffusion appears to be by means of cation vacancies. To nullify or reduce the effects of ferrous ions in decreasing the resistivity, we can use ions such as Ti^{4+} and Sn^{4+}. The mechanism here is to localize the valence electrons of the ferrous ions by combining one Fe^{2+} with one quadrivalent ion (Ti^{4+} or Sn^{4+}) so that the net effect is the same as two +3 ions. The same technique appears to work for the disaccommodation as reported by Knowles (1974). Dense, high permeability ferrites tend to have low vacancy concentrations and therefore, disaccomodation is not a severe problem. However, when low losses are required, the high oxygen fires often used produce a large number of cation vacancies and it is in these materials that the worst problems with disaccommodation are experienced.

Chemistry Dependence of Low Field Losses Loss Factor

Earlier, we stated that the low field losses are composed of three different contributions;

    1. Hysteresis Losses

2. Eddy Current Losses
3. Anomalous or Residual Losses

Hysteresis Losses are those that occur even at D.C. or low frequencies. They can be visualized as the area inside of a D.C. hysteresis loop. For minimization of these losses, low anisotropy and magnetostriction are required as in the maximization of permeability. The chemical implications of these factors have already been discussed. Other requirements are microstructure related (grain size, porosity). The anomalous losses are actually a catch basin for all other effects. They include such things as ferromagnetic resonance losses.

The eddy current losses become an important factor as the ferrite is pushed to higher and higher frequencies. We have spoken of the use of excess iron to reduce magnetostriction but incorporating iron does have its drawbacks. For example, eddy current losses become too high at high frequencies reducing the performance considerably. Consequently, in these high frequency applications (as in the range in excess of 10 MHz., one usually switches to Ni or NiZn ferrite where the iron excess may be reduced to zero. Sometimes, even a deficiency of iron is indicated. The high frequency characteristics are improved by the increase in the resistivity. To illustrate the great impact of iron content in these materials, the resistivity of nickel ferrite drops from 10^9 to 10^3 ohm-cm abruptly at 50 mole percent iron oxide composition as the iron content is increased (Figure 12.17) (van Uitert,1955). The reason for this great decrease of resistivity or increase in conductivity is an electron hopping mechanism from Fe^{++} to Fe^{+++} ions. With regard to Mn-Zn ferrite, minor additions are common in commercial practice for reduction of losses. Early in the exploitation of ferrites, C. Guillaud (1957) discovered that the addition of CaO to the ferrite in small quantities produced significant decreases in losses of Mn-Zn ferrite. The improvement in losses can be seen in Figure 12.18 where the μQ product (which is the reciprocal of the loss factor) is plotted against the calcium content. This decease in losses is attributed to the large reduction in the eddy current component of the losses.

The other oxide which later was found to improve properties and appears to work in conjunction with CaO, but at a significantly lower concentration is SiO_2.(Akashi, 1961) More recently, materials such as SnO_2 and TiO_2 (Stijntjes, 1970) have also lowered losses and have increased temperature stability. The addition of cobalt oxide through the anisotropy compensation previously described lowers the losses in Ni ferrite especially at high frequencies. See Figure 12.19 (Lescroel and Pierrot,1960)

Tetravalent and Pentavalent Oxide Substitution

The tetravalent oxides that can be used in more than trace amounts are TiO_2 and SnO_2. The pentavalent ones are V_2O_5 and Ta_2O_5. In the lattice, the main function is to combine with a Fe^{++} such that the one valence electron from Fe^{++} can travel or resonate between the Fe^{++} and the Ti^{++++}. The net result is the equivalent of a Fe^{+++} and a Ti^{+++}.

$$Fe^{++} + Ti^{++++} \rightarrow Fe^{+++} + Ti^{+++}$$

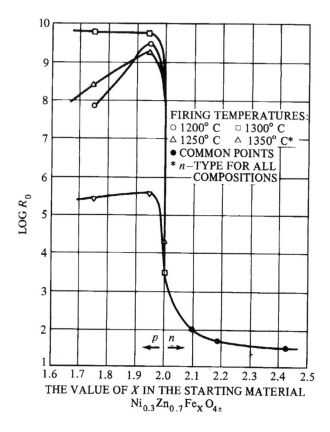

Figure 12.17- Variation of the resistivity of nickel zinc ferrites as a function of Fe_2O_3 content in the starting material. From van Uitert(1955)

In another sense, the electron from the Fe^{++} has been localized in the vicinity of the TiO_2 and thus is prevented from being free to conduct (by electron hopping to Fe^{+++}). However, the effect of the Fe^{++} is till operable to reduce the magnetostriction as discussed earlier. This is one way to use the beneficial effects of the ferrous ions without paying the penalty in loss increase. The same is true for SnO_2 and GeO_2. Sugano (1972) has given an equivalence of the TiO_2 and SnO_2 and its effect on the Fe^{++} content. The use of TiO_2 and SnO_2 was proposed by Roess (1969,1970) as a substitution for some of the iron to move the secondary maximum. He found it increased the resistivity and lowered the losses primarily at higher frequencies. Stijntjes (1970) also reported on the permeability and conductivity of Ti-substituted MnZn ferrites. Giles and Westendorp (1977) proposed the simultaneous substitution of Ti^{+4} and Co^{+2} in MnZn ferrites. Buethker(1982) reported on a high frequency power material incorporating Ti and Co. At ICF4, Stijntjes(1985) further used the same additions to develop a power ferrite for use at 500 KHz. At ICF5, he compared the power losses of the Ti-Co substituted material ($Mn^{+2}_{0.715}Zn^{+2}_{0.204}Co^{+3}_{0.006}Ti^{+4}_{0.03}Fe^{+2}_{0.105}Fe^{+3}_2O_4$) with those of a similar unsubstituted material. The

CHEMICAL ASPECTS OF FERRITES

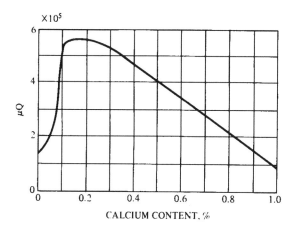

Figure 12.18- Effect of Ca addition on the µQ product of a MnZn ferrite. From Guillaud (1957)

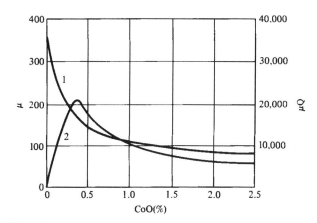

Figure 12.19- Effect of Co additions on the permeability and µQ product of a NiZn ferrite containing 25% NiO. From Lescroel and Pierrot (1960)

reduction in the substituted material is shown in Figure 12.20 Note that the eddy current losses are only a small part of the losses even at the higher frequencies. Johnson(1989) examined the similar Co-Ti substitution in MnZn ferrite single crystals. He found that the magnetostriction coefficients, λ_{100} and λ_{111} as well as B_s and the Curie point, T_c, were the same as those found in the unsubstituted case.

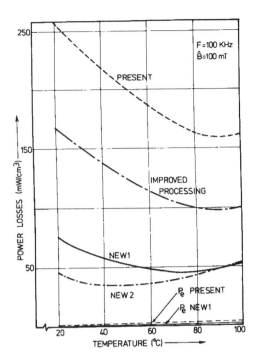

Figure 12.20- Comparison of core losses of several MnZn ferrite materials, showing the effects of improved processing and a new chemistry involving Ti + Co additions. From Stijntjes, T.G.W. and Roelofsma, J.J., Advances in Ceramics, Vol.16, p.493 (1985)

However, K_1 and K_2 depended on the Co content, especially at lower temperatures. Visser (1989) also studied the beneficial effects of the Ti-Co substitution in single crystals and polycrystalline MnZn ferrites and by the observation of the temperature variation of disaccomodation, was able to separate the permeabilities due to wall movement and rotational processes. He found the Ti-Co substitution reduced the wall permeability and enhanced the rotational permeability. The result was a decrease in the damping losses due to reverible wall movement and a decrease in the hysteresis losses due to irreversible wall movement.(See Figure 12.21). At 100 KHz, in the polycrystalline doped material the experimental total power loss consisting of mainly hysteresis and damping losses was almost equal to the calculated damping losses. This factor led to a lowering of the overall losses. Hanke (1984) described a power ferrite with Ti-Sn substitutions with the composition, $Mn_{.631}Zn_{.266}Ti_{.022}Sn_{.01}Fe_{2.071}O_4$. For use at higher frequencies, Sano (1989) in describing the conditions for development of TDK's H7F material included the use of TiO_2 for control of the magnetic anisotropy. In addition to the Ti addition, he cites the need in H7F for addition of adequate amounts of CaO and SiO_2. The comparison of the losses of this material with their best previous power material, H7C4, is shown in Figure 12.22. In the new material, H7F, the CaO, SiO_2 and TiO_2 concentrations are higher than in H7C4. Near the grain boundaries, H7F has a 56% lower

CHEMICAL ASPECTS OF FERRITES 255

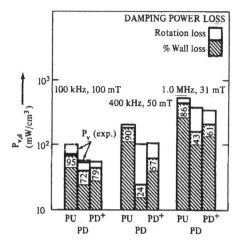

Figure 12.21-Separation of wall losses and rotational losses in several MnZn ferrites at three different frequencies, PU is undoped, PD has a doping of Ti and Co and PD+ has 1.1 times the amount of Co as PD. The dotted lines show some experimental results. From Visser, Roelofsma and Aaftink, Advances in Ferrites, Volume 1- Oxford and IBH Publishing Co.,New Delhi, India, 605 (1989)

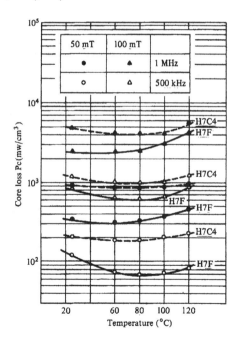

Figure 12.22-Improved core loss characteristics at higher frequencies of the new TDK H7F over the older H7C4. From Sano, Morita and Matsukawa, Advances in Ferrites, Volume 1- Oxford and IBHPublishing Co.,New Delhi, India, 595 (1989)

CaO Auger intensity than H7C4 (See Figure 12.23) but it has triple the grain boundary area. The ratio of CaO addition in H7C4 to H7F is 1:2.

When all these factors are considered, he concludes that there is little difference between the two with respect to Ca accumulation near the grain boundary. Ishino (1992) proposed a low loss ferrite for high frequency switching power supplies. Core loss at 1 MHz. was improved by sintering at 1150°C. with 0.02 to 0.04 mole percent Ta_2O_5 added. The temperature property of core loss for an iron-rich sample was optimized by sintering in a higher than equilibrium partial pressure

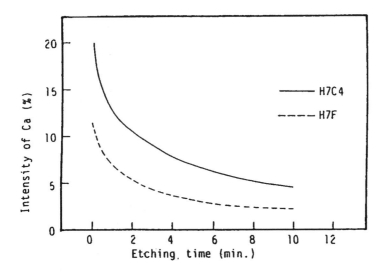

Figure 12.23- Auger intensity of Ca ions as a function of etching time for the TDK H7C4 and the new H7F materials. From Sano, Morita and Matsukawa, Advances in Ferrites, Volume 1- Oxford and IBH Publishing Co.,New Delhi, India, 595 (1989)

Yamamoto(1997) also reported using small amounts of Ta_2O_5 to increase the resistivity of MnZn ferrites while large amounts caused a decrease in resistivity without significant grain growth.

Losses at High Power Levels

Some of the same factors that minimize low field losses can also help reduce high field losses. The one difference involving chemistry appears in the need for higher saturation materials in power applications. Power supply designers would prefer to operate at as high a flux level as possible. This makes the core smaller and lowers nonmagnetic related losses such as wire losses. Almost always, the wire losses are much higher than the core losses. The combination of core losses and winding losses heats the wound core, that has two poorly conducting paths to remove the heat. These are 1), by conduction to the outer surface of the ferrite and 2) through the insulating wall of the coil into the copper windings which themselves are heat generators. The increased temperature drops the saturation induction of the material, which may increase core losses and heat the core even further, dropping the saturation again and so forth. If this situation continues unchecked, the core can go into saturation with the magnetizing current rising sharply. The result is a thermal runaway that is the most severe problem in power supply design.

At lower frequencies, the losses are lower and designs at moderate flux levels are possible. A higher saturation to start with provides a cushion for the expected temperature rise. Operation at 50-60 percent of saturation is usual. Within the limitations of core losses, the higher the saturation the higher the possible flux level of operation. In addition, since power supplies may be expected to run hot, the minimum in the core loss should be designed to be near the intended temperature of operation. The minimum of core loss often occurs at the secondary permeability maximum. Therefore in combining the chemistry needs for high saturation with those for optimizing the high temperature loss minimum, a satisfactory power ferrite can be developed. However, even with the loss minimum, protection against further core loss increases at temperatures higher than the minimum must be provided. Therefore, the presence of a broad core loss minimum is desired. At operation at very high frequencies (>500 MHz.) the operation flux density must be so low that the needs for a high saturation material are not too important.

In the Manganese-Zinc ferrites, the greatest increase in usage is in the application as power materials. It is not surprising that the chemistry studies reported are in that area. Roess(1989) examined the compositions of several of the commercially available low- power-loss (LPL) materials and compared them to the compositions of several commercial very-high permeability (VHP) materials. He notes that the range for the VHP materials is much smaller than the corresponding range for LPL materials, making the former more sensitive to major chemistry. The LPL materials can be made with low loss by a variety of different techniques at somewhat different compositions. Figure 12.24 shows the ranges for the two types. Note that the Fe_2O_3 content includes possible additions of TiO_2 and SnO_2 as part of the substitutions that may improve the quality of LPL materials. He proposes a compromise material that has a permeability of 5000 and a saturation induction of 5000 Gausses with a Curie Temperature in excess of 200°C. Usually, with a high perm material, the temperature dependence near the secondary permeability maximum is very steep (See Figure 12.25,Curve A) making it difficult to keep the5000 perm over a wide temperature range needed for power ferrites. Obtaining a curve like that of Curve B is quite difficult to achieve. Ochiai (1985) gives the range of useful compositions for a power material in Figure 12.26. In addition to overall major element chemical composition, Berger (1989) found that discontinuous variations on a microscale resulted in higher power losses. With major constituents, there may be strains and variations of anisotropy and magnetostriction from grain to grain.

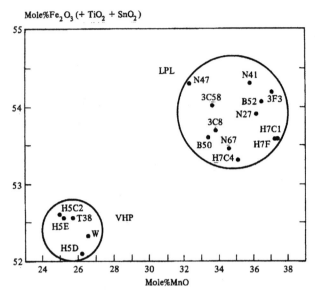

Figure 12.24-Comparison of ranges of composition of commercially-available MnZn ferrite materials of the very-high permeability (VHP) and low-core-loss(LCL) varieties. Note that some of the iron may be substituted with TiO2 or SnO2. From Roess,E., Advances in Ferrites, Volume 1- Oxford and IBH Publishing Co.,New Delhi, India, 129 (1989)

Otsuka (1992) found that, in addition to the increase in resistivity of MnZn ferrite and decrease in power loss with the use of silica and calcia , hafnium oxide also enriched the grain boundary layers increasing the resistivity even furthere. They also noted that, with increasing frequency, the temperature at which the minimum core loss occurred shifted downward. This fact is attributed to the positive dependence of the eddy current loss, P_e, on temperature and the increasing effect of P_e on P_c, the core loss with frequency. On the basis of these findings, they developed a new material, B40, with a power loss one half to one third of the current product.

Kwon (1997) studied the effect of vanadium pentoxide additions to MnZn power ferrites. Small amounts of V_2O_5, in combination with silica and calcia additions, gave a fine grain structure and improved the magnetic properties, especially core loss. The V_2O_5 formed a liquid film at the boundary, inhibited grain growth and reduced the eddy current losses. If too much it it is used, hysteresis loss becomes dominant and the total core loss can increase.

Takahashi(1992) studied the effect of non-stoichiometric oxidation degree on properties of MnZn power ferrites. The non-stoichometry was adjusted by composition, oxygen partial pressure or by addition of TiO_2. The temperature of the Secondary Permeability Maximum(SPM) and the temperature of minimum power loss decreased with ferrous ion content.

Inoue (1992) studied the mechanism of core loss in single crystal and polycrystalline MnZn ferrites. The core loss was divided into hysteresis and eddy current losses from their frequency dependence. Although the eddy current loss in the

CHEMICAL ASPECTS OF FERRITES

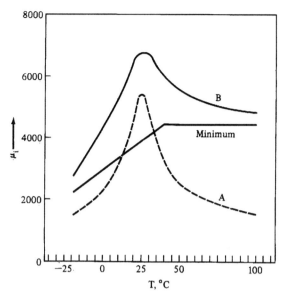

Figure 12.25- The temperature dependence of permeability of two MnZn ferrite materials showing a rather sharp peak at the secondary permeability maximum in the case of material A. This drop-off particularly above room temperatures makes it unsuitable for power use which needs the minimum value shown. Material B is desirable but hard to produce. From Roess,E., Advances in Ferrites, Volume 1- Oxford and IBH Publishing Co.,New Delhi, India, 129 (1989)

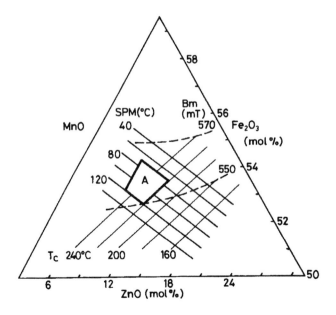

Figure 12.26- Desireable range of composition for a MnZn ferrite power material. From Ochiai,T. and Okutani,K. at ICF4, Advances in Ceramics, Vol.16 p.447.(1985)

Chemistry Considerations for Hard Ferrites

In optimizing hard ferrites for permanent magnet applications, a compromise is struck between obtaining a high remanence value or a high coercivity. Aside from the choice of either barium or strontium ferrites, a good deal of the optimization involves microstructural effects. For example, the remanence, B_r, is increased greatly by the attainment of higher densities. This subject will be dealt with in greater detail in the second part of this chapter. However, as previously stated, the microstructural features are often strongly influenced by the chemistry. Even with regard to the major metal oxides, the chemistry affects the density. In common pratice, the ratio of BaO to Fe_2O_3 is usually maintained at 5.6 rather than the 6 required by the stoichiometry ($BaO.6Fe_2O_3$). The microstructure may also be influenced by the presence of certain minor elements such as Si. Kools (1985) has reported that in strontium ferrite, silica is an active ingredient that definitely alters the microstructure. Depending on the ratio of SiO_2 to SrO, quite different microstructures are obtained. Bi_2O_3 (or Na_3BiO_3) is another additive that is used to improve density. The effects of Sb_2O_3, TiO_2, As and P have also been investigated.

A new hexaferrite with mixed β-alumina and magnetoplumbite structure was synthesized by Nariki(1989)

Saturation Induction of Microwave Ferrites and Garnets

For microwave applications, both spinel ferrites and garnets are used. The ferrites in general have higher saturations than the garnets the former reaching as high as 5000 Gausses while the garnets are limited to about 1950 Gausses. However, other saturation-related properties must be considered for this application, primarily the temperature stability of the saturation value. Higher temperature stability of the satuation can be accomplished in several ways;

1. Use of a material with a high Curie point-Since the temperature course of the saturation magnetization is a parabolic function, the closer to the Curie point, the steeper will be the slope and the more temperature-sensitive the saturation will be. Therefore it the T_c is high, we will be operating on a flatter portion of the curve.

2. Use of a material with a compensation point. Some materials, primarily garnets, have compensation points with a flat portion between the compensation point and the Curie point. If the rare earth ions are chosen as the primary ion or blended with other rare earths so that the temperature of operation is on the flat part of the curve, greater temperature stability can be obtained.

Spinel ferrites for microwave application usually have saturations of about 2000-2400 Gausses. Nickel ferrite has a saturation of about 3200 but this may be lowered by the use of Al. Lithium ferrites have a saturation of 3600 and that can be reduced through the use of Ti^{4+} or increased with Zn^{+2}. Magnesium- manganese ferrites have $4\pi M_s$ values of about 2400 Gausses.

With the garnets,The prototype, yttrium-iron garnet,YIG, has a saturation of 1790 Gausses. Al or rare earth substitutions can reduce this to values of from 1200 to 300 Gausses. The saturations can be increased somewhat to about 1900 through the use of In(indium) or Ca-Zr substitutions.

Chemistry Dependence of Microwave Properties

As mentioned earlier, the line-width, ΔH, in ferrimagnetic resonance is broadened by anisotropy and porosity. The porosity effect will be discussed later. The spinel ferrites have higher anisotropies than the garnets and consequently higher line-widths. The anisotropy in the spinels is mostly due to Fe^{3+} ions on the octahedral sites. Substitutingf Al^{3+} for the Fe^{3+} will lower the anisotropy and the line-width in MgMn ferrites. In the case of Ni ferrite the negative anisotropy can be compensated and the line-width reduced by the addition of Co, a method previously employed in non-microwave applications. In the case of lithium ferrite, Ti is used to reduce octahedral Fe^{3+} ions and also lower ΔH. With regard to the rare earth garnets, low ΔH is obtained with yttrium-iron garnet (YIG) because the magnetic moment is due to an excess of Fe^{3+} ions on the tetrahedral rather than octahedral sites. Substitution for Fe^{3+} with Al^{3+} in YIG will lower the value of $4\pi M_s$ while maintaining the low line-width. In general, lowering the Curie temperature by substitution of Fe^{3+} ions will reduce the anisotropy and the line-width. For low anisotropy garnets, the only other rare earth ion substituted for Y in YIG is Gd which has no orbital moment and therefore doesn't increase anisotropy. Extremely low linewidths are obtained with single crystal garnets that are formed into spherical samples and polished to a very smooth. finish.

For high power microwave operation, we have mentioned the need for a high h_{crit} that ,in turn, requires a large ΔH_k or spin-wave linewidth. One method of obtaining this is by the addition of magnetic ions of the so-called "relaxing" variety . These include most of the rare earths (with the exception of Gd, previously mentioned). The ones with the greatest effect are Ho and Dy. Another "relaxing" ion is Co^{2+}. There is also an increase in ΔH that is a disadvantage so a compromise is usually struck. If Co^{2+} is used, there is usually an equimolar addition of a tetravalent ion such as Si^{4+} or Ge^{4+} to obtain the equivalent of a +3 valence needed in the garnet crystal structure for charge balance.

In the case if the YIG filter, the saturation and line width are temperature dependent. In the case of very sensitive filters with very narrow line widths, small changes in temperature may detune the system from resonance. Therefore, the temperature must stay constant or the material must be changed to one which is less sensitive but more stable. The same situation applies to high power circulators where the $4\pi M_s$ may vary sharply at high power levels. As a result, forced cooling may be necessary. In addition, a material with a lower $4\pi M_s$ may be used.

Han (1989) found that in the microwave material, calcium-vanadium-bismuth iron garnet that chemistry variations from grain to grain with the attendant change in $4\pi M_s$ increased the value of ΔH. Line broadening was attributed to incomplete solid reaction. Han (1992) also reported on the effect of In^{3+} and Zr^{4+} on lowering the temperature coefficient of M_s for Ge:BiCaVIG. Ge promotes the completion of reaction and densification. . Excellent properties were obtained by the combination. Dionne, G.F.(1997) reported on microwave ferrites for cryogenic applications. Fe^{3+} ions were replaced by diamagnetic substitutions such as Al^{3+}, Ga^{3+}, In^{3+} and Zn^{2+} to tailor the magneto-elastic properties. The use of Gd^{3+} was avoided because of magnetic loss characteristics of the rare earths. Replacement of the Y was done by the use of Bi or of Ca+ V.

Ferrites for Memory and Recording Applications

In the previous chapter, we have discussed the need for a high squareness ratio, B_r/B_s, in memory applications. Wijn(1954) has written that for high squareness or the attainment of a square hysteresis loop, the anisotropy, K_1, should be large and negative while the magnetostriction, λ_{111} or that measured in the body diagonal direction (the preferred axis in most ferrites) should be close to zero. Several varieties possess square-loop properties naturally while in others, they can be induced by thermal or mechanical treatments. Most of the ferrites that can have square loops contain Mn including MgMn (mentioned under microwave applications), MnCu, MgMnZn, MnCuZn, MnCuNi, and MnLi. The ferrites with the best square loop properties contain about 40-45% Fe_2O_3 rather than the 26-38% predicted by Goodenough(1957). The square-loop ferrites introduced by Albers-Schoenberg (1954) is one of the new classes of ferrites developed in the United States. Zn increases the squareness but decreases H_c and as we have previously noted, it also decreases the Curie point. Cd has been substituted for Zn (Eichbaum 1959) because it gave a lower coercive force but unfortunately was very toxic. In many cases involving memory cores, good properties were obtained by quenching from high temperature that was easy to do without cracking the small cores. Memory cores practically disappeared because of the emergence of semiconductors for random access memories and disk memories for storage systems.

For recording media, the large volume of tape and disk memories use γ-Fe_2O_3 or more recently Co- doped γ-Fe_2O_3. Another "ferrite" in common usage is CrO_2. A more recent development in recording media is in the area of so-called perpendicular recording, in which the particles of ferrite are arranged perpendicular to the surface of the tape or disk instead of the normal longitudinal or parallel- to-surface recording. A ferrite being used for this purpose is a special variety of barium ferrite. Additions of Co^{2+} and Ti^{4+} are sometimes made to improve the properties.

Rare Earth Garnets for Bubble Domain Devices

The rare earth garnets used for low level microwave applications were chosen for their cubic symmetry and low anisotropies. The garnets for bubble domain applications, on the other hand, will require that the structure be anisotropic so that the bubble domain structure can be supported.

The first bubble domain materials were orthoferrites, typified by the compound, $YFeO_3$ (Bobeck,1967). Extensive research on rare-earth garnets substituted with non-magnetic ions yielded materials with improved properties for these applications. A strong advantage was their greatly improved methods of fabrication along the lines of mass- produced semiconductor techniques. The garnets were prepared as films grown epitaxially on non-magnetic substrates. The universally used substrate is GGG or Gadolinium Gallium Garnet. Growing epitaxially means that the crystal structure and the unit cell dimensions of the substrate and the film are so close to each other that there is a natural extension of growth from substrate to the magnetic garnet film. In the garnets chosen for bubble usage, the anisotropy field, $H_a = 2K_1/M_s$, is larger than $4\pi M_s$. To reduce the magnetostrictive effects (due to distortion of the lattice),neighboring rare earths in the periodic table are often used as their ionic radii are similar. Some common effective composition of garnets used for bubbles include $Er_2Eu_1Ga_{.7}Fe_{4.3}O_{12}$ and $Er_1Eu_2Ga_{.7}Fe_{4.3}O_{12}$. The gallium is added to reduce the $4\pi M_s$. The anisotropy field is about 4000 oersteds. Structural perfection of the film is important. Film preparation will be discussed in the chapter on processing.

References

Abe, M.(1989), Itoh, T.,Tamaura, Y. and Yoshimura, K., Advances in Ferrites, Vol 2- Trans-Tech Publications, Aedermannsdorf, Switzerland, 1131

Buethker,C.(1982) Presented at American Ceramic Society(Electronics Subsection, Sept. 13,1982, Boston, Mass.

Chiba,A.(1989) and Kimura,O.,Advances in Ferrites, Vol. 1 Oxford and IBH Publishing Co., New Delhi, India, 35

Date, S.K.(1989), Deshpande, C.E., Kulkarni, S.D., and Shroti, ibid, 55

Dormann, J.L.(1989) and Nogues,M., Presented at ICF5, Bombay, Jan 10-13,1989, Paper A8-O2

Franken, P.E.C.(1980) and Stacey, J. A. Ceramic Society, 63,

Giles, A.D.(1977) and Westendorp, F.F.,J. Phys. Suppl.34, 38 p.47

Goldman, A.(1989), Advances in Ferrites, Volume 1- Oxford and IBH Publishing Co.,New Delhi, India, 13

Han, Z.Q.(1989),ibid.,Advances in Ferrites, Volume 2, Trans-Tech Publishing Co.,Aemannsdorf, Switzerland, 995

Hanke,I.(1984) and Neusser, IEEE Trans. Mag MAG-20,#5,Sept. 1984, p.1512

Jha, V.(1989), and Banthia,A.K., ibid, 61

Johnson, M.T.(1989), ibid ,399

Kim, T.O.(1989), Kim, S.J., Grohs, P.,Gronnenberg, D. and Hempel,K.A., Presented at ICF5, Bombay, Jan10-13, Paper C1-04

Kimura, O.(1989), ibid,169

Kools, F.(1989) and Henket,B., ibid,417

Manoharan, S.S.,(1989) and Patil,K.C., Advances in Ferrites,Volume 1- Oxford and IBH Publishing Co.,New Delhi India, 43

Matsumoto, K (1989), Nakagawa, S. and Naoe,M., ibid, 545

Matsuyama,K.(1989), Shimizu, S.,Watanabe,K.,Hirota,K.,and Kugimiya, K., ibid, 565

Mochizuki,T.,(1985), Sasaki, I. and ,Torii,M. ,Presented at ICF4, Advances in Ceramics, Vol.16, p.487

Narayan, R.,(1989), Tripathi,R.B. and Das, B.K.,267

Nariki,S.(1989), Ito,S.,Fujiwara,S. and Yoneda,N., 121 Neyts,R.C.(1989) and Dawson, W.M., ibid,293

Postuplski, T.(1989) ibid, 639

Rambaldini, P.(1989) Advances in Ferrites, Volume 1- Oxford and IBH Publishing Co.,New Delhi, India, 305

Rikukawa, H.,(1987), Sasaki,I. and Murakawa, K. Proceedings Intermag 1987

Ries,H.B.(1989), Advances in Ferrites, Volume 1- Oxford and IBH Publishing Co.,New Delhi, India, 155

Roelofsma, J.J.(1989) and Kools,F.X.N.M., Proc.Inst. ECERS, Maastricht, Holland, June 18-23,1989

Roess,E.(1969), German Patent 1300860, Issued 7 Aug. 1969

Roess, E.(1970, Phy. stat. sol.(a) $\underline{2}$, K185

Roess, E.(1986), Proc. 3rd Int. Conf. on Phys. of Mag. Mat., Szczyrk- Bita, Poland, Sept. 9-14, 1986

Roess, E.(1987), ERA Report 87-0285

Roess, E.(1989), Ruthner, M.J.(1989) Advances in Ferrites, Volume 1- Oxford and IBH Publishing Co., New Delhi, India, 129

Ruthner, M.J.(1989), ibid, 23

Saimanthip, P.(1987) and Amarakoon, V.R.W., Abstract for Paper 8-E-87,, presented at the Annual Meeting of the Am. Cer.Soc., April 21,1987, Pittsburgh, Pa.

Stijntjes, T.G.W.,(1970), Klerk, J. and Broese van Groenou, Philips Res. Repts. $\underline{25}$, 95

Suresh, K.(1989) and Patil, K.C., Ruthner, M.J.(1989) Advances in Ferrites, Volume 1- Oxford and IBH Publishing Co., New Delhi, India, 103

Tamaura, Y.(1989), Itoh, T. and Abe, M., ibid, 83

Ventkataramani, N. (1989), Srivastava, C.M., and Patni, M.J., ibid, 435

Visser, E.G.(1989), Roelofsma, and Aaftink, G.J.M., ibid, 605

Wagner, U.(1980), J. Magnet. and Mag. Mat. $\underline{19}$, 9

13 MICROSTRUCTURAL ASPECTS OF FERRITES

INTRODUCTION

In the last chapter, we dealt with the important properties of ferrites in terms of the chemistry and the cation site arrangement in the crystal lattice. Even in polycrystalline ferrites, these properties are related to those that we would obtain on single crystal ferrite material. Chemistry control is important here or several reasons, namely:

1. To attain proper saturation
2. To minimize anisotropy
3. To minimize magnetostriction
4. To avoid foreign ions that can strain the lattice

In our discussion of intrinsic properties, we are assuming that the lattice chemistry is homogeneous. In the finished ferrite, a gross inhomogeneity even with the correct bulk chemistry can, of course, be quite harmful. However, if we enter the final sintering step with a compact of a non- homogeneous powder, we may correct the chemical inhomogeniety by the use of an extended heat treatment and the proper atmosphere. This approach is not generally used, however, and chemical inhomogeneities present before sintering often lead to an inferior product that is relegated to the scrap heap. Microstructural defects due to the chemical inhomogeneities may be the ultimate reason for failure. The source of the microstructual defects may not be obvious when we examine the finished ferrite inasmuch as the chemical aberration was present early in the processing scheme and has long since disappeared. Knowledge of the effects of chemical inhomogeneities and minor impurities on microstructure development during sintering is therefore indispensable. Then, in turn, we can examine how these ceramic micro- structures affect the magnetic properties.

Although chemical-microstructural interactions can have deleterious effects, we can sometimes take advantage of chemical aspects to improve microstructural and magnetic properties. It is becoming quite evident that microstructural problems are the most serious obstacles in obtaining high- quality reproducible ferrites. The pure chemistry of the lattice appears simple compared to the myriad of possible ceramic effects and inconsistencies, known and unknown, that can come into play.

materials and very low losses especially at higher frequencies. Limitations on the desireable properties at high frequencies may be more easily overcome by control of microstructure than by choice of chemistry. In the last chapter, the choice of chemistry was made on the basis of the optimizing the intrinsic properties needed for a specific application. Similarly, the microstructure of a ferrite can also be designed to optimize the features that will improve operation at the particular frequency, flux level, mode of operation and stability required in the final component or device. For example, for some applications, we may require remarkably high permeability at moderate frequencies. Other devices may require low losses at higher frequencies. The microstructures best suited for each case will be different.

The Effects of Grain Size on Permeability
Let us refer back to domain consideration. If only high magnetic permeability is desired without regard to high frequency losses, the presence of grain boundaries will act as impediments to domain wall motion. The fewer the number of grain boundaries present, the larger the grains and the higher the permeability (See Figure 13.1). Generally, this principle is consistent with metallic magnetic materials. As we noted earlier, there are several differences here between the two types of materials.. The main difference is that metals have few of the "dirty" features found in ferrites. Inclusions are few, porosity is extremely slight and grain boundaries are relatively clean and unobtrusive. As by Tebble & Craik(1969) have pointed out, in high permeability metallic materials such as permalloy (an 80% Ni-Fe alloy), grain size is not an important factor for high permeability. The reason is that domain walls appear to be able to move across the grain boundaries easily. In ferrites, where the grain boundary is thicker, the same unhindered movement does not occur. Again, the lack of a purification scheme in processing, the presence of pores and inclusions, as well as greater chemical inhomogeniety prevent the attainment of the very high permeabilities, which may extend up to 100,000 in metals.

The earliest work on correlating grain size with permeability was done by Guillaud and Paulus (1956) on Mn-Zn ferrites (See Figure 13.2). Although the initial permeabilities only reached about 4,000, this was considered high for the time, although not according to present standards. The general trend is obvious. Guillaud (1957) related the inflection at 5 microns to a change from domain rotation as a mechanism to wall movement above 5 microns. Guillaud considered the limitation at about 20 microns to be due to the presence of pores included in the interior of grains. In later work, Guillaud (1960) reported similar results for $NiFe_2O_4$ (Figure 13.3). The trend to 14 microns follows the same pattern with the consequent dropoff of permeability again related to included porosity.

Many other workers such as Beer & Schwartz (1966, Perduijn and Peloschek (1968) and Roess (1966)(Figure 13.4) have confirmed the linear dependence of grain size to permeabilities of 40,000 and grain sizes to 40 microns. As will be seen later, precaution has to be taken to limit included porosity. Since permeabilty depends on so many factors in addition to grain size, the slope of the line relating

MICROSTRUCTURAL ASPECTS OF FERRITES

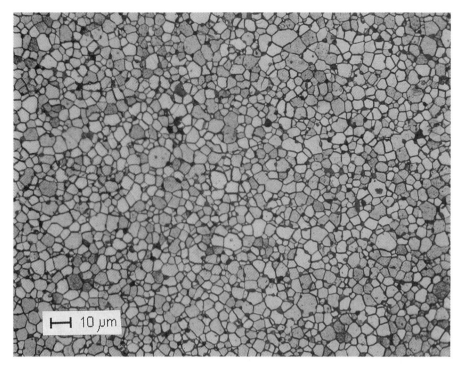

Figure 13.1a-Microstructure of a high-frequency MnZn power ferrite,(250 X)-Permeability = 2300. Photo courtesy of Magnetics.

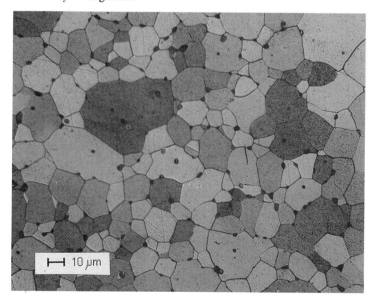

Figure 13.1b-Microstructure of a high permeability ferrite, (250 X), Permeability = 16,000
Photo courtesy of Magnetics

grain diameters versus permeability will change with the composition of the basic ferrite. Heister (1959) reports that permeability rose from 1000 to 4000 as grain size increased from 3.5 to 30 microns with the porosity also decreasing. At still larger grain sizes, the permeabilities drop presumably because of porosity. The evidence is clear that, if pores can be suppressed or located at the grain boundaries, the permeability will increase with grain size. Roess(1971) found that contrary to expectations, the permeability versus temperature curve for a ferrite with a uniform grain size had a broader secondary permeability maximum than one with a mixture of smaller and larger grains.

Exaggerated Grain Growth in Ferrites

The relationship between grain size and permeability will generally be linear only if the grain growth is normal, that is if all the grains grow pretty much at the same time and same rate. This leads to a rather narrow range of final grain sizes. If, indeed, some grains grew very rapidly, they would trap pores, which as we have seen, can limit permeability by pinning domain walls. When conditions permit this type of grain growth to occur with many included intragranular pores, it is called exaggerated or discontinuous grain growth.

Drofenik (1985,1986) has recently reported results that indicate that distances between pores account for variations in permeability. Samples with giant grains and included porosity owing to exaggerated grain growth still had higher permeabilities

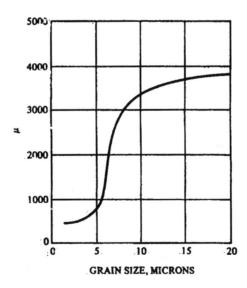

Figure 13.2- Permeability of a MnZn ferrite as a function of grain size in microns (Guillaud)

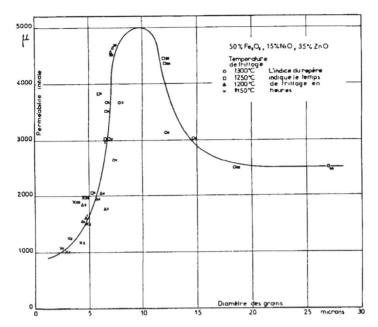

Figure 13.3 - Permeability of a NiZn ferrite as a function of grain size (Guillaud 1960)

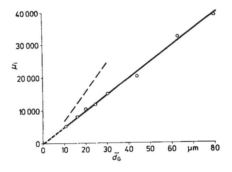

Figure 13.4 Permeability of High Permeability MnZn Ferrites as a function of Grain Size (Roess 1966)

than those with normally grown grains provided the distances between pores were the same. Drofenik concludes that the large grained samples are less sensitive to grain boundary effects and thus the μ versus T curve is more peaked. Yan and Johnson (1978) have studied exaggerated grain growth in 20 different elements (See Table 6.1). Figure 13.5 shows the growth of some large grains in a Ti-rich high frequency power material. In his study of high permeability ferrites, Roess (1971 found that in the same type of ferrite, the one with exaggerated grain growth and much included porosity had a permeability of 2000 while one with about the same grain size that grew normally had a permeability of 40,000.

Table 13.1
Tendencies of Metal Oxides to Produce Exaggerated Grain Growth

METAL OXIDES WITH NO EFFECT	METAL OXIDES WITH SLIGHT EFFECT	METAL OXIDES WITH STRONG EFFECT (Liquid Phase)	METAL OXIDE WITH STRONG EFFECT (No Liquid Phase)
Y_2O_3	Fe_2O_3	CaO	TiO_2
La_2O_3	Mn_2O_3	SrO	SiO_2
ZrO_2	MnO_2	V_2O_5	
ZnO		Nb_2O_5	
CdO		PbO	
Al_2O_3		CuO	
GeO_2		Sb_2O_3	
SnO_2			

Source: M. F. Yan and D. W. Johnson, 1978, *J. Am. Ceram. Soc. 61*, 342. Reprinted by permission of the American Ceramic Society.

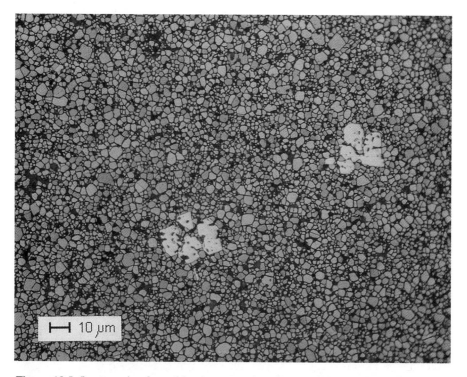

Figure 13.5- Large grains formed by exaggerated grain growth in Ti-rich grains in a high frequency power material. Photo courtesy of Magnetics.

Kimura et al (1977) prepared samples of magnesium ferrite with and without the exaggerated grain growth. They found that the presence of the exaggerated grain growth retarded the cation redistribution rate because of the decreased surface area of the grain boundary. The redistribution process presumably nucleates at the boundaries and the reduced area leads to retardation.

Drofenik (1985) reported on a mechanism other than impurities associated with the presence of exaggerated grain growth. In his study, the growth of the giant grains was brought about by zinc loss from the surface with its consequent acceleration of grain boundary movement. Thus, a smaller sample that had a larger volume fraction affected by zinc loss had larger grains and a higher permeability than a larger sample. Figure 13.6a shows the large grains only on the outer layers of a larger sample whereas Figure 13.6b shows giant grains throughout the thickness. The link with zinc loss is shown in Figure 13.7 in which the smaller core of Figure 13.6b is placed on a layer of zinc oxide during firing. The resulting microstructure contains fine grains in the inner portion of the core.

Duplex Grain Structures

A duplex structure is an undesirable type of microstructure that lowers the permeability and increases the losses. This type of structure has some very large (giant) grains in a matrix of fine grains. It is most often due to segregation of a particular impurity such as SiO2 which produces rapid grain growth locally while other undoped areas are unaffected. Figure 13.1a shows a normal microstructure compared to one (Figure 13.8) with the duplex structure. Yoneda (1980) has shown that incomplete binder burnoff can cause a duplex structure. His study revealed the appearance of the duplex structure accompanied by a degradation of the magnetic properties including a decrease in permeability and an increase in the losses.

Effect of Porosity on Permeability

As we have already noted, porosity is an important microstructural feature limiting the movement of domain walls. Again, this factor is not encountered as frequently in metals. Pores and other imperfections would appear to pin domain walls especially inside of the grain. However, as Globus (1972) and others point out, domain wall bulging would still permit wall movement even while the endpoints were tied down Such a mechanism is illustrated in Figure 13.9 . The growing of large grains creates a problem: many pores are swept over by the grain boundary and remain within the large grains. intragranular porosity is more deleterious than the intergranular.

Guillaud(1957) showed that although permeability in nickel zinc ferrites decreased with grain size up to 15 microns, it decreased thereafter. He contended that this decrease was due to included porosity.

Brown and Gravel (1955) demonstrated a decrease in permeability with porosity for a Ni and a NiZn ferrite. Porosity was varied by control of the firing temperature that was different for the two materials. Grain size was not controlled. However, since permeability did not decrease at higher temperatures when we would expected grain sizes to be larger, the included porosity does not seem to be as much a factor in the Brown-Gravel study as it was in Guillaud's work.

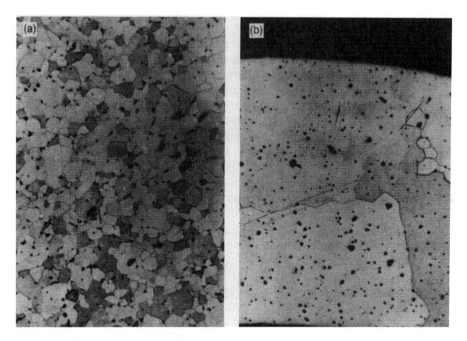

Figure 13.6- Microstructures of a MnZn ferrite; a) O.D. = 36 mm. b) O.D. = 4 mm. From Drofenik (1985

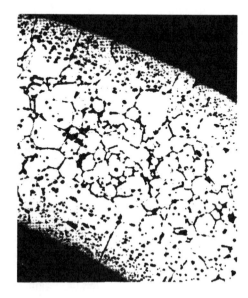

Figure 13.7-Microstructure of MnZn ferrite in Figure 13.6b, but fired on a bed of ZnO. (From Drofenik 1985)

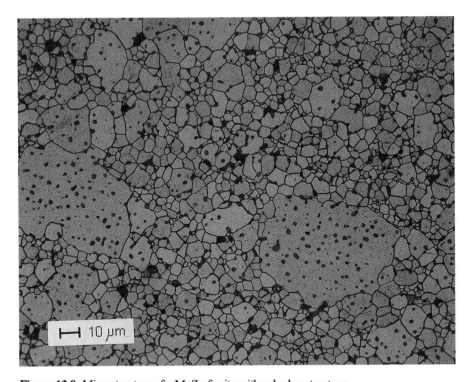

Figure 13.8- Microstructure of a MnZn ferrite with a duplex structure

Economos 1958, noted the increase in permeability with increasing density in $MgFe_2O_4$ but his study did not consider the grain size effect.

Separation of Grain Size and Porosity Effects
It is very difficult to identify the effects that are due to grain size and those that are due to porosity since both grain growth and densification occur simultaneously. Globus and Duplex (1966)first attempted the separation of these two effects on nickel ferrite and yttrium iron garnet(Globus 1971). In the nickel ferrite study, Globus and Duplex first made the ferrite material by "technological" preparation that was assumed to be conventional ceramic processing. The results are shown in Figure 13.10. The 1971 study used pure powder(no iron pickup in milling) which was classified according to particle size by sedimentation. Porosity was controlled by varying compaction pressures. The results are shown in Figure 13.11, which shows that he was able to obtain about the same susceptibility, (μ-1), for a large variation in porosity and for the same porosity, a large variation in susceptibility. Globus believes that the poor quality of the powder (as in the first experiment) is often blamed on porosity while the better powder in the second experiment gives a truer picture.

Igarashi (1977) used other technique to perform the separation on nickel ferrite. He used a variation of binder content to vary porosity at constant grain size and hot pressing to vary grain size at constant density. Thus, he was able to get two independent variations in each of grain size and porosity (Figure 13.12) He then plotted permeability against each parameter separately (Figures 13.13 and 13.14). He found the following relationship;

$$\mu \propto D^{1/3} \qquad [6.1]$$
where D= diameter of the grain

His theoretical analysis shows the relationships u_{app} $D^{1/2}$ but as he points out, this relationship refers to domain walls fixed at grain boundaries. Fixing the walls by included pores may account for the difference. Igarashi explains the porosity on the basis of its demagnetizing effect on permeability:

$$\mu_{app} = (1-p)(u_1-1)/1+N(u_1-1) \qquad [6.2]$$
μ_1 = permeability at p = 0
N = demagnetizing factor

μ_1 was obtained by extrapolation of the experimental values. Then N was calculated for both levels of grain size and turns out to be proportional to the porosity.

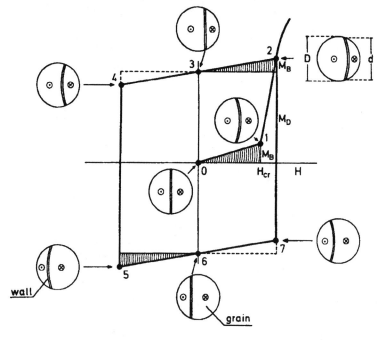

M_B Bulging magnetization
M_D Displacement magnetization

Figure 13.9 - Diagram of magnetization of a ferrite showing domain wall bulging and displacement (Globus 1972)

MICROSTRUCTURAL ASPECTS OF FERRITES

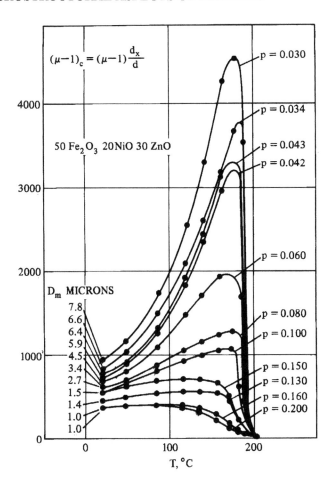

Figure 13.10-Permeability versus temperature curves of NiZn ferrites as a function of porosity and grain size using normal ceramic processing. From Globus © 1966, IEEE

The solid lines of Figure 13.15 are calculated by the formula showing the exactness of the fit of the experimental points.

Li, 1986, examined samples of Ni-Zn ferrite from 2 commercial sources and found great scatter in grain size dependency at small grain sizes and better linearity at large grain sizes. He attributes this to the Globus model in which wall motion depends linearly on grain size and domain rotation is independent of grain size. Ocurrence of wall motion at the larger grain size, might explain the observation.

Studies on the separation of porosity and grain size effects for the Mn-Zn ferrites are difficult to perform, possibly because of the greater complexity such as that caused by variable valence ions.

Effect of Grain Size and Porosity on B-H Loop Parameters

Thus far, we have been primarily concerned with initial permeability or very low level properties such as those needed in telecommunication filters and inductors.

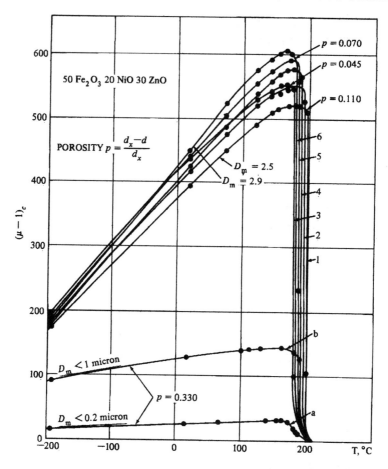

Figure 13.11- Permeability versus temperature curves of NiZn ferrites as a function of porosity and grain size using special powder processing (Globus, 1966)

applications. The coercive force is probably the property most sensitive to porosity and grain size. Smit & Wijn (1954) showed the variation of H_c with porosity in Ni Zn ferrites. The increase in coercive force with porosity was linear (as would be expected) at low porosity and deviates at higher levels. This effect may be caused by the fact that the high-porosity samples contain smaller (possibly single-domain) particles, which have higher coercive forces. The coercive force of mixed Ni-Zn ferrites of varying porosity can be correlated fairly well with Néel's theoretical

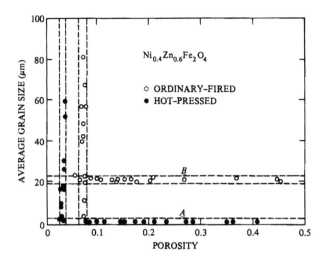

Figure 13.12- Combinations of porosities and grain sizes obtained in NiZn ferrites by special techniques (Igarishi 1977)

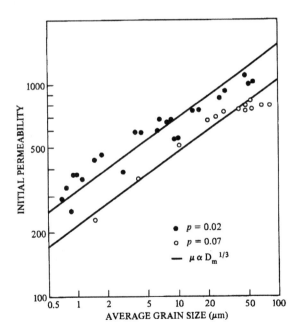

Figure 13.13- Permeability versus grain size for 2 different porosities in NiZn ferrites (Igarishi 1977)

mathematical model treating the demagnetizing influence of non-magnetic materials in cubic crystals. In this method, the domain processes are considered rotational. Economos,(1958) showed that coercive force decreases in Mg Ferrite as the porosity decreases.

With regard to saturation, we would certainly expect the saturation magnetization to increase with decreased porosity because of the increase in density or the packing of more magnetic material in a specified volume. Indeed, Smit & Wijn (1954) show a decided increase in saturation with increased density. For square loop materials, Schwabe & Campbell, (1963) have shown that the thresh-hold field (close to H_c) varies with the grain size for lithium ferrite. This is normally the case with H_c being inversely proportional to the grain diameter.

Igarashi (1977) in his study of NiZn Ferrite in which he separated porosity and grain-size effects concluded that porosity changes the B-H loop as shown in Figure 13.16. B_{max} would then be independent of grain size, but varying with porosity as (1-p) $4\pi M_o$. (M_o is the magnetization extrapolated to zero porosity). H_c is then assumed proportional to 1/r (r being the grain radius) and independent of porosity. Figures 6.17 and 6.18 show the correlation of the experimental results with theoretical curves, as well as the correlations of the postulated and experimental B_r values. B_r is assumed to be independent of grain size. B_r would then be equal to $4\pi M(1-p)$ until the demagnetizing field became equal to the coercive force and then decreased linearly. The independence of H_c from porosity and B_r from grain size are two sur-

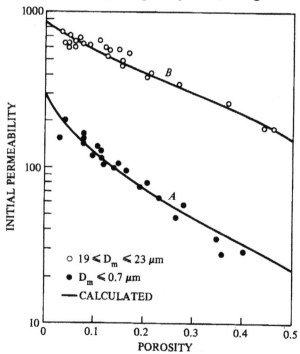

Figure 13.14- Permeability versus porosity for 2 different grain sizes in NiZn ferrites (Igarishi 1977)

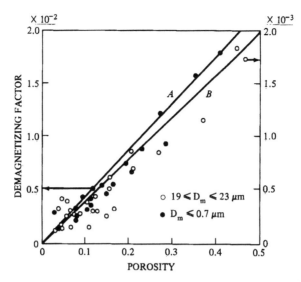

Figure 13.15- Demagnetizing factor,N, versus porosity for 2 different grain sizes in NiZn ferrites (Igarishi 1977)

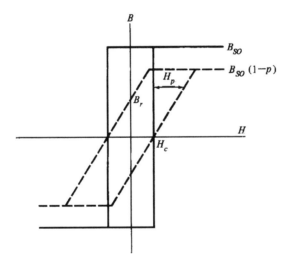

Figure 13.16- Change of hysteresis Loop of a NiZn ferrite due to an increase in porosity(Igarishi 1977)

prising results of this study. Whether this model works for other systems remains to be seen.

For commercial NiZn and MnZn ferrites, Li (1986), studying magnetic recording applications, did indeed find that saturation and remanence were independent of grain size and H_c varying in a 1/d manner. For YIG, Globus (1972) also showed that remanence was independent of grain size and coercivity inversely proportional.

Grain Boundary Considerations

Early workers in ferrite materials such as Guillaud developed the correlations of microstructure to magnetic properties with scientific equipment such as X-rays which were then considered state of the art. Most of the results obtained then are still valid today. However, with the advent of new methods of microstructural analysis such as Scanning Electron Microscopy (SEM), Auger Microscopy (AM), and Secondary Ion Mass Spectroscopy (SIMS), a wealth of new information became available. Stuijts (1977) in the concluding remarks of his paper on ceramic microstructures in 1966 stated "I have the impression that in one decade, unexpected external influences (energy crises) and instrumental developments (TEM-Auger spectroscopy, Auger spectroscopy) can have a dominant effect on actual developments". Time has confirmed his impression, especially in the investigation of grain boundary phenomena. The previous results on grain size may, in some respects, simply reflect the presence of more or less grain boundary area. Even porosity may be strongly related to grain boundaries since the latter is certainly one mechanism that can remove porosity. Grain boundaries may have even a

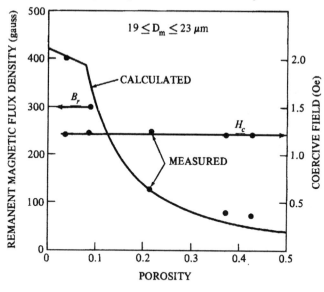

Figure 13.17 Remanence, B_r, and coercive force, H_c, as a function of porosity in NiZn ferrites and the calculated (solid) curves (Igarishi 1977)

broader impact than either consideration. The thickness and chemical composition of the grain boundary are two of the most critical factors in determining magnetic properties of ferrites.

Early Studies of Grain Boundaries

Guillaud (1957) examined eddy current losses in MnZn ferrites found the resistivity of a single crystal lower than that observed across a grain boundary. He concluded that in the bulk material the resistivity of the boundaries predominated. He had also discovered the importance of Ca^{++} as a useful additive in reducing eddy current losses. Using radioactive Ca^{++} and a technique called autoradiography, Guillaud proved conclusively that the Ca segregated at the grain boundaries. In this technique, the Ca is added to the ferrite before processing, having been tagged with a radioactive isotope. After sintering, the polished sample is placed in close contact with a photographic film for some time. The radioactive Ca will then expose the film in the places where it is located and in effect take a picture of itself. The resulting photomicrograph shows the segregation of Ca at the grain boundaries. The eddy current losses of the sample with Ca were less than one-tenth of those without Ca.

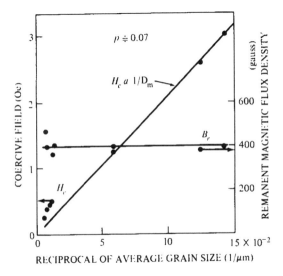

Figure 13.18 Remanence, B_r, and coercive force, H_c, as a function of grain size in NiZn ferrites and the calculated (solid) curves (Igarishi 1977)

Evidently, the grain boundary can be modified by firing and this has profound effects on the intergranular strength and magnetic properties. Heister (1959) in sintering Mn Zn ferrites found that fractures occurred at grain boundaries at 1260° C. and across grain boundaries at 1365°C. The material with pronounced grain boundaries gave the lowest losses whereas the one fired at the higher temperature gave the highest permeability.

Akashi(1961) showed that combining the CaO addition with judicious amounts of SiO_2 could increase the resistivity and lower losses. This effect was again shown to be due to an increase in grain boundary resistivity. Kono(1971) describes the oxides that can constitute the grain boundary as acidic (SiO_2, B_2O_3, P_2O_5 or and As_2O_5) or basic (Na_2O, CaO, FeO, and MnO). If an acidic oxide is at the boundary, it will extract a basic oxide from the crystal (FeO or MnO). This lowers the resistance of the boundary especially with the multiple valences of the metal ions involved. Such intergranular layers are semiconductive. If an acidic oxide is mixed with a strongly basic oxide such as CaO, the less basic oxide such as FeO will be pushed out of the liquid phase at the boundary. The result will be a much higher resistivity due to the stable valences of the ions at the boundary.

Electron and Auger Microscopy of Grain Boundaries
Stjntjes (1971), using electron microprobe analysis showed the high concentration of Ca and Si at the grain boundaries (Figure 13.19) as well as the depletion of Mn, Zn and Fe. Other combination which gave similar increases in resistivity and lowered losses were B_2O_3 + CaO, ZrO_2 + CaO, ZnO_2 + CaO + SiO_2. Another substituent used in the same paper was TiO_2 which was reported to act by asomewhat different mechanism. Additional identification of the composition of the grain boundary was made in a Ti-substituted Mn-Zn Ferrite by Franken (1978)

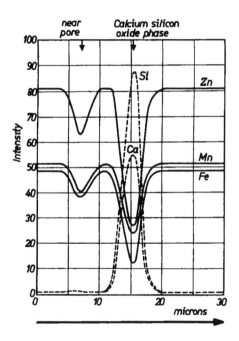

Figure 13.19 Electron microprobe intensities for several elements at the grain boundary of a MnZn ferrite (Stijntjes 1970)

using Auger electron spectroscopy. A Ti gradient was found at the grain boundaries that extended for some thickness. By sputter etching, the composition profile could be determined as a function of distance from the grain boundary. The grain boundary itself contained a CaSiTi-rich layer of about 2 um thickness with the Ti extend-

MICROSTRUCTURAL ASPECTS OF FERRITES

ing further into the grain.

Tsunekawa (1979) examined the microstructure and properties of several commercial grade Mn-Zn ferrites by high-voltage high-resolution transmission electron microscopy. These techniques followed X-ray analyses to be performed on areas with a radius on the order of 100A. The grain sizes varied from 7u to 35u. Pore distribuition varied considerably, but very surprisingly was both intragranular and intergranular in the highest permeability (18,000u). In addition, the grain size appeared to be larger in the lower perm ferrite (3590u). Tsunekawa and his coworkers attribute the superiority of the grain size to processing control and higher purity powders that eliminate glassy phases and stress/strain gradients at grain boundaries. In the lowest perm material (μ=1180) they found a glassy phase at the grain boundary along with Ca + Si segregation. The author postulated a $CaSiO_3$ glassy phase, and in addition, found that the lattice parameter was increased near the grain boundary, but not in the highest perm material. They conclude that all this indicates stresses and strains which affect magnetostriction and anisotropy and therefore μ. Ca and Si segregation was also not observed in the high perm case.

In further work using AES and TEM Franken (1980) found the enrichment of the grain boundary never exceeded 22 atomic percent Ca or 8 atomic percent Si. The influence of Ca concentration on resistivity is best explained by oxidation of the grain boundaries because of segregation during cooling. The secondary phases occur mainly at multiple grain junctions. Franken found no glassy phase, but only one crystalline & two amorphous phases. There seems to be some discrepency between the two findings, but since the raw materials used and the processing were vastly different, the differences found could be real.

Sundahl (1981) used Auger spectroscopy and X-ray stress measurements and related them to magnetic permeability and permeability-frequency spectra of Mn-Zn ferrites. He concluded that variations in cooling conditions could cause pronounced variations in magnetic properties primarily through changes in the grain boundaries brought about by the introduction of a uniform microstress throughout the core. Sundahl and his coworkers could not correlate permeability variation with segregation levels of Ca and Si. They noted zinc depletion at the grain boundaries similar to that which occurs at the surface of the core. Lattice parameter changes induced by the zinc depletion causes the microstresses. They did not explain the increase in Fe level at the boundaries.

Ghate (1981) in a second paper in the same book in which Sundahl; et al presented their findings, reviewed the boundary phenomenon in MnZn soft ferrites. For low-loss ferrite, Ghate states that grain boundaries influence properties by;

1) Creating a high resistivity intergranular layer
2) Acting as a sink for impurities which may act as a sintering aid and grain growth modifiers.
3) Providing a path for oxygen diffusion which may modify the oxidation state of cations near the boundaries.

The permeability of high-permeability ferrites is affected by the interactions of pores and boundaries, the type of the boundary phase and the microstresses.

Chiang & Kingery (1983) used STEM (scanning transmission electron microscopy) with an ultrathin window that can monitor lower energy lines. This means that the oxygen can also be detected. Ferrites were prepared with two Ca levels, one with residual Ca levels (500 ppm) and one with added $CaCo_3$ to bring the Ca level to 2180 ppm. The average grain size was about 7 microns. The authors compared sections near the surface with a section at the center of the sample. The silica segregation was equal and low in both samples, but the Ca segregation was greater in the higher Ca level sample. Surface sections showed less Ca segregation than the interior, which was caused by inward migration of Ca during the sintering cycle. Oxidized grain boundaries were seen near the surface in the 500 ppm Ca sample, but not in the high Ca sample. Chiang and Kingery conclude that the grain boundary diffusion of oxygen occurs in the former sample when the cooling atmosphere is oxidizing. Increased Ca segregation in the latter lowers the boundary diffusion of oxygen.

Yan (1986) reporting on a oral presentation by Ghate (1982) shows the core loss per cycle for four different ferrites, single crystal, hot pressed, poly- crystalline with no additive, and polycrystalline with .07% CaO. At higher frequencies, the highest loss was in the single crystal followed by the hot pressed, and the polycrystalline. The lowest was the polycrystalline with added CaO, Auger analysis shown the segregation of Ca at the grain boundaries due to its larger ionic radius (.99Å versus .64. to 8Å for the major elements).

A recent paper by Lavel 1986 using STEM and tungsten microelectrodes revealed vitreous phases at the triple grain junctions where these elements were observed in the glassy phase . When no glassy phase was present, they saw only Ca. The authors calculated the resistivity of the grain boundary to be 10^3 -10^5 Ω-cm. The bulk resistivity is usually 1 Ω-cm. The voltage drop across the electrodes is proportional to the distance between them or to the number of grain boundaries transversed.

Imperfections

In addition to grain boundaries, ceramic imperfections can impede domain wall motion and thus reduce the permeability .Among these are pores, cracks, inclusions, second phases, as well as residual strains. Imperfections also act as energy wells that pin the domain walls and require higher activation energy to detach. Néel (1949), in his study on soft steels, found that the coercive force increased linearly with the volume of non-magnetic inclusions (Figure 13.20). The same effect should be operative in ferrites. Stresses are microstructural imperfections that can result from impurities or processing problems such as too rapid a cool. They affect the domain dynamics and some workers feel they are responsible for a much greater share of the degradation of properties than we would expect.

High Frequency Materials

For high frequency materials, the eddy current effects become quite important. The skin-depth into which alternating magnetic flux can penetrate, becomes thinner as the frequency rises. Beyond this depth, eddy current loss reduces the effectiveness of the material to support the frequency increase. To minimize these losses, thin sheets, fine particles, or small grains are needed. Processing steps are taken to maintain small grains by limiting grain growth and even producing a significant porosity. It is found that in this case, eddy currents are reduced. Apparently, there is a compromise between the large- grain dense ferrites for high permeability at low frequencies and the small grain, porous ferrites for low losses at high frequencies. The role of the grain boundary has become all-important at high frequencies. If only rotational processes are involved, there is an incompatibility of high permeabilities at low frequencies and low losses at high frequencies. From Snoek's 1948 theory, the following relationship exists;

$$f(\mu-1) = 4/3\pi M_s, \qquad [6.3]$$

That means that the product of frequency and permeability reaches a limiting value. If we examine permeability versus frequency for various perm ferrites Figure 10.5, we find the higher the permeability the lower the frequency of the roll off

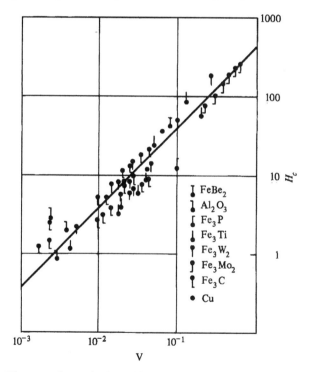

Figure 13.20- Electron microprobe intensities for several elements at the grain boundary of a MnZn ferrite (Stijntjes 1970)

which is caused by increased losses. Another way of looking at this problem is to examine the complex permeability μ' and μ'' as a function of frequency (Figure 4.3).

Note the μ'' representing the losses get larger at lower frequencies the higher the permeability of the ferrites. To overcome the limitations of Snoek's Law, a material with a preferred plane rather than a preferred direction was developed (Ferroxplana). Even though the permeability is about the same as nickel ferrite, the frequency of operation is many times higher.

It should be apparent now that microstructural requirements range from the very high perm ferrites for mainly lower frequencies to the lower permeability materials which are useful at higher frequencies.

To lower the losses at higher frequencies it is necessary to reduce the eddy current losses that become more important there. We can be achieve this condition by increasing the crystal lattice resistivity or by Ca addition and microstructural control utilizing the grain boundary resistivity in this case. Because of both skin depth and resistivity considerations, fine-grained materials with pronounced grain boundaries are needed.

De Lau 1969 found that NiZnCo ferrites could be improved by reducing grain size. The greatest effect was at frequencies just below the ferromagnetic resonance frequency. Grains below 1u in size were produced by a hot pressing technique. De Lau explained these effects in terms of a further reduction in domain wall motion due to diminished grain size.

Several authors have called attention to the need for more uniform, pore-free grains. Buthker (1982) compares the then present MnZn ferrite with a new one with Ti and Sn substitutions. In addition, Ochiai (1985) compares material from several sources of raw materials and purities and concludes that the material with the lowest losses is obtained from a high-purity material prepared using a new process involving co-spray roasting (see Chapter 14). Ochiai believes that the uniform pore-free grain structure is responsible for the improved performance.

Ishino (1987) summarized the requirements for low-loss ferrite materials such as the MnZn types at frequencies up to 1 MHz. He considered the following combination of chemical and microstructural factors to be important:

1. Suppression of electron hopping from Fe^{+2} to Fe^{+3} within the grains. Ishino prefers accomplishing this through the use of Ti^{+4} ions.
2. Insulating films surrounding the grain boundaries. This can be done by additions of Ca and Si.
3. Small, homogeneous grain size. If the grain size is large, the fraction of the grain occupied by the domain wall will increase, also increasing eddy current and hysteresis losses.
4. Reduction of pores for increased density. This factor decreases demagnetizing influences and increases flux density. Residual pores should be at grain boundaries.

Sakaki (1986) made an interesting comparison of high permeability ferrites and power ferrites for higher frequencies. After examining the behavior of the Steinmetz coefficients for both materials are and found they were different. The high perm material behaved more like Si-Fe grain -oriented material with few thin grain boundaries. The power material consisted of isolated areas surrounded by insulating material that magnetized somewhat independently. Sakaki found that domain wall motion in the power material was confined to each grain, whereas, in the high-permeability material, it was partly to motion in grains and partly to motion in aggregates. The incidence of aggregation increases the eddy current losses that are important at higher frequencies.

As in the case of the chemical aspects of ferrites, the microstructural concerns of MnZn ferrites were centered around the power materials. Ochiai(1985) has attributed the superiority of the H7C4 material to a large degree on the attainment of a fine but uniform microstructure. This sentiment was echoed by Sano(1988,a,b, 1989). In the development of the new H7F material, the grain size was reduced to less than half of that for H7C4. The fundamental studies directed to this material suggests that the average grain size be no larger than 5 microns for the higher-frequency 500 KHz operation. He reports that despite the fine grain size, the density should be as high as possible, which requires special processing.

Stijntjes (1989)shows that the presence of large numbers of inclusions can seriously decrease the performance of a power ferrite as measured by his PF_{200} factor. In Figure 13.21, the difference between a MnZn ferrite with few inclusions (B) and one with many is given. On the same figure is shown the effect of stresses

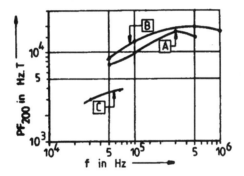

Figure 13.21 The performance factor PF_{200} of a MnZn ferrite with few inclusions (B), with many inclusions(A), and one being a highy stressed reoxidized ferrite with Fe_2O_3 segregation,(C).From Stijntjes,T.G.W., Advances in Ferrites, Volume 1-Oxford and IBH Publishing Co.,New Delhi,India, 587 (1989)

(C) caused by reoxidation, probably during the cool. Micrographs indicate the appearance of a second phase of $-Fe_2O_3$ on the surface and microcracks within the grains. Poor mechanical strength is also associated with this condition(Neyts 1989). The PF_{200} of a MnZn ferrite with Ti and Co additions with a good microstructure (E) and one with a poor microstructure (F) are compared in Figure 13.22

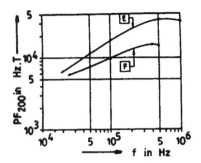

Figure 13.22- Performance factor, PF_{200}, of two MnZn ferrites, Material F has a poor microstructure and Material E has a fine and uniform grained microstructure. From Stijntjes,T.G.W., Advances in Ferrites, Volume 1-Oxford and IBH Publishing Co.,New Delhi, India, 587 (1989)

Sano (1989) attempted to increase the density of the TDK H7F material by additionally subjecting the conventionally fired cores to vacuum sintering or hot isostatic pressing, which are known methods of increasing density. In the vacuum sintering, the pressure was reduced between 900°C. and 1120°C. and the vacuum set at 133Pa. or 267 Pa. at a holding temperature of 1150°C. . The hot isostatic pressing (HIP) was done at 1100°C.for 2 hours at 98 Pa. argon pressure. The higher the vacuum, the fewer pores there were. In the HIP process, densification occurred with fewer pores. The grain size also remained about the same. In both the vacuum sintered and HIP samples, there was a segregation of Ca near the surface and diminished Ca in the core. There is also reduced Ca at the grain boundaries. This Ca migration is thought to increase internal stresses and reduce performance. At the vacuum sintering at 133 Pa., there was a 20% reduction of core losses over the H7F because the higher density and lower hysteresis losses overrode the Ca segregation effect. For the other samples, the two effects cancelled with no net improvement. Higher Ca additions and higher vacuum are suggested for further improvement.

Sakaki (1985) found that when the surface layer of a MnZn ferrite was removed, great reductions in hysteresis losses and increase in permeability were achieved. In industrial applications, he feels low loss materials can be made by ensuring that surface oxidation does not proceed too far.

Znidarsic (1992) was able to modify the microstructure of low loss MnZn ferrites using tantalum oxide (Ta_2O_5) additions. Below the solid solubility temperature, it is an excellent grain boundary inhibitor leading to high grain boundary resistivity and low eddy current losses. At a higher sintering temperature, a drastic difference occurs where the tantalum oxide dissolves in the lattice and induces an excess of cation vacancies. This is accompanied by discontinuous grain growth, intragranular porosity , low grain boundary resistivity and high eddy current losses. Znidarsic (1997) combined the tantalum additions with sintering at 1280C. in a low oxygen atmosphere to decrease the average grain size, increase the grain boundary resistivity and thus get lower power loss.

Mochizuki (1992) used A.ES (Auger emission spectroscopy) and TEM (transmission Electron Microscopy) to examine the relations between microstructure and core loss in MnZn ferrite for power applications. A suitable particle size for the operational frequency, a high sintered density and a minimum thickness of the grain boundary were found to lower losses. With regard to Ca-Si additions, attention must be paid to the melting glassy phase. The core loss was reduced 10% when using a proper heat patterns and additives affecting the melting point of the high resistivity glassy phase. Hafnium oxide was mentioned as one of the additives.

Otobe (1992) found that the Ca layer at the grain boundary became larger as the sintering temperature and heating rate increased. If the grain size is minimized, the core loss is decreased by a thin Ca layer. The grain boundary thickness varies with the amount of liquid phase formed in the grain boundary during sintering. Thus, it is necessary to determine the amount of Ca and Si to add to reduce the core loss. Other additives such as sodium carbonate and lanthanum oxide were used to modify the melting points of the grain boundary elements thus improve the magnetic properties.

Otsuki(1992) studied the nanostructural factors that affect the power loss in MnZn power ferrites. The Fe^{2+} content which decreases with oxygen potential in the sintering atmosphere affects the resistivity, eddy current loss and temperature dependence of the hysteresis loss. The eddy current loss is proportional to $1/\rho$ at the temperature of minimum core loss. At the sintering temperature, Ca dissolves in the lattice while the solubility limits for silica depend on sintering temperature and oxygen partial pressure. The segregation of additives in cooling depends on the change of their solubilities with the oxygen potential. Vanadium pentoxide (V_2O_5) was listed as one additive used.

Otobe (1997) found that phosphorus in MnZn ferrites for power applications produced larger grains and higher core losses. The P had a large solubility in the ferrite along with a low melting point (lower than silica) and generated a great deal of liquid phase even in small additions.

Lebourgeois(1997) optimized chemistry, additives and microstructure to develop a a new high frequency material having low losses in the 0.5-2 MHz. range. Ca and Si additions were combined with TiO_2 substitutions and a grain size of 5-10 microns. The parameter of interest is $(\mu_s-1) \times f_r$ where f_r is the resonant frequency. Although microstructural dependent, this term may be considered a intrinsic limit for a material. Experimental values of 5 GHz. have been reported for high-frequency MnZn ferrites. At a frequency of above 1 MHz., the resonance-relaxation loss will contribute to the core loss. Below this frequency, the main contributions to core loss are the hysteresis and eddy current losses if the material has a high resistivity. When the grain size is small, the parameter is increased . When the complex permeability spectrum of a conventionally processed material is compared to one of a coprecipitated material,certain differences appear. The resonance frequency of the conventional material is at 5 MHz. while that for the coprecipitated material occurs at 10 MHz. The grain size of the former material is between 6-7 microns while that of the latter is 1-2 microns. The permeability of the coprecipitated aterial is lower than that of the conventional material but there is a 30% reduction of core losses at 1 MHz. an 50 mT. Since the saturation and the dynamic resistivity are the same in both, Lebougeois concludes that the resonance relaxation is lower in the case of the coprecipitated material. Under 3 MHz, the permeability of the conventional material is higher, but above this frequency the reverse is true. If the sample core is bias longitudinally with the flux direction, the resonant frequency is 5 MHz. while one biased perpendicular to the flux direction has a resonance frequency of 10 MHz. At this frequency a MnZn power ferrite presents better dynamic properties when the magnetization mechanisms are mainly by spin rotation. Using some of the findings of the study a new material was developed for use up to 2 MHz.

Perriat (1992) found a magnetostrictive effect in explaining the non-linearity of permeability and grain size I low loss ferrites. Although not present in high permeability ferrites because of their low magnetostriction, magnetostrictive effects have great importance in properties of MnZn ferrites involving anisotropy. Aside from permeability, hysteresis losses and induction at a given field are also affected.

Otsuki (1992) found that, in addition to the increase in resistivity and decrease in power loss in a MnZn ferrite with silica and calcia, the addition of hafnium oxide (HfO_2) also enriched the grain boundary layers increasing the resistivity. They also noted that, with increasing frequency the temperature at which the minimum core loss, P_c, occurred shifted downward. This is attributable to the positive dependence of the core loss, P_e with temperature and the increasing importance of P_e to P_c. On the basis of these findings, they developed a new material, B40, which has one third to one half of the power loss at 1 MHz. of current materials.

Yamamoto (1997) used hydrothermal synthesis to produce MnZn ferrites with a grain size of 1.5 microns. This was achieved with a high density by the use of fine particle additives, planetary milling and low temperature sintering. The core loss was significantly lower than a commercial ferrite. Boerekamp (1997) made the comparison between the power loss behavior of MnZn ferrites having a average grain size ranging from 4-16 microns. For fine grains, the power loss at induction levels of 50 mT rose anomalously as did the Steinmetz coefficient. An explanation is given involving irreversible rotational loss in mono-domain grains.

Yamada (1992) studied the frequency dependence of electrical resistivity and power loss for MnZn ferrites. The sample without additives had a linear dependence ofg power loss to the square of the frequency. The additions of silica and CaO result in a reduction in power loss by the increase in resistivity but the increment of power loss with frequency is pronounced. This is attributed to dielectric loss and dimensional resonance loss.

Lebourgeois (1992) analyzed the low and high level losses in MnZn and NiZn ferrites. The low level losses are produced by domain wall displacements at low frequencies (below 10 MHz.) and magnetization rotation at high frequencies. For high level and high frequency (f >500 KHz.) core losses of fine-gained MnZn (<5 microns) and NiZn ferrites are mainly hysteresis and relaxation losses and cannot be separated. Eddy current losses represent a small part of the core loss considering the temperature dependence of the dielectric loss and the $\mu_s f_r$ product.

Visser (1992) made a new interpretation on the permeability of ferrite polycrystals. For small grains, a model was developed to account for the proportionality between permeability and grain size emphasizing the role of the low-permeability grain boundary. The work will be extended to the case of large-grained high permeability MnZn ferrites. For NiCuZn ferrites used for chip inductors, Nakano(1992) found that Ag used for the conductors accelerated the densification of the ferrite and promoted the dissociation of the Cu from the ferrite which causes discontinuous grain growth.

Lebourgeous (1992) developed a new MnZn low-loss power ferrite for frequencies up to 1 MHz. The raw materials impurities and reactivity must be controlled. The calcining process is done in an oxidizing atmosphere to make milling easier. He used isostatic pressing with no binder. In the firing, after reaching the proper oxidation degree, the ferrite is cooled under equilibrium conditions. Although good high frequency properties were obtained, the method, Lebourgeous concedes is hardly possible for production.

Considerations of Microstructure for Microwave Ferrites

Although the mode of applications is somewhat different for microwave ferrites than high frequency non-microwave ferrites, one requirement remains the same, but even to a greater degree. This is the need for very high resistivity. In microwave fields, the dielectric losses become extremely important. Low dielectric losses are often found in materials with high resistivity. We have spoken of the need for chemistry control to achieve this resistivity. Depending on the application, the use of grain boundary resistivity is not as applicable here as it was in the power materials. For very low line widths, a major requirement of microwave ferrites is high density or low porosity. As a result, single crystal ferrites always give the lowest linewidths. In polycrystalline materials, pores exert demagnetizing influences that seriously broaden the linewidth. The density should be upwards of 97-98% (Van Uitert 1956). Van Uitert's use of copper was one method of increasing the density of Nickel Ferrite.

Baba(1972) achieved the same object by using Bi in Li ferrite. He achieved densities greater than 99% with addition of .005 ions Bi per formula unit and firing as low as 1000°C. The low firing temperature also minimizes oxygen loss and lithium volatility. For high power microwave applications, Suhl (USPat. 2,883,629) found that in order to avoid the increased loss in microwave ferrites at very high power levels, using very fine-grained material would move the spinwave linewidth, H_k) and therefore the instability threshhold (h_{crit}) to much higher power levels. These observation has been verified by Malinofsky (1961) Green (1964) and several others in commercial and government-supported contract reports. Paladino (1966) reported the hot pressing of several Ni-Co and Ni-Mn powders to different grain sizes and found large decreases in dielectric losses for very low grain size (below 5u). Critical fields for the best samples were 192 Oe versus 24 Oe. for conventionally fired materials. There is, of course, the problem of attaining fine grains and also maximum density and hot pressing is one of the best means in accomplishing this objective. Inui(1977) has reviewed the effects of grain size in ferrites for microwave applications. He specifically refers to two extremes of useful grain size. One is the fine grained material for high power applications and the other is the very large grained, "single-crystal-like" material. The latter refers to CaVSn garnets that have rather low line-widths. As we might expect, growing these very large grains with very few pores presents some manufacturing difficulties.

For the garnet materials, Nicolas (1980) believes that the use of dopants offers better possibilities because of the cost and delicacy of hot pressing. Because of this, fine grained ferrites are not used often in practice.

Microstructural Considerations of Hard Ferrites

In the case of hard ferrites for permanent magnets, similar dichotomies arise with regard to fine grain size and high density. We have previously mentioned that the two important properties of permanent magnets are the coercive force H_c and the remanence B_r. The criterion of quality is the maximum energy product (B H) max. The remanence is a strong function of the chemistry, density and orientation, neither of which is microstructurally related. Attaining a high remanence in hard ferrites is just a matter of packing as much total magnetic moment with the right orientation in a specific volume. The coercive force, on the other hand, is microstructurally dependent as we have seen. Once magnetized, the magnetic domain structure must be resistant to demagnetization. Since one method of demagnetization (or magnetization) occurs through domain wall motion, the absence of domain walls would eliminate this mechanism. If the grains contain only one domain, this object would be accomplished and demagnetization could only occur by domain rotation. In a uniaxial anisotropic material, this process is more difficult than wall movement. Single domain size in most hard ferrites is about 1u so that is the size of grain preferred. In actual fact, attainment of single domain size in all grains is is not achieved, and some degradation due to poly-domain particles is encountered.

In the sintering of hard ferrites, the problem then is again to achieve fine grain size and high density simultaneously. The practical result is generally a compromise depending on the requirements. Another microstructural feature of the barium ferrites is that the grains usually grow in platelets with the thickness aligned in the c or preferred direction. To take advantage of shape anisotropy, we would preferably have the c axis be in the long axis of the particle. Recently, using non-conventional processing to produce media for perpendicular magnetic recording, barium ferrite particles have been produced with the c axis in the long direction.

Kools (1985) has found that SiO_2 in the right proportion when added to Sr- rich Sr ferrite, inhibited grain growth by forming second phases at the grain boundaries.Schippan and Hempel(1965) found that if fine milled barium ferrite particulates were etched in HCl, surface defects were removed, thereby reducing the grain growth during sintering, the coercive force would be increased.

In the area of hard ferrites, Besenicar (1989) studied the influence of sintering conditions and morphology of the starting strontium ferrite powder on the microstructural development. When the Sr ferrite powder had a wide grain size distribution, anomolous grain growth was promoted. A CaO addition promotes anisotropic grain growth and higher orientation during sintering

Influence of Material Properties on Thermal Conductivity

Earlier, in the section on core losses in power ferrites, we discussed the problem of removing the heat generated in the ferrite owing to the ferrite's low thermal conductivity . Hess (1985) has found that a higher conductivity ferrite requires the production of a ferrite with high density and a homogeneous microstructure along with a sufficient Ca concentration at the grain boundaries. The difference in thermal conductivity between the lowest and highest material reported was on the order of 35%. Thus, skillful ceramic engineering practice can produce an optimized ferrite from the thermal conductivity point of view.

Thomas(1989) stressed the importance of high resolution electron microscopy in characterizing ferrites. In the recording head application, grain boundary problems can be avoided by using single crystals or properly oriented polycrystals of desired composition and structure. He reports on the development of a new "Y"-type hexagonal "Ferroxplana" material, $Ba_2Cu_{.8}Zn_{1.2}Fe_{12}O_{22}$, prepared by coprecipitation from which well-oriented polycrystals can be prepared.

Resistive and Capacitive Effects in Grain Boundaries

Berger(1989) studied the relations between grain boundary structure and hysteresis losses in MnZn ferrites for power applications. At low B levels,(< .1 Tesla) there is much attention paid to decrease Eddy Current losses so grain boundary resistivity has been the dominating factor. While workers such as Thomas(1989) feel that the high resistivity is due to the $CaO-SiO_2$ glassy boundary layer, others such as Franken(1980) feel that the CaO segregation causes a depletion of the Fe^{+2} at the grain boundary and the consequent resistivity increase. For high B aplications, the Eddy Current losses can be effectively suppressed by the above mechanisms, so that the hysteresis losses become the limiting factor. The use of Ti^{+4} and Co^{+2} has been used to reduce these (Stijntjes 1984).

Materials with similar composition and grain size but cooled differently can have different power losses, so differences in local structure, specifically intergranular microstructure is considered a contributing factor. Berger made up 12 samples of a MnZn ferrite with the same composition (See Table 13.2) and grain size but cooled the under varying oxygen partial pressures at both high and low temperatures and also with different cool rates. See Table13.3. He then measured various material properties such as ac and DC resistivity, power losses, permeability and power losses versus temperature. In addition measurements of local resistivity at grain boundaries were measured. Other observations include microstructure studies by TEM (Transmission Electron Microscopy), energy- dispersive X-ray microanalysis and EELS (Energy Loss Spectroscopy). The correlation of the samples described by Tables 13.2 and cooled as shown in Table 13.3 with the various electrical with magnetic properties are given in Figure 13.23 There is no correlation of power losses with resistivity but a strong correlation with permeability. There is also no correlation between losses and bulk Fe^{+2} content. It can be concluded that the power loss differences are mainly due to hysteresis losses rather than Eddy current losses. The electron micrographs showed some grain boundaries which had coincidence of the atomic layers(lattice fringes) and some that did not show coincidence. The energy-dispersive X-Ray showed some boundaries with Ca segregation and some without it .The ones showing segregation also showed by EELS a decrease in the Fe^{+2}/Fe^{+3} ratio at the grain boundaries. The

Table 13.2
Composition of Samples used in Cool Rate and Atmosphere Studies

Component	Mole %
Fe_2O_3	51.46
MnO	36.47
ZnO	10.12
TiO_2	1.64
CoO	0.31
SiO_2	0.012 wt. %
CaO	0.023 wt. %

From: Berger, M.H., Laval, J.Y., Kools,F. and Roelofsma,J., Advances in Ferrites, Vol 1- Oxford and IBH Publishing Co.,New Delhi, India, 619 (1989)

samples that did not show Ca segregation also did not show changes in the Fe^{+2}/Fe^{+3} ratio. From the series of voltage drop measurements across the grain boundaries(indicating G.B. resistivity), two different groups of resistivities are distinguished. The histograms of these distributions are given in Figures13.24a and 13.24b. The ones in Figure 13.24a are the voltage drops for those showing high power losses (as in samples 5,6,11, and 12). Note that less than 30% of the grain boundaries had voltage drops(indicating resistivities) of greater than 20 mV. For those in Figure 13.24b, which showed low magnetic losses more than 50% of the samples showed voltage drops of greater than 20 mV. Berger feels that the study shows:
1. A drop in the Fe^{+2} content accompanies Ca segregation at boundaries.
2. Correspondence of the statistical behaviour of the grain boundaries and the bulk hysteresis losses
3. A possible explanation is that Ca segregation and GB reoxidation decrease lattice stress level and therefore lower effective anisotropy

Berger concludes that to minimize hysteresis losses, one must consider the distribution of grain boundaries as well as post sintering cooling. It is then necessary to keep the grain boundary coincidence small and maintain a high oxygen partial pressure during cooling.

In studies on the Eddy current losses in power ferrites, Sano(1988a,b) found that the conventional dependence of these losses on the reciprocal of the resistivity must be modified at higher operating frequencies. A term involving the grain size must be included to obtain good correlation across the whole frequency range up to 1 MHz. The resulting equation is;

$$P_e = k(d^2/\rho) \qquad [13.3]$$

where d^2 = the cross section area of the grain

The correlation coefficient for this equation is given in Figure 13.25 for the frequency range. The rationale behind this need for parameters in addition to the

Table 13.3
Sample Numbers of MnZn Ferrites Cooled at Two Cool Rates and Various Atmospheres at High and Low Temperatures of Cool

	p_{O2}(at HT)	High		Medium		Low	
	p_{O2}(at LT)	High	Low	High	Low	High	Low
Cool Rate	Fast	1	2	3	4	5	6
	Slow	7	8	9	10	11	12

p_{O2}(at HT) = Oxygen Partial Pressure at High Temperature (>900°C.) of Post Sintering Treatment
p_{O2}(at LT) = Oxygen Partial Pressure at Low Temperature (<900°C.) of Post Sintering Treatment
From: Berger, M.H., Laval, J.Y., Kools,F. and Roelofsma,J., Advances in Ferrites, Vol 1- Oxford and IBH Publishing Co.,New Delhi, India, 619 (1989)

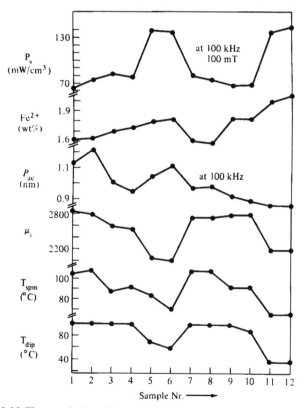

Figure 13.23-The correlation of the various electrical and magnetic properties of samples with the same chemistry and grain size(Table 13.1) but with the variation in their post-sinter (cooling) conditions as shown in Table 13.2. From Berger, M.H.Laval, J.Y., Kools,F. and Roelofsma,J., Advances in Ferrites,Vol 1- Oxford and IBH Publishing Co.,New Delhi, India, 619 (1989)

resistivity may be seen in the frequency dependence of the resistivity as shown in Figure13.26. The drop in resistivity of the four finer-grained materials as well as the approach to a constant resistivity of all of the materials is explained by the Koop's model involving the combination of resistive and capacitive effects at high frequencies. At the highest frequencies, the grain boundaries that we have assumed to be resistive take on other effects. The capacitive reactance given as;

$$X_C = 1/2\pi fC \qquad [13.4]$$

When f becomes large, the reactance becomes small, that is, it offers very little "resistance" to current flow. Therefore, the conventional resistance at the grain boundary is overshadowed by the capacitive effect. In that case, only the lattice resistivity will govern the bulk material rather than the grain boundary resistivity

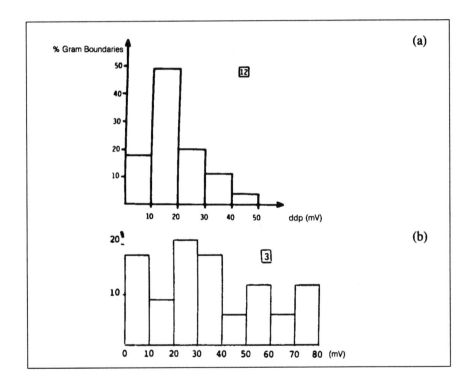

Figure 13.24- Histogram showing the statistical distribution of the percentages of grain boundaries having given voltage drops across them shown on the horizontal scale. The voltage drops are proportional to the grain boundary resistivitiesThe histogram in Figure 13.25 is typical of samples having high power losses (Figure 13.24) while those in Figure 13.13b are typical of samples with low power losses. From Berger, M.H., Laval, J.Y., Kools,F. and Roelofsma,J.,Advances in Ferrites,Vol 1- Oxford and IBH Publishing Co.,New Delhi, India, 619 (1989)

MICROSTRUCTURAL ASPECTS OF FERRITES

that we assumed at lower frequencies. Sano attributes the dependence of Eddy current losses to grain size to switching mobility that is a function of the microstructure. Stijntjes (1989) has examined this phenomenon in a simplified but less rigorous manner. The ferrite microstructure is assumed to consist of grains of low resistivity separated by grain boundaries of high resistivity. (Figure 13.27A) The electrical schematic representation of this situation is given in Figure 13.27B.

Irvine (1989) examined the electrical and magnetic behavior of a NiZn ferrite between 5-13MHz. Two resistive components were observed associated when measuring the impedance of the sample. These are associated with the bulk and grain boundary properties. Figure 8.27 is a complex impedance plot as a capacitor. The semicircular plot represents the grain boundary properties. The capacitance calculated from the plot is 13 nF-cm^{-1}. The high frequency intercept (at low Z' (Z_R)) does not pass through zero. A smaller semicircle should appear at high frequencies between the observed one and the origin corresponding to the bulk properties. The capacitance calculated for this semicircle is 5 pF-cm^{-1}.

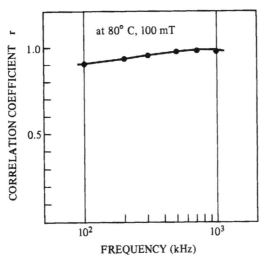

Figure 13.25- The frequency dependence of the resistivity of 5 different MnZn ferrite materials. Note that at the higher frequencies, they all become nearly equal. This is due to the decrease in capacitive reactance at the grain boundary at the high frequencies, making the lattice the resistivity-controlling factor. From Sano, Morita and Matsukawa, Proc. PCIM, July 1988.

Kim (1992) stressed the importance of the grain boundary resistivity with chemistry(Ca and Si) and processing but to operate at higher frequencies, conventional grain boundaries may become insignificant. In this case, the frequency where the grain boundary resistivity vanishes is roughly proportional to $1/2\pi RC$. Then;

$$RC = (\rho l/A)(\varepsilon A/l) = \mu\varepsilon \quad [13.5]$$

So to increase the operating frequency, the product of μ x ε must be optimized by chemistry control. Thicker grain boundaries are preferred to increase the resistance (not resistivity) if the product remains unchanged. Somewhat the same approach is reported by Tung(1997) for reducing the power loss in Mn Zn ferrites at frequencies over 1 MHz. under the constraint that the product of f x B = 25 KHz. He finds that at these frequencies, the power loss is affected not only by the hysteresis and eddy current losses but also the dielectric losses. At the middle frequency of 1.2-3 MHz., the dielectric loss dominates the core loss. He suggests that decreasing the grain boundary capacitance is the appropriate way to improve the properties of MnZn ferrites in this frequency range. By the use of high amounts of calcia and silica, sintering at 1250 C., annealing at 1100 C. and cooling to 800C. in 10 hours, the core loss in this frequency range dropped from 172.9 to 145.1(lower Ca and Si and shorter cooling time). The grain boundary resistance increased while the grain boundary capacitance dropped. At frequencies below 1 MHz., the core loss under the above listed constraint is proportional to 1/f. Above 3 MHz, the loss is independent of the measured frequencies.

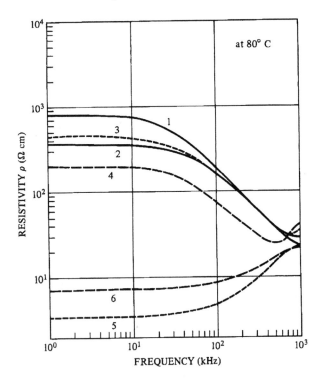

Figure 13.26 The frequency dependence of the resistivity of 5 different MnZn ferrite materials. Note that at the higher frequencies, they all become nearly equal. This is due to the decrease in capacitive reactance at the grain boundary at the high frequencies, making the lattice the resistivity-controlling factor. From Sano, Morita and Matsukawa, Proc. PCIM, July 1988.

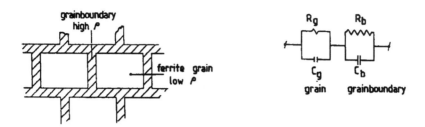

Figure 13.27-(Left)Schematic of the action of the grain boundary on the resistance of a ferrite (Right)The electrical circuit equivalence of a grain and grain boundary. At high frequencies, the action of the capacitor associated with the grain boundary becomes ineffective and a.c.-wise "shorts out".From Stijntjes, T.G.W., Advances in Ferrites, Volume 1-Oxford and IBH Publishing Co.,New Delhi,India, 587 (1989)

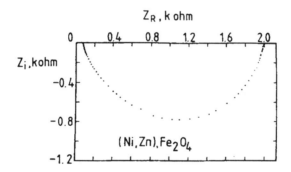

Figure 13.28- Complex impedance (as a capacitor) curve for a NiZn ferrite. The semicircular curve is characteristic of the grain boundary capacitance. The high frequency end (low Z_R or Z')does not coincide with the origin. A small semicircle should appear between this end and the origin. This semicircle would represent the capacitance due to the bulk of the material.From Irvine, et al, Advances in Ferrites, Volume 1, Oxford and IBH Publishing Co.,New Delhi, India, 221 (1989)

Phase Transformation and Oxidation

Leroux (1989) examined the phase and oxidation transformation during the sintering of a MnZn ferrite powder under different atmospheres. After preparing a ferrite green powder by standard ceramic techniques, the powder was pretreated by heating to 800°C. in O_2 or to 1100°C. in N_2 and quenched. Then TGA (Thermogravimetric Analysis) was carried out on the original and treated powders to 1000°C. in air. A low temperature peak was related to the Mn^{+3}/Mn^{+2} transition and the upper one to the Fe^{+3}/Fe^{+2} transition. The original powders were heated in and the upper one to the Fe^{+3}/Fe^{+2} transition. The original powders were heated in air, N_2, or O_2 and quenched at temperatures between 500°C. and 1000°C. In all cases, a second phase of Fe_2O_3 first increased in XRD intensity, peaked and decreased at the higher temperatures. The active oxygen behaved similarly. The magnetization decreased to a very low value and then increased sharply at higher temperatures with the sample heated in N_2, showing the increase at a much lower temperature (800°C.) than the others(1000°C. During the heating cycle, when Fe_2O_3 precipitated, no oxides of Mn or Zn were found by X-ray, leading Leroux to propose that valence changes and cation vacancies were created in the spinel phase when Fe_2O_3 precipitated. The study shows the importance of the oxidation reactions during the heatup portion of the sintering step. It also points out that from a practical consideration, the oxidation state of the calcine will be important in determining the heatup behavior in the final sintering. Kim(1989) studied the morphology of the precipitation of hematite in MnZn single crystals in air. He found that the hematite precipitated as thin platelets on the <111> habit plane. Because pores formed along of cations and cation the precipitate, the mechanism for diffusion apparently involves the counterdiffusion vacancies.

Shigematsu (1989) has confirmed the precipitation of hematite from FeZn ferrites at temperatures from 673°C. to 1373°C. with a maximum amount of precipitate at around 1073°C. The precipitate appeared on the surface first and advanced to the inner layers. Sugimoto (1989) reviewed progress in the development of amorphous ferrite materials. He finds it very difficult to produce these oxides with ferromagnetic features until the best technology is found to produce an amorphous oxide with more than 50 mole percent of Fe_2O_3 or of dispersing the magnetic clusters in a glass. On the practical side, the elimination of microcracks must be avoided and forming into the right shape possible.

SUMMARY

The last three chapters have described the crystal structure, chemistry and microstructure of ferrite materials. This study should go a long way in dictating the manner in which ferrites are processed. The next chapter will build on our knowledge and review the processing steps that have been used by conventional means and by non-conventional means. The chapter will discuss powder preparation, pressing and firing steps. In addition, there will be sections on preparation of ferrite films that has gained much attention of late. Lastly, there is a section on single crystal growth.

References

Akashi, T. (1961), Trans. J. Inst. Metals, 2, 171
Baba, P.D. (1972), Argentina, G.M., Courtney, W.E., Dionne, G.F., and Temme, D.H., IEEE Trans. Mag., 8 (1),83
Beer, A. (1966) and Schwartz, T., ibid, 2,470
Berger, M.H.(1989)Laval, J.Y., Kools,F. and Roelofsma,J., Advances in Ferrites, Vol 1- Oxford and IBH Publishing Co.,New Delhi, India, 619

Aedermannsdorf, Switzerland, 163
Brown, F. (1955) and Gravel, C.L., Phys. Rev., 97, 55
Chiang,Y. (1983) and Kingery, W.D.,Advances in Ceramics, 6, 300
De Lau, J.G.M. (1969), Proc. 1969 Intermag Conf.
Drofenik, M. (1985) Besenicar,S. and Limpel, M. Advances in Ceramics, 16, 229
Drofenik, M. (1986) Am. Ceram Soc. Bul., 65, 656
Economos, G. (1958) Ceramic Fabrication Processes, John Wiley, New York, 201
Franken, P.E.C.(1978) IEEE Trans. Mag., 14, 898
Franken, P.E.C.(1980) J. A. Cer. Soc., 63, 315
Ghate, B.B.(1981) Advances in Ceramics, 1, 477
Ghate, B.B.(1982) Sundahl, R.C.,and Nguyen,T.V.; Paper 56-Be-82F,Ceramic Bull., 61, 809
Globus, A. (1966) and Duplex,P. IEEE Trans. Mag., 2, 441
Globus, A. (1972) and Guyot, M., Phys. Stat. Solid. 52, 427
Green, J.J.(1964) Waugh, J.S. and Healy, B.S., J. Appl.Phys. Supp., 35, 1006
Guillaud, C. (1956) and Paulus, M., Comptes Rend., 242, 2525
Guillaud, C. (1957) Proc. IEE, 104, Sup.# 5, 165
Heister, W. (1959) J. Appl. Phys., 30, 22S
Hess, J.(1985) and Zenger, M.. Advances in Ceramics, 16,501
Igarashi, H. (1977) and Okazaki, K., 60, 51
Inui, T. (1977) and Ogasawara, N., IEEE trans. Mag., 13, 1729
Irvine, J.T.S,(1989), West, A.R., Huanosta, A. and Valenzuela, R, Advances in Ferrites, Volume 1, Oxford and IBH Publishing Co.,New Delhi, India, 221
Ishino, K. (1987) and Narumiya, Y, Ceramic Bull., 66, 1469
Kim, M.G.(1989) and Yoo, H.I.,Advances in Ferrites, Volume 1- Oxford and IBH Publishing Co., New Delhi, India,109
Kim, Y.S.(1992) Ferrites,Proc.ICF6, Jap. Soc. Of Powder and Powder Met. Tokyo,37
Kimura, T. (1977) Yoneda, M., and Yamaguchi, T., J. A. Cer. Soc., 60, 180
Kools,F. (1985) Advances in Ceramics, 15, 177
Kono, H. (1971) Ferrites, U. Of Tokyo Press, Tokyo, 137
Laval, J.Y. (1986) and Pinet, M.H., J. de Phys., 47 (Sup. #2), C1-329
Lebourgeois R. (1997), Ganne, J.P. and Loret, B.,Proc. ICF7, J. de Physique, 7 C-1, 105
Le Roux, D.(1989),Onno,P. and Perriat,P., Advances in Ferrites, Volume 1- Oxford and IBH Publishing Co., New Delhi, India, 95
Li,S.X. (1986) IEEE Trans. Mag., 22, 14
Malinofsky, W.W. (1961) and Babbitt, R.W., J. Appl. Phys., 32, 237S
Mochizuki, T.(1992) Proc.ICF6, Jap. Soc. Of Powder and Powder Met. Tokyo, 53
Morell, A,(1989) Eranian, A, Peron, B. and Beuzelin, P. Advances in Ferrites, Volume 1- Oxford and IBH Publishing Co.,New Delhi, India, 137
Neyts, R.C.(1989) and Dawson, W.M. Advances in Ferrites, Volume 1- Oxford and IBH Publishing Co.,New Delhi, India, 293
Nicholas, J. (1980) Ferromagnetic Materials, edited by E.P. Wohlfarth, North Holland Pub. Co. Amsterdam, 243
Ochiai, T.(1985) Presented at ICF4, Advances in Ceramics,Vol. 16, 447

Otobe, S.(1997) Hashimoto, T., Takei, and Maeda, T. Proc. ICF7, J. de Physique, $\underline{7}$ C-1,127

Otobe, S (1992) and Mochizuki, T. Proc.ICF6, Jap. Soc. Of Powder and Powder Met. Tokyo, 329

Otsuka, T.(1992) Otsuki, T.,Sato, T. and Maeda, T.,ibid, 317

Otsuki, T.(1992) ibid,659

Paladino, A.E. (1966) Waugh, J.J., and Green, J.J., J. Appl. Phys., $\underline{37}$, 3371

Perduijn D.J. (1968) and Peloschek, H.P., Proc. Br. Cer. Soc., $\underline{10}$, 263

Perriat,P.(1992) Lebourgeois, R.. and Rolland, J.L. Proc.ICF6, Jap. Soc. Of Powder and Powder Met. Tokyo, 827

Roess, E. (1966) Electronic Component Bull., $\underline{1}$, 138

Roess, E. (1971) Ferrites, U. of Tokyo Press, Tokyo, 187

Roess, E. (1985) Advances in Ceramics, $\underline{15}$, 38

Sakaki, Y. (1986) and Matsuoka, T.,IEEE Trans. Mag., $\underline{22}$,623

Sakaki, Y.(1985) and Matsuoka,T.,IEEE Translation, J. on Magnetics in Japan,Vol. TJMJ-1,36,Sept.1985, p.772

Sano,T.(1988a),Morita, A. and Matsukawa, A.,PCIM, July,1988,p.19

Sano, A.(1988b),Morita, A.,and Matsukawa, A.,Proc. HFPC, San Diego, CA.,May 1-5,1989

Sano,A.(1989), Morita, A. and Matsukawa,A., Advances in Ferrites, Volume 1- Oxford and IBH Publishing Co.,New Delhi, India, 595

Schippan, R. (1985) and Hempel, K.A., Advances in Ceramics, $\underline{16}$, 579

Schwabe, E.A. (1963) and Campbell, D.A., J. Appl. Phys., $\underline{34}$, 1251

Shigematsu, T.(1989), Kubo, T. and Nakanishi, N., Advances in Ferrites, Volume 1- Oxford and IBH Publishing Co.,New Delhi, India, 89

Smit, J.(1954)and Wijn, H.P.J., Advances in Electronics and Electron Physics, $\underline{6}$, 69

Snoek, J.L. (1948) Physica, $\underline{14}$, 207

Stijntjes, T.G.W. (1971), Broese van Groenou, A., Pearson R.F., Knowles, J.E., and Rankin, P., Ferrites, U.of Tokyo Press, Tokyo, 194

Stijntjes, T.G.W.(1985), Presented at ICF4, Advances in Ceramics, Vol. $\underline{16}$,493

Stijntjes,T.G.W.(1989), Advances in Ferrites, Volume 1-Oxford and IBH Publishing Co.,New Delhi, India, 587

Stuijts, A.L.(1977) Ceramic Microstructures, Proc. 6th Int. Mat. Symp., R.M. Fulrath and J.A. Pask,Editors, Westview Press, Boulder Colorado,

Sugimoto, M.(1989) Advances in Ferrites, Volume 1-Oxford and IBH Publishing Co. ,New Delhi, India, 3

Suhl, H. (1956) Proc. IRE, $\underline{44}$, 1270

Sundahl, R.C. Jr.(1981) Ghate, B.B., Holmes, R.J., and Pass, C.E., Advances in Ceramics, $\underline{1}$, 502

Thomas,G. (1989) Advances in Ferrites, Volume 1-Oxford and IBH Publishing Co. ,New Delhi, India, 197

Tsunekawa, H.(1985),Nakata,A.,Kamijo,T.,Okutani,K.,Mishra R.K., and Thomas, G., IEEE Trans. Mag., $\underline{15}$,1855

Tung, M.J.(1997) Tseng, T.Y.,Tsay, M.J. and Chang, W.C. Proc. ICF7, J. de Physique, $\underline{7}$ C-1, 129

Van Uitert, L.G.(1956) Proc. IRE $\underline{24}$, 1294

Yan, M.F. (1978) and Johnson, D.W., J.A. Cer. Soc., $\underline{61}$, 342

Yan, M.F. (1986) J. de Phys., 42, Suppl. #2, C1-269
Yoneda, N. (1980) Ito, S. and Katoh, I, Ceramic Bull., 59, 549
Znidarsic, A.(1992) Limpel, M.,Drazik, G. and Drofenik, M. Proc.ICF6, Jap. Soc. Of Powder and Powder Met. Tokyo,333

14 FERRITE PROCESSING

INTRODUCTION

The crystallographic and chemical influences on ferrite properties were presented in Chapters 11 and 12 and the ceramic or microstructural aspects in Chapter 13. The problem now is to process or prepare the ferrites, keeping in mind the chemical and physical requirements previously described.

A goal common to all the ferrites is the formation of the spinel structure. The starting materials are conventionally oxides or precursors of oxides of the cations. This process involves the interdiffusion of the various metal ions of a preselected composition to form a mixed crystal. Nonconventional powder processing in a liquid medium may produce intermediate, finely divided mixed hydroxides or mixed organic salts to assist the subsequent diffusion process.

The formation of the ferrite could be made at $100°$ C. or lower since precipitation and digestion methods have produced fine ferrite powders at these temperatures. A classic example of this technique involves the preparation of magnetite or ferrous ferrite by direct coprecipitation and heating of the aqueous suspension of the mixed hyroxides $\{Fe(OH)_2 + Fe(OH)_3\}$. Except in the case of recording media, copier powders or ferrofluids, ferrite powders are not generally the finished products. In most cases, the powders must be consolidated into a body with a microstructure appropriate for the application. The two requirements can be met by carefully controlling two processing steps:

1) Powder preparation
2) Sintering

These steps are quite closely intertwined. The characteristics of the powder will strongly affect the quality of the product after sintering. Any remaining inadequacies in the powder can be corrected by extended times or higher temperatures in the sintering step, but usually at the cost of deterioration of other properties. The optimum combination, therefore, is a coordinated process in which powder making and sintering enhance each other.

POWDER PREPARATION-RAW MATERIALS SELECTION

The process of selecting raw materials will be governed by how critical the properties desired are, the type of equipment and processing used and economic considerations. As pointed out earlier, if high permeability, high quality materials are needed, the purity should be very high (Less than .1 percent total metal impurities.) Ferrites for consumer applications require highly competitive costs but since their specifications are much less stringent than telecommunication or microwave applications, less pure materials may be used. One exception to this general rule involves a company making a lower grade ferrite material out of a higher grade of iron oxide because the powder production process could be simplified (the calcining step could be skipped,) so that the overall powder cost is lower. The particle size requirements of the raw materials depend on the process equipment used. For example, if we use ball milling to blend the original mix, a step which coincidently reduces the particle size, we can use a lower cost, coarser raw material. Dry blending without milling will require a finer particle size. In general, a finely divided raw material commands a premium price. Although the raw materials vendor may specify the average particle size, a more meaningful criterion is the specific surface area (SSA), which is usually given in terms of square meters per gram (m^2/g). By assuming a particle geometry and the known material density, we can calculate the average particle size.

The reactivity of the raw materials is an operational parameter that is well understood but difficult to measure quantitatively. Chol (1968) has defined the reactivity of iron oxide in MnZn ferrites by the lowest temperature that the first spinel phase (zinc ferrite) forms. Of course, the lower this temperature, the higher the reactivity. The finer the iron oxide, the higher the reactivity. In Chol's study with two oxides of the same surface area, the one with the spherically-shaped particles had higher reactivity than the one with needle-shaped particles. This difference is thought to be caused by the increased packing efficiency of the former powder. Iron oxide accounts for the highest percentage of the raw materials for spinel ferrites, and therefore, has the highest potential for affecting the properties. The two most important quality criteria are purity and the specific surface area. The main sources of iron oxide for ferrites in order of increasing cost are;

1. Natural or refined iron ore (hematite,etc.)
2. Spray roasted iron oxide from HCl regeneration of steel-mill pickle liquor.
3. Synthetic iron oxide from decomposition of copperas(ferrous sulfate hydrate)
4. Carbonyl iron oxide derived from the oxidation of carbonyl iron

Kohno (1992) reported how the some of the characteristics of spray-roasted iron oxide can be controlled by the operation of the HCl regeneration roaster. Such properties as particle size, oxide diameter (Sub Sieve Sizer) ,pressed compact density and chloride content can improved by selection of conditions in the roaster. Another way of recovering iron oxide from hydrochloric and sulfuric acid waste pickle liquor is by a crystallization method is described by Yamazaki(1992). Impurities such as N, P and Cr are removed by the process. For the sulfuric process, ferrous sulfate is produced which is a raw material for magnetic recording media.For low-cost hard ferrite production, Narita(1992) proposes the use of refined iron ore containing less than 0.2% silica and 0.15% alumina. The magnetic properties are said to be more stable against variation in calcining temperature and variations in the raw materials than spray roasted iron oxide

Shrotri (1992) prepared nickel ferrite with four different iron oxides. Comparative evaluation showed the ferrite made with a γ-Fe_2O_3 had better electrical properties. Khurana (1992) reported the use of a non-spray roasted iron oxide(natural)purified only by physical means. The silica was high at .2% and the alumina was .1% and the surface area was low. With additions of CaO and vanadium pentoxide, power ferrites with moderately acceptable properties were made. Zhang (1997) was able to use iron oxide from open-hearth furnace dust to make Ba and Sr hard ferrites. Maximum energy products of up to 0.91 MGO were obtained.

The per pound cost of these oxides runs from a few cents to several dollars. With regard to purity, the heavy metals that can enter the lattice and cause strains are highly undesirable. The alkali metals such as Na and K may lower resistivity and hurt high frequency properties. A final consideration of purity concerns the presence of oxides that alter the sintering action by fluxing or producing liquid phase- sintering. These oxides include SiO_2 and V_2O_5. The properties and approximate costs of some iron oxides are given in Table 14.1.

Table 14.1
Properties and Costs of Commercial Iron Oxides

Oxide Grade	% Fe_2O_3	% SiO_2	% Cl^-	S.S.A.*	Cost/lb.
Premium, Carbonyl, Calcined $Fe_2(SO_4)_3$	99.5	.007	----	5 m^2/gm	$1-5
Purified Spray-Roasted	99.3	.007	.15	4 m^2/gm	$0.60
Medium Spray-Roasted	99.2	.015	.2	3.5 m^2/gm	$0.25
Low Grade Spray Roast, Refined Ore	95-99	>.1	.2	2 m^2/gm	$0.05-- $0.10

* S.S.A. = Specific Surface Area

In the case of iron oxide, a large enough quantity should be purchased from one lot so that once the peculiarities of that lot are determined and the properties optimized, a large amount of consistent material can be made without additional testing. Some companies will actually make a ferrite from a sample bag before purchasing a large quantity for production.

With regard to raw materials for ferrites, Ruthner(1989) examined the availability of spray-roasted iron oxides. He finds that, up to the year 2000, there will be sufficient supply to meet the increased demand. He feels that investments in plants not linked to steel mills must be considered on an individual basis. Chiba

(1989) disclosed a new method for production of highly pure iron oxide using a solvent extraction method. Jha (1992) reported on the use of natural ferric oxide for the manufacture of soft ferrite components.

For MnZn ferrites the most costly raw material is usually the manganese component. As is true for the iron oxide, many different grades are available. However, although the iron oxide varies mostly in its source, in the instance of manganese, the difference is mostly in the manganese compound itself. Analogous to the iron oxide case, Chol(1968) defines the reactivity of the manganese raw material according to their rates of solution in zinc ferrite at 1000 K. Table 7.2 lists the available materials The choices range from low- cost Mn ore to the reagent

Table 7.2
Sources and Costs of Manganese for Ferrites

Source	Percent Mn	Cost/lb.	Features
MnO_2	62	$0.60	Coarse Particles
Mn_2O_3	69.6	$1.00	High Surface Area
Mn_3O_4	70.5	$1.50	High Surface Area
Mn	99.8	$0.80	Pyrophoric if fine
$MnCO_3$	44	$3.30	Reagent Grade
MnO- Reduced Ore	62	$0.25	High Impurity Content

grade $MnCO_3$. The reagent grade has the advantage of producing newly formed or nascent surfaces on decomposition of the carbonate at a relatively low temperature. Being more costly than the other forms, it is used in rather critical applications where cost is not too important (such as in recording head ferrites). Mn metal is a moderately low cost source but must be ground fine that unfortunately leads to explosion hazards because of its pyrophoricity. CaO is usually added as $CaCO_3$.

Calculation of Weights and Raw Materials

From the theoretical point of view, the composition of a ferrite is calculated in terms of mole percent. Thus for a ferrite whose formula would be $Mn_{.5}Zn_{.5}Fe_2O_4$ or $0.5MnO.0.5ZnO.Fe_2O_3$ the corresponding mole percent would be

MnO	25 mole %
ZnO	25 mole %
Fe_2O_3	50 mole %

These mole percentages are then converted to weight percentage by traditional chemical calculations. Two cautions must be observed in the calculations. First, if the source of the particular oxide is not in the form shown above, then a conversion factor must be applied and the weight of that component must be increased or decreased accordingly. Second, when the raw material is not 100% definable ($MnCO_3$ which may not be exactly stoichiometeric), then the assay given as the metal must be used in the conversion factor. Others may be expressed as percent oxide (i.e. 99.5% Fe_2O_3). In addition to moisture, other impurities may be present to account the assay. Here again the assay is used to calculate the weight of each component.

One deviation from the mole percent usage occurs in the case of minor additions such as $CaCO_3$, SiO_2, SnO_2 and TiO_2, which are generally given as weight percent. Another correction frequently inserted in the calculation involves the compensation for the iron picked up during the grinding operation. This correction can be determined by experience and involves reducing the amount of Fe_2O_3 in the raw materials by the anticipated pick up which is on the order of 0.5 mole % Fe_2O_3.

Major element chemistry may be determined on the ferrite powder or the fired ferrite by Wavelength-dispersive X-Ray Fluorescence Analysis. The powder or the ground up fired ferrite part is mixed with a standard weight percentage of an organic binder and pressed into a disk that is used as the sample for the X-Ray. Minor elements such as Ca, Si, Ti and Sn can be analyzed as their solutions by atomic absorption spectroscopy or by ICP(Inductively-coupled plasma or DC plasma spectroscopy. A study by Reynolds (1981) shows extensive deviation of the major chemical analysis of the same MnZn ferrite powders in a round robin among some major ferrite producers. In some cases the variation was as much as one percent. There was much better agreement in the results of multiple analyses within one laboratory. Thus, in the interest of ferrite product reproducibility, analytic standards should be used based on the chemistry of the best electrical performance even though the absolute chemistry may not be exact. The reasons for the varied results between labs are related to differences in sample preparation, analytical procedures and instrumentation.

Weighing and Blending

For exacting applications, the accuracy of the weighing should be ±0.1percent.. In production practice, weighing hoppers are used so that there is no material lost in the transfer. Load-cell scales are frequently used. Digital readouts and hard copy recording of the weights are advisable.

Blending may be done by several different means, wet or dry and with or without coincident grinding. Traditionally, ball milling for ferrites is done wet are as dry milling may produce troublesome packing of the powder on the walls and balls. Continuous ball mills and rod mills have also been used. A variation of the ball mill is the attritor in which the axis of the mill is vertical and whose balls that are smaller than those used in a ball mill (1/4" diameter) are driven by vanes rotated from the central axis. Milling times are shorter as the balls can be driven faster than the tumbling experienced in a ball mill. Continuous attritors are also available. Chol (1968) in his study of optimization of high-quality soft ferrite manufacture has concluded that vibratory mill produces more homogeneous mixtures than those obtained with a ball mill. In the same paper, he also finds that Mn and Fe oxide components become more homogeneous with milling time for up to 12 hours of milling. However, the Zn component reached the maximum homogeneity after about 4 hours of milling. At that point, the homogeneity decreased with further milling presumably because of some agglomeration of the ZnO. In a similar study, Auradon (1969) found that the permeability of the ferrite produced increased with milling time but that the uQ product maximized at four hours of milling time.

Many dry blenders are available and most of them are based on high shear mixing. Dry blending, though a simpler process, does not produce the intimacy of a ball mill. In addition, it permits much air to be incorporated into the powder, rendering it fluffy and thus more difficult to calcine.

Between the dry and the wet processing exists an intermediate semi-wet or moist processing. Mullers that mix powders with about 10 percent moisture illustrate the action of two heavy rollers compressing and blending the powders and also producing some milling. Scrapers are used to clean the walls. Another semi-wet process involves the use of the drum blender which may be combined with a pelletizer producing spherical pellets of varying size (1/4" - 1/2").

If no intermediate calcining step is used in the process, ball milling, or attritoring is the recommended blending process. The binders and other additives can be incorporated in the ball mill and the slurry spray dried or granulated. If intermediate calcining is used, the material produced by wet blending must be dried before calcining. For ball milled materials, the slurry is dewatered by filter pressing and the cake is dried and broken up. Another option is spray drying but this requires the addition of a binder. In mulled powders, the wet mix may be dried in a continuous belt dryer. The material produced by a pelletizer may be dried and calcined or calcined directly. In the conventional powder processing area, Ries (1989) reported on the increased use of new grinding and pelletizing equipment. Pelletizing is a compromise between the wet powder mixing and grinding such as wet ball milling and the strictly dry processing. Ries(1992) describes the use of the pelletizer to mix, grind and pelletize ferrite powders. Microspheres can also be made as a substitute for spray drying. Automated equipment using a pelletizer combines several operations. Durr (1997) describes the production of ferrite granules with a vacuum hot steam process using an Eirich vacuum-tight pelletizer. Superheated steam is the drying medium under slight vacuum. Zaspalis (1997) developed a technique for quantitatively measuring the homogeneity of the green mixture of a MnZn power ferrite involving an SEM using Energy Dispersive X-Ray Analysis(EDAX). This evaluation is correlated to the calcined and sintered material homogeneity and ultimately to the power loss. As expected, the ferrite with the greatest green homogeneity had the lowest core loss.

Calcining

Calcining involves heating the blended material to an intermediate high temperature (900 - 1100° C.for ferrites; 1200° C. for garnets). In general, the calcining temperature will be about 100 to 300° below the final firing temperature. The purpose of the calcining (if used) is to start the process of forming the ferrite lattice. This process is essentially one of interdiffusing the substituent oxides into a chemi-

FERRITE PROCESSING 311

cally and crystallographically uniform structure. The driving force for the interdiffusion is the concentration gradient. As the individual oxides interdiffuse, some ferrite is created at the interface. This completed phase reduces further diffusion since the concentration gradient is no longer there to act as a driving force. The material in the center of each of the oxide particles experiences difficulty diffusing through the ferrite since the diffusion distances become larger. If the material is then broken up exposing the inside of the particles, the driving force for diffusion is again established. Since some shrinkage occurs in calcining, one advantage of calcining is to reduce the shrinkage in the final sintering. This allows better control of the final dimension in cases where this control is necessary. In addition, calcining helps homogenize the material, which obviously is advantageous. In some cases of premium materials such as microwave or recording head ferrites, double calcining is used with intermediate millings.

The calcining can be continuous or by batch. Batch calcining is not economically attractive, however, so that in practice continuous calcining is most often used. In some cases, the material to be calcined is placed in boats or containers and sent through a continuous belt or pusher kiln. Commercially, most calcining is done in rotary calciners which can be either gas or electric fired (See Figure 14.1). The material is fed into a rotating tube inclined at a pre-designed angle for proper dwell time and economical through put. The proper operation considers rotation speed, angle of incline, heat input, temperature, and depth of fill.

Figure 14.1- Rotary calciner for ferrites, Courtesy of Magnetics, Division of Spang and Co.,Butler, Pa.

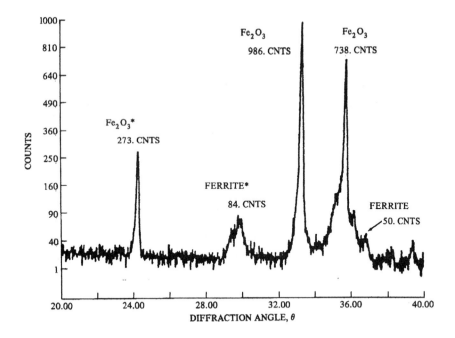

Figure 14.2- X-Ray diffraction pattern of a MnZn ferrite that is undercalcined

During the process of the calcining, the powder coarsens considerably and the color changes from red to gray or black. In air, the manganese changes from its original form to Mn_2O_3. There are also varying degrees of ferrite formation depending on the feed material and the temperature and residence time. In a batch calciner, the controls are simple. In a rotary calciner, it is possible to control residence time by the rotational speed and the tilt angle of the rotating tube.

The degree of calcining action may be studied by the X-Ray diffraction determination of the residual Fe_2O_3 content as suggested by Chol (1968). In addition, the degree of spinel formation can be determined simultaneously. For reproducibility of calcining, this method provides a good control. Figure 14.2 shows the diffraction spectrum of a MnZn ferrite that is undercalcined while Figure 14.3 is one of an overcalcined powder. Control of the degree of calcining will produce the proper shrinkage in the final sintering step and will also leave sufficient reactivity or driving force to permit good densification in the final firing. Ruthner (1992, 1997) has described a vertical calciner with a very short residence time. MnZn, NiZn and strontium ferrites were calcined in this equipment. A pilot plant has been set up in Switzerland. Economic and quality advantages are claimed. Kang, D.S.(1992) has studied the variation of microstructure and magnetic properties in a microwave ferrite with the calcination temperature. This temperature determines the amount of spinel formed and crystallite size thus affecting the driving force of sintering. The differences in linewidth was due to porosity and the coercive force depended on the density.

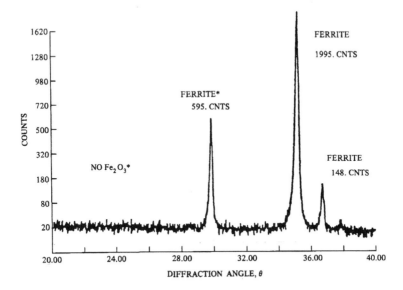

Figure 14.3- X-Ray diffraction pattern of a MnZn ferrite that is overcalcined

Milling

After calcining, the material which has been coarsened must be broken up by ball mills or attitors. The amount of milling will determine the particle size distribution, which, in turn, will influence the homogeneity of the compact going into the final firing as well as the microstructure after the sintering process. The optimum particle size is generally on the order of 1 micron or smaller. The percent solids in the milling operation is critical for proper grinding and subsequent processing. If the solids content is too high, as the grinding proceeds, the particle size will decrease and the slurry will become quite viscous, reducing grinding and slurry-handling facility. If the solid content is too low, however, the ensuing spray drying operation will produce a very fine spray-dried powder that is not desirable. To keep the solid content high and prevent viscosity problem, deflocculants such as gum arabic are added. Operationally, the viscosity of the slurry is often measured with paddle-type viscometers to monitor the combined effects of the degree of grinding, solids content and binder percent

Granulation or Spray Drying

After milling, the slurry must be converted to pressable powder. We can emply several additives to facilitate this step.. First, we add a binder which is used to give strength to the pressed compact. Many such binders are available from various gums, cellulose derivatives and poly vinyl alcohol. The polyvinyl alcohol is commonly used, but unfortunately it produces rather hard spheres in spray drying. Consequently, we may use plasticizers such as polyethylene glycol to soften the particles. Sometimes, lubricants such as zinc stearate are added. Harvey & Vogel (1980)discussed the use of a combination of two different binders (polyvinyl alcohol and polyamine sulfone) which have improved burnoff characteristics. Vogel (1979) has also studied the use of different dispersants or deflocculants as compared to gum arabic. Deflocculants are added before the ball milling but binders and plasticizers are added a short time before the end of the milling.

Granulation allows the water from the slurry to evaporate while avoiding separation of the binder by constant agitation. The dry cake must then be broken up into the proper aggregate size. Generally, the powder is screened to assure this condition. Whereas granulation produces rather dense particles, the flow characteristics of the powder are not as good those of spray drying. In addition, spray drying is a more efficient operation.(Figure 14.4, Ferrite Spray Dryer)

Several different varieties of spray dryers are available. Some have double-fluid nozzles, others have single-fluid nozzles, and still others are the centrifugal types. The first two are generally made in the shape of a tall tower with the slurry sprayed upward with (double) or without (single) the aid of a compressed air jet. The last type consists of a squatty cylinder in which the slurry is fed into a whirling centrifugal wheel (in the center of the top section) flinging the particles outward. The atomized slurry is always dried by a blast of preheated air producing a cyclonic air flow. Since the drops of slurry are dried in mid-air, the resultant particles are essentially spheres that make the resulting powder free flowing and capable of filling pressing dies very reproducibly. Unfortunately the spheres produced are generally somewhat hollow. Mano(1992) examined the PVA characteristics and behavior during the preparation of the slurry and the spray drying. By varying the slurry preparation, the volume of PVA deposited between the particles will change affecting the pellet strength and compaction ability. The PVA distribution is a more important factor than particle strength and pressure conductivity. The effect of chemical additives on characteristics of ferrite slurries

Pressing

Most ferrite powders are die-pressed. Presses for die-pressing of ferrites can be of the hydraulic or mechanical variety. For simple shapes such as toroids or E cores, presses with single-level lower punches may be used. For complicated shapes such as pot cores, presses with secondary lower punches be employed (Figure 14.5). Depending on the pressed density needed and the characteristics of the powder, pressures from several thousand psi to as high as 20,000 psi may be used. Die-pressing produces compacts which may have density gradients caused by the friction of the powder along the wall. To reduce this problem, external lubricants such as zinc-stearate may be used. In designing the dimensions of the die, consideration must be given to the shrinkage characteristics of the powder to be pressed including the binder content, particle size, pressed-density, and degree of calcination. The conditions of the final fire are alsoimportant. All of these factors help determine the final dimensions of the part. Obviously, in order to produce parts with low tolerances on dimensions, shrinkage considerations are critical. The previously

Figure 14.4- Ferrite spray dryer for ferrite production. Courtesy of Ferroxcube, Division of Amperex Corp, Saugerties, New York

Figure 14.5- Cutaway cross-section of a complex die for pressing muli-level ferrite parts such as pot cores. Photo courtesy of NV Plilips Elcoma-BG Materials, Einhoven, Netherlands

for spray drying was examined by Alam(1992). The optimum system was one with a solid loading of 27 volume percent and containing tri-ammonium citrate (0.085%), PVA (1.75Wt.%), and PEG(Polyethylene glycol) (1.0 Wt. %)with a slurry density of 1.85 g/cc a flow time of 16 sec. and a sediment density of 1.81 g/cc.

In some special cases, the ferrite powder can be isostatically-pressed, either cold or hot. Isostatic pressing produces compacts that are more uniform than die-pressed parts. This is due to the uniform pressure on all surfaces of the compact. Cold isostatic presses used the transmission of pressure through a liquid medium against a rubber or plastic bag containing the powder. Of course, there is little control of dimensions so the final part must be machined before or after firing. Hot isostatic pressing (HIPing)is generally used for producing dense parts for critical application such as recording heads or microwave ferrites. An additional advantage of hot isostatic pressing is the attainment of fine grained dense parts. Instead of using a metal can as a container for HIPing metal powders, the alternative procedure for ferrites is to conventionally press and sinter the body in such a way as to produce a sealed or "case-hardened" surface on the ferrite. The ferrite itself thus provides the case and the compact is then isostatically hot-pressed. Oudemans (1968) at the Philips Laboratories in Eindhoven has developed a scheme for continuous hot pressing of ferrites.

Broussard(1989) studied the influence of oxidation state on the green strength in MnZn ferrites. He found that in the burn out of the PVA binder, the surface of the powder was modified with respect to oxidation state and surface area in a manner that may not be reversible. Parameters that resulted in high mechanical strength included those which lead to the important oxidation-reduction ractions including a) High surface area for increased reactivity b) low initial degree of oxidation so that oxidation rate is enhanced. A slow heating rate to 500°C. is recommended.

Morell(1989) used PEG (Polyethylene glycol) in place of water to plasticize the PVA(polyvinyl alcohol) binder used in the preparation of ferrite slurries. The amount of previously used water was hard to control.With the proper grade of PEG, the same flexion strength produced with the water was obtained.Rambaldini(1989) examined the compressibility and mechanical strength of the PEG- plasticized PVA mentioned above. He found that pressing improves with increased PEG addition. However, the "active" binder(PVA) decreases with the addition of PEG so green strength is reduced when PEG is added in larger ratios. Mano(1989) used mercury porosimetry in a green ferrite body to measure the pore structure(including total pore volume, pore diameter, and trapped mercury volume.

Alam (1992) studied the conditions affecting the pressing characteristics of soft ferrites. The properties of the spray-dried powder granules, complexity of shape, pressing pressure influence the open and closed porosity, green density, pressure transmission, shrinkage and ultimately the quality of the sintered ferrite.

Roess (1989) reported that for the LPL(Low Power Loss) ferrites, there is a special requirement for low prices even for the quality materials. He cites the need for a great deal of automation. However, because of the nature of ferrites, the use of a straight CIM (Computer Integrated Manufacturing) is not practical for ferrites. Such a system does not allow for the versatility of the many different sizes, shapes and materials in a single ferrite plant. Instead, Ross advocates the use of CHIM

(Computer-Human-Integrated-Manufacturing) Under these circumstances, the automation is combined with some batch operations and a system of process control with adjustments and corrections as needed. He cites the problems of making 3000 different types of cores in 20 different materials. Automation advances cited by Roess include the use of rotary presses, pressing 6-8 times as many parts per hour as conventional presses at the same cost per press. Thus, in the pressing of 145 EE-55 cores per minute, 4 tons of ferrite are used per hour.

NONCONVENTIONAL PROCESSING

Today, the large majority of ferrite powders are made by the process described earlier in this chapter, the conventional ceramic process. Most non-conventional processes involve producing the powder by a wet method Among these processes are:

1. Coprecipitation
2. Organic Precursors
3. Co-Spray Roasting
4. Freeze drying
5. Activated Sintering
6. Fused Salt Synthesis
7. Sol-Gel Synthesis
8. Hydrothermal Synthesis
9. Mechanical alloying
10. Plasma Spraying

Coprecipitation

Numerous examples of coprecipitation are reported in the literature. Few cases of actual commercial usage are known. The precipitates can be in the form of hydroxides, oxalates, or carbonates all of which, of course, can be thermally decomposed to the corresponding oxides. The precipitation can be accomplished chemically or electrolytically. The main advantages of coprecipitation are:

1. Greater homogeneity
2. Greater reactivity
3. High purity - no grinding
4. Fine particle size
5. Elimination of calcining

Akashi (1971) reported achieving extremely good magnetic properties in commercial materials using coprecipitated hydroxide powders. Goldman (1977) obtained materials with loss factors on the order of $1. \times 10^{-6}$ from coprecipitated carbonate-hydroxide powders. Yu (1985) reported forming mondisperse (individual unagglomerated particles) coprecipitated spheres by precise control of the precipitation and aging processes (Figure 14.6). Economos (1959)

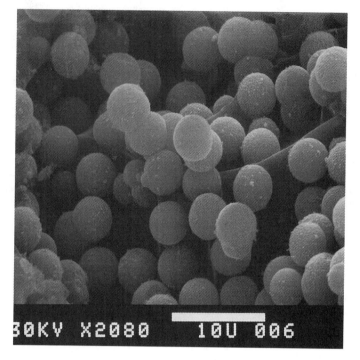

Figure 14.6 - Electron micrograph of a coprecipitated MnZn ferrite powder.

coprecipitated nickel-Iron hydroxides with tetramethylammonium hydroxide and decomposed and reacted them to form nickel ferrite. With hydroxide precipitation, ammonia is the precipitant of choice since it leaves no inorganic cation residue as is the case with some soluble metal hydroxides such as NaOH or KOH. However, for Ni, Co, and Zn, ammonia is not recommended due to complex formation which inhibits precipitation. However, with regard to the substituted ammonias (amines), the organic analogs of ammonia, the presence of methyl groups sterically hinders complexing and thus the hydroxide precipitation is accomplished. Goldman (1977) reported the use of diethylamine for Ni and Ni-Zn ferrites. Knese (1992) prepared NiZn ferrite using ammonia at a constant pH of 7.5-8.5 and characterized it by Moessbauer Spectroscopy. No chemical analysis is given but the author believes that complexing of the Ni and Zn occurred.

Aluminum-doped γ-Fe_2O_3 was prepared by Filho (1992) by coprecipitation of the sulfates. It was studied by X-Ray diffraction, magnetization and Moessbauer Spectroscopy. The ferroxplana material, Zn_2Y, was prepared by Kim (1992) by calcining coprecipitated δ-FeOOH, $Fe(OH)_2$ and $BaCO_3$ at 1200°. Above 1250° C., the ferroxplana decomposes.

Coprecipitation can also be accomplished by electrolytically forming the oxides or hydroxides. Beer (1958) used a continuous scheme of electrolytic coprecipitation. Grenier (1997) proposed a new way of preparing highly oxidized orthoferrites by an electrochemical reaction in air in alkaline solution.

Wickham (1954) coprecipitated the oxalates of iron combined with Co, Ni, or

Zn by the addition of ammonium oxalate to the metal sulfates. He then decomposed the resulting mixed oxides at approximately 500°C. The oxides formed were then fired to ferrites at higher temperatures. Bo (1981) reported MnZn ferrite preparation by an oxidation process. At one time, large commercial production of ferrites was made using the oxalate process. The disadvantage is the high cost of the oxalate which cannot be recovered. Goldman (1975) using $(NH_4)_2CO_3$ as the precipitant reported a scheme of recovering the NH_3 and CO_2. Another disadvantage of coprecipitation is the costly processing involving large volumes of water. Lithium ferrite is difficult to coprecipitate as there are few insoluble Li salts. Micheli (1970) reported using Li stearates combined with the hydroxides of the other elements. In the area of non-conventional processing, Kim(1989) used coprecipitation to make the ferroxplana material Zn_2Y. Coprecpitation was also used by Date(1989) to prepare ultrafine particles of strontium ferrite.

Fine particles by Coprecipitation
Since the advent of nanocrystalline metallic strip materials, there has been a complementary research effort to develop corresponding nanocrystalline ferrite materials. Two considerations propel the effort. First, the material research has shown that nanocrystalline metallic materials behave quite different than normal-sized grained material. Second, the need for power materials for increasingly higher frequencies would be aided considerably by finer-grained ferrites. Coprecipitation is certainly one method of achieving fine-grained ferrite particles.

Gomi(1992) prepared fine particles of Bi and Ce substituted yttrium iron garnet by coprecipitating using sodium or ammonium hydroxides and the nitrates of the metals dissolved in water and ethanol(50 volume ratio for YIG and Bi:YiG and only water for Ce:YIG). A pH of 10 was used to assure complete precipitation. After drying the first two precipitates were annealed for crystallization for from 3 minutes to 23 hours at 600-1200° C. The Ce:YIG was annealed for 20 minutes in a hydrogen-nitrogen atmosphere. They had average particle sizes as low as 40 nm. Fine particle nickel ferrite prepared by coprecipitation by Michalk (1992)showed anomalous non-collinear or canted spins. The canting depended on the annealing conditions rather than the fine particles. High magnetization CoZnFe ferrites were synthesized by Sato(1992a)by coprecipitation having particle sizes of 12 nm. A new sensitive detection method was used to detect these particles. The detection limit of this device was 1.6×10^{-8} emu. Ultrafine particles of zinc ferrites were reported by Sato (1992b) using coprecipitation. A large magnetization of 73 emu/g was found for 8 nm. zinc ferrite at 4.2K. Similarly ultrafine particles of cadmium ferrite were produced by Yokoyama (1992) by coprecipitation with NaOH. The solution was boiled at 100° C. filtered and dried at 60° C. The saturation magnetization was much higher than for bulk cadmium ferrite. From the particle size dependence of magnetization, a magnetic inactive layer on the surface is postulated. Thin platelets of hematite were prepared by Iwauchi (1992) byprecipitation of ferrous sulfate with sodium hydroxide, and oxidation of the ferrous hydroxide to magnetite in air at 70° C. Samples with thicknesses of 5-100 nm. were produced. Needlelike particles of α-Fe_2O_3 without micropores were attained by Kiyama (1992) by autoclaving in NaOH solution. Polyhedra-shaped magnetite for magnetic toner applications was made by

Koma (1992) by precipitating ferrous hydroxide from ferrous chloride and sodium hydroxide in nitrogen. A fixed amount of air was introduced and the ferrous-ferric ratio monitored by chemical analysis. Preferable results as a toner were obtained. Bagul (1992) prepared active strontium ferrite powders by coprecipitation. The nitrate solutions of the metals were precipitated with sodium hydroxide and sodium carbonate. Submicron sized particles were formed. The effect of residual sodium content on process parameters was studied.

Perriat (1997) studied the oxidation-reduction reactions in finely divided coprecipitated spinels. Each cation oxidizes in a specific range of temperature when there are more than one oxidizable cation present

Organic Precursors

Many laboratory preparations have also reported in the literature involving organic precursors. Commercial application is hampered by both the fire hazard and the high cost of the process. Two rather common organic complexing compounds used in organic precursors to ferrites are the acetylacetonates and the 8-Hydroxy quinolines. Hirano (1985) produced cobalt ferrite by the hydrolysis of the metal acetylacetonates. Suwa (1981) also prepared Zn ferrite from similar precursors.

Earlier Wickham (1963)using double acetates, recrystallized them from pyridine as pyridinates and then decomposed them.

Busev (1980) used the 8-Hydroxquinoline derivatives to produce Co ferrite and achieved good compositional control and controlled particle size.

Hiratsuka (1992) synthesized fine acicular particles of MnZn ferrite from acicular α-FeOOH and the acetylacetonates of Mn and Zn. The mixture was sintered in a nitrogen atmosphere between 880 and 1080° C. for 10 hours and then quenched in air. The mixture as coated on a polyester field under a field of 7 kOe. Such a technique may be useful in preparing magnetic recording media. In a later paper, Hiratsuka (1997)extended this work to include crystal oriented particles when the annealing temperature was extended to 1250° C. for 4 hours in air and cooling in a nitrogen atmosphere. Kodama (1992)formed high-vacancy-content magnetites by reduction with dextrose of strongly alkaline solutions of ferric tartrate. In addition, high-vacancy-content zinc ferrites were made in the same manner without the dextrose reduction. The particle size of these ferrites was less than 20 nm. The preparation of nickel ferrite from their formate precursor was made by Randhawa (1997). The magnetization was high at 4440 Gausses showing potential to function at high frequencies. Tachiwaki (1992) used hydrolysis of iron and yttrium isopropoxides to form an amorphous material which was converted to yttrium orthoferrite by calcination.

Fine particle MgMn ferrite with a surface area of 50 m^2/g was made via the decomposition of the organic precursor hydrozinium metal hydrazine carboxylate (Manoharin 1989). Suresh (1989) prepared high density MnZn ferrites by blending the individual $MnFe_2O_4$ and $ZnFe_2O_4$ powders which were prepared from the decomposition of the relevant organic precursor hydrozinium metal hydrazine carboxylates as given above. Sintered at 1000°C. for 24 hours in nitrogen gave 98% of theoretical density as well as the required saturation magnetization.

Another organic precursor method was poposed (Bassi 1989)for making alkaline earth ferrites. He used ferric malonate and alkaline earth(Ca, Mg, Sr, and Ba) malonate solutions, concentrated them and coprecipitated using acetone. The resulting precipitate was decomposed to the respective ferrites at lower temperatures than those used to decompose the oxalate complexes,

Co-Spray Roasting

In the co-spray roasting process, the metals are added as dissolved salts (usually chlorides) in an aqueous medium. The solution is sprayed into a large heated reaction vessel where the metal salt is hydrolyzed and in the case of iron, then oxidized. The acid (HCl) is recovered and the mixed oxide is accumulated at the bottom of the roaster.

Ruthner (1970) and Akaski (1973) have described co-spray roasting. Co-spray roasting refers to the simultaneous spraying of more than one component (i.e.,Fe2O3, MnO). Ochiai (1985) reported the use of co-spray roasting in commercial ferrite materials and cited the following advantages:

1. Increased homogeneity
2. Elimination of Calcining
3. Good magnetic properties
4. Economical process

The most important recent development in ferrite powder processing is in the introduction of co-spray roasting as a commercial method. Ochiai (1985) reported at ICF4 that the TDK H7C4 power material was superior partially due to the use of spray-roasted powder. The same reason was advanced by Sano (1989) for the new TDK H7F material for use at higher frequencies. The TDK process is described as using Fe_2O_3 and Mn_2O_3 from the spray roasting process, mixing them with ZnO and milling to an average particle size of .85 microns. The powder was pressed to a density of 3.0 g/cm^3 and sintered under controlled firing conditions. Goldman (1989) spoke to the economics of co-spray roasting process and the need for large capital investments for purification schemes and large spray roasters. On the basis of the size requirements for efficiency, he predicted that for the bulk of the quality power ferrites, a few companies with large powder preparation plants would dominate the future power ferrite market. The possibility of special companies just producing powder for the various ferrite producers seems remote because of the need for many different compositions.A process similar to co-spray roasting called spray firing was reported by Wagner (1980).

Wenckus & Leavitt(1957) described a variation of spray roasting called flame spraying was by but no commercial application was ever made. Here the metal salts are dissolved in an alcohol solution and sprayed and reacted in an alcohol-oxygen flame. Very fine ferrite particles were produced in this manner.

Freeze Drying

The use of freeze drying has been reported by Bell Lab workers(Schnettler 1970). The metal ions are dissolved in an aqueous medium and sprayed into a very cold organic liquid, producing fine, frozen spheres. The solvent is then extracted and the resulting frozen spheres then freeze dried to leave mixed crystal precursors. These spheres were subsequently decomposed to form ferrites..

Activated Sintering

Wagner (1980) has reported a variation of spray roasting in which a slurry of the reactant ferrite materials are sprayed into a reaction vessel so that calcining takes place rapidly. No commercial application of this process is reported.

Fused Salt Synthesis

In this method, the oxides of salts of the component metals of the ferrite are dissolved in a low melting inorganic salt that aids in the reaction of the oxides in the molten state. After the reaction is complete, the soluble salts are dissolved in water leaving the residual ferrite powder. In this process, the fine milling of the powders is avoided, but the ferrite powder is very fine after washing indicating the low solubility of the Fe_2O_3 or other oxides in the molten salt. Wickham (1971) produced Li ferrite by using $LiSO_4$-Na_2SO_4 as the salt mixture with Li_2CO_3 and Fe_2O_3 as the ferrite raw materials. Using the same process, he also produced the ferrites of MgZnCo and Ni . Kimura (1981) also reported forming the same ferrites from a Li_2SO_4-Na_2SO_4 fused salt. The particles were very fine, being in the micron- to submicron-sized range. The particles may be so fine that achieving reasonably fired densities may be difficult.

Sol-Gel Synthesis

This is a very new technique in which small colloidal particles are first formed in solution usually by hydrolysis of organic compounds. They then link to form a gel or are formed into ceramic particles. Iron-excess manganese zinc ferrites with B_2O_3 additions were synthesized by Sale (1997) using the citrate gel-processing route. The boric oxide addition promoted grain growth and altered the magnetic properties. Chien (1992) used the same method to study the effect of V_2O_5 additions to iron-deficient MgMnZn ferrites. Small additions inhibited grain growth but higher amounts (> 0.25%) promoted it. Ochiai (1992) prepared ultrafine MnZn and hard ferrite powders by the amorphous citrate process. The nitrate solutions of the metals were mixed with citric acid to bind the constituent ions in solution. Ethylene glycol was also added as a dispersant. The solution was heated in an oil bath. The water and nitric oxides evaporate at 90° C. After two hours the reaction is complete and gelation begins. Further heating of the gel at 120° C. results in complete evaporation of the ethylene glycol. The precursor was burned out at 500° C. for 4 hours. The burnt-out material was calcined, wet milled pressed in a die and sintered. For MnZn ferrite, spinel formation was complete at 1100° C. Magnetic properties were similar to those produced conventionally.

Sol-Gel techniques were used to make microwave absorbers from waste ferrite raw materials(Jha 1989). Saimanthip and Amarakoon (1987) made use of a novel sol-gel reaction to achieve a uniform distribution of CaO and SiO_2 in MnZn ferrite powders. The ferrite powder was initially prepared chemically. It was the suspended in a solution of tetra-orthosilicate dissolved in ethyl alcohol. This was then partially hydrolyzed with hydrochloric acid. Aqueous calcium acetate was then added and the solution stirred. The CaO-SiO_2 film coated each particle. The excess liquid was drained and the powder dried. The fired ferrite (with a composition of $Mn_{.52}Zn_{.44}Fe_2O_4$) coated with the Ca-Si material show much better microstructure than that prepared from the uncoated powder. The electron micrograph in Figure 14.6a (lower picture-uncoated) shows many intragranular pores which are absent in Figure 14.6a (upper picture-coated). Because uniform distribution of the Ca-Si is so important in MnZn ferrites, this method may prove useful.

Hydrothermal Synthesis

Hydrothermal synthesis involves the aqueous reaction of constituents under high temperature and pressure in a sealed reaction vessel. Hasegawa (1992a) used hydrothermal synthesis to control the size of hematite and magnetite particles. Conditions affecting the nucleation mechanism are deemed responsible. Hasegawa (1992b) also reported using the method to control the size of MnZn ferrites. The size of the partcles processed at 160-300° C. were larger than those made by other wet processes. A narrow distribution of average particle sizes between 0.3-8 microns was obtained. Lucke (1997) used the hydrothermal process route to make Mn Zn ferrites. Greater homogeneity and higher sintered density produce 20% higher initial permeability than conventionally prepared material.

Mechanical Alloying

In metallurgical circles, the use of mechanical alloying and mechanochemistry has existed for some time. The method has been extended to ceramics over 25 years ago on simple oxides such as magnesia or titania. In ferrites, zinc ferrite has been made from ferric oxide and zinc oxide. The properties of mechanical activation by prolonged milling can be different from those made conventionally. Kaczmarek (1997) reviewed work on simple and complex iron oxides produced by this process. Potential applications in soft and hard ferrites are reviewed.

Completion of Non-Conventional Processing

The non-conventional process always produces powder that can be converted to the oxides. Often, calcining is not necessary except where the powder is so fine that it may be difficult to press. Non-conventional powder can then be treated similarly to conventionally produced powder.

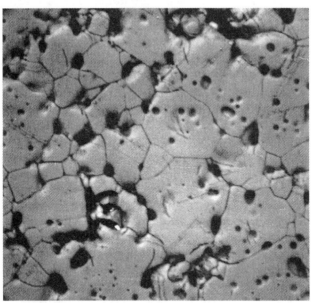

Figure 14.6a- Electron micrographs of two MnZn ferrites. Material for upper figure was made from a powder that was treated to produce a $CaO\text{-}SiO_2$ coating by a sol-gel process while the ferrite powder for material in lower figure was uncoated. The latter shows many intragranular pores which are absent in the sample made from coated powder. From Saimanthip and Amarakoon

Plasma Spraying

Varshney (1992) used plasma spraying with a Metco plasma flame spray system to prepare a NiZnCo ferrite from commercially produced powder. The method is suggested for making net shapes of complex geometries. It eliminates cumbersome machining and material waste. It also eliminates stresses and associated core losses. The results can be improved by hot isostatic pressing (HIP'ing)

POWDER PREPARATION OF MICROWAVE FERRITES

The use of microwave ferrites represents only a tiny fraction of the amount of the total ferrite consumption. However, because of the critical nature of the application and its connection with military space and aircraft radar, its importance is disproportionate to its small usage. Elements that produce exaggerated grain growth (i.e. silica) are excluded and particle size is controlled to favor small grain size. For conventional firing, allowing a very fine particle size to go into the final fire may be undesirable as very large grains may result. Coprecipitation has been used to get the purity and as a feed material for hot pressing which is another technique sometimes used in microwave ferrite materials.

Wolfe & Rodrigue (1958) produced yttrium iron garnet by precipitating the hydroxides from the mixed nitrates and pre-sintering them. The fired samples had densities up to 98%. Goldman (1975) produced Ni and NiZn ferrites for microwave applications using diethylamine and the mixed sulfates.

HARD FERRITE POWDER PREPARATION

The powder for hard ferrites is made by conventional ceramic processing as described for spinels. The problems of stoichiometry are not as critical however, as in the case of the spinels. In the processing, it is important to obtain the proper particle size so that the grain size after firing will be on the order of 1 micron. If too small a grain size is obtained, the particles may be superparamagnetic. If the grain size is too large, the existence of domain walls will decrease the coercive force. By proper choice of milling times and sintering schedules we can obtain the optimum properties. Even here, we have a trade off of coercive force and remanence in which differing processing techniques will favor one over the other.

Another variation of processing permanent magnets is in the preparation of oriented or anisotropic magnets. Here the singe-domain particles are aligned in a magnetic field during the pressing operation. The procedure can be done either wet or dry. If it is done wet, the milled slurry is poured into the die, and the field is turned on to orient the particles in line with the axis of the die. When pressure is applied, the water is squeezed out of the slurry through a porous plug in the die. The time to press these magnets is much longer than that for the conventional non-oriented type but the properties are vastly superior. A hydraulic press used to form the oriented ferrite magnets is shown in Figure 14.7.

Additions of SiO_2 to improve density and remanence may be used, but the amount must be low enough to prevent growth of large grains. This addition of SiO2 is especially useful in strontium ferrite. Other additives used include PbO (Pb ferrite has the same crystal structure) and CaO.

On the non-conventional side, barium ferrite powder has been produced by a fast reaction sintering method similar to that described for spinel ferrites.

Takada & Kiyama (1970) also produced Ba ferrite by coprecipitating the nitrates of Fe-Ba with NaOH and then subjecting the product to a hydrothermal treatment between 150-300°C.

SINTERING SPINEL FERRITES

The firing of ferrites is closely tied in with the powder properties and thus to the powder processing methods. There are also definite variations in the sintering cycle, depending on the final properties desired as dictated by the application. In discussing the sintering process, we will first review the individual ferrite group requirements and discuss the firing cycle that best accomplishes the purpose. We will then review the types of kilns available to accomplish this.

The purposes of the sintering process are;
1. To complete the interdiffusion of the component metal ions into the desired crystal lattice.
2. To establish the appropriate valencies for the multi-valent ions by proper oxygen control.
3. To develop the microstructure most appropriate for the application.

Sintering Nickel Ferrite

Nickel Ferrites may be the least complicated ferrite from the processing point of view. The nickel remains essentially in the Ni^{++} state so that air is almost always used as the firing atmosphere. Kedesy and Katz (1953) using X-ray diffraction found that nickel ferrite began to form at a temperature of about 700° C. or about 100° higher than that for zinc ferrite. In the case of nickel ferrite, however, its continued formation was much slower. At 1100° C., the nickel ferrite was the major phase but the crystal structure and microstructure are poorly defined as evidenced by its magnetic measurements. At 1250° the reaction was complete. The formation of nickel ferrite was accelerated by a lower oxygen pressure that destabilizes an oxygen-rich nickel oxide and makes it more reactive. O'Bryan (1969), using fine coprecipitated powders, studied the microstructure control in nickel ferrous ferrite with an Fe/Ni ratio of 4:1. Both densification and grain growth increased with increasing temperature but the effect of oxygen partial pressure depended on the Fe^{2+} content. However, grain growth occurred much more rapidly in lower oxygen atmospheres. At 12 hours, the grain size in nitrogen was twice as large as that in air (See Figure 14.8). Ni ferrite is generally used at very high or microwave frequencies. Obtaining a dense material with fine grain size requires either long-time low temperature fires with additives or pressure-assisted short-time fires (hot pressing or hot isostatic pressing).

Sintering Nickel-Zinc Ferrites

The presence of zinc complicates the sintering process because high temperature coupled with low oxygen firing will cause zinc loss. High density is important for high permeability, but so is zinc conservation. Tasaki (1971) described two alternative firings to achieve high density.

Figure 14.7- Hydraulic press for forming anisotropic hard ferrite, Courtesy of Dorst-Maschinen und Anlagenbau, Kochel Am See Federal Republic of Germany

1. low sintering temperature excluding O_2 (vacuum, argon, nitrogen)
2. high sintering temperature in pure oxygen to reduce zinc loss.

Accordingly, other properties correlate along with density.

1. Lattice constant is greater for O_2, smaller for vacuum
2. Curie temperature is greater for vacuum, smaller for O_2
3. Resistivity is greater for O_2, smaller for vacuum
4. Weight loss is greatest for O_2, smaller for vacuum

Morell (1980) has described a method that may be used for rapid sintering of NiZn ferrites in air at temperatures from 1300-1460° for times from 10-95 minutes. She showed that sintering cycle times can be reduced by factors on the order of 100. Optimum conditions are 1460° in air for 12 minutes.

Stuijts (1971) has observed that maximum density is found for compositions in which the divalent ion is in excess (less than 50 mole % Fe_2O_3). His study avoided additives were since ions that increase sintering velocity also increase grain size. Preparation of powder that was not agglomerated prevented bridging and voids in the compact. Using these techniques, ferrites were sintered with porosities approaching .1 percent.

Inazuka (1992) fabricated NiZnCu ferrites with low sintering shrinkage. The ferrite powders were mixed with a small amount of Pb-based low softening point glass powder as a sintering additive. Sintering shrinkages of less than 1.5% were obtained.

Fetisov (1997) investigated the oxidation kinetics of non-stoichiometric NiZn ferrites. The oxidation curves can be explained by taking into account two types of defect formations.

Rao (1997) has reviewed the sintering of nickel-zinc ferrites for high frequency switching power supply applications. Changes in density, grain size and microstructure of these ferrites were correlated to zinc loss at elevated temperatures. Densification and grain growth were found to be Arrhenius controlled rate processes.

Shu (1992) investigated the effect of atmosphere in the sintering of a NiZn ferrites with different stoichoimetry. The controlling species on densification were oxygen vacancies rather than cation vacancies. The initial permeability was affected by the anisotropy constant, sintered density and second phase. The resistivity is reduced by decreasing iron excess or sintering in high oxygen both leading to lower Fe^{2+} levels. The low temperature sintering of NiZnCu ferrites was investigated by Nakamura (1997) by usual ceramic technique. The sintered density and permeabil
ity were strongly affected by the size of the starting oxide powder and the presintering temperature. The best conditions were fine oxide powders and an 800° C. calcine for the high permeability ferrite. The sintered density was 4.5 g/cc and the permeability at 10 MHz was 200 achieved at a relatively low sintering temperature of 900 ° C. This condition is suitable for multilayer chip inductors. Satoh (1992) analyzed the interaction between dielectric and ferrite in a monolithic SMD-LC filter. A NiZnCu ferrite was used. To reduce the Cu-peak at the interface, the use of glass frit in the ferrite to obtain a dense microstructure and a higher green density of the ferrite were suggested.

Sintering Manganese-Zinc Ferrites

Since manganese-zinc ferrites represent the predominant type of ferrite used commercially, we will discuss the sintering process in greater detail. In common with the NiZn ferrites, Zn loss is a problem especially when high density is required. Other serious problems encountered are (1) stabilization of the Fe^{2+} content consistent with the excess iron oxide present and (2) the variable valence of the Mn.

FERRITE PROCESSING

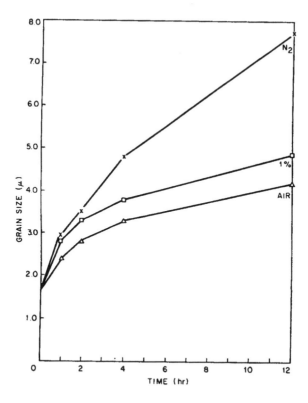

Figure 14.8- Variation of grain size versus time in nickel ferrous ferrites for 3 different atmospheres (O'Bryan 1969)

Much has been written on the subject of firing manganese-zinc ferrites. In addition to the problems already mentioned, further considerations on the type of fire used are:

1) Use of additives for sintering aids or grain boundary modifiers (Ca,Si)
2) Inherent impurities in the raw materials
3) Size of the sample since degree of both O_2 equilibrium and Zn loss depend on the diffusion path or the smallest dimension.

In the next section, we will discuss some of the sintering variables that must be controlled and their influence on the attainment of the chemical, crystallographic and microstructural properties required for specific applications (outlined in Chapter 6).

Roess(1989) in discussing the increased efficiency needed for power ferrites, cites increasing the load on pusher plates and increase in pusher velocity. This has increased the firing capacity of a 30 meter long pusher kiln by 400 tons per year to

1000 Tons per year.

Sano(1988, a,b,1989) reports that, for the attainment of the good microstructure and high density needed in the TDK H7F material, a very careful sintering process must be used. This includes a low temperature fire below 1200°C. and control of the heating rate and ambient atmosphere between 900°C and 1100°C in the heat-up stage. This method is effective in obtaining a finer microstructure and controlling the grain and pore growth as well as the oxygen discharge by ferrite formation. In Sano (1988,a,b), several firing conditions were used in a ferrite containing 54 mole % Fe_2O_3, 37% MnO, and 9% ZnO in connection with different additives. The atmosphere was controlled during the cool to maintain phase equilibrium. Thus;

Application	Additives	Firing T	O_2
Low loss	CaO, SiO_2, Ta_2O_5	1200°C.	2%
High B	CaO, SiO_2, V_2O_5	1270°C.	4%
High μ	CaO, Bi_2O_3	1340°C.	6%

For low loss materials, firing at or above 1270°C. gave abnormal grain growth while a 1200° fire gave a uniform 4.6 micron grain size. It showed the lowest core loss, highest resistivity and medium hysteresis losses. The largest grain size gave the highest hysteresis losses but this was due to porosity. For the high B materials, no abnormal grain growth was found at any of the temperatures. The grain size increased with temperature. B_s and density increased with temperature. The core loss was slightly higher than the low loss material and the resistivity was lower.

Rikukawa(1987) showed a method of computer simulation for materials development called MAGSYS. A key use of this system was to model the sintering conditions. Figure 14. 8a gives the comparison for the experimental results and the simulated results for the effect on the loss factor of the temperature A and the atmosphere parameter in the equation;

$$\log P_{O_2} = a - b/T \qquad [7.1]$$
where a= atmosphere parameter
b= constant = 14,540
P_{O_2}= Oxygen partial pressure

Nagata (1992) used a plasma activated sintering method (PAS)to a make fine grained MnZn ferrite for high frequency use. After PAS, the ferrite was annealed at 970° C. for 6 hours. With decreasing grain size, the high frequency core loss was reduced even though the lower frequency properties were degraded.

Hold or Soak Temperature and Duration

These related variables may range from 1100 to 1450°C for the hold temperature and usually from one hour to about eight hours in duration of the hold. A notable exception is the fast fire reported by Morell (1980) who fired MnZn ferrites in cycle

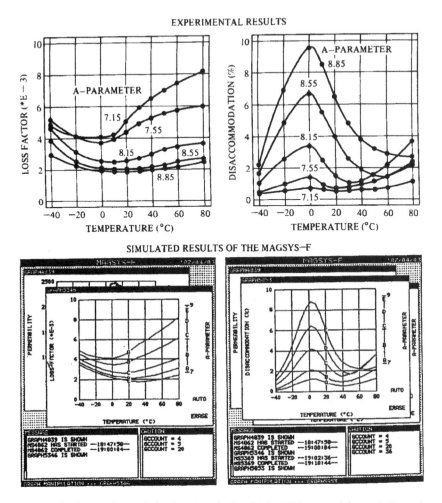

Figure 14.8a-Comparison of simulated analysis of MAGSYS materials modeling scheme with experimental results. From Rikukawa and Murakawa.1987 © IEEE

times from 16-33 minutes on a moving Pt-Rh(platinum-rhodium)belt. The optimum hold temperature was 1300°C. It would then seem that long firing times are not necessary for small cores (14.8 mm. pot cores). Obviously, capacity, core size, equipment and economic considerations limit the practical utilization of this approach. There has however, been an overall drive to decrease firing time (usually accompanied by higher temperature over the past 13 years) coinciding curiously with the energy crisis of 1974. Many manufacturers found that cycle times could be cut in half by judicious choice of raw materials, additives, processing techniques and firing parameters.

The optimum temperature and time conditions should be the minimum needed to achieve the required homogeneity and microstructure. Excessive temperature or time past this point degrades the ferrite by atmospheric contamination, exaggerated grain growth and increased porosity.

Chol 1969 has shown that the soak temperature and time are the main independently variable parameters since the eqluibrium atmosphere is generally prescribed by the hold temperature. The rate of disappearance of porosity is proportional to the remaining porosity and hence a compomise between densities and economic factors determine the optimum firing conditions.

Hirota (1987), studying a Na-doped MnZn ferrite has shown the several features of the microstructure, particularly the 1) grain size, 2)presence of an Fe_2O_3 precipitate and 3) presence of a duplex structure varied with the final sintering temperature. In addition, the grain size and density of the fired ferrite was dependent on the rate of temperature increase. At high heating rates, grain growth may be inhibited by the "impurity drag" effect. High densities are obtained at low and medium Na contents (<0.1%). The lower Na additions do not greatly affect saturation magnetization and permeability, whereas disaccomodation and resistivity are improved. At high Na contents of about 1%, the density is alway low even with the fine grain size due to the fast heat up.

Atmosphere Effects

For MnZn ferrites, unlike the Ni & NiZn ferrites, the atmosphere control may be the most crucial variable in the sintering process. The work of Blank (1961), Macklen (1965), Slick (1970), and Morineau (1975) have stressed the importance of maintaining the equilibrium oxygen atmosphere above Mn-Zn ferrite to obtain the appropriate Fe^{++}/Fe^{+++} ratio as outlined in Chapter 5. This objective is now taken as a matter of course in ferrite production. Because oxygen diffusion rates in the ferrite are still rapid enough during the first stages of the cool, it is still possible at that stage to control the oxidative state of the ferrite through the atmosphere. As the ferrite is further cooled, the possibilities of equilibration are reduced with the decreased diffusion rates.

Manufacturers have developed elaborate schemes of automatically controlling the atmosphere during the initial cooling period. This period extends from the firing temperature to about 900-1000°C. where the diffusion rates are slow. In order to produce high-quality ferrites it is most critical that the proper temperature-atmosphere equilibrium be determined and maintained for the specific MnZn ferrite produced. A typical log P_{O2} vs $1/T(K)$ is shown in Figure 14.9 with the boundaries in which the Mn Zn ferrite phase is stable. Blank (1961) proposed a universal equilibrium diagram for all ferrites with certain portions of it being appropriate for specific ferrites.

Macklen (1965) showed that for a specific soak temperature, the density increased as the oxygen partial pressure in the atmosphere decreased. The effect of atmosphere on grain size is more complex and will be treated later under specific applications.

FERRITE PROCESSING

A complicating factor in the atmosphere during the sintering of Mn Zn ferrites is the accompanying los of Zn especially at high temperatures and low oxygen partial pressures. Special precautions must be taken to deal with this problem.

Sintering of MnZn Ferrites for Specific Applications

In the case of MnZn ferrites, three major categories will be treated separately. The groups are;

1) Low loss linear materials for telecommunications
2) High permeability materials for sensitive flux sensors or wide-band transformers
3) High frequency power materials

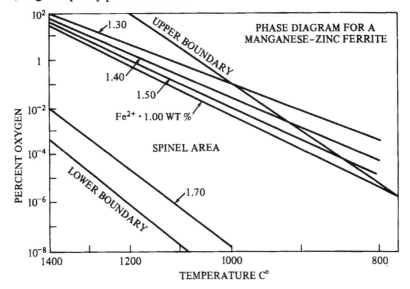

Figure 14.9- Equilibrium oxygen partial pressure, P_{O2}, as a function of temperature for MnZn ferrites (Blank 1961)

The first and third group may have subgroups depending on whether the application frequencies are low or high.

The following problems are common to all these groups to a greater and lesser extent:

2. Prevention of Zn loss
   1. Maintenance of the proper ferrous iron content.
   3. Attainment of appropriate microstructure including avoidance of discontinuous grain growth.
   4. Attainment of the desired magnetic properties.

We usually solve the final and overriding problem through compromise since the process for attaining the best in one property often conflicts with the optimum in another.

Perriat (1992) developed a thermodynamic and mechanical model of MnZn ferrite sintering which permits him to calculate the stresses in the cores during sintering. With it, he can predict the influence of oxidation degree and specific area of the calcined powder. Considering the chemical reactions may help explain crack formation.

Rikukawa (1997) used computer-aided design (CAD) to make an analysis of the oxygen concentration in a tunnel kiln. Tani (1992) made a model of the gas flow in a ferrite sintering kiln. There was a downward flow at the gap between refractories and an upward flow near the side wall. Okazaki (1992) investigated the reaction between MnZn ferrites and the alumina refractory setters. Zn migrated from the ferrite into the refractory and the zinc depleted layer extends to 200 microns in depth. The zinc loss induces residual stresses and deteriorates the magnetic properties.

Stijntjes (1992) presented an excellent review of the processing and crack formation in MnZn ferrites. Causes for cracking are listed as;

1. Haberstoh Effect- Deficiency of oxygen in the calcined powder leading to shrinkage of the compact at 600° C.
2. Inhomogeneous densification of compact
3. Binder burnout cracks
4. Reduction cracks
5. Reoxidation cracks
6. Impurity bulges

Low Loss Linear Ferrites

These materials are almost always used as gapped pot cores, RM cores or similar telecommunication cores. Cores meant for medium high frequencies (100 KHz) are generally fired at about 1200-1300°C in an appropriate equilibrium oxygen content within the confines of Blank's curve and cooled maintaining the proper relationship between O_2 and temperature as dictated by the same curve. A wide latitude exists as to where to stay within the limits of this curve. Herein lies the compromise of factors such as permeability, losses, temperature-coefficient, disaccommodation and frequency response. Oxidizing fires represented on the upper portion of the curve, generally lead to low Eddy Current losses, but high disaccomodation (low cation vacancy level). On the other hand, reducing fires lead to high permeability, high losses, and low disaccomodation. For high frequency cores of this type, the fire is usually made at a low temperature (1100-1200° C.) leading to finer grain size and higher porosity. This yields lower permeability, but low losses at high frequencies.

High Permeability Materials

Because of their need for high density and large grain size high permeability materials are fired at rather high temperatures (1300-1400°C). At these temperatures, loss of Zn is a serious problem and appropriate atmosphere precautions are taken. Many researchers have suggested high O_2, high temperature firing, with an attempt to reestablish the O_2 equilibrium after the hold or during the cooling process. When large grains are needed, the growth of the grains are not so rapid that the grain boundary will sweep across pores leaving a large intragranular porosity, pinning domain walls and reducing permeability. Shichijo(1971) prepared vacuum sintered MnZn ferrite of high density and high permeability. He attributed the high density to removal of the gas from the green (unfired) compact so that porosity was removed and densification proceeded. The vacuum sintered ferrite was then fired in an equilibrium atmosphere and high permeability ferrites (23,000 perm) were obtained. The outside surface suffered from zinc loss in vacuum sintering, but when this surface (about 1 mm.) was removed, the permeability rose to 35,000.

Chol (1969) has proposed a solution to the attaining of high permeability ferrites in spite of the contradictory demands of high density, controlled Fe^{++} and no zinc loss. He recommends a high temperature soak at high O_2 (pure O_2) levels to grow large grains and promote open porosity. This is followed by a second soak under low oxygen to densify and attain proper Fe^{++} concentration.

Tsay (1997) described the manufacture of high permeability MnZn ferrites by using atmospheric protection to prevent zinc loss. Control of oxygen partial pressure and flow rate can reduce zinc loss 20-35%.. The permeability was increased from 12,000 to 15,000 using this method. In a tube furnace, the yield of 15,000 permeability ferrite was increased when a boat containing Zn oxide powder was placed near the ferrite.

Power Ferrites

Thus far, we have been concerned with permeability requirements of high density and large grain size. However, in recent years with the growing requirement for high frequency power materials, the need for other microstructural and chemical properties has arisen. For example, high resistivity materials with prominent grain boundaries and small grain sizes are used. Once the chemistry (including Fe^{++}) is established for high saturation and moderate permeability, good density is still needed for saturation purposes (although not as critically as in the high perm case). Buthker, 1982 has reported that for a low loss power material, if an equilibrium oxygen curve is used, the resistivity drops dramatically and pores appear inside of grains. Ti^{++++} coupled with an oxidative firing is used to control the Fe^{2+} content. However, even when additives are not used, the resistivity and the high frequency loss factor and core losses are lower under these firing conditions. In discussing power materials for 400-600 KHz switching power supplies, Mochizuki(1985) used a strongly oxidizing atmosphere during the cooling process to increase the resistivity of the internal grains. Uniformity of grain size was improved by sintering at a high heating rate. Rikukawa (1985) used a special atmosphere control for the sintering of power ferrites. This process will be described in the section of microstructural development..

Lin (1986) has reviewed the possibility of annealing sintered ferrites to improve the electrical resistivity by Fe^{++} reduction. The method, though successful, severely reduced permeability. As an alternative as an alternative Lin suggests control of atmosphere during the earliest stages of sintering.

The firing of power ferrites is again related to the frequencies of operation. For lower frequencies, the permeability is intermediate (2000-3000) and a firing temperature of about 1200-1300°C with equilibrium oxygen pressure is generally used. Temperature and time should be sufficient to form the proper ferrite lattice yielding the highest B_{max} available. This was not as critical in the telecommunication ferrites, but is necessary for low core loss in power materials. For high frequency power materials, grain size must be kept small again and lower firing temperatures are also used with lower O_2 pressures. The saturation magnet-ization may have to be reduced to obtain lower core losses.

Hon (1992) fired MnZn power ferrites in a two stage sintering process. The first part was an isothermal one at 900° C. for 30 minutes that he feels is superior to ones at 950-1100° C. The second step is sintering at 1335° C. for one hour. There were fewer pores in the lower temperature first step process even though the grain size was the same.

Chemical Mechanism of Sintering in Mn-Zn Ferrites

After the initial heating period in which the binder is burned off (room temperature to 500°), the first chemical change occurs in the formation of zinc ferrite starting at about 600°C and continuing to about 800°C. During this time, the Mn will assume its equilibrium form of Mn_2O_3. Shortly after 800°C, MnZn ferrite will begin forming slowly through dissolution of the Mn into the zinc ferrite. Since some of this reaction has already occurred during the presintering step, only the unreacted material may be involved. After 1000°C, the reaction to form the final ferrite structure is increased. At the firing temperature and with the equilibrium oxygen partial pressure, the appropriate amount of Fe^{++} is formed to produce some Fe_3O_4. Excess O_2 will cause precipitation of Fe_2O_3.

Microstructural Development

The previous section dealt with satisfying the chemical requirements of the ferrite during the sintering process. However, the microstructural requirements must be considered concurrently with the chemical needs. Satisfying both sometimes presents a conflict situation. For example, in the case where the O_2 partial pressure is adjusted for Fe^{2+} equilibrium, the presence of a dense microstructure (which may be desireable) will impede the diffusion of oxygen during the cool.

Rikukawa(1985) has examined the influence of oxygen partial pressure on the grain growth and densification process. His study used the expression of Blank (1961) namely;

$\log P_{O2} = a - b/T$ [7.1]
  where a = atmosphere parameter
  b = constant = 14,540
  P_{O2} = Oxygen partial pressure

To the constant, b, he assigned the value given by Morineau and Paulus (1975) and varied the a parameter, which they called the atmosphere parameter. During the heat-up, the density and grain size varied as shown in Figure 14.10. We see that the grain size is constant as the P_{O2} is varied but the density increases with lower values of O_2 concentration. Morineau and Paulus attribute this to the increased number of anion vacancies (or fewer cation vacancies) which increases the oxygen diffusion.

At high temperatures, if the oxygen content is relatively high (large value of a), more cation vacancies are produced aiding the ionic diffusion needed for densification. For an atmosphere during the heat-up period of .2 atmospheres,

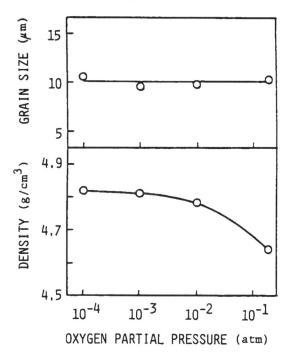

Figure 14.10- Variation of grain size and density as a function of oxygen partial pressures during the heat up portion of the fire (Rikukawa 1985)

Figure 14.11 gives the dependence of grain growth and densification during the high temperature holding period. Grain growth is accelerated as the oxygen is increased past the equilibrium value. In such cases grain growth accompanies densification and both increase under those circumstances. Figure 14.12 shows the increased resistivity and lower loss factor as the Fe^{2+} concentration is lowered due to the increased oxygen parameter.

In Chapter 13, we mentioned that Ishino (1987) listed several chemical and microstructural requirements for low-loss MnZn ferrites. These requirements included the formation of a thin insulating film at grain boundaries, a small and homogeneous grain size and reduction of porosity. To accomplish these aims, Ishino suggested the sintering cycle given in Figure 14.13 is. Also shown in this figure are the grain structures at each sintering stage and the concentration gradient of the Ca and Si at the grain boundaries. This program differs from the firing profile given in Figure 14.14 which is used for high permeability ferrites. In the latter, the grain size is much larger and there is no detectable Ca or Si at the grain boundaries. With regard tothe firing temperature and time , the recommended process for the low-loss ferrite is by slow grain growth involving low temperatures and longer times with a high impurity content (Figure 14.15). This leads to small grains, thicker grain boundaries and higher resistivity. The fast grain growth fire at higher

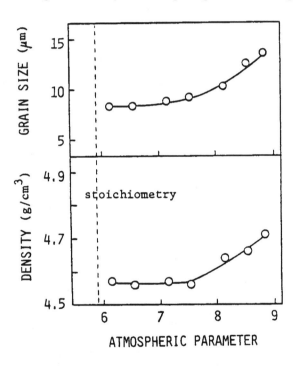

Figure 14.11- Variation of grain size and density as a function of atmospheric parameter during high temperature hold (Rikukawa 1985)

temperatures and longer times coupled with a with low impurity content leads to high permeability and high losses.

Shichijo(1971) used vacuum sintering in preparing MnZn ferrites. He achieved high densities but the vacuum-sintered product was oxygen-deficient requiring an additional anneal in higher O_2. Zinc loss was serious and the outside skin had to be removed by etching in a mixture of hot sulfuric and phosphoric acids. The process consisted of vacuum sintering at 1200-1300 for 2 hours producing a pore-free fine-grained structure. It was the easy to grow larger grains from this matrix in the higher oxygen anneal at 1250 to 1400°. Crystal formation occurs at 650 to 1000°, densification at 900 to 1250°, and oxidation and grain growth between 1250 to 1375°.

Ferrite Kilns

Firing may be performed in one of many different types of furnaces or kilns. Box furnaces, tube furnaces, and elevator furnaces may be used for periodic firing. The continuous types include pusher kilns (see figure 14.16), roller hearth kilns, and

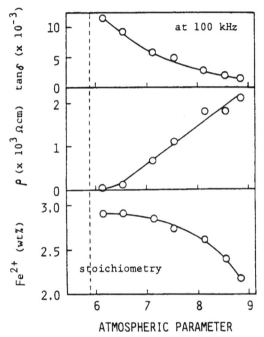

Figure 14.12- Variation of tan δ, resistivity, ρ, and Fe^{2+} content as a function of atmospheric parameter during high-temperature hold (Rikukawa 1985)

sled kilns with provisions such as elevators or offsets to separate the binder burnoff from the firing sections.

In the past, because of uncertainty in reproducing temperature and atmosphere profiles, and because of variations in temperatures and atmospheres at different parts of a kiln, variation in properties in a periodic kiln may be greater than in a continuous kiln. However, other considerations such as maximum temperature, equipment cost, and operation cost were involved.

The first section of a continuous kiln as well as the first period of time in a periodic kiln is devoted to the binder burnoff. The percent binder and the size of the part will determine these parameters. For binder burnoff, the atmosphere is oxidizing (for example, air). After burnoff, the kiln is gradually raised to the final firing temperature. At this point, consideration must be made for the atmosphere in equilibrium with the particular ferrite at a particular temperature. The higher the temperature, the higher the equilibrium partial pressure of oxygen, pO_2. Studies by Blank,(1961) Slick,(1971) and Paulus,(1975) have dealt with the treatment of this oxygen-temperature equilibrium. The most critical step of the firing is the cooling portion as this is the region where the oxidation states of Fe, Mn, and other multivalent ions are fixed. The firing is accomplished by controlling the O_2 partial

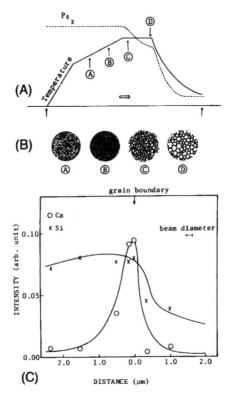

Figure 14.13-(A)- Temperature and atmosphere profile for a firing of low-loss MnZn ferrite; (B)-Grain size change during fire; (C)- Ca and Si concentrations at a grain boundary. (Ishino 1987)

pressure, p_{O2}, as the temperature is lowered. Even if the proper equilibrium is established during the high temperature hold, improper cooling may negate the previous oxidation state. Conversely, if the hold is not equilibrium, proper equilibrium may still be established during the cool if sufficient time is used.

Recently, an improved design of a periodic elevator kiln was developed which is reported to overcome some of the objections given previously. First, the design permits rather rapid firing cycles on the order of one day with a large load. Second, the spatial variations of atmosphere and temperature are reportedly reduced signifi-

FERRITE PROCESSING

cantly. It remains to be seen whether these new kilns will receive wide acceptance. Several photographs of such a kiln are shown in Figures 14.17 and 14.18.

Kijima (1997) described the operation of an atmosphere-controlled roller-hearth kiln for firing MnZn ferrites. In the roller-hearth kiln, the total sintering timeis less than 11 hours which is less than half the time in a pusher kiln. With good control of the oxygen profile, both high permeability and low core loss ferrites

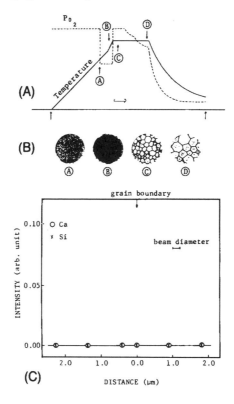

Figure 14.14- (A)-Temperature and atmosphere profile for a firing of a high-pemeability MnZn ferrite); (B)-Grain Size change during fire; (C)-Ca and Si Concentrations at a grain boundary)(Ishino 1987)

can be sintered simultaneously. The properties of these materials match those of the highest level in mass production.

Firing of Microwave Ferrites and Garnets

As previously noted, nickel ferrites are generally fired in air. Garnets, in most cases are fired at higher temperatures (1350-1400° C.) than ferrites. Because of the high sintering temperatures and since the ions involved all have the same valence

(+3), the atmosphere is very often pure oxygen at high temperatures with a possible switch to air during the cool. In addition, since the quantities are usually quite small in comparison to ferrites, and the selling prices are much higher, the firing is usually done in periodic box-type furnaces.

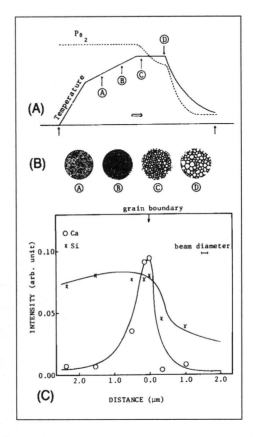

Figure 14.15- Permeabilities and microstructural features of MnZn ferrites formed by firing according to the profiles given in Figures 14.13 and 14.14. (Ishino 1987)

Firing of Hard Ferrites

As in garnets, the ions in hard ferrites do not change valencies at their sintering temperatures and so for economy considerations, the atmosphere is most often air. Firing is often done in continuous kilns of the pusher variety.

Finishing Operations for Ferrites

The mechanical finishing operations carried out on the fired ferrite include

FERRITE PROCESSING

1. Tumbling of toroids to remove sharp edges
2. Flat grinding of mating surfaces of pot cores, E-cores and so on.
3. Grinding of the center post of a pot core or the center leg of an E-core to obtain a required gap.

Figure 14.16- A modern pusher type kiln for firing ferrites (Courtesy of Bickley Furnaces, Inc.)

These operations will be described further when we deal with these components.

Ferrite and Garnet Films

The processing of ferrite materials discussed thus far was for polycrystalline bulk materials. Earlier, we mentioned the use of garnet films for magnetic bubble materials. Films, thin or thick, have also been produced in ferrites by a variety of

344   HANDBOOK OF MODERN FERROMAGNETIC MATERIALS

Figure 14.17- A new elevator type kiln for firing ferrites-Elevator down. (Photo courtesy of Ferroxcube, Div. of Amperex Corp, Saugerties, N.Y.)

Figure 14.18-Elevator kiln for firing ferrites-Elevator up (Photo courtesy of Ferroxcube, Div. Of Amperex Corp., Saugerties, N.Y.

different methods. The earliest of these used was that of vapor deposition of the metals (as alloys of the ferrite metallic composition) followed by oxidation (Banks 1961). The films were porous, polycrystalline and approximately 1000 Å thick. Many common ferrites were produced by this method and were single phase spinel in crystal structure. However, in the rare earth garnets such as YIG the widely differing melting points and vapor pressures of the iron and rare earth metals in the evaporation cannot produce stochiometric films with this method unless the iron content of the alloys is adjusted accordingly. Ferrite films have also been made by decomposition of the appropriate hydroxides (Lemaire 1961) or of the metallo-organic precursors (Wade 1966). The substrates are often aluminum oxide or fused quartz because of the high temperatures required. Single crystal ferrite films have also been deposited (Gambino 1967).

While many of the early ferrite films were studied for microwave applications, they were also researched for their potential as recording media . Reactive evaporation on glass and aluminum substrates produced mixtures of γ-Fe_2O_3 and Fe_3O_4 (Bando 1978,1981, Inagaki 1976 and Hoshi 1981).Nishimoto (1981) evaluated the read-write characteristics of these films.. Yamazaki (1981) prepared cobalt ferrite films by a similar method using successive evaporations of Fe and Co layers followed by annealing and oxidation . Morisako (1985) reported hexagonal Z-type ferrite films on Si substrates prepared by a sputtering technique. Double layering with interlayers of aluminum nitride or aluminum oxide was also done. Matsuoka (1985) described the deposition of barium ferrite films for perpendicular magnetic recording media.. Here, the barium ferrite was deposited with the easy (c) axis oriented perpendicular to the film plane . First a c-axis oriented ZnO film was deposited, followed by a (111) axis oriented MnZn ferrite and finally, a c-axis oriented Ba ferrite film.

Thin films with of MnZn ferrite with (111) plane orientation were successfully grown by Matsumoto (1989) by sputtering at a substrate temperature of $100^{\circ}C$.

Films of MnZn and Co ferrite for microwave magnetic Integrated Circuits (MMIC) were formed on Cu or PET plastic substrates by an electroless plating process followed by oxidation (Abe 1983,1984 and 1985). Multilayered ferrite-organic films on glass or GaAs substrates for the same applications were reported using a spray-spin coating method. Solutions of $FeCl_2$, $NiCl_2$, and $ZnCl_2$ were sprayed onto a heated substrate spinning at 300 rpm together with an oxidizing solution of $NaNO_2$. A buffer layer of dextran was used to separate the ferrite layers. Fe_3O_4 layered structures were produced similarly (Abe 1987a and 1987b and 1987c). Abe (1992) further reviewed uses of his plating process. Zhang (1992) reported on the effect of oxygen plasma treatment on magnetite and NiZn ferrite films using the spin-spray plating method. The oxygen plasma treatment increased the number of nucleation sites of ferrite and enhanced adhesion of the films to the substrate. Abe (1997) listed further improvements in his plating method including the addition of ultrasound power to the aqueous solution. Further applications included the use of Fe_3O_4/CdS backlayers for perpendicular magnetic recording

The earliest garnet films were reported by Mee (1969) and Pulliam (1971) using the method is known as chemical vapor deposition CVD. In this method, the film is deposited onto a substrate from the vaporized salts (almost always chlorides) of the metals of the ferrite or garnet composition i the presence of water vapor. The deposition is accomplished by hydrolysis(or reaction with water) to form the oxides. The reaction for the formation of YIG can be represented by;

$$5FeCl_3 + 3YCl_3 + 12H_2O \rightarrow 24HCl + Y_3Fe_5O_{12}$$

Prior to using this method, the material to be used for bubbles was made by growing single crystals and cutting very thin sections of the appropriate crystallographic orientation. Although the CVD process successfully replaced the previous cumbersome method, it, too, was replaced by a method that has essentially lasted until this day.

Liquid Phase Epitaxy

The newer process is termed LPE (for liquid phase epitaxy) which was first reported for other systems by Linares (1968). It was first used on garnets by Van Uitert (1970) and in bubbles by Shick (1971). Almost immediately, Levinstein (1971) refined the process to one similar to high-volume integrated circuit technology. Liquid phase epitaxy is accomplished in a solution of the component oxides in a solvent of flux usually composed of molten PbO and Bi_2O_3. The substrate is cut from a single crystal of a non-magnetic material (commonly gadolinium gallium garnet). Its lattice parameter is similar to the garnet being deposited and with its surface plane is that desired in the garnet film. When the substrate is placed in the molten solution of the garnet plus flux and the temperature is lowered, the garnet film will deposit epitaxially or in the same orientation as the substrate. Levinstein's improvement calls for the substrates(and there can be a very large number of them)to be dipped in the liquid and then to be pulled out when the films were deposited. The details of the process are described by Blank (1973). Another improvement in the preparation of bubbles was involves using ion implantation to strain the uppermost layers of the film and thus improve magnetic properties (Pulliam 1981)

Il'yashenko(1992) studied the decrease of growth-induced anisotropy by annealing on Bi-substituted garnet films formed by LPE. Additions of calcium suppressed the decrease. The ratio of the garnet-forming rare earths played an imporrtant part in the annealing process. Uchida(1992) investigated the effect of growth conditions on the surface morphology in $(BiTbNd)_3Fe_5O_{12}$ films on GGG(Gadolinium Gallium Garnet) grown from a Bi_2O_3-PbO-B_2O_3 flux. It took 50 hours to grow a 600 micron thick film with a mirror surface. Using Moessbauer spectroscopy, Itoh (1992) studied the effect of H^+ ion bombardment on LPE-grown films of bubble garnet films. Some Fe^{3+} ions were changed to Fe^{2+} ions as a result. Vertesy (1997) investigated the stress-dependent magnetic parameters of epitaxial (YSmCa) garnet films. Lattice distortion was found to have a significant effect on the uniaxial anisotropy. Padlyak (1997) also grew bubble garnets, $(Y,Sm,Lu, Ca)_3(FeGe)_5O_{12})$ by LPE on GGG and studied the FMR spectra, domain structure and basic magnetic properties of the films. The Ca is added for charge compensation for the Ge to maintain the trivalent ion average. Annealing at 1100-1200°C.created a magnetic transition layer between the film and the substrate.

Sputtered Films

Gu (1992) studied the effect of annealing on r-f magnetron sputtered Co ferrite films. With a 500 ° C. anneal, the magnetization increased rapidly. At a 400 ° C. anneal, a wasp-waisted hysteresis loop was obtained. Using facing-targets sputtering(FTS), and Co ferrite and ZnO targets, Matsushita (1992) produced Co ferrite-ZnO multilayer films. Hoshi (1997a) used FTS to make films of YBaCuO and indium oxide films. With DC magnetron sputtering, Morisako(1997) attempted to form Ba ferrite with the easy (c) axis perpendicular to the film plane which is desirable in perpendicular magnetic recording. The effect of the post- deposition anneal on the magnetic properties was studied. Unfortunately preferential orientation was not observed. Hoshi (1992b) also used FTS to produce Ba ferrite on a thermally-oxidized silicon substrate. On annealing a film deposited in a Ar atmosphere cracks formed after annealing above 850 ° C. but not in an Ar-O_2 atmosphere. Deficiency of oxygen in the Ar atmosphere caused the cracks and suppressed the hexagonal phase on annealing. Hoshi(1997c) again using FTS was unable to obtain for barium ferrite on a carbon substrate due to diffusion of the C into the film on deposition. He then tried coating the carbon with various materials such as Ba ferrite, silica,alumina, ZnO metals(Pt and Cr) and Si_3N_4. With the silicon nitride, the film showed X-Ray peaks of Ba ferrite and a smooth surface. It also had better adhesion than the other oxides. Nakagawa (1992) employed FTS to successively deposit films of Ba ferrite and the YBa_2Cu_3O superconductor material on amorphous substrates. The c axis was perpendicular to the film plane. He suggests that this film can be grown on any substrate. Complex films of Bi_2O_3-Fe_2O_3-$PbTi(Zr)O_3$ were deposited by Kajima(1992) using RF reactive sputtering in an attempt to achieve ferroelectric and ferromagnetic properties simultaneously. A ferroelectric loop was obtained and the film was ferromagnetic when annealed above 500° C raising the possibility of both effects in a restricted region of the ternary system. For the first time Ramamurthy(1997) was able vary the texture to get perpendicular or in-plane anisotropy in a deposited Sr ferrite film. The perpendicular anisotropy is used for perpendicular recording media. At low power levels and annealing at >800°C. , the orientation was perpendicular while at high power levels under the same annealing conditions, the orientation was in-plane. Dash (1997) produced LiZn ferrite films by sputtering on fused quartz using targets of varying Zn content. The films were amorphous when annealed at 750° C. The lattice constant of the film was the same as that of the bulk. The coercive force was higher than the bulk and decreased with zinc content. Gomi(1997) used RF sputtering to deposit α-Fe_2O_3 and Cr_2O_3 on GGG(Gadolinium Gallium Garnet)using ceramic targets of the same composition. With the iron oxide, the deposition went by two steps, the initial growth at ambient followed by further growth at 500° C. while with the chromium oxide, it was only on deposition. Noma (1997) studied the effect of adding Xe (Xenon) to Ar-O_2 sputtering atmosphere using a Ba ferrite target. The heavy bombardment to the surface of the growing films influence the crystallization and decomposition in films and excellent magnetic characteristics are obtained. The $4\pi M_s$ was 4.7 KG. Okuno (1997) prepared Co ferrite thin films by using a primary 1000 eV Ar-ion sputtering and a secondary 200 eV Ar-ion beam for ion bombardment of the growing surface. The perpendicular anisotropy of the <111> oriented film was considered causaewd by the compressive stress characteristic to ion bombardment. Murthy (1997) studied internal friction in NiZn ferrites. Nanocrystalline Co-Fe-Hf-O films for high frequency operation were deposited by Hayakawa (1997) using RF sputtering in Ar-O_2 atmosphere. The films as deposited were amorphous and nanocrystalline(3-4 nm.) The real permeability μ' is 160 almost constant to 1 GHz., μ'/μ'' is 61 at 100 MHz. The high frequency attributes arise from the high resistivity of 13 $\mu\Omega$m and large anisotropy field of 4.8 kA/m.

Talhaides (1992) compare sputtered films of Co-Mn cation deficient spinel ferrites with submicronic powders of the same composition. Even though strict comparisons between the powders and the films are difficult, the main result of the study lies in the fact that sputtered ferrite films is assumed to made up of finely divided grains. As a result, the different thermal treatments used to improve properties of fine powders may be applied to thin films.

Laser Ablation

Omata(1992) used lasr ablation to deposit NiZn ferrite film using sintered NiZn ferrite as a target. By annealing at 400° C., the inherent flux density was restored and the composition of the film was the same as the target. Again using laser ablation, Masterson(1992) prepared Ba ferrite and iron oxide films. The Ba ferrite deposition was on a sapphire substrates from a sintered B ferrite target. The Ba ferrite films were annealed to >850° C. in oxygen and had perpendicular magnetic anisotropy. The typical thicknesses of the films were 0.25 microns and were measured interferometrically or by etching off the films and measurig the mass change. The composition of the iron oxide films varied between magnetite and α-Fe_2O_3. The magnetite films were oxidized to γ-Fe_2O_3.

Chemical Vapor Deposition

Torii (1992) deposited Fe_3O_4-γ-Fe_2O_3 films by plasma-assisted MOCVD from a mixed gas containing iron acetylacetonate and oxygen on soda-lime gas, fused silica, sapphire or silicon substrates. The advantages were the high deposition rate and the low temperature formation of the ferrites with good crystallinity. Fujii (1992) achieved <100> orientation and columnar structure perpendicular to the substrate surface in Co ferrite films using plasma-assisted MOCVD. Pignard(1997) used a new process of injection-MOCVD to deposit Ba ferrite films with the c axis perpendicular to the substrate. He used solid organic precursors of the iron and barium dissolved in an organic solvent. Droplets of a few microliters of the solution were injected and vapors of these precursors were formed by heating. Ito (1997) prepared (Zn,Fe)Fe_2O_4 by MOCVD applying a novel evaporation method. The metal oxides(MO) were evaporated in a single evaporation vessel and the vapors carried to the reaction vessel to react with oxygen. High crystallinity was obtained even at 500°C. without annealing at a higher temperature.

Sol-Gel Method

Matsumoto (1992) prepared magnetoplumbite type hexagonal lead ferrite films by the sol-gel method. The single phase films were formed by calcining at 600° C. Amorphous films of Bi_2O_3-Fe_2O_3-$PbTiO_3$ were deposited by Miura (1992) and Fujii (1997) by the glycol-gel process. Previous attempts using sputtering involved lots of pinholes after annealing. By the present technique, optically flat films without microcracks and pinholes could be made. For the liquid precursors, the nitrates of the Bi, Fe and Pb together with the isopropoxide of the Ti were dissolved in ethylene glycol. The sol-gel reaction proceeds by heating the solution to 80° C. in a nitrogen flow. The gel was coated on glass plates using the spinning disc method. The films remained amorphous even after a 700° C. anneal. The films exhibited ferromagnetic and ferroelectric properties simultaneously. The sol-gel method was also used to prepare Ba ferrite films by Cho who studied the crystallographic, morphology and magnetic characteristics. The metal nitrates were dissolved in ethylene glycol and coated onto silica substrates. They were dried at 250° C. and heated at 800° C.

Reactive Evaporation

Komachi(1992) prepared thin films of Co-CoO by reactive evaporation. The perpendicular anisotropy varied with changing oxygen pressure and deposition rate. When the deposition rate was kept constant, the content of CoO in the film and the perpendicular magnetic anisotropy appeared with increasing oxygen pressure. Yano(1992) studied the magnetic anisotroppy of obliquely evaporated Co-0 films on a 10 micron polyethylene pterophthalate(PET) using a web-coater vacuum evaporation apparatus. Thin films of c-axis oriented Ba and Sr ferrites crystal alumina substrates were prepared by Fujii(1997) using reactive evaporation. Subsequently, Ba ferrite/Sr ferrite multilayered films were tried. There were three evaporation sources, two for Ba and Fe were heated by electron beams and one for Sr was by a crucible heater. From the magnetization measurements, all films had perpendicular anisotropy. However, the M_s for the Ba ferrite film was only 55% of that for the bulk and the Sr ferrite film had 80% of the bulk. The multilayered films combined the advantages of the individual films. Tanaka(1997) used reactive evaporation to prepare Cu ferrite films.

Molecular Beam Epitaxy (MBE)

Chang (1992) deposited YIG by the molecular beam epitaxy method. After first depositing single films of Fe_2O_3 and Y_2O_3 on a GGG substrate, the YIG was deposited simultaneously using two molecular beams and the same substrate. The film was similar to one grown by LPE. Yamazaki (1992) studied Fe_3O_4-γ-Fe_2O_3 intermediate thin film with TEM. Fe films were deposited on vitreous silica substrates at room temperature using an electron beam heating unit. These were oxidized to form α-Fe_2O_3 and reduced in methanol to form convert Fe_3O_4 films. The Fe_3O_4-γ-Fe_2O_3 intermediate films were prepared by annealing the Fe_3O_4 films at 240° C. for various times.

Those who prepare single crystals from the melt encounter problems of 1)high temperature gradients producing strains 2) High melting points of the ferrites causing equipment difficulties 3) High oxygen partial pressures for these temperatures.

Figure 14.19 A large commercial Bridgman furnace for growing large ferrite single crystals. Courtesy of Philips Electronic Components, Elcoma Division, Eindhoven, The Netherlands

Use of a flux can help alleviate some of these problems. Some of the fluxes used for ferrites are;sodium carbonate(Kunsman 1962), sodium tetraborate or borax (Galt 1950, 1951), and PbO (Remeika 1958).

The problems of growth from the melt are much greater in the garnets than in ferrites.The reason is that the melting points are higher and also that the garnets (unlike the ferrites) do not melt normally but incongruously. Thus ,the melt composition is different than the liquid, thereby precluding a stoichiometric crystal. The flux method is used almost exclusively in the garnets. . The flux originally was PbO (Neilsen 1958). It was later changed to a mixture of PbO and PbF_2, where PbF_2 lowered the viscosity of the melt. Linares (1962) has also used a mixture of BaO and B_2O_3 as a flux.

Wolczynski (1992) reported on predictions to the nature of microsegregation for MnZn ferrites monocrystallization using the open Bridgman technique.The melting zone for the method of continuous addition of the ferrite can be adjusted to

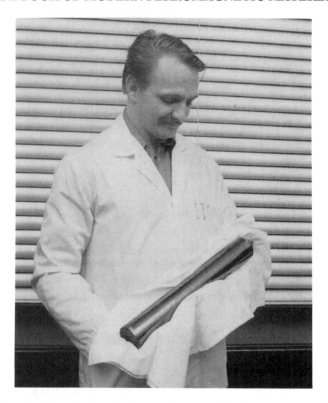

Figure 14.20- A large ferrite single crystal grown in the apparatus shown above. Courtesy of Philips Electronic Components, Elcoma Division, Eindhoven, The Netherlands

the required nature of segregation. Guzik(1992) expands on this topic by describing the conditions for unidirectional solidification of a MnZn ferrite. The dis placement rate of the crucible has been determined to be equal to 1.89×10^{-4} cm/sec.and the temperature gradient imposed on the solid liquid interface was about 15K/cm. Single crystals of MnZn ferrites with trace additives were grown by Matsuyama(1989) for VTR heads.Crystals with good properties were obtained and the segregation was found to be insignificant. Furukawa(1992) grew YIG single crystals for optical isolators with highly uniform transparency by the Floating Zone Method. Mn doping was the most appropriate method to control the valency of the iron. Various Mn-Si single crystals were grown by Okada (1992) using a high temperature tin solution method. Nasagawa (1992) was able to prepare a MnZn single crystal grown by applying metallic potassium dissolved in ethyl alcohol to a parallelpiped of a polycrystalline MnZn ferrite, contacting a seed crystal grown by the Bridgman method to the surface of the polycrystal and heating them near 1370° C. In this way, a single crystal over 20 mm long was made. Kozuka (1992) prepared single crystals of MnZn ferrite using a HIP (Hot Isostatic Pressed) method using a solid-solid reaction. The Bridgman technique was used to obtain seed crystals. The

HIP method consists of contacting a polycrystal which shows continuous grain growth with a seed crystal and heating the contacted body under solid phase. They obtained single crystals with homogeneous composition and no inclusions.

References

Abe, M. (1983) and Tamaura, Y.,Jap.J. Appl. Phys, 22, L511,
Abe, M. (1984) and Tamaura, Y.,J.Appl. Phys. 55, 2614
Abe, M. (1985) and Tamaura, Y.,Advances in Ceramics, 15, 639
Abe, M. (1987a) ,Itoh, Y, Tamaura,Y., and Gorni,M.,Presented at Conference on Magnetism and Magnetic Materials, Chicago, Nov.9-12,1987,Paper EP-05
Abe, M. (1987b) and Itoh, Y., Presented at Intermag Conference, Tokyo, Apr. 14-17,1987 Paper HD-07
Abe, M. (1987c) and Tamaura, Y.,ibid, Paper GA-1
Abe, M.(1992) Ferrites, Proc. ICF6, Jap. Soc. Powder and Powd. Met. Tokyo, 472
Abe, M.(1997) Proc. ICF7, J. de Physique, IV, Vol.7, C-1, 467
Akashi, T. (1971),Kenmoku, Y.,Shinma, Y., and Tsuji, T.Ferrites,Proc. ICF1, University of Tokyo Press, Tokyo, 1971 ,183
Alam, M.I (1992) and Jain, S.K.,Ferrites, Proc. ICF6, Jap. Soc. Powder and Powd. Met. Tokyo, 159
Auradon, J.D. (1969), Damay, F. and Chol, G.R., IEEE Trans. Mag., 5, 276
Bagul, A.G.(1992) Shrotry,J.J.,Kulkarni, Deshpande, C.D. and Date, S.K. Ferrites, Proc. ICF6, Jap. Soc. Powder and Powd. Met. Tokyo, 109
Banks, E. (1961) Riederman, N.H., and Silber, L.M.,J.Appl. Phys., 32, 44S
Bando, Y. (1978) Horii, S. and Takada,T. J.Appl. Phys., 17 ,1073
Bando, Y. (1981) Mishima,T, Horii, S., and Takada, T., Ferrites, Proc. ICF3, Center for Academic Publ., 602
Bassi, P.S. (1989) Randhawa, B.S. and Kaur, S., Advances in Ferrites, Volume 1- Oxford and IBH Publishing Co.,New Delhi, India, 67
Beer, H. (1958) and Planer, G.V., Br. Commun. Electronics, 5, 939
Blank, J.M. (1961) J.Appl. Phys.., 32, 378
Blank, J.M. (1973)and Neilsen, J.W., J. Crystal Growth, 17, 1973
Broussard, M.(1989) Abouaf, M., Perriat, P. and Rolland, J.L. Advances in Ferrites, Volume 1- Oxford and IBH Publishing Co.,New Delhi, India, 75
Bo, L. (1981) and Zeyi, Z., IEEE Trans. Mag. 17, 3144
Busev, A.(1980), Koroslelov, P.P.,and Mikhailov, Inorg. Mat. (USA) 16,1259
Buthker, C. (1982), Roelofsma, J.J. and Stijntjes, T.G.W.,Ceram. Bull., 61,809
Chang, N.S.(1992) and Nonomura, Y. Ferrites, Proc. ICF6, Jap. Soc. Powder and Powd. Met. Tokyo,413
Chiba,A.(1989) and Kimura,O.,Advances in Ferrites, Vol. 1 Oxford and IBH Publishing Co., New Delhi, India, 35
Chien, Y.T. (1992) and Sale, F.R. Ferrites, Proc. ICF6, Jap. Soc. Powder and Powd. Met. Tokyo, 301
Cho, W.D.(1997) Byeon, T.B.,Hempel, K.A.,Bonnenberg D., Surig, C. and Kim, T.O. Proc. ICF7, J. de Physique, IV, Vol.7, C-1,499

Chol, G.R. (1968), Damay, F., Auradon, J.P. and Strivens, M.A., Electrical Commun., 43, 263

Chol, G.R. (1969), Auradon, J.P. and Damay, F., IEEE Trans. Mag. 5, 281

Czochralski, J. (1918), Z. physik. chem. 92, 219

Dash, J.(1997) Ventkataramini, N., Krishnan, R., Date, S.K.,Kulkarni, S.D.Prasad, S. , Shringi, S.N.Kishan P.and Kumar, N. Proc. ICF7, J. de Physique, IV, Vol.7, C-1,477

Date, S.K.(1989), Deshpande, C.E., Kulkarni, S.D., and Shroti, Advances in Ferrites, Vol. 1 Oxford and IBH Publishing Co., New Delhi, India, 55

Durr, H.M. (1997) Proc. ICF7, J. de Physique, IV, Vol.7, C-1, 57

Economos, G. (1959) J.A.Cer. Soc., 42, 628

Fan, F. (1997) and Sale, F.R. Proc. ICF7, J. de Physique, IV, Vol.7, C-1, 81

Fetisov, V.B.(1997) Kozhina, G.A.,Fetisov, A.V.,Fishman, A.Y. and Mitofanov, V.Y. ICF7, J. de Physique, IV, Vol.7, C-1, 221

Filho, M.F. (1992) Mussel, W.,Qi,Q and Coey, J.M.D. Ferrites, Proc. ICF6, Jap. Soc. Powder and Powd. Met. Tokyo,1267

Fujii, E(1992) Torii, H. and Hattori, M. Ferrites, Proc. ICF6, Jap. Soc. Powder and Powd. Met. Tokyo, 468

Fujii, T.(1997) Kato, H.,Miura, Y.and Takada, J. Proc. ICF7, J. de Physique, IV, Vol.7, C-1, 485

Fujii, T.(1997) Wada, T., Tokunaga Y., Kawahito, K., Inoue, M.,Kajima, A,Jeyadevan B. and Tohji, T. Proc. ICF7, J. de Physique, IV, Vol.7, C-1, 493

Furukawa, Y.(1992) Fujiyoshi, M., Nitanda, F., Sato,M.and Ito, K. Ferrites, Proc. ICF6, Jap. Soc. Powder and Powd. Met. Tokyo, 378

Galt,J.K. (1950), Matthias, B.T.,and Remeika, J.P., Phys. Rev., 79,391

Galt, J.K. (1951), Yager, W.A., Remeika, J.P., and Merritt, F.R.,Phys. Rev., 81, 470

Gambino, R.J. (1967) J.Appl.Phys. 38, 1129

Goldman, A. (1975) and Laing, A.M.,.Presented at Electronics Section Meeting, American Ceramics Society,

Goldman, A. (1977) and Laing, A.M., J.de Phys. 38, Colloque C1, C-297

Goldman, A.(1989), Advances in Ferrites, Volume 1- Oxford and IBH Publishing Co.,New Delhi, India, 13

Gomi, M. (1987) Satoh,E, and Abe, M., Presented at Conference on Magnetism and Magnetic Materials, Chicago, Nov. 8-12,1987 Paper EP-10

Gomi, M. (1992)Serada, S and Abe, M., Ferrites, Proc. ICF6, Jap. Soc. Powder and Powd. Met. Tokyo, 999

Gomi, M.(1997) Toyoshima, H. and Yamada, T. Proc. ICF7, J. de Physique, IV, Vol.7, C-1, 481

Grenier, J.C.(1997) Wattiaux, A, Fournes, L.,Pouchard, M. and Etourneau, J., Proc. ICF7, J. de Physique, IV, Vol.7, C-1, 49

Gu, B.X.(1992) Zhang, H.Y,Zhai, H.R.,Lu, M., Zhang, S.Y.,Miao, Y.Z., and Xu, Y.B. Proc. ICF6, Jap. Soc. Powder and Powd. Met. Tokyo, 425

Guzik, E.(1992) and Wolczynski, W. Ferrites, Proc. ICF6, Jap. Soc. Powder and Powd. Met. Tokyo, 340

Harvey, J.W. (1980) and Johnson, D.W., Ceramic Bull., 59, 637

Hasegawa, F.(1992) Watanabe, K. and Nakatsuka, K., Ferrites, Proc. ICF6, Jap. Soc. Powder and Powd. Met. Tokyo, 112

Hayakawa, Y.(1997) Ohminato, K.,Hasegawa, N. and Makino, A. Proc. ICF7, J. de Physique, IV, Vol.7, C-1, 495

He, H.(1992) Jiang, D., and Su, J. Proc. ICF6, Jap. Soc. Powder and Powd. Met. Tokyo,489

Hirano, S. (1985) Watanabe, J. and Naka,,S.,Advances in Ceramics, 15, 65

Hiratsuka, N.(1992) Sasaki, I.,Fujita, M. and Sugimoto, M. Ferrites, Proc. ICF6, Jap. Soc. Powder and Powd. Met. Tokyo,980

Hirota, K. (1987) and Inoue, O., A. Cer.Soc. Bul., 66,1755

Hon, Y.S.(1992) and Ko, Y.C. Ferrites, Proc. ICF6, Jap. Soc. Powder and Powd. Met. Tokyo, 305

Hoshi, Y. (1985) Koshimizu, H., Naoe, M., and Yamazaki, S.,Ferrites, Proc. ICF3, Center for Academic Publ. Japan, 593

Hoshi, Y.(1992) Tezuka, T. and Naoe, M. Proc. ICF6, Jap. Soc. Powder and Powd. Met. Tokyo,432

Hoshi, Y.(1992), Speliotis, D.E. and Judy, J.H. ICF6, Jap. Soc. Powder and Powd. Met. Tokyo,444

Hoshi, Y.(1992), Speliotis, D.E. and Judy, J.H. ICF6, Jap. Soc. Powder and Powd. Met. Tokyo,440

Il'yashenko, E.I.(1992) Vasil'chikov, A.S.,Gaskov, N., Watanabe, N, Ohkoshi, M., and Tsushima, K., Ferrites, Proc. ICF6, Jap. Soc. Powder and Powd. Met. Tokyo, 485

Inagaki, N. (1976), Hattori, S., Ishii, Y., and Katsuraki, H., IEEE Trans. Mag., 12, 785

Inazuka, T.,(1992) Harada, S. and Kawamata, T., Ferrites, Proc. ICF6, Jap. Soc. Powder and Powd. Met. Tokyo, 362

Ishino, K. (1987) and Nurumiya, Y., Ceramic Bull., 66, 1469

Ito, S.(1997) Mochizuki, T., Chiba, M., Akashi, K. and Yoneda, N. ICF7, J. de Physique, IV, Vol.7, C-1,491

Itoh, J.(1992) Toriyama, T., and Histake, K., Proc. ICF6, Jap. Soc. Powder and Powd. Met. Tokyo, 496

Iwauchi,K(1992) Kiyama, M. and Proc. ICF7, J. de Physique, IV, Vol.7, C-1, 491

Jha, V.(1989), and Banthia,A.K., ,Advances in Ferrites, Vol. 1 Oxford and IBH Publishing Co., New Delhi, India, 61

Kaczmarek, W.A.(1997) and Ninham,B.W. Proc. ICF7, J. de Physique, IV, Vol.7, C-1, 47

Kajima, A.(1992) Ideta, K., Yamashita, K.,Fujii, T.,Nii,H., and Fujii, I., Advances in Ferrites, Vol. 1 Oxford and IBH Publishing Co., New Delhi, India, 452

Kang, D.S.(1992) Kim, H.S., You``B.D., Paik, J.G. and Kim, S.J. Ferrites, Proc. ICF6, Jap. Soc. Powder and Powd. Met. Tokyo, 1302

Kedesdey, H.H. (1953) and Katz, G., Cer. Age., 62, 29

Kijima, S.(1992) Arie, K.,Nakashima, S., Kobiki, H.,Kawano, T., Soga, N. and Goto, S. Proc. ICF7, J. de Physique, IV, Vol.7, C-1, 65

Kim, M.G.(1989) and Yoo, H.I.,Advances in Ferrites, Volume 1- Oxford and IBH Publishing Co., New Delhi, India,109

Kim, T.O.(1992) Kim, S.J., Grohs, P., Bonnenberg, D. and Hempel, K.A., Ferrites,

Proc. ICF6, Jap. Soc. Powder and Powd. Met. Tokyo, 75

Kimura, T. (1981), Takahashi, T., and Yamaguchi, T.,Ferrites, Proc. ICF3, Center for Acad. Publ., Japan, 27

Kiyama, M.(1992) Nakamura T., Honmyo, T. and Takada, T. Ferrites, Proc. ICF6, Jap. Soc. Powder and Powd. Met. Tokyo, 79

Knese, K.(1992) Michalk, C., Fisher, S., Scheler, H. and Brand, R.A. Ferrites, Proc. ICF6, Jap. Soc. Powder and Powd. Met. Tokyo, 163

Kobayashi, S. (1971),Yamagishi, I.,Ishii, R. and Sugimoto, M., Ferrites, Proc. ICF1, University of Tokyo Press, Tokyo, 326

Kodama, T.(1992) Mimori, K., Yoshida, T. and Tamaura, Y. Ferrites, Proc. ICF6, Jap. Soc. Powder and Powd. Met. Tokyo, 118

Kohno, A.(1992) Takad, K. and Yoshikawa, F., Ferrites, Proc. ICF6, Jap. Soc. Powder and Powd. Met. Tokyo, 140

Koma, S.(1992) Yoshida, S., Oka, K. and Suzuki, A. Ferrites, Proc. ICF6, Jap. Soc. Powder and Powd. Met. Tokyo, 71

Komachi, N.(1992) Namikawa, T. and Yamazaki, Y. Ferrites, Proc. ICF6, Jap. Soc. Powder and Powd. Met. Tokyo, 405

Kozuka, Y..(1992) Naganawa, M., Ouchi, R.., Imaeda, M. and Matsuzawa, S. Ferrites, Proc. ICF6, Jap. Soc. Powder and Powd. Met. Tokyo,354

Kunnman, W. (1962) Wold, A. and Banks, E., J.Appl.Phys. $\underline{33}$, 1364S

Khurana, Y. (1992) Ferrites, Proc. ICF6, Jap. Soc. Powder and Powd. Met. Tokyo,167

Lemaire,H. (1961) and Croft,W.J., J.Appl.Phys., $\underline{32,46S}$

Levinstein, H.J. (1971), Licht, S.,Landorf, R.W., and Blank, S.L., Appl. Phys. Lett., $\underline{19,}$ 486

Lin, I. (1986), Mishra, K. and Thomas, G., IEEE Trans. Mag., $\underline{22,}$ 175

Linares, R.C. (1962), J.A.Cer. Soc., $\underline{45,}$ 307

Lucke, R.(1997) Schlegel E. and Strienitz, R. Proc. ICF7, J. de Physique, IV, Vol.$\underline{7,}$ C-1, 63

Macklen, E.D. (1965), J.Appl. Phys. $\underline{36,}$ 1072

Mano, T.(1992) Mochizuki, T. and Sasaki, I. Ferrites, Proc. ICF6, Jap. Soc. Powder and Powd. Met. Tokyo, 152

Mano, T.(1989) Mochizuki, T. and Sasaki, I. Advances in Ferrites, Volume 1- Oxford and IBH Publishing Co., New Delhi, India, 143

Manoharan, S.S.,(1989) and Patil,K.C.,, Advances in Ferrites, Volume 1- Oxford and IBH Publishing Co., New Delhi, India, 43

Masterson, H.J.(1992) Lunney, J.G. and Coey, J.M.D. Ferrites, Proc. ICF6, Jap. Soc. Powder and Powd. Met. Tokyo,397

Matsumoto, M.(1992) Morisako, A. and Haeiwa, T. Hoshi, Y.(1992), Speliotis, D.E. and Judy, J.H. Ferrites, Proc.ICF6, Jap. Soc. Powder and Powd. Met. Tokyo,460

Matsuoka, M. (1985) and Naoe, M.,Advances in Ceramics, $\underline{16,}$ 309

Matsushita, N(1992) Koma, K.,Nakagawa, S. and Naoe, M. Ferrites, Proc. ICF6, Jap. Soc. Powder and Powd. Met. Tokyo,428

Mee, J. E. (1969) Pulliam, G.R., Archer, J.L. and Besser, P.J., IEEE Trans. Mag. $\underline{5,}$ 717

Micheli, A.L.(1970) Presented at Intermag Meeting 1970

Miura, H.(1992) Yamaguchi, K.,Kajima, A and Fujii, T. Ferrites, Proc. ICF6, Jap. Soc. Powder and Powd. Met. Tokyo,456

Mochizuki, T. (1985) Sasaki, I., and Torii, M., Advances in Ceramics, 16, 487

Morineau, R. (1975) and Paulus, M.,IEEE Trans. Mag. 11, 1312

Morisako, A. (1985) Matsumoto, M., and Naoe, M., Advances in Ceramics, 16, 349

Morisako, A. (1992) Matsumoto, M., and Naoe, M. ICF6, Jap. Soc. Powder and Powd. Met. Tokyo,436

Morrell, A. (1980) and Hermosin, A., Cer. Bull., 59, 626

Morell, A,(1989) Eranian, A, Peron, B. and Beuzelin, P.Advances in Ferrites, Volume 1- Oxford and IBH
Publishing Co.,New Delhi, India, 137

Murthy S.R.(1997) Proc. ICF7, J. de Physique, IV, Vol.7, C-1,489

Nakagawa, S.(1992) Matsushita, N. and Naoe,M., Hoshi, Y.(1992), Speliotis, D.E. and Judy, J.H. ICF6, Jap. Soc. Powder and Powd. Met. Tokyo, 448

Narita, Y.(1992) Ogasawara, S.,Ito, T. and Ikeda, Y Ferrites, Proc. ICF6, Jap. Soc. Powder and Powd. Met. Tokyo, 143

Nasagawa, N.(1992) Yamanoi, Aso, K., Uedaira, S. and Tamura, H. Ferrites, Proc. ICF6, Jap. Soc. Powder and Powd. Met. Tokyo, 350

Neilsen, J.W. (1958) and Dearborn, E.F., Phys. and Chem. Solids, 5, 202

Neilsen, J.W., (1958), J.Appl.Phys., 31, 51S

Nishimoto, K. (1981) and Aoyama, M., Ferrites, Proc. ICF3,Center for Acad. Publ. Japan, 588

Noma, K.(1997) Matsushita, N., Nakagawa, S. and Naoe M. Proc. ICF7, J. de Physique, IV, Vol.7, C-1,487

O'Bryan, H.M. (1969), Gallagher, P.K., Montforte, F.R., and Schrey, F., A.Cer.Soc. Bull., 48,203

Ochiai, T. (1985) and Okutani, K., Advances in Ceramics, 16,447

Ochiai, H.(1992) Ferrites, Proc. ICF6, Jap. Soc. Powder and Powd. Met. Tokyo, 93

Okada, S.(1992) Kudou, K. and Lundstom, T. Ferrites, Proc. ICF6, Jap. Soc. Powder and Powd. Met. Tokyo, 389

Okazaki, Y.(1992) Kitano, Y. and Narutani, T. Ferrites, Proc. ICF6, Jap. Soc. Powder and Powd. Met. Tokyo, 313

Okuno, S.N.(1992) Hashinoto, S and Inomata, K., Proc. ICF6, Jap. Soc. Powder and Powd. Met. Tokyo, 417

Omata, Y (1992) Tanaka, K. Nishikawa, Y. and Yoshida, Y Ferrites, Proc. ICF6, Jap. Soc. Powder and Powd. Met. Tokyo,393

Oudemans, G.J. (1968), Philips Tech. Rev., 29, 45

Padlyak, B.(1997) Proc. ICF7, J. de Physique, IV, Vol.7, C-1,503

Papakonstantinou, P.(1997) Teggart, B. and Atkinson, R. Proc. ICF7, J. de Physique, IV, Vol.7, C-1,475

Perriat, P.(1992) Ferrites, Proc. ICF6, Jap. Soc. Powder and Powd. Met. Tokyo, 321

Perriat, P.(1997) Gillot, B. and Aynes, D. Proc. ICF7, J. de Physique, IV, Vol.7, C-1, 43

Pignard, S.(1997) Seneteur, J.P. Vincent, H., Kreisel, J. and Abrutis, A. Proc. ICF7, J. de Physique, IV, Vol.7,C-1,483

Pulliam,G.R. (1971),Heinz, D.M., Besser, P.J.,and Collins, H.J., Ferrites, Proc.ICF1, University of Tokyo Press, 315

Pulliam, G.R. (1981), Mee,J.E., and Heinz,D.M., Ferrites, Proc. ICF3, Center for Acad. Publ.,Japan, 449
Rambaldini, P.(1989) Advances in Ferrites, Volume 1- Oxford and IBH Publishing Co.,New Delhi, India, 305
Randhawa, B.S.(1997) and Singh, R., Proc. ICF7, J. de Physique, IV, Vol.7, C-1, 89
Ramamurthy, B.(1997) Acharya, S. Prasad, S.Ventkataramini, N. and Shringi, S.N.
Rao, B.P. (1997) SubbaRao, R., and Rao, K.H. Proc. ICF7, J. de Physique, IV, Vol.7, C-1, 241
Remeika, J.P. (1958) U.S.Patent 2,848,310
Ries, H.(1992) Ferrites, Proc. ICF6, Jap. Soc. Powder and Powd. Met. Tokyo, 146
Reynolds, T.G. (1981) Ferrites, Proc. ICF3, Center for Acad. Publ. Japan, 74
Ries,H.B.(1989), Advances in Ferrites, Volume 1- Oxford and IBH Publishing Co.,New Delhi, India, 155
Rikukawa, H.,(1987), Sasaki,I. and Murakawa, K. Proceedings Intermag 1987
Rikukawa, H. (1985) and Sasaki, I.,Advances in Ceramics, 16,215
Rikukawa, H. (1997) and Sasaki, I. Proc. ICF7, J. de Physique, IV, Vol.7, C-1, 133
Roess,E.(1997) and Ruthner, M.J.(1989) Advances in Ferrites, Volume 1- Oxford and IBH Publishing Co.,New Delhi, India, 129
Ruthner, M.J.(1989),ibid,23
Ruthner, M.J. (1971),Richter, H.G. and Steiner, I.L.,Ferrites, Proc. ICF1, U. of Tokyo Press, Tokyo, 75
Ruthner, M.J.(1992) Ferrites, Proc. ICF6, Jap. Soc. Powder and Powd. Met. Tokyo, 40
Ruthner, M.J.(1997) Proc. ICF7, J. de Physique, IV, Vol.7, C-1, 53
Saimanthip,P.(1987) and Amarakoon, V.R.W.,Abstract for Paper 78-E-87,, presented at the Annual Meeting of the Am. Cer.Soc., April 21,1987, Pittsburgh, Pa.
Sano,T.(1988a),Morita, A. and Matsukawa, A.,PCIM, July,1988,p.19
Sano, A.(1988b),Morita, A.,and Matsukawa, A.,Proc. HFPC, San Diego, CA.,May 1-5,1989
Sano,A.(1989), Morita, A. and Matsukawa,A., Advances in Ferrites, Volume 1- Oxford and IBH Publishing Co.,New Delhi, India, 595
Sato, (1992a) Fujiwara, K. , Iijima, T., Haneda, K. and Seki, M. Ferrites, Proc. ICF6, Jap. Soc. Powder and Powd. Met. Tokyo, 961
Sato, (1992b) Fujiwara, K. , Iijima, T., Haneda, K. and Seki, M. Ferrites, Proc. ICF6, Jap. Soc. Powder and Powd. Met. Tokyo, 984
Schnettler, F.J. (1971) and Johnson, D.W., ibid, 121
Shichijo Y. (1971) and Takama, E., ibid, 210
Shick, L.K. (1971), Neilsen, J.W., Bobeck, A.H., Kurtzig, A.J., Michaelis, D.C., and Reetskin, J.P., Appl. Phys. Lett. 18, 89
Slick, P.I. (1971) Ferrites, Proc. ICF1, U. of Tokyo Press,Tokyo, 81
Smiltens, J. (1952) J. Chem. Phys., 20, 990, Stockbarger, D.C. (1936), Rev. Sci. Inst., 7, 133
Stijntjes, T.G.W.(1985), Presented at ICF4, Advances in Ceramics, Vol. 16,493
Stijntjes,T.G.W.(1989), Advances in Ferrites, Volume 1- Oxford and IBH Publishing Co.,New Delhi, India, 587

Stijntjes,T.G.W.(1992) Roelofsma, J.J. Boonstra, L.H. and Dawson, W.M., Ferrites, Proc. ICF6, Jap. Soc. Powder and Powd. Met. Tokyo, 45
Stuijts, A.L. (1971), Ferrites,Proc. ICF1, U. of Tokyo Press, Tokyo, 108
Sugimoto, M. (1966), J. Appl. Phys. Jap., 5, 557
Suresh, K.(1989) and Patil, K.C., Advances in Ferrites, Volume 1- Oxford and IBH Publishing Co.,New Delhi, India, 103
Suwa, Y. (1985), Hirano, S., Itozawa, K., and Naka, S., Ferrites, Proc.ICF3, Center for Acad. Publ. Japan, 23
Tachiwaki, T.(1992) Takano, H., Hirota, K. and Yamaguchi, O. Ferrites, Proc. ICF6, Jap. Soc. Powder and Powd. Met. Tokyo, 122
Tailhades, Ph.(1992) Chassaing, I., Bonino, J.P.,Rousset, A. and Mollard, P. Proc. ICF6, Jap. Soc. Powder and Powd. Met. Tokyo, 421
Takada, T. (1971) and Kiyama, M., Ferrites Proc. ICF1, U. of Tokyo Press, Tokyo, 69
Tanaka, T.(1997) Chiba, N., Okimura, H.and Koizumi, Proc. ICF7, J. de Physique, IV, Vol.7, C-1, 501
Tani, M.(1992)and Sawada, I., Proc. ICF6, Jap. Soc. Powder and Powd. Met. Tokyo, 156
Tasaki, J. (1971) and Ito, T., ibid, 84
Tsay, M.J.(1997) Tung, M.J.,Chen, C.J. and Liu, T.X., ICF7, J. de Physique, IV, Vol.7, C-1, 71
Torii, H.(1992)Fujii, E. and Hattori, M. Ferrites, Proc. ICF6, Jap. Soc. Powder and Powd. Met. Tokyo, 464
Uchida, N.(1992) Yamasawa, K.,Oido, A.,Maruyama, S. and Nakata, A. Yamazaki, Proc. ICF6, Jap. Soc. Powder and Powd. Met. Tokyo, 493
Van Uitert,L.G. (1970), Bonner, W.A., Grodkiewicz, W.H.,Pictroski, L. and Zydzik, G.J., Mat. Res. Bull.,5, 825
Verneuil, M.A. (1904), Ann. chim. et phys., 3, 20
Vertesy, G.(1997) Proc. ICF7, J. de Physique, IV, Vol.7, C-1, 479
Vogel, E.M. (1979) Ceram. Bull., 58, 453
Wade, W. (1966) A. Cer. Soc. Bull., 45, 571
Wenckus, J.F. (1957) and Leavitt, W.Z., Conf. on Magn. and Mag. Mat., 1957, Boston, Mass.,IEEE Publ. T-91,526
Wickham, D.G. (1961),Ferrette, A,Arnott, R.J.,Delaney, E.and Wold, A., J.Appl. Phys., 32,905
Wickham, D.G. (1954) MIT Lab. for Ins. Res. Rept. 89
Wickham, D.G. (1960) J.Inorg. Nucl. Chem., 14,217
Wickham, D.G. (1971) Ferrites, Proc. ICF1, U. of Tokyo Press Tokyo, 105
Wolczynski, W.(1992)and Guzik, E., Ferrites, Proc. ICF6, Jap. Soc. Powder and Powd. Met. Tokyo,337
Wolf, W.P. (1958) and Rodrigue, G.P.,J.Appl. Phys., 29, 105
Yamazaki, Y. (1981) Namikawa, T., and Satou, M., Ferrites, Proc. ICF1, U. of Tokyo Press, Tokyo, 606
Yamazaki, Y. (1992) and Matsue, M., Ferrites, Proc. ICF6, Jap. Soc. Powder and Powd. Met. Tokyo, 136
Yamazaki, Y. (1992) Okuda, K., Komachi, M. Sato, M and Namikawa, T. ibid, 401
Yano, A (1992) Ogawa, y and Kitakami, O. Ferrites, Proc. ICF6, Jap. Soc. Powder

and Powd. Met. Tokyo, 409

Yokoyama, M.(1992) Sato, M., Ohta, E. and Seki, M., Ferrites, Proc. ICF6, Jap. Soc. Powder and Powd. Met. Tokyo, 998

Yu, B.B. (1985) and Goldman, A., Ferrites, Proc. ICF3, Center for Acad. Publ., Japan, 68

Zaspalis, V.T.(1997) Mauczok, R.,Boerekamp, R. and Kolenbrander, M. ICF7, J. de Physique, IV, Vol.7, C-1, 75

Zhang, Y. (1992) Jiang, G., Zu, G., Deng, G. and Chang, Q. Ferrites, Proc. ICF6, Jap. Soc. Powder and Powd. Met. Tokyo, 184

Zhang, Q.(1992) Itoh, T., Abe, M. and Tamura, Y. Ferrites, Proc. ICF6, Jap. Soc. Powder and Powd. Met. Tokyo, 481

15 FERRITE INDUCTORS AND TRANSFORMERS FOR LOW POWER APPLICATIONS

INTRODUCTION

We have dealt with magnetic units without regard to the operational parameters such as the current, I, and the voltage, E. The relationship with the current is simple since we can derive equations for the magnetic fields produced by electrical currents. These equations vary with geometry, but for a coil around a long bar or a toroid, the equation is;

$$H = .4\pi NI/l \text{ (cgs) or } NI/l \text{ (MKSA)} \quad [15.1]$$

where N = turns
  I = Current (amperes)
  l = length of coil (cm. in cgs, m in MKSA)

This, then, is the dependence of H on the magnetization curve and hysteresis loop with the current in the coil that is really the input to the magnetic circuit. To discuss the relation of the output portion of the magnetic circuit, or the induction, B, to an operational parameter, it is necessary to introduce the important Faraday equation of magnetic induction;

$$E = -Nd\phi/dt \quad [15.2]$$

or using the BH loop parameters;

$$E = -Nd(BA)/dt = -NAdB/dt \quad [15.3]$$

Integrating and setting $1/t = \omega$ (angular velocity) $= 2\pi f$ and multiplying by .707 to get the rms sine wave voltage we get the final equation (cgs).

$$E = 4.44 BNAf \times 10^{-8} \quad \text{(volts)} \quad [15.4]$$

The corresponding equation for a square wave voltage is;

$$E = 4BNAf \times 10^{-8} \quad \text{(volts)} \quad [15.5]$$

The negative sign in the upper equations indicates that the polarity of the induced voltage is in opposition to the voltage causing the change in flux. Since we do not know how the flux changes with I, we introduce a parameter, L, inductance, which is defined as

$$L = Nd\phi/dI \quad \text{(henries)} \quad [15.6]$$

Combining this equation with the induction equation, we get;

$$E = -LdI/dt \quad [15.7]$$

INDUCTANCE

In simple resistive circuits, the relation between voltage, E, and current, I, is made with Ohms Law;

$$E = IR \quad [15.8]$$

Where R= Resistance, ohms

This equation holds in DC circuits and even in AC circuits which have no components other than resistances. When there are other components, namely inductors or capacitors in the circuit, there is a voltage drop across these components that is the analog of resistance, R. We can calculate this from the equation at the end of the last section;

$$E = -LdI/dt \quad [15.9]$$

Integrating and setting 1/t again equal to $\omega = 2\pi f$, we get;

$$E = (2\pi fL)I \quad [15.11]$$

We can now define our equivalents to R in inductors and capacitors;

$$E = X_L I \text{ for inductors} \quad [15.12]$$

and

$$E = X_C I \text{ for capacitors} \quad [15.13]$$

where

$$X_L = \text{Inductive Reactance, ohms} \quad [15.14]$$

and

$$X_C = \text{Capacitive Reactance, ohms} \quad [15.15]$$

In the inductor, sine wave voltage leads the current by 90° (see Figure 15.1). This occurs because the current creates a magnetic field in the magnetic material that stores the energy until the current, and consequently the field, is reversed. Thus a retardation of current is accomplished. In the capacitor, the opposite is true and the voltage is stored on the plates of the capacitor. Here the current leads the voltage by 90°. If the resistance is considered 0°, then inductive reactance and capacitive reac-

FERRITE INDUCTORS AND TRANSFORMERS FOR LOW POWER

tance are vectors and have magnitude and direction. As such, they can be added or subtracted vectorially. Thus equal X_L and X_C will cancel and the circuit will be purely resistive. If all three X_L, X_C, and R are present, there will be a remaining X after subtraction of the small of X_L and X_C from the larger and this residual X will be vectorally combined with R. See Figure 15.2. This vector sum is known as the impedance, Z, which represents the combined equivalent of R in an ac circuit.

$$Z = \sqrt{R^2 - (X_C - X_L)^2} \qquad [15.16]$$

We shall naturally be primarily concerned with L, the inductance. We would like to correlate the inductance to the magnetic properties of the component. The inductance was defined by the expression;

$$L = N \, d\phi/dI \qquad [15.17]$$

where 0 is the magnetic flux as previously described. We can also write;

$$L = N \, d(BA)/dI \qquad [15.18]$$

Now if the loop traversal is symmetrical, dB is $\pm B_{max}$ and $dI = I_{max}$ so that the expression reduces to

$$L = NBA/I \qquad [15.19]$$

Now:

$$B = \mu H = .4 \, \pi \mu NI/l \qquad [15.20]$$
where l = magnetic path length, cm

so that:

$$L = .4 \, \pi \mu N^2 A/l \qquad [15.21]$$

We see then that L is a parameter of the core dependent on the dimensions, the number of turns, N, of the winding and the permeability, μ, of the material. Therefore, for the same core and winding, L is proportional to μ. Herein lies the link between our bulk magnetic properties of the material producer and the operational properties of the designer. If the magnetic circuit is closed(no air gap), the cross section is known and is uniform as it is in a toroid, and the magnetic path length is known, then measurement of the inductance electronically (See section on measurements) and substitution in equation 15.21 allow calculation the permeability.

364   HANDBOOK OF MODERN FERROMAGNETIC MATERIALS

EFFECTIVE MAGNETIC PARAMETERS

The toroid was described as a closed magnetic circuit with uniform cross section. Even in a toroid, however, the magnetic path length varies from the circumference formed by the ID and to that formed by the OD. The mean length is often taken as

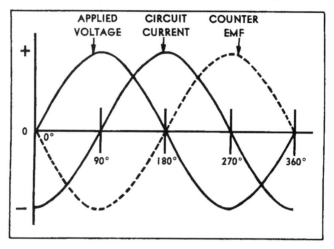

Figure 15.1-Phase relationship between applied voltage and circuit current in an inductor. The voltage leads the current by 90°. The induced voltage appears as a counter-emf that is 180° out of phase with the applied voltage. (Courtesy of Philco Corp.)

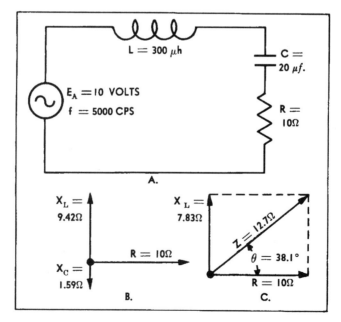

Figure 15.2- Vectorial addition of inductive and capacitive reactances and resistance to form impedance, Z.(Courtesy of Philco Corp.)

Pulsed Laser Deposition

Papakonstantinou (1997) prepared Bi and Ga substituted DyIG films (~350 nm. thickness) by pulsed laser deposition on single crystal GGG, Y-stabilized zirconia and Si substrates. Epitaxial growth was obtained only on the GGG substrates. In all cases, the films exhibited perpendicular magnetic anisotropy.

Single Crystal Ferrites

Single Crystals of ferrites and garnets are indispensible for certain types of fundemental magnetic measurements such as anisotropy, magnetostriction and ferrimagnetic resonance linewidth. Single crystals of ferrites are also used commercially primarily for recording heads. Of course the garnets films produced by LPE are really single crystals as well. Most of the methods used in growing ferrite and garnet single crystals have been applied previously to other systems. These include;

1. Czochralski Method (1918)- The crystal is pulled slowly from the molten ferrite on a rod which usually contains a seed of the crystal material.
2. Bridgman Method - Later modified by Stockbarger (1936)- The component oxides are melted in a crucible made of an inert, high-melting-point-material such as platinum and passed slowly through a sharp temperature gradient.
3. Flame Fusion Method (also called the Verneuil (1904) method-A fine powder of the oxides are dropped though a high temperature flame such as one of hydrogen and oxygen. The molten drops fuse onto a pedestal (sometimes containing a seed crystal) which is gradually lowered as the crystal grows. This method is used successfully to grow the non-magnetic gadolinium gallium garnet used as the substrate for magnetic bubble material.
4. Flux Method- In this method, the component oxides are dissolved in a solvent or flux at a high temperature. The temperature is then lowered slowly and the crystals start forming as the ferrite or garnet solubility is lowered. When the the crucible holding the crystals and flux is cooled, the crystals are separated from the flux by selective acid extraction.
5. Hydrothermal Method- Here, the component oxides are dissolved in an autoclave containing an aqueous solution at a rather high temperature and pressure. This method although used very successfully for quartz, is rarely applied to ferrites.

Smiltens (1952) was one of the first one to grow ferrite single crystals. He used the Bridgeman method for Fe_3O_4. Later investigators grew single crystals of Mn ferrite, Co ferrite and finally MnZn and the NiZn ferrites (Ohta 1963). Sugimoto (1966) designed a modified Bridgeman furnace operating at pressures up to 20 atmospheres and temperatures up to 1800° C. An induction coil was used as the heat source and a Pt-Rh crucible was used as the susceptor since the high resistivity of the ferrite prevents it from absorbing the radio-frequency power. At present, extremely large Mn Zn single crystals are grown commercially. For example, Kobayashi (1971) reported crystals of MnZn ferrite as large as 60 mm in diameter and 150 mm in length. See Figures !4.19 and 14.20.

FERRITE INDUCTORS AND TRANSFORMERS FOR LOW POWER

the circumference of the average diameter [$le = \pi(d_o + d_i)/2$]. Where there is a large variation between the OD and ID, the average value is invalid and a more complex method involving integration of all the paths is necessary. The situation on other shaped components is usually not as simple. First, the circuit may have an air gap (intentional or that formed by mating surfaces). The permeability of the magnetic circuit will be

$$\mu_e = \mu_o/\{1 + \mu_o l_g/l_m\} \qquad [15.22]$$

where; μ_e = Effective permeability of gapped structure
μ_o = permeability of the ungapped structure
l_g = length of gap
l_m = length of magnetic path

It is very important for us to appreciate the impact of this relationship especially in high permeability materials. For example, let us take the case of an EP core of 10,000 permeability material with no intentional gap. A separation of only 1 micron(or .00004 inches) will reduce the effective permeability to about 6700.

The effective permeability, μ_e is actually the permeability of an equivalent ungapped structure having the same inductance and same dimensions.
If there is a varying cross section of the component (such as a pot-core), then special methods are available for determining the effective length, le, the effective cross section, A_e, and effective volume, V_e, of these shapes by combining the contributions of each varying section.

MEASUREMENT OF EFFECTIVE PERMEABILITY

The effective permeability can be measured by several different methods. Impedance Bridges which separate the inductive and resistive components of an impedance are generally used for ferrites. The effective permeability is given by;

$$\mu_e = L_s l_e /.4\pi N^2 A_e \qquad [15.23]$$

These measurements will be discussed in a later chapter.

Effective Permeability-Relation to Other Magnetic Parameters

If we know the magnetic parameters for an ungapped circuit, then for any gapped structure in which the u_e is known we can calculate the corresponding parameters for that core.

The following relations hold:
TC or TC_e = TF x μ_e [15.24]
DA or DF_e = DF xμ_e [15.25]
where:

TC or TC_e = Temperature coefficient (Gapped)
TF = Temperature factor (ungapped)
DA or DF_e = Disaccomodation coefficient (Gapped)
DF = Disaccomodation factor (ungapped)

and

$$\tan \delta_e = \tan \delta \times \mu_e \qquad [15.26]$$

Inductance Factor, A_L

Another characterization of the inductance of a component is the inductance factor A_L. It is defined as the inductance of the core in henries per turn or millihenries per 1000 turns. This factor for a specified core can be used to calculate the inductance for any other number of turns, but we must remember that since L varies as N^2, so does A_L.

$$L_N = A_L N^2/(1000)^2 \text{ for L in millihenries} \quad [15.27]$$
$$L_N = A_L N^2 \qquad \text{for L in henries} \qquad [15.28]$$

The standard A_L values for pot cores are chosen from the International Standards Organization R5 series of preferred numbers. In this system, the antilogs of .2, .4 .6 .8, 1.0 and their multiples of 10 are selected numbers. Thus common A_L's are 16, 25, 40, 63, 100, 160, 250, 400 and so on. Some companies include A_L's from the R10 series which include antilogs of .1, .5 an.9 and have A_L's of 125, 315 and 800.

MAGNETIC CONSIDERATIONS: LOW-LEVEL APPLICATIONS

For low level applications such as those in channel filters, the hysteresis loop traversed will be a minor loop and is said to be in the Rayleigh or linear region where the permeability is described as the initial permeability. The induction change is usually not more than about .05 Teslas (50 gausses). The ferrite component can serve as an inductor supplying the inductance part of the LC network or it can act as a transformer for matching impedances, isolating or coupling. Because the induction is so low, there is no great need for a high saturation magnetization as we will require in the high level applications to be discussed next. However, low level inductors or transformers have other requirements . These include medium to high initial permeability, low loss factor at relevant frequencies, high stability of inductance to changes in operating conditions (temperature, AC signal level, DC bias) and low disaccommodation (decrease of permeability with time).

LC-Tuned Circuits-Channel Filters

Now X_L and X_C are related to the component's inductance and capacitance respectively but in addition are frequency dependent. This frequency is the angular frequency expressed in radians ($\omega = 2\pi f$) thus,

$$X_L = \omega L = 2\pi f L \qquad [15.29]$$

FERRITE INDUCTORS AND TRANSFORMERS FOR LOW POWER

and

$$X_C = 1/\omega C = 1/2\pi fC \quad [15.30]$$

where L = Inductance in henrys
C = capacitance in Farads

This means that in a series LC circuit, the higher the frequency, the higher the X_L and the lower the circuit current or signal. On the other hand, the lower the frequency, the higher the X_C, and again the lower the circuit current. At a certain frequency, where both canceled, the circuit would be purely resistive with only the resistive losses limiting the circuit current or signal. We would reasonably expect these circuits to provide a means of passing certain frequency bands while rejecting others. A major application is in the telephone transmission circuits in what are called channel filters.

The Q of a magnetic component is defined by the equation:

$$Q = X_L/R_s = 2\pi fL/R_s \quad [15.31]$$

where;

R_s = Series resistance (ohms)

This ratio or quality factor represents a type of efficiency of output or inductance to loss ratio. Earlier, we spoke of the complex permeability components, μ', and μ'' with regard to the permeability spectrum. We can now speak of them in an operational or circuit sense. Therefore another correlation between materials parameters and applications can be made.

In fact, the following relationship holds;

$$Q = \mu'/\mu'' = 1/\tan \delta \quad [15.32]$$

and in addition:

$$\mu Q = \mu/\tan \delta \quad [15.33]$$
$$L.F. = 1/\mu Q = \tan \delta /\mu \quad [10.34]$$

The selectivity of a specific frequency (or bandwidth) is given by:

$$\Delta f/f = 1/Q \quad [15.35]$$

where:

Δf = Bandwidth

That is to say that the higher Q of the component, the narrower will be the frequency bandwidth or the more useful a component will be to filter or separate out frequencies above or below a specific band (See Figure 15.3)

Telephone signals are transmitted by impressing the voice frequencies on specified carrier frequency bands that have limited bandwidths for each of the calls. By using a large number of adjacent bandwidths, we may send many messages over the same line. The selectivity of sending or modulation and receiving or demodulation of these different signals will be determined by the narrowness of the bandwidth. The Q then becomes a very important property and is usually categorized by the vendor according to frequency and inductance. When the many transmissions or messages are sent on the same telephone transmission line, lack of this selectivity leads to an annoying problem known as "cross talk" in which a telephone user hears a neighbor's conversation on his or her line. For the same reason, it is important that the inductance (or permeability) of a material should not vary significantly with changes in temperature, drive level, superimposed DC signal, and time. Magnetic components in telephones and transmission lines can be subject to extreme temperatures. Most telephones have superimposed DC at times for ringing purposes. Time wise, some components are expected to operate for about 20 years without great difference in properties. Figures 15.4 and 15.5 show the variations of some ferrite materials with temperature and frequency. Earlier we saw that obtaining the basic materials that show these different temperature and frequency dependencies is a matter of chemistry and processing control. Note that in Figure 15.5, there are materials in which the permeability versus temperatures slopes are positive and one material in which it is flat. The resonant frequency of a LC circuit is obtained when the effects of inductance and capacitance cancel. At that frequency, the following condition exists;

$$f_{res} = 1/2\pi\sqrt{LC} \quad [15.36]$$

where f_{res} = resonant frequency
L = inductance
C = capacitance

To maintain constancy of the resonant frequency, the LC product must be constant. Many capacitors such as the polystyrene types have temperature coefficients in the usable range which are slightly negative. If the inductor material T.F. has the same, but opposite (positive) slope, the net effect will be cancellation or no variation. If the capacitor is a silver mica capacitor which has a flat T.C., then a flat T.F. inductor material is used. The T.F. can be adjusted by either movement of the secondary permeability maximum by chemistry control as shown in a Chapter 6 or by additives such as TiO_2 or CoO.

Loading Coils
Loading coils are used in transmissions lines to add inductance that offsets the increase in capacitance that occurs over long distances. The "loading coils" are placed about every 1/2 mile and have inductances of 88 or 118 mH. In the United States., where transmission lines are mainly above ground and temperature variation and lightning may change the inductance of a material, moly-permalloy powder cores are used extensively since ferrites do not possess the same degree of stability(See Figures 15.4 and 15.13).. In most of the rest of the world, where transmission lines are below ground, ferrite is the material used for cost considerations. This is another application of a low level inductor (ferrite pot core) which needs low losses and good stability. The Q and needed stability can be attained by insertion of an air gap into the magnetic circuit.

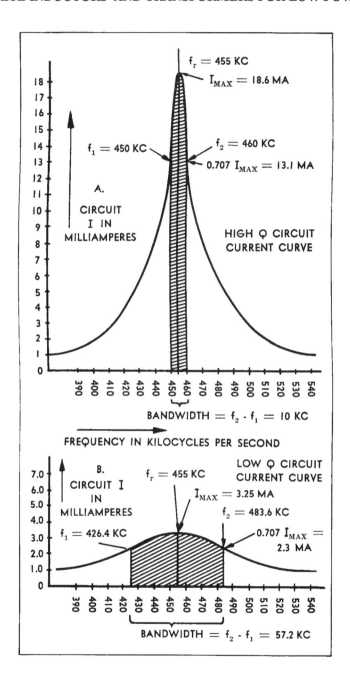

Figure 15.3- Bandwidths in a high Q and low Q LC Circuit (Courtesy of Philco Corp.)

Graph 2 — Initial Permeability (μ_i) vs. Temperature

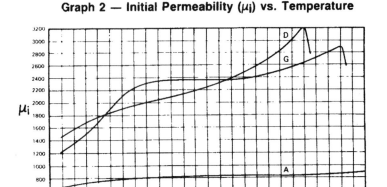

Figure 15.4- Variation of permeability with temperature for several ferrite materials for telecommunication application.(Courtesy of Magnetics,Division of Spang and Co., Butler, Pa.)

Pot Core Assembly

For many inductor and transformer applications, the component of choice is the ferrite pot core. The pot core has evolved over many years from a cup shaped core with a separate center post and cover to the present form consisting of two halves with the center post an integral part of each half. In most cases, the halves are identical; a notable exception is the one used as a tone-generator in touch-tone telephones. In this case, one half is a full-round core whereas the other one is a slab-sided core. A breakdown of a ferrite core assembly is shown in Figure 15.6. Note that it consists of;

1. Two ferrite halves
2. A plastic bobbin for winding the coil
3. A tuning slug
4. A metal clamp for securing the halves

Each pot core half, in turn, consists of a outer cylindrical shell called the skirt, a disk -shaped base called the floor and a cylindrical center post. The two halves mate on the two skirts and center posts. The center-posts will frequently have an axial hole to accommodate an inductance adjustor. The small space between the mating surfaces constitutes an air gap which reduces the core effective permeability to a value below that of the material's permeability. For reasons that are discussed in the following sections, an additional air gap often is formed in the magnetic circuit. This gap is ground on the mating surfaces of the center posts. If the gap is small, it will be ground on one side only, but if there is a large gap, it is usually split between the two halves so as to center it with respect to the coil.

FERRITE INDUCTORS AND TRANSFORMERS FOR LOW POWER

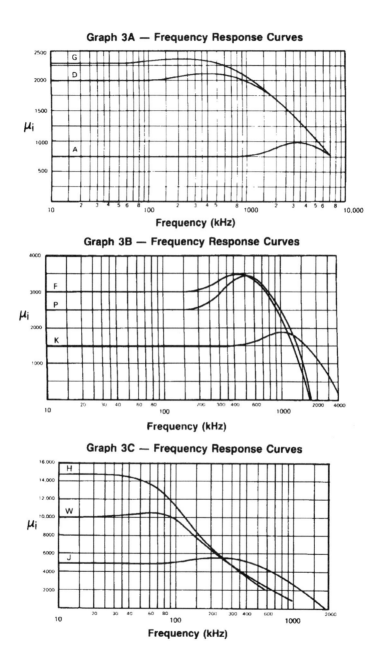

Figure 15.5- Variation of permeability with frequency for several ferrite materials for telecommunication applications. (Courtesy of Magnetics, Division of Spang and Co., Butler,Pa.)

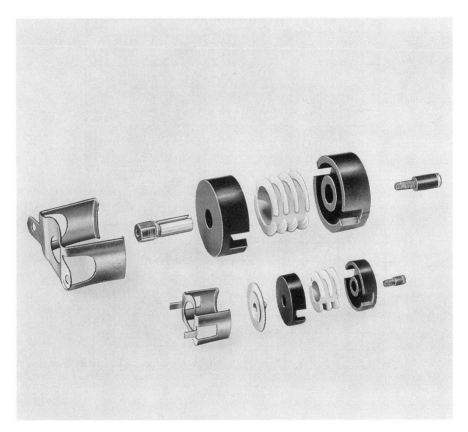

Figure 15.6- A pot core assembly with accessories (Courtesy of Magnetics, Division of Spang and Co., Butler, Pa.)

Although the pot core is a more costly ferrite shape than some of the other simple structures such as E cores, it provides several features not found in the others.

1. The presence of a nearly enclosed structure which shields the coil from unwanted external signals.
2. A mechanism for fine tuning the inductance with the use of the ferrite adjustor.
3. The ability to obtain high Q's by air gap and winding optimization,.
4. Improved stability of temperature and time by use of the air gap.
5. By using different combinations of materials, sizes and air gaps for the same inductance, optimization for the most critical requirements.
6. Lower cost of bobbin winding used versus toroidal winding.

FERRITE INDUCTORS AND TRANSFORMERS FOR LOW POWER

Figure 15.6a- Ferrite pot cores for telecommunications (Courtesy of Magnetics, Division of Spang and Co., Butler, Pa.)

Pot Core Shapes and Sizes

Pot cores come in a large variety of International Standard sizes from a small 9x5 core to a 42x29 core. In addition, there are non-standard sizes as small as 3.3x2.6 and as large as 70x42. The conventional method of expressing the size is to give the outside diameter first(in mm.)followed by the total height (mm) of the two halves. The International Standard adopted by most countries of the world on the dimensions and tolerances of pot cores is IEC (International Electrotechnical Commission) Publication 133 Third Edition, (IEC 1985). Similar standards for other shapes are published by the same organization. (Figure 10.6a) In addition to the older and more traditional round pot core shape, several new variations have been developed over the last few years to conform to changing component packaging concepts. For example, the shape of the touch-tone core was dictated by the need for many leads created by the several windings in the core. Therefore, one half of this core is slab-sided for larger lead openings. Four-holed cores instead of the original two were adopted for the same reason as well as the possible use of fractional numbers of turns.

The need for more compact packing of components on printed circuit boards led to the development of RM and R cores. In addition to their hexagonal shape for

374 HANDBOOK OF MODERN FERROMAGNETIC MATERIALS

closer packing, the use of lead terminal pins attached to the bobbin and directly insertable into PC boards eliminated the projections in round cores. The large wire openings in these cores reduce the shielding effect but that is often a small price to pay. A different type of clamp is used (See Figure 15.7) and often, the core halves are glued.

The dimensions for RM cores are given in IEC Publication 431 (IEC 1983). A slightly varied construction is found in the R cores. The major difference between RM and R cores lies in the fact that the angle of the wire opening is 90° in the R cores and 110° in the RM core.

The numbering system of the RM cores is based on the grid system for holes on printed circuit boards. There are 10 grids to an inch (25.4mm) The RM number corresponds to the number of grids that a side of the square that contains the core. Thus an RM4 core would fit in an are of 4X4 grids (0.4X 0.4 inches) or about 10 x 10 mm.

Figure 15.7- Variations of pot cores-RM Cores (hexagonal) and EP Cores (square). Courtesy of Magnetics, Division of Spang and Co.,Butler,Pa.)

Effective Magnetic Parameters of Pot Cores

In normal use, when the pot core is assembled from two halves, the flux path produced by the winding will travel through the center post, across one base (called the skirt) up the cylindrical side, and across the other base back to the center post. The cross- sectional area of the parts of this path vary so each section's core constant, A_e/l_e, is calculated separately and then combined to give an effective core constant. The method of determining these core constants is given in IEC Publication 205 and 205B (IEC 1966 and IEC 1974). Often a good approximation for the core constant is obtained by calculating a constant using the dimensions of the center post.

Specifications for Pot Core Design

In the design of ferrite pot cores, several operating parameters are usually listed (Magnetics 1987 and Philips 1986). These include;

1. Frequency of operation, f.
2. Inductance, L.
3. Minimum Q at the operating frequency
4. Applied alternating voltage
5. Maximum dimensions
6. Required stability with respect to temperature
7. Required stability with respect to time
8. Range of adjustment
9. Maximum current through coil
10. Cost

These factors are usually not of equal importance and as it is true for many electronic designs, the choice of the pot core is often a compromise. The choice is considerably accelerated by past experience. However, there are design aids available from several sources. Two books by Snelling (1969, 1982) are the best available on inductor and transformer design. The reader is strongly advised to consult them for greater detail. Useful guide lines are also available from vendor catalogs and design manuals (TDK 1990, Philips 1986, Siemens 1986, Magnetics 1987, Thomson 1983, Ferroxcube 1986, Tokin 1977, Fair-Rite 1987)

Designing for Inductance in Pot Cores

We have said that there are several variables that can be combined to obtain a pot core of required inductance. First, the inductance depends on the material permeability. If inductance is the only or major consideration, then the material with the highest permeability would be the choice. However, if the losses or consideration for core stability or variability are important factors, then the material with the lowest loss factor at the operating frequency or the one with the proper temperature factor is used. The loss factor is a good criterion because the losses are normalized per unit of permeability. After we select the material, we choose a trial core with the highest A_L available in the material to be used. Since A_L depends on core size and gap, there will be several core sizes that can have the right inductance. Considering inductance only, the core with the smallest size with that A_L is chosen from the list.

A good way of checking that core for practical winding convenience is to use the graph of inductance vs A_L versus number of turns. Such a graph is shown in Figure 15.8 (Magnetics 1987). This type of graph may be split up into several plots with smaller ranges. The point of highest A_L and the desired inductance is matched with the number of turns needed. From the vendor's core information, the available winding space is found. The winding used should be that which completely fills the bobbin. If the bobbin is not filled, the expected inductance decreases. A graph

376 HANDBOOK OF MODERN FERROMAGNETIC MATERIALS

showing this reduction as a function of percent of bobbin winding area filled is given in Figure 15.9. By combining the number of turns and winding area, we calculate the appropriate wire size used to fill the bobbin. Another way of finding the turns is through the use of a graph of turns versus wire size specifically for the pot core sizes (see Figure 15.10). If the number of turns is too high or the AWG (American Wire Gauge) wire size is too large (i.e. wire diameter is too small), a larger core with the highest A_L is chosen and the test for turns and wire size is repeated. If the wire size is too small (i.e., diameter is too large), a smaller A_L is used. This process is repeated until the smallest core with the highest A_L is found that can be economically wound.

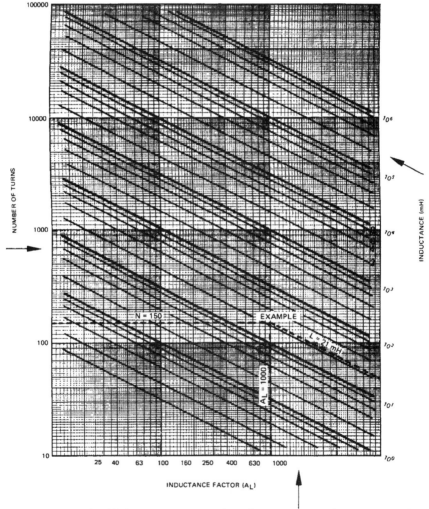

Figure 15.8- Graph of inductance versus number of turns and A_L value.(Courtesy of Magnetics, Division of Spang and Co., Butler,Pa.)

Designing a Pot Core Inductor for Maximum Q

Where maximum Q is desired, it is especially important to choose the material with the lowest losses (lowest loss factor). After the material is selected, the vendors' Q curves for the various cores and A_L's should be scanned for those cores that show peaks at the operating frequency. One such curve is shown in Figure 15.11. A alternative convenient method is through the use of Iso-Q curves (Figure 15.12) These give the Q contours as a function of frequency and inductance. In this case both Q and inductance can be optimized by finding the cores whose point of specified frequency and inductance comes closest to the center point of the contours (point of maximum Q). Using either method, we can choose the core with the maximum Q is. If only a minimum Q is specified, then the smallest core meeting the Q spec is chosen. If the Q curves are used, we must check the core chosen for any inductance requirements. After we find the core with the proper Q and inductance, we must determine the number of turns and wire size as outlined in the previous section. From the inductance of the core chosen, the value of the capacitor needed for resonance is calculated from the equation given earlier (Equation 15.36)

Since Q is a ratio of inductance to losses and since inductance is larger in larger cores (A_e/l_e is larger), we would expect larger cores to have higher Q's and indeed this is the case. However, large cores and high A_L's usually have their Q maxima at lower frequencies. In addition, large cores are more costly to buy and wind.

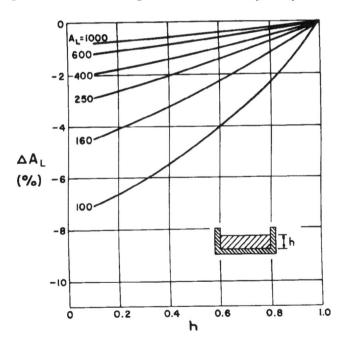

Figure 15.9- Deviation of inductance of a pot core caused by partially-filled bobbin. "h" is the fraction of bobbin filled. (Courtesy of Magnetics, Division of Spang and Co.,Butler, Pa.)

Designing Ferrite Inductors for Stability Requirements

In this case, in addition to picking a material based on the lowest loss factor at the operating frequency, another choice may be involved depending upon the type of capacitor used in the LC circuit. If the capacitor has a negative temperature coefficient and is linear (polystyrene type), then a ferrite material is chosen that has a positive linear TF in the temperature range of operation. On the other hand, if the capacitor has a temperature coefficient that is close to zero (silver-mica type), then a ferrite material is chosen which has a flat temperature- inductance curve. The temperature factor in this case should be centered about zero. Vendors' material specifications always include the temperature factors. After the material is chosen, the individual core listings should be scanned for cores and A_L's whose temperature coefficients coincide with the specified limits. From this list, the cores with the maximum Q at the operating frequency are chosen. The smallest of these is checked for required inductance and then, the turns and wire size are determined. Several trials with different cores and A_L's may be necessary to find the core which best meets the specifications. If the vendor does not list the temperature coefficients for each core and A_L, these can be calculated from the formula;

$$TC_{eff} = TF \times \mu_e \qquad [15.37]$$

Several other stability requirements may be of concern. For example, inductance stability with time is an important factor in some frequency-selective applications such as telephony. The material property involved is the disaccomodation factor. As in the case for temperature stability, the choice of material should include this factor. The disaccomdation coefficients DC_e (or DF_e) for the individual cores may be consulted in a manner similar to the temperature coefficients. The DC_e may also be calculated from the μ_e.

$$DC_e = DF \times \mu_e \qquad [15.38]$$

The process of choosing the core based on time stability is similar to the process used for the temperature case. To approximate the reduction in permeability after demagnetization or exceeding the Curie temperature, the following equation can be used;

$$L_1/L_2 = DC_e \log t_2/t_1 \qquad [15.39]$$
where t_1 = time after demagnetization of 1st measurement
t_2 = time after demagnetization of 2nd measurement

Since the largest changes per unit time take place in the first few decades after manufacture or demagnetization, the cores (especially those with higher disaccommodations) should be stabilized by aging before final adjustment of inductance is made.

FERRITE INDUCTORS AND TRANSFORMERS FOR LOW POWER

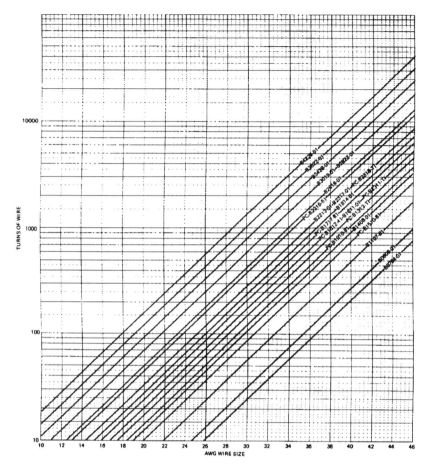

Figure 15.10-Number of turns versus wire size for a number of pot core Sizes (Courtesy of Magnetics, Division of Spang and Co., Butler, Pa.)

Saita (1992a, b) has substantiated the aging effect (disaccommodation) over 20 years of a detection and tuning circuit containing inductors and capacitors. They reveal that the change in inductance based on disaccommodation and vacancy displacement theory has been proven predictable.

FLUX DENSITY LIMITATIONS IN FERRITE INDUCTOR DESIGN

The initial permeability found in vendors' catalogs is valid in the linear or Rayleigh range of the magnetization curve. If the core chosen for a particular frequency operates outside of this range, the design may not be right. To determine whether the flux density is in the linear range (200 Gausses is a conservative limit), it may be calculated for sine wave excitation by;

$$B = E_{rms} \times 10^{-8}/4.44BNA_ef \text{ Gausses} \quad [15.40]$$

380 HANDBOOK OF MODERN FERROMAGNETIC MATERIALS

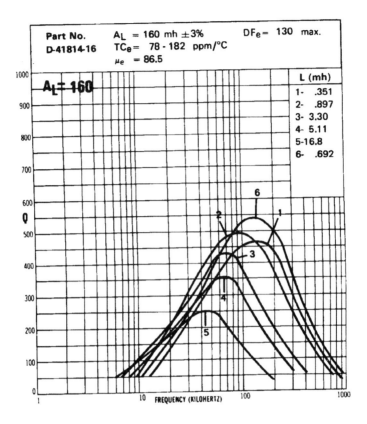

Figure 15.11-Q Values of an 18 x 14, 160 A_L, pot core as a function of frequency and inductance (Courtesy of Magnetics, Division of Spang and Co., Butler, Pa.)

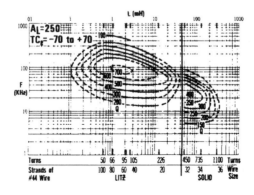

Figure 15.12 Iso-Q contours for a 22 x13 pot core as a function of frequency and inductance (Courtesy of Magnetics, Division of Spang and Co., Butler, Pa.)

FERRITE INDUCTORS AND TRANSFORMERS FOR LOW POWER

If there is an additional D.C. bias (See next section), it must be added to the flux density equation to give;

$$B = (E_{rms} \times 10^{-8} / 4.44 N A_e f \times 10^{-8}) + N I_{DC} A_L / 10 A_e \quad [10.41]$$

If the flux density is too high, a larger core (larger A_e) can be used to lower B.

DC Bias Effects in Ferrite Inductor Design
Sometimes, as in telephony, there will be a superimposed DC voltage present in the LC filter circuits. This is an additional consideration in the design of the core. The presence of a DC current in excess of a cutoff value will result in a drop off of A_L owing to operation at a higher point on the magnetization curve. The vendor usually supplies characteristic curves of A_L versus D.C. bias. The curve will sometimes be given for each individual core. An example of such a curve is given in Figure 15.13.

Inductance Adjusters for Ferrite Pot Cores
As mentioned earlier, an axial hole is often formed in the center post for the purpose of inserting an inductance adjustment mechanism. In a core such as the Touchtone core, a threaded ferrite screw core is screwed into a threaded plastic insert that runs the length of the hole. In most other adjustment mechanisms a ferrite tube is molded into a plastic holder containing threads on the bottom end and a notched head to accommodate an adjusting screwdriver tool. For an individual pot core size there may be one or more types of adjusters with varying adjustment ranges. The vendors' catalogs will contain a graph showing the percent change of inductance as a function of turns on the thread. One such adjuster is shown in Figure 15.6.

SURFACE-MOUNT DESIGN FOR POT CORES
With the increased used of printed circuit board, two different designs have been used to insert the cores so the leads could be soldered into the printed circuit in a manner attractive for automated manufacturing. The older method is called the pin-through-hole(PTH) method. This involves a bobbin that has pins that insert into pre-arrangfed holes in the PC board for mounting and also connecting leads from the core to the board circuit. The new method which is more automatic and cost-saving is the surface-mount design. In this method, the leads and mounting strips are attached to metallic pads of a special SMD bobbin. The core with bobbin is then placed on the PC board so these pads are located directly over corresponding pads on the board. Some of the pads are leads carrying the signal. For telecommunications usage, the cores using the SMD method are mostly RM cores whereas the standard pot cores use the PTH method for attaching to PC boards. Magnetics (Magnetics 1996) does list a small 9X5 pot core with SMD.

LOW LEVEL TRANSFORMERS
The pot cores for filter applications discussed thus far are used to supply an inductance in an LC tuned circuit. One winding was involved and the inductor was considered in series with the rest of the circuit. However, there are other low-level applications in which the ferrite is used as a transformer with two or more windings involved.

382 HANDBOOK OF MODERN FERROMAGNETIC MATERIALS

The transformer may be used for several purposes;
1. To isolate one circuit from another, for example, to isolate a D.C. component.
2. To couple one part of the circuit to another.
3. To change voltages or currents.
4. To match impedances between various parts of the circuit.

In this case, the ferrite component is the instrument of transformation of the voltage and current parameters of the input to those of the output. Aside from the losses in the core, no net energy is produced or absorbed in the process. The relation of voltages and currents in a transformer are well known;

$$E_1/E_2 = N_1/N_2 \qquad [15.42]$$

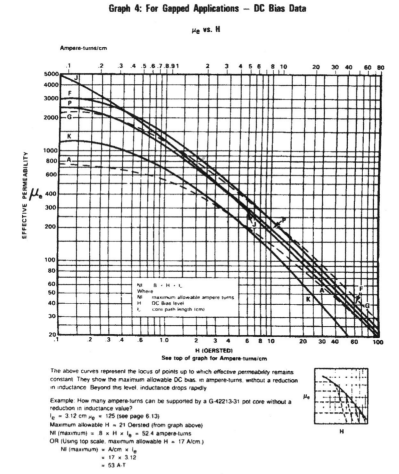

Figure 15.13- Effective permeability of several ferrite materials as a function of D.C bias (Courtesy of Magnetics, Division of Spang and Co., Butler, Pa.)

FERRITE INDUCTORS AND TRANSFORMERS FOR LOW POWER

$$I_1/I_2 = N_2/N_1 \qquad [15.43]$$

It can be shown that in matching impedances;

$$Z_1/Z_2 = N_1 2/N_2 2 \qquad [15.44]$$

There are two classifications of low level ferrite transformers;

1. Broadband transformers
2. Pulse transformers

The first is used to transmit a signal (usually of a sine wave) over a wide frequency range. The second is to transmit, with little distortion, a square wave generally for digital applications. For circuits such as radio or television it is necessary to tune all the frequencies over the broadcast range. The transformers in the circuit must function over a wide range and thus are called wide or broad band transformers. The width of the band can vary by a factor of 10 to 100 depending on the application. Under those conditions, the transformer is specified to operate between two frequencies, the upper one or high frequency cutoff, f_2, and the lower or low frequency cutoff, f_1. One criterion used to gauge the effectiveness of a transformer is the insertion loss, which is a measure of the power in the load without the transformer

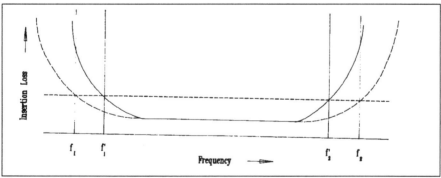

Figure 15.14- Insertion loss of a wide band transformer as a function of frequency (Courtesy of Fair-Rite Products, Walkill, N.Y.)

compared to the power with the transformer. The specification usually calls for the insertion loss to be less than a certain amount at the two frequencies, f_1 and f_2. (See Figure 15.14. A treatment of the components of the insertion loss is given in Snelling (1969). An equivalent circuit model of a transformer is given in Figure 15.15. Without the various equivalent resistances and reactances shown, the model would be considered an ideal transformer. At first glance, we would be especially concerned with the parallel components of the impedance since these could shunt some of the power from the output windings. At low frequencies, the shunt inductive reactance, X_p is low while the frequency effect on R_p is negligible. To reduce the ratio of R_p to L_p should be as low as possible. This can be seen from the equation for the low frequency insertion loss or attenuation.

$$A = 10\log\{1+ (R_p/2\pi fL_p)^2\} \quad [15.45]$$
where; A = Attenuation or Insertion Loss

To minimize the R_p/L_p ratio, we need a high permeability ferrite. The mid-band losses are mainly due to the winding losses. It is therefore difficult to use a large

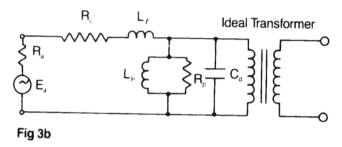

Fig 3b

Figure 15.15- Equivalent circuit of a transformer showing parallel components (Courtesy of Fair-Rite Products, Walkill, N.Y.)

number of turns to satisfy the low frequency cutoff requirement as the midrange winding losses would be too high. For the high frequency cutoff requirement, the leakage inductance and shunt capacitance cause attenuation problems.

In wide band transformers, we would like a maximum coupling of primary and secondary flux. Unfortunately, there is a loss of flux called leakage flux where some of the primary flux is lost through the air. We also encountered leakage flux in the case of the permanent magnet. The degree of coupling between primary and secondary fluxes is known as the coupling coefficient. This coupling coefficient should be as high as possible.

In order to minimize the leakage flux or maximize the coupling coefficient, the modern design engineer uses as high a permeability ferrite as is available which can maintain reasonable losses at the frequencies involved. High permeability toroids that minimize leakage flux can be chosen but for winding economy, an ungapped type of core is often used. Since the inductance depends not on the material permeability but on the effective permeability, the effect of the gap should be minimized. Therefore, in addition to grinding the mating surfaces, they are often lapped with a very fine diamond grit to obtain a mirror finish. To aid in this operation, the core should have a moderately large surface such as those obtained in EP or H cores. Pot cores and E-cores also be used with highly-polished mating surfaces in permeabilities from 10,000 to 15,000.

Ferrite Pulse Transformers

Ferrite cores, especially small toroids, are widely used in pulse transformers. This application requires transmission of a square wave with little distortion. The shape of a typical square wave voltage pulse is shown in Figure 15.16. During the time that the voltage pulse is on, the current is ramping up as is the flux density. In the case of a square wave, the ΔB is given in terms of the applied voltage, E, and the pulse width, T. For a specific core area and number of turns, the equation is;

$$B = ET \times 10^{-8}/NA_e \quad [15.46]$$

From the value of ΔB, the corresponding value of H, the magnetizing field can be determined from the vendor's curves on the material properties. From this value of H, and the l_e of the trial core, the excitation current can be determined from;

$$H = .4 \pi NI_p/l_e \quad [15.47]$$

If E, T, N are given and a ΔB is assumed, the effective dimensions of the core can then be given as;

$$l_e/A_e = 0.4 \pi N^2 I_p \Delta B \times 10^8/ET\Delta H \quad [15.48]$$

The cores corresponding to various values of l_e/A_e are listed by the vendor. The pulse permeability is given by;

$$\mu_p L_p = ET/I_p \quad [10.49]$$

The pulse transformers used in digital data processing circuits will usually have ΔB's on the order of 100 Gausses and are always little toroids inserted in small TO-5 cans for use on PC boards. Higher power pulse transformers may use pot cores or E cores that may be gapped to prevent saturation. All of the frequency related problems encountered in wide-band transformers are present in the pulse transformer, but here, it is evidenced by pulse attenuation (droop). As in the previous case, the permeability should be as high as possible, but when high pulse repetition rates or fast rise times are used, permeability concerns may be compromised for lower losses. Permeabilities of about 5-7000 are frequently used for small toroids.

FERRITES FOR LOW-LEVEL DIGITAL APPLICATIONS

Thus far, this chapter has dealt with the use of ferrites for low level analog applications. In the early 1990's and continuing till the present time there has been a remarkable change in communications technology. It has been driven by the proliferation of the following new developments

1. Personal and business computers
2. Fax transmission
3. Internet access
4. PBX(Private Branch Exchange)
5. LAN (Local Area Network)
6. Image Scanners
7. ATM Machines
8. Cellular phones

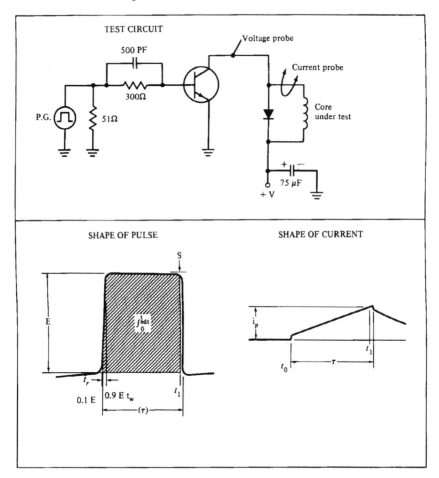

Figure 15.16- Shapes of voltage and current Wave forms in a ferrite core pulsed with a square wave (Courtesy of TDK Corp., Tokyo, Japan)

Most of these developments use digital circuitry rather than analog with a great deal of transmission of data as opposed to voice. Much of the resonant circuits which used tunable ferrite pot cores have been replaced by semiconductor devices called

FERRITE INDUCTORS AND TRANSFORMERS FOR LOW POWER

SLIC's. Touch-Tone cores have gone the same way. The magnetic material requirements for the new digital circuitry, as expected, have also changed. In the last section, we spoke of pulse and wideband transformers. The digital circuitry depends heavily on materials for pulse applications. The magnetic component needs for digital applications include;

1. Interface transformers for linking various devices (Fax, computer) central-offices
2. EMI(Electromagnetic Interference) suppression devices
3. Efficient D.C. power supplies(SMPS)

The materials and components for EMI suppression will be examined in the Chapter 18 while those for the SMPS will be discussed in Chapter 19. Those for the interface transformers will be discussed here in the low-level applications section.

ISDN COMPONENTS AND MATERIALS

The principal new telecommunications technology employing the new digital technique is the ISBN or Integrated Service Digital Network that is now being standardized in several countries and internationally. In this method, the speed and bandwidth is increased greatly by the use of digital transmission through regular telephone lines. A schematic of the system is shown in Figure 15.17. There are several interfaces which require the use of ferrite cores for coupling and impedance matching functions. These interface transformers include;

1. S_o Interface-This couples the terminal equipment (ISDN-capable Fax, telephones and adapers) with th ISDN network termination. The ISDN-capable terminal equipment is designated TE1, the non-ISDN is TE2 and the adapter TA. NT-1 is the network terminator. The S_o(S/T in the figure) also includes an interface converting from S to T.

2. U_{ko} Interface- (Here shown as U) connects the local central office with the network terminator, NT1.

3. U_{pn} and U_{po}-Interface is used as the link with the terminals in the PBX and can send signals for several kilometers.

4. S_{2m} Interface- This is used to connect PBX's with the central office if a high data transmission rate of 2 Mbits/s is needed.

Materials and Components for ISDN Interface Transformers

For the S_o interface, the important magnetic material properties are high inductance, low leakage inductance and low DC resistance. Hess (1996) states that these requirements can be met with very high perm materials such as Siemens T-38 or T-42 which have permeabilities of 10,000 and 12,000 respectively.See Figure 15.18. For the component, the RM-5 core is somewhat standard. Toroids are also used. The trend is to use 2 S_o toroids and a data line transmission choke in one module. The cores are usually potted but other methods may be developed. Both PTH (Pin through hole) or SMD (Surface Mount Design) devices are used.

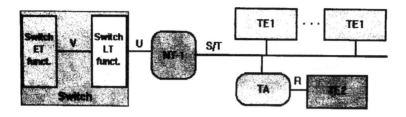

Figure 15.17-Schematic of an ISDN network with the various interfaces

	T38	T42	T44
Initial permeability μi	10000	12000	15000
Saturation Flux density (10KHz/400 A/m/25°C) [mT]	380	400	400
Hysteresis material constant ηB [mT*10^{-6}]	<1.4	<1.4	<2.0
Curie Temperature[°C]	>130	>130	>130
DC-resistance Ωm	0.1	0.1	0.01

Figure 15.18- Materials for the S_o interface for ISDN

For the U_{ko} interface, special matching transformers to handle the transmission codes and also the DC current needed for emergency telephone calls in case of system failure. A filter ferrite such as Siemens N48 is used in the shape of an RM6 that can handle the DC in a gapped core. The low loss factor permits a large signal operational range.

For the U_{pn} and U_{po} interface transformers, the range of transmission is short so the requirements are not as severe as for U_{ko} material except for the high DC current needed. Two power ferrites, an E core with N27 for 90 mA DC or a RM 41p with N67 for 140 mA max. D.C.

For the S_{2m} interface, Since no D.C is involved and low leakage is needed for the high frequencies, the transformers can be E6.3 cores in T38 material.

LOW PROFILE FERRITECORES FOR TELECOMMUNICATIONS
As pointed out by the author in an earlier book, the increased use of PC (printed circuit boards on which to place the magnetic components (ferrites) has required the use of low-profile cores. These must fit between the stacked boards with a space of only about one half inch between them. This requirement has given rise to a large number of core shapes that have been reduced in height. One particular shape that has been especially targeted for this reduction is the RM core. Aside from the un-

FERRITE INDUCTORS AND TRANSFORMERS FOR LOW POWER

gapped variety without adjusters that are used in some coupling or power functions(see chapter on high power applications), there are also some gapped cores in some RM sizes that accommodate adjusters. The Siemens catalog(Siemens 1997) shows the height of the standard RM5 reduced from 10.5 mm to 7.8 mm in the low profile case.(See Figure 15.4). In the RM6 Philips shows a similar reduction in height from 12.4 mm to 9.0 mm. The low-profile variety is also found in the RM8 and RM10. As in other RM cores, simple clamps can be used to fasten the two halves of the assembly. In addition, the smaller sizes are available in surface-mount design(SMD).

MULTI-LAYER CHIP INDUCTORS AND LC FILTERS

Continuing the advancements in magnetic core technology through miniaturization and surface-mount-design has led to thre development of the SM chip inductor and the SM chip LC filters. Earlier (in 1977) SM chip components were resistors, ceramic capacitors and transistors for use first in radio sets and then in tuners in TV's VCR's and other consumer items. In 1983 the first chip inductor was introduced by TDK. The smallest size listed is one in which the length is 2.0 mm and the width is 1.25 mm and the thickness is .85 mm. The length of the electrode is 0.5 mm. TDK lists the advantages as;

1. Easily handled by mounting machine.
2. Highly dense and versatile
3. Reliable due to monolithic construction
4. No crosstalk due to magnetic shielding
5. Excellent solderability and heat resistance.

The method of forming the ferrite with enclosed conductor turns is shown in Figure 15.19. The process describing the individual steps is as follows;

A. Ferrite paste is printed.
B. Conductive paste is printed. S is the starting point of the conductor.
C. Ferrite paste is printed onto ½ of the inductor.
D. Conductive paste is printed to be connected to the pattern C.
E. Ferrite paste is printed onto opposite side of C.
F. Final conductive paste is printed to opposite pattern E
   F is the finish of the conductor.
G. The last ferrite paste is printed.
H. Cofiring
I. External electrode

In addition to the multilayer chip inductors, TDK combines inductors with a capacitors to form LC filter networks in the same type of miniature surface-mountable format

References

Ceramic Magnetics (1986) Ferrite Catalog,Ceramic Magnetics Inc., 87 Fairfield Rd., Fairfield, NJ, 07006

DeMaw, M.F. (1981) Ferromagnetic Core Design and Application Handbook,Prentice-Hall, Inc., Englewood Cliffs, NJ, 07632

Fair-Rite (1987) Linear Ferrite Catalog, Fair-Rite Products, Wallkill, NY, 12589

Ferroxcube (1986) Linear Ferrite Materials and Components,Seventh Edition, Ferroxcube, Division of Amperex Electronic Corp., Saugerties, NY 12477

IEC (1966) IEC Document 205,International Electrotechnical Commission,1 Rue de Varembe, Geneva, Switzerland

ICE (1974) IEC Document 205B

IEC (1983) IEC Document 431

IEC (1985) IEC Document 133

Magnetics (1987) Magnetics Ferrites, Magnetics, A Division of Spang and Co., P.O.Box 391, Butler PA 16003

Philips (1986) Philips Data Handbook, Book C4,Philips Electronic Components and Materials Division, P.O. Box 218, Eindhoven, The Netherlands

Siemens (1986) Ferrites, Data Book,1986/7, Siemens AG, Bereich Bauelemente, Balanstrasse 73,D-8000,Munich, 80 W.Germany

Snelling, E.C.(1969) Soft Ferrites, Properties and Applications, Iliffe Books Ltd. London

Snelling, E.C. (1984) Ferrites for Inductors and Transformers, Research Studies Press, Letchworth,England

Steward (1987) Ferrite Cores, D.M. Steward, P.O.Box 510,Chattanooga, TN 37401

TDK (1986) Ferrite Cores for TV and Radio, TDK Corp. 13-1Nihonbashi 1-chome,Chuo-ku, Tokyo,103, Japan

TDK (1987) Ferrite Beads

Thomson (1983) Soft Ferrites, Ferrinox Booklet 13B, LCC Cofelec Department,50 Rue J.P. Timbaud,92400Courbevoie, France

16 SOFT MAGNETIC MATERIALS FOR EMI SUPPRESSION

INTRODUCTION

In Chapter 15, the soft magnetic materials for low level applications in transformers and inductors were chosen for their high Q, i.e. low losses per unit of inductance. This characteristic gave them greater frequency specificity in LC filter circuits and low losses in transformers. Their use in analog circuits in telephony, radio and other telecommunications applications satisfied the requirements and had few problems. The introduction of transistors, IC's and digital circuitry prompted changes in these magnetic materials again with few problems. However, in the past twenty years, with the proliferation of complex digital applications in computers, LAN and other networks, there has literally been a pollution of conducted and radiated electromagnetic interference (EMI) which has threatened to disturb severely the operation of much sensitive equipment. In fact, even earlier during World War II, there was a "critical mass" of delicate communication systems that could be tolerated before a breakdown in the operation would occur. The advent of computers, digital data transmission, cellular phones, Internet and satellite transmission have forced the governments of many countries to make restrictions on the EMI that could be generated by various electrical and electronic devices. In the early seventies, several companies started making ferrite and powder core components to assist manufacturers and designers of the offending equipment in taking steps to reduce these emissions. In addition, the victims of the disturbing emissions could also use the same components to attenuate the incoming EMI. One step for reducing the incoming radiated emi was discussed in Chapter 8 on magnetic shielding. A second is in the use of LC filters of several types. The third is in the use of ferrite beads and other shapes to encircle the wires and cables and by absorbing the offending "noise", reduce it. This chapter will deal with the magnetic materials and components used to combat EMI.

THE NEED FOR EMI SUPPRESSION DEVICES

Some of the EMI generated is from natural means but the largest amount is from man-made devices. Table 14.1 lists the most common generators of electromagnetic interference. In Table 14.2, another list gives thew most common victims of EMI. In some cases, the same ones listed as generators are also victims. There is EMI that is conducted through the wires and there is EMI that is radiated through the air. The EU (Formerly the EEC European Economic Community) has ruled that, after 1996, all products marketed in the EU must either meet the EMI Directive 89/336 or EMI

Table 16.1- From Woody (1994)

Generators of Electromagnetic Interference

Natural Phenomena	Manufactured

• Atmospheric – Primarily Storm Discharge
• The Sun
• Remote Stars
• Cosmic Noise
	• Communication
	  –Equipment
	  –Operation
	  –Transmission
	• Radar
	• Electric Power
	  –Generation
	  –Transmission/Distribution
	• Electrical Machinery
	  –Operations
	  –On/Off Line
	• Ignition (Spark)
	  – Systems
	• Digital Systems
	  –Processors
	  –Computers
	  –Controls
	• Nonradar Navigational Aids
	• Lighting
	  –Fluorescent

Table 16.2- From Woody (1994)

Victims of Electromagnetic Interference

Broad Band Amplifiers – Digital Circuits
Low Level Sensors – Ordnance
Communications Systems
Power Controls
Computer Equipment
Control Processors
Radar
Navigational Aids
Weapons Systems
Automotive Systems
Life Support and other Medical Systems

regulations of that country. That directive states that any equipment marketed in the EC;

a) must not interfere with radio or telecommunications equipment.
b) must be immune to emi emissions.

Proof of compliance must be demonstrated by the EMI district commission (CENELEC). Much of the EU requirements were spelled out in a document called CISPR 22. CISPR is a committee if th IEC (International Electrotechnical Committee). The CISPR 22 Emission Limits on Class A and Class B devices are given in Tables 16.3 and 16.4. They are similar to but not the same as FCC Title 47,

SOFT MAGNETIC MATERIALS FOR EMI SUPPRESSION

Table 16.3
CISPR 22 Emission Limits for Class A Devices

Radiated Emissions (30 meters)

Frequency (MHz)	uV/m	dB(uV/m)
30 - 230	31.6	30
230 - 1000	70.8	37

Conducted Emissions

Frequency (MHz)	uV QP (AV)	dB(uV) QP (AV)
0.15 - 0.5	8912.5 (1995)	79 (66)
0.5 - 30	4467 (1000)	73 (60)

Table 16.4
CISPR 22 Emission Limits for Class B Devices

Radiated Emissions (10 meters)

Frequency (MHz)	uV/m	dB(uV/m)
30 - 230	31.6	30
230 - 1000	70.8	37

Conducted Emissions

Frequency (MHz)	uV QP (AV)	dB(uV) QP (AV)	
0.15 - 0.5	1995-631 (631-199.5)	66-56 (56-46)	(limit varies line
0.5 - 5	631 (199.5)	56 (46)	
5 - 30	1000 (316)	60 (50)	

Part 15 Subpart B listed below. It is applicable to residential, commercial and light industrial cases. It also includes Information Technology and Electronic Data Processing equipment. In earlier occasions, the criterion for EMI interference for the victim was "susceptibility", which is a negative way of looking at the problem. Now, it is called "immunity" which is the opposite but has a more positive connotation. This puts some burden of the responsibility on the victim. In the U.S., the

Table 16.5
FCC Emission Limits for Class A Digital Devices

Radiated Emissions (10 meters)

Frequency (MHz)	uV/m	dB(uV/m)
30 – 88	90	39
88 – 216	150	43.5
216 – 960	210	46.5
>960	300	49.5

Conducted Emissions

Frequency (MHz)	uV	dB(uV)
0.45 – 1.705	1000	60
1.705 – 30	3000	69.5

Table 16.6
FCC Emission Limits for Class B Digital Devices

Radiated Emissions (3 meters)

Frequency (MHz)	uV/m	dB(uV/m)
30 – 88	100	40
88 – 216	150	43.5
216 – 960	200	46
>960	500	54

Conducted Emissions

Frequency (MHz)	uV	dB(uV)
0.45 – 30	250	48

SOFT MAGNETIC MATERIALS FOR EMI SUPPRESSION

Table 16.7- MIL-461 Standard Frequencies and Applications Section

MIL-STD-461D

Conducted Emission

CE101 - 30 Hz to 10 kHz (Power and Interconnecting Leads)
CE102 - 10 kHz to 10 MHz (Power and Interconnecting Leads)
CE106 - 10 kHz to 40 GHz (Antenna Terminals)

Conducted Susceptibility

CS101 - 30 Hz to 50 kHz (Power Leads)
CS114 - 10 kHz to 400 MHz (Power and Interconnecting Control Leads)
CS115 - Spikes, 2ns x 30ns (Power Leads)
CS116 - 10 kHz to 100 MHz (Damped sinusoidal transients, Cables)

Radiated Emissions

RE101 - 30 Hz to 100 kHz (Magnetic Field)
RE102 - 10 kHz to 18 GHz (Electric Field)
RE103 - 10 kHz to 40 GHz (Spurious and Harmonics)

Radiated Susceptibility

RS101 - 30 Hz to 100 GHz
RS103 - 10 kHz to 40 GHz (1000 Hz Square Wave Modulation)
RS105 - Electromagnetic Pulse Field Transient

FCC, among other devices, regulates in FCC Rule-Title 47, Part 15, Subpart B (shown in Tables 16.5 and 16.6), "unintentional radio-frequency devices" which include any unintentional radiator that generates or uses timing pulses at a rate exceeding 9000 pulses per second and uses digital techniques. This, of course, would include every personal computer, point-of-sale terminal, modem, printer and many electronic games. Manufacturers of this equipment must have the product measured and approved for radiated and conducted emissions before advertising or selling the product. There are two categories to this rule, namely Class A and Class B. Class A devices are for use in industrial and business climates. Class B items are those for home use. Those for Class B are more stringent. For military equipment, MIL STD-461 (Table 16.7) is applicable for items from tools to workstations and actually sets limits on radiated and conducted emissions and immunities for the same. The corresponding test procedure is given in MIL-STD-462 . A generalized set-up for radiated and conducted EMI testing is given in Figure 16.1 and 16.2.

396 HANDBOOK OF MODERN FERROMAGNETIC MATERIALS

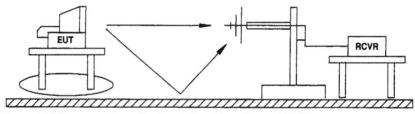

Figure 16.1-Test setup for radiated emissions From Steward(1995)

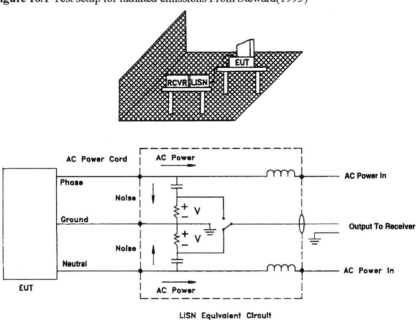

Figure 16.2- Test setup for conducted EMI From Steward(1995)

MATERIALS FOR EMI SUPPRESSION

The materials available for EMI suppression applications essentially are of two types. The most widely used would be soft ferrites and the other less widely used one would be powder cores. Although the principal operational frequency may be quite low (line or mains frequency, 50-60 Hertz), it is not primarily that frequency which is designed for in EMI suppression. It is rather the interference or disturbance frequency that mostly determines the choice of material used. This frequency can be high frequency ac or square or other digital waveform in the high Kilohertz or Megahertz region. The secondary consideration would be the operational frequency in that the material must pass the lower frequency with sufficient inductance. This means that, at low frequencies, the material must behave as a fairly good inductor but at high frequencies, it must be quite lossy. The frequencies involved in this application preclude many of the materials listed earlier (metals) except for the ferrites and iron powder cores. Other possible new materials which will be listed later are the High Flux NiFe powder cores and the Sendust (Fe-Al-Si) powder cores.

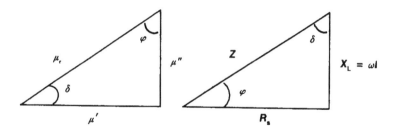

Figure 16.3-(Right)-Vector relationship between X_L, R and Z. (Left)-Vector relationship between μ', μ'' and μ_e

The ferrites used in EMI applications are of 2 generic types. i.e., NiZn and the MnZn. The powder cores would be of the higher permeability type (35-200 perm). This would limit their use to the lower frequency applications, i.e., up to a few hundred Khz. When one looks at the technical papers on ferrite materials for applications such as low-level transformers and inductors or for high level transformers and inductors, there is a large number of these giving close details on the chemistry and microstructure. Papers on the chemistry of EMI suppression materials are notoriously absent. Part of the reason for this anomaly may be due to the recent recogni-

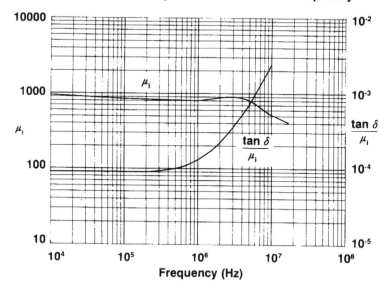

Figure 16.4- Frequency dependence of μ' and Loss Factor ($\tan\delta/\mu'$) for a EMI suppression ferrite.

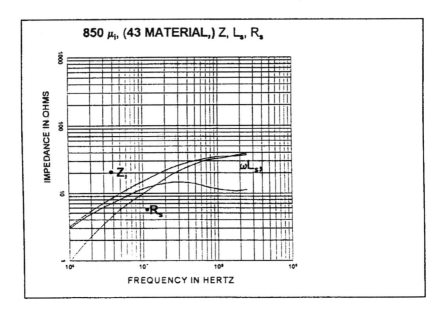

Figure 16.5-Frequency dependence of X_L, R, and Z for an EMI suppression material. From Parker (1994) Courtesy Fair-Rite Products

tion of their usefulness in combating EMI. However, more to the point, it maybe due to the fact that many companies have specific proprietary chemistries and processing to obtain the good EMI material. Another factor may be the small number of companies who do a sizeable business in this application. It is pretty clear that, in the U.S., two companies have dominated the field. They are Fair-Rite Products and Steward. This is somewhat surprising since a large market for these EMI components is in Europe or for products made elsewhere but sold in Europe.

Since the frequencies for many EMI suppression applications are in the higher Megahertz band, the ferrites for EMI suppression are mainly of the NiZn variety. The use of MnZn material has been limited to the high KHz. or, at most, the very low MHz.region. The MnZn materials that are often used for noise filters at lower frequencies are the high permeability (5000-10,000 perm) ferrites. These materials have been described in an earlier chapter. A recent MnZn material typified by the Philips 3S4 material has been introduced to compete with the NiZn material in the 800-1200 perm range and at somewhat higher frequencies. The resistivity of this material is said to be 2 orders of magnitude than other MnZn materials.

The NiZn materials normally contain copper for lower temperature firing and Mn for increased resistivity. The permeabilities for many of these materials are in the 500-1200 perm range. There is also a NiZn material with cobalt additions for higher frequency operation. This material is of lower perm (125). In addition to the chem-

SOFT MAGNETIC MATERIALS FOR EMI SUPPRESSION

istry, of EMI suppression material, there is also concern for the microstructure.

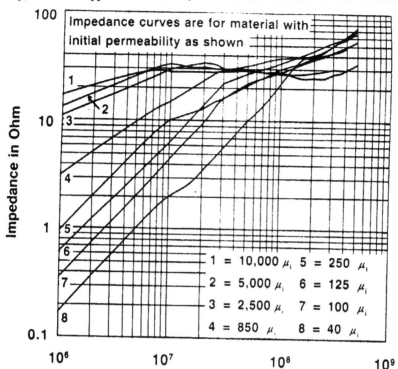

Figure 16.6- Impedance versus frequency curves for a wide range of EMI suppression materials. Courtesy Fair-Rite Products

When high perm MnZn material is used, large grains are conducive to high perm and high inductance needed to pass the lower frequencies but cause permeability fall off and rapidly increasing losses at somewhat higher frequencies where the suppression is desired. I would guess that the higher frequency MnZn material has increased resistivity due mainly to the prominent grain boundary structure. For the higher frequency, finer grain size and various degrees of intentional porosity aid in passing the medium high frequencies while attenuating the very high frequencies. Obviously, there is an interplay between chemistry and microstructure to obtain the desired result. Another consideration in these two property-determining factors is the influence of DC bias on the EMI suppression. Much of this information is obtained by experience and experiment.

FREQUENCY CHARACTERISTICS OF EMI MATERIALS

The function of the EMI suppressor core must meet two general needs, namely;

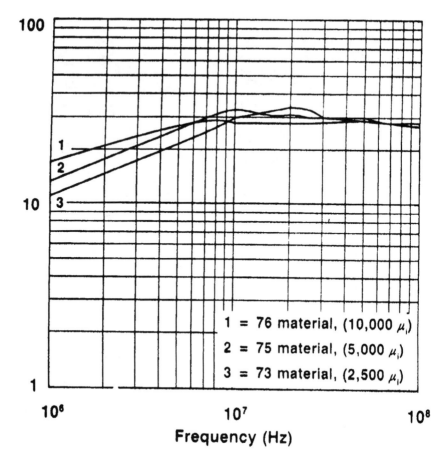

Figure 16.7 Impedance versus frequency curves for EMI suppression in the 1-10 MHz range. Courtesy Fair-Rite Products

1. The passage of the lower frequency with sufficient inductance.
2. The suppression of the higher frequencies.

One governing rule of EMI suppression is Snoek's Law;

$$f = \gamma M_s / 3\pi(\mu-1) \text{ Hz.} \quad [16.1]$$
Where f = Resonant frequency
M_s = Saturation Magnetization
γ = Gyromagnetic Ratio

This equation can also be approximated by

$$f = B_s/\mu_i \text{ MHz.} \quad [16.2]$$

SOFT MAGNETIC MATERIALS FOR EMI SUPPRESSION

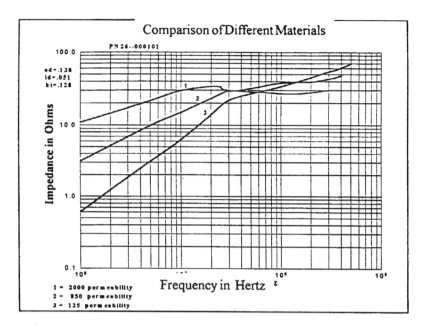

Figure 16.8-Impedance versus frequency curves for EMI materials in the 10-100 MHz. Range. From Parker (1994) Courtesy Fair-Rite Products

This shows what we have already surmised, that is, when f is low, the permeability will be high. When f is high, μ must be low. Thus, the higher the frequency of operation, be it ferrite or powder core, the lower the perm, and the converse. It also follows that the higher the perm, the lower the frequency of fall off and vice versa.

THE MECHANISM OF EMI SUPPRESSION

Thus far in our discussion of EMI suppression, we have dealt with material parameters such as permeability and resistivity, In actual operation of the device, we must change over to the equivalent component parameters such as inductance and resistance. The most important parameter for rating Emi suppression is the impedance, Z, which may be regarded as the a.c. resistance of a material to current flow at a particular frequency. The impedance is defined as;

$$Z = \sqrt{R^2 - (X_C - X_L)^2} \quad [16.3]$$

Unless we are using an LC filter network, we may focus our attention primarily on the R and XL parameters. In this case;

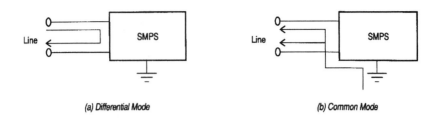

(a) Differential Mode   (b) Common Mode

Figure 16.9- Diagram of differential mode and common mode currents in a switched mode power supply. In the differential mode, when a ferrite core is placed over both conductors, the net flux is 0 and the induced voltage is 0. In the common mode current, there is net flux and the induced voltage is not 0.

$$Z = \sqrt{R^2 - X_L^2} \quad [16.4]$$

The vector relationship between these 3 parameters is shown in Figure 14.3. The inductive reactance, X_L is given by

$$X_L = 2\pi f L \quad [16.5]$$
Where L = inductance.

At low frequencies, X_L will dominate the impedance value as R is comparatively low. At these frequencies, some EMI suppression occurs by inductive reactance. However, at the frequency when the permeability (as represented by μ') drops sharply, as shown in Figure 16.4, X_L will also decrease as R (represented by loss factor, LF) increases. Thus, the main source of the suppression at higher frequencies is the absorption of the these frequencies by resistive heat. This interplay of the three parameters is shown in Figure 16.5. The best material for suppression in a particular frequency range may be inferred by the peak or upward trend of the impedance. Figure 16.6 shows a family of impedance versus frequency curves for a wide range of material permeabilities. The inverse relation of frequency of impedance peaks and material permeabilities is demonstrated. A localized version of this curve for the low frequency range (1-10 MHz.) is shown in Figure 16.7. Figure 16.8 shows a similar curve for higher frequencies.

COMPONENTS FOR EMI SUPPRESSION
Typical ferrite materials for EMI suppression applications are shown in Table 16.8 with the pertinent magnetic parameters. From the frequency of the noise to be suppressed, the appropriate material is chosen.

Before our discussion of the actual components used for EMI suppression, it is useful to look at the circuitry involved along with the currents both intentional and unintentional. The latter, of course are the EMI interference or noise currents.

A Simple Model of CM and DM Currents

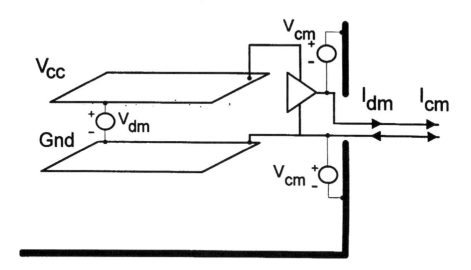

Figure 16.10- Common and differential mode currents in an EMI suppression circuit. From Lee Hill,(1994) MMPA Soft Ferrite Users Conference, Feb. 24-25, 1994, Rosemont, IL

Basically, there are two types of EMI currents, namely the common-mode and the differential currents. These are contained in a pair of wires leading to and from the load. The first of these is the differential currents which is the same as ordinary intended or designed circuitry as shown in Figure 16.9. The differential EMI currents then flow in the same direction as the intended currents. If a current probe is placed around the pair of conductors, no current flow will be detected for either the intentional or unintentional(EMI) currents. In the case of the common-mode EMI currents, they flow in the same direction in both conductors. Now, while the differential intended currents will cancel, the non-intentional (EMI) currents willnot and there will be a current indicated with a probe. Another way of describing the two types of current flow is shown in Figure 16.10 that shows the voltages producing the currents. In the differential case, the voltage is between the high voltage line and the neutral line while in the common-mode case, it is the voltage between both the high voltage and neutral lines to ground.

Common-Mode Filters
Since there are two types of EMI currents, there are two types of filters to handle them. The common mode filter uses a single core enclosing both conductors. The intended signal will be passed by cancellation of the two opposing currents. The EMI currents will produce magnetic flux in the core which because of the higher

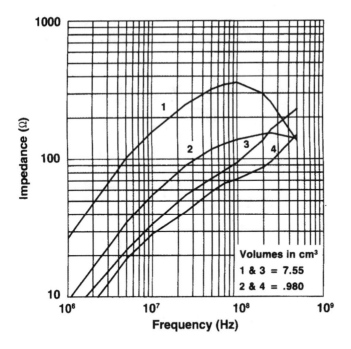

Figure 16.11-Impedance versus frequency curves for two pairs of cores each pair with the same volume but with different impedance curves. From Parker (1994) Courtesy Fair-Rite Products

frequency will be attenuated. The common-mode filter is then transparent to differential mode currents and attenuates common-mode currents. One simple method of determiming whether EMI noise is due to common mode or differential mode currents is by placing a ferrite core over both conductors, If the interference is removed or lessened, the noise was common mode. If no improvement, differential mode currents were the cause. Applications using the common-mode configuration are;

1. High current DC power filtering
2. Reducing noise on high speed differential lines

Very small common-mode currents(as low as 8µA) can cause failure in radiated or conducted EMI tests. For a differential mode filter, the equivalent current would be 19.9 mA, showing the greater sensitivity of the common-mode currents. Probably the simplest method of EMI suppression is the use of a ferrite core alone placed right over the device leads or on a PC board to prevent parasitic oscillations or attenuate unwanted signal pickup. The most common component of this type is the shield bead (sometimes with leads) but other shapes to accommodate cables either flat or round are available. For multiple lines, discs and plates are also used.

Ferrite beads are available to slide over conductors or they are available with preformed leads. The choice of amount of impedance is not always dependent on the volume of the core. Figure 16.11 shows 4 cores of the same material. Two sets have the same volumes but have different impedances. In fact, Figure 16.12 shows cores having volumes twice and tenfold volumes with similar impedance

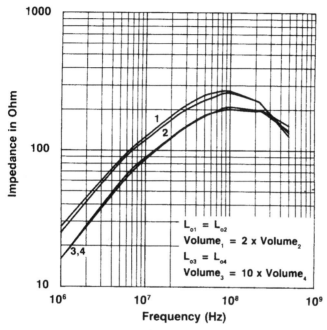

Figure 16.12-Impedance versus frequency curves of two sets of cores with similar impedance curves but core volumes that vary by factors of four and ten. From Parker (1994) Courtesy Fair-Rite Products

characteristics. A useful representation of the impedance as a function of dimensions which is valid below the resonance frequency is the use of the toroidal L_o which is the air core inductance. If we define ;

$Z = K/L_o$
Where $L_o = .046 \times N^2 \times \log_{10}(OD/ID) \times h \times 10^{-8}$
Where the dimensions are in mm.
Then; $Z = .046 K \times N^2 \times \log_{10}(OD/ID) \times h \times 10^{-8}$
and $K = Z/L_o$ ohm/H x 10-8, $N = 1$

If one determines the K by measuring the impedance and dimensions of one core of a material, the one can approximate the impedance of another size core. One important revelation given by this equation is the importance of specific dimensions. The impedance varies as only the log of the OD/ID ratio but directly with the height or length of the bead. In most cases of usual OD/ID ratios, an increase in length is more predominant in impedance determination. Figure 16.13 shows a curve of impedance versus K or Z/L_o for a higher perm (2500) material. Figure 16.14 and 16.15 show similar curves for a medium perm and a low perm (high frequency) material. The approximation of impedance can be made from the K for each material (at the specific frequency) and the dimensions. Also shown on each curve is the phase angle (the angle whose sine is X_L/R_s). Pure inductors have phase-angles of 90° and for pure resistors it is 0°. Thus the course of phase angle is an indication of the frequency response of this ratio. Note that, in the curve for the highest perm material, the phase angle goes from about 80° to 10° in the frequency range 1-30 MHz. In the curve for the medium perm material, the phase angle goes from 75° to 0° for frequencies from 1-100 MHz. For the lowest perm material, the phase angle goes from 90° to 10° from 10-400 MHz. The behavior of Z/L_o is also reflective of the permeabilities of the material.

Stability of Ferrite Suppression Cores

The impedance of a ferrite suppression core will be degraded with temperature and D.C. bias. The variation of impedance for three ferrite materials with temperature at 25 Khz. is shown in Figure 16.16. As might be surmised, the most stable material is the lowest perm NiZn material. Surprisingly, the Highest perm MnZn ferrite is next followed by the middle perm NiZn material. The same type of curve for 100 MHz is shown in Figure 16.17. The variation is similar but shows greater reduction.

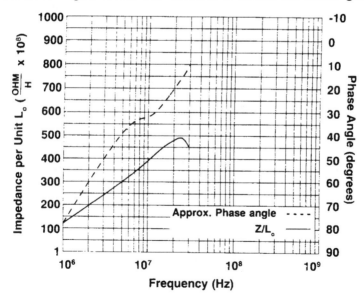

Figure 16.13-Impedance per unit L_o(Z/L_o) or K and approximate phase angle for a higher perm (2250) material. Courtesy Fair-Rite Products

SOFT MAGNETIC MATERIALS FOR EMI SUPPRESSION

Another important stability characteristic is DC bias since DC is often present. Figures16.18a,b and c give the DC bias effect on impedance for the three different permeability materials at the pertinent frequency ranges. For the MnZn material (a) even at the two lower frequencies, the maximum bias before a large drop is only about 2-3 Oersteds. For the higher perm NiZn (b)the drop-off is slower and as much as 10 Oersteds can be tolerated. For the low perm NiZn(c) at higher frequencies, there is still sufficient impedance at 15 Oersteds. In the common-mode filter, there must be by-pass capacitors to ground. For safety reasons, the value of the capacitor cannot be too high. For a resonant circuit, this restriction limits the minimum inductance that the core can carry. This value is 1000 µH.

Aside from the simple ferrite core type of suppressor, there is also the coil suppressor, often a wound toroid or slug. In this case, the extra turn produce a capacitive reactance, which combines with the inductance to form a type of LC filter. The ferrite suppressor has lower impedance but greater bandwidth. The coil has higher impedance and better frequency control. With the ferrite core, the performance is determined by the length of material while in the coil, the performance is determined by the winding as well as the core dimensions. Cost-wise, the coil approach is more expensive.

DIFFERENTIAL MODE FILTERS

In the case of the differential EMI filter, the offending currents do not cancel and thus, each line must be protected separately. Since the full circuit current will pass through this filter, the component used for the common-mode would not be appropriate since the higher permeability material would cause the core to saturate. Instead a core with a low effective permeability must be used. The two choices are;

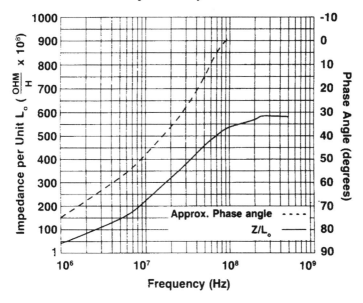

Figure 16.14-Impedance per unit L_o(Z/L_o) or K and approximate phase angle for a medium perm (850) material. Courtesy Fair-Rite Products

1. A gapped ferrite core
2. A powder core of either iron, high flux MPP or Sendust (Kool-Mu or MSS core)

The differential-mode filter is used in
1. Low current DC and Ac filtering
2. Removing HF Noise from low speed signals
3. Shaping rise times of high speed signals

METAL POWDER CORES FOR EMI SUPPRESSION

While common-mode filters are mostly used in unbalanced circuits where the currents return to ground, the differential-mode filter is used primarily in balanced systems. Consequently, putting a ferrite toroid around both wires would not cause any flux change in the core and so not suppress the EMI. The solution in this case is to put suppressor cores on each of the wires. However, this means that the full ac (and D.C.) signals would pass through the suppressor core. While the common-mode ferrite core can be used in high current power filters, the differential-mode suppressor ferrite core is only used in low current power filters. With the differential-mode or in-line filters, the core losses could be a problem (except for D.C.), the main problem would be core saturation. To prevent this, a core with low permeability is needed. Either a gapped ferrite core or a powder core can be used. The gapped

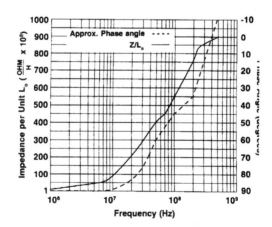

Figure 16.15-Impedance per unit L_o (Z/L_o) or K and approximate phase angle for a lower perm (125) material. Courtesy Fair-Rite Products

SOFT MAGNETIC MATERIALS FOR EMI SUPPRESSION

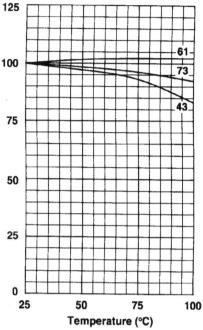

Figure 16.16-Percentage variation of impedance with temperature for three different EMI suppression materials at 25KHz. Courtesy, Fair-Rite Products

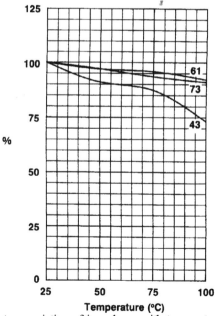

Figure 16.17- Percentage variation of impedance with temperature for three different EMI suppression materials at 100 KHz. Courtesy Fair-Rite Products

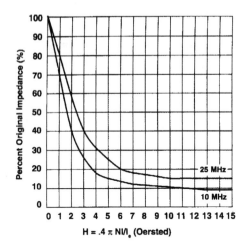

Figure 16.18a-Percent reduction in impedance with DC bias for a 2500 perm MnZn ferrite EMI suppression material Courtesy Fair-Rite Products

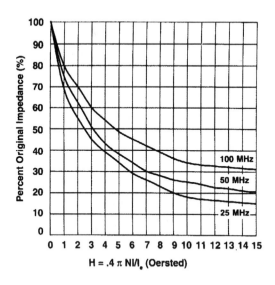

Figure 16.18b-Percent reduction in impedance with DC bias for a 850 perm NiZn ferrite EMI suppression material Courtesy Fair-Rite Products

SOFT MAGNETIC MATERIALS FOR EMI SUPPRESSION

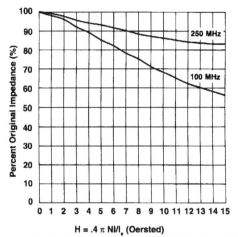

Figure 16.18c-Percent reduction in impedance with DC bias for a 125 perm NiZn ferrite EMI suppression material Courtesy Fair-Rite Products

The iron powder cores, unlike those listed in the chapter on low level telecommunications applications are of the higher permeabilities (from 70-100 perm). They are usually of the hydrogen-reduced variety. Before discussing the magnetic properties of powder cores for EMI suppression, we must point out that, while permeability has been our criterion for EMI ferrite suppression ability, with powder cores, vendors do not specify impedance. Since the application of these cores involves high flux densities and often D.C. bias, the magnetic properties usually listed are;

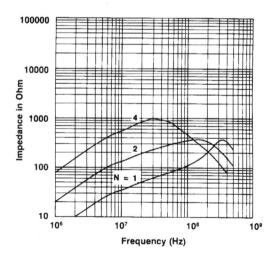

Figure 16.19- Impedance vs frequency characteristics of a ferrite EMI suppression core when the turns are doubled and quadrupled. From Parker (1994) Courtesy Fair-Rite Products

1. Permeability versus Flux Density
2. Permeability versus DC Bias
3. Core Losses
4. Permeability versus Frequency
5. Permeability versus Temperature
6. D.C. Energy storage curves

Another important property for an application as a D.C. Choke is the variation of energy stored versus D.C. current. The energy storage criterion is given by ½ LI^2. For the iron powder cores, the high saturation of about 20,000 Gauss is suited for this application. Curves displaying the variation of permeability with flux density and DC bias for iron powder cores are shown in Figures 16.21-22. Permeability versus frequency is shown in Figure 16.23. Core loss and Energy storage curves are shown in Figures 16.24-25.

The NiFe powder cores listed as High Flux cores are different from the MPP cores listed in the chapter on low level applications of powder cores. These NiFe cores are 50% Nickel-50% Iron. The have about twice the saturation (15,000)of the MPP cores and thus are much better for the present application. Cores of this material are available in permeabilities of 200, 160, 147, 125, 60, 16 and 14. The variations of permeability with flux density and D.C. bias, frequency and temperature for different permeabilities of this material are shown given in Figures 16.23-24. The core losses are given in Figure 16.25. As expected, the stability is inversely propor-

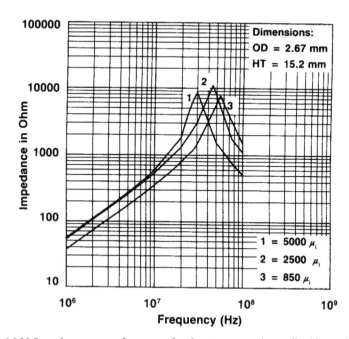

Figure 16.20-Impedance versus frequency for slug-type cores in a coil with varying material

SOFT MAGNETIC MATERIALS FOR EMI SUPPRESSION 413

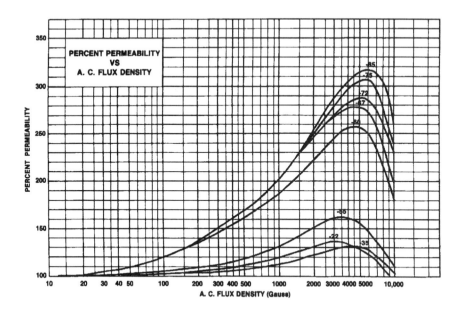

Figure 16.21- Permeability vs ac flux density for several different permeability iron powder cores. Curve numbers are the permeabilities. From Pyroferric

tional to the permeability but the high frequency core losses are proportional to the frequency. Cost-wise the High Flux powder cores are more expensive than the iron-powder cores, but somewhat less expensive than the MPP cores. The Sendust cores are marketed under the trade names of Kool-Mu and MSS materials. It is a new application of an old material having been described by Matsumoto in 1936 and patented in 1940 (Matsumoto 1940). Sendust is a ternary alloy containing about 6% aluminum and 9% silicon. Its attraction is that it is close to a zero anisotropy-zero magnetostriction material. Its brittleness and difficulty in producing it have limited its use in the past to recording head material due to its great hardness. When used in powder cores, its brittleness helps in the comminution process. The high saturation of this material (on the order of about 10,000 Gauss provides much more energy storage than MPP cores or gapped ferrites. The cores come in permeabilities of 60, 75, 90 and 125. Figures 16.26-28 show the permeability variations of different permeabilities of this material for flux density, D.C. bias and temperature. The core loss of the 125 perm material is given in Figure 16.29.

414 HANDBOOK OF MODERN FERROMAGNETIC MATERIALS

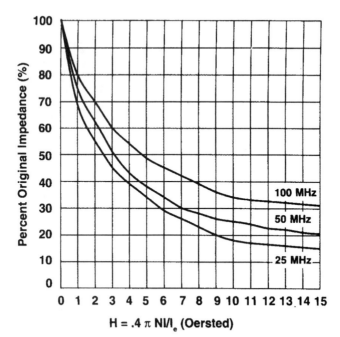

Figure 16.22-Percent reduction in permeability with DC bias for several different permeability iron powder cores. Curve numbers are the perms. From Pyroferric

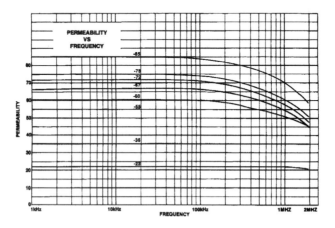

Figure 16.23-Variation of permeability with frequency for several different permeability iron powder cores From Pyroferric

SOFT MAGNETIC MATERIALS FOR EMI SUPPRESSION 415

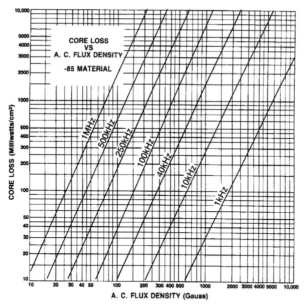

Figure 16.24- Core Losses of a 85 perm iron powder core. From Pyroferric

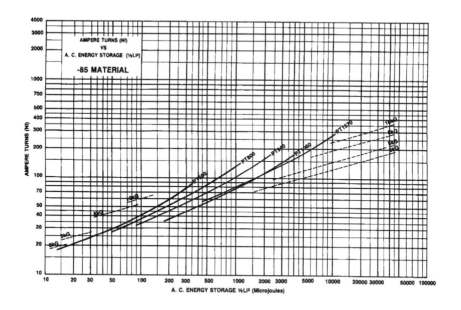

Figure 16.25- Energy storage curves for 85 perm iron powder cores. From Pyroferric

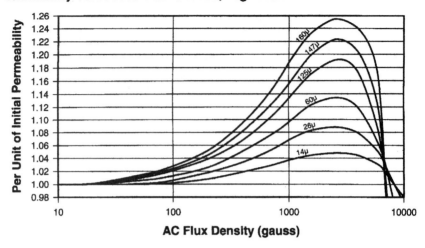

Figure 16.26a-Permeability versus flux density for different permeability NiFe High Flux Powder Cores. From Magnetics 1998

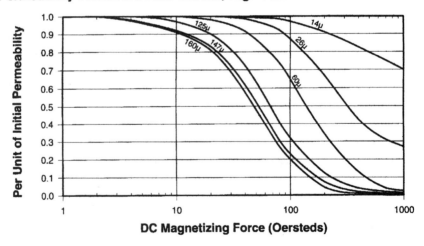

Figure 16.26b-Permeability versus DC bias for different permeabilty NiFe High Flux powder cores. From Magnetics 1998

SOFT MAGNETIC MATERIALS FOR EMI SUPPRESSION

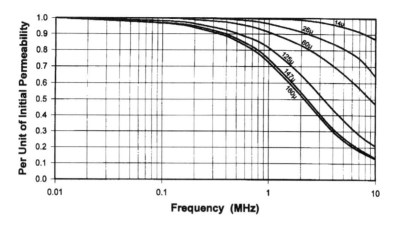

Figure 16.26c-Permeability vs frequency for different permeabilty NiFe high flux powder cores. From Magnetics 1998

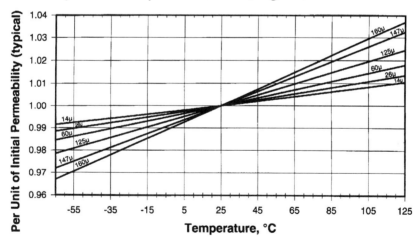

Figure 16.26d-Permeability vs temperature for different permeabilty NiFe high flux powder cores. From Magnetics 1998

Core Loss Density Curves, High Flux 147μ / 160μ

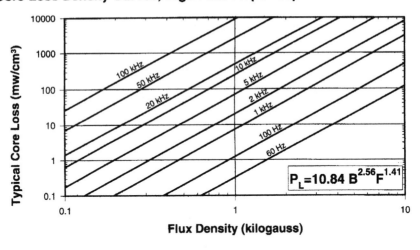

Figure 16.27-Core loss curves for different permeabilty NiFe high flux powder cores. From Magnetics 1998

Permeability versus AC Flux Curves, Kool Mμ

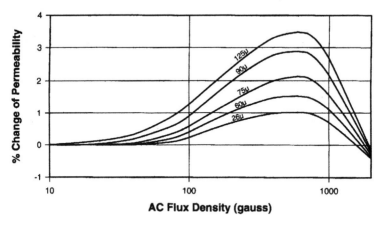

Figure 16.28a- Permeability vs flux density for several different permeability Kool-Mu (Sendust) Cores. From Magnetics 1998

SOFT MAGNETIC MATERIALS FOR EMI SUPPRESSION 419

Permeability versus DC Bias Curves, Kool Mµ

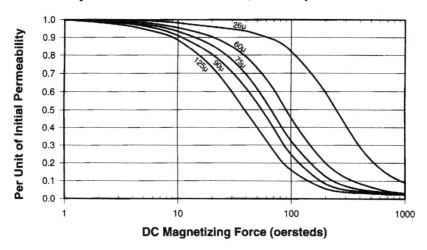

Figure 16.28b- Permeability vs DC bias for several different permeability Kool-Mu (Sendust) cores. From Magnetics 1998

Permeability versus Frequency Curves, Kool Mµ

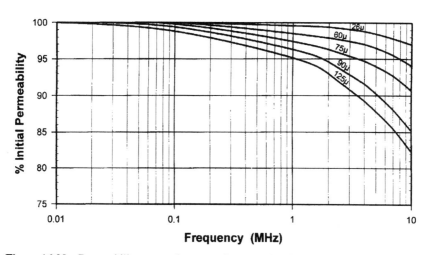

Figure 16.28c- Permeability versus frequency for several different permeability Kool-Mu (Sendust) cores. From Magnetics 1998

Permeability versus Temperature Curves, Kool Mµ

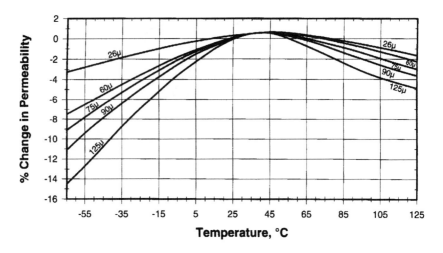

Figure 16.28d- Permeability versus temperature for several different permeability Kool-Mu (Sendust) cores. From Magnetics 1998

Core Loss Density Curves, Kool Mµ

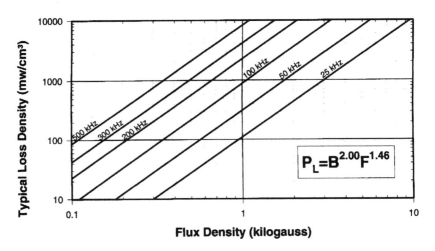

Figure 16.29- Core loss Curves for 125 perm Kool-Mu (Sendust) powder cores

The core losses are significantly lower than the iron powder cores. However, the Sendust cores are somewhat more expensive. Cores with O.D's from .140 inches to 2.25 inches are available.

AMORPHOUS- NANOCRYSTALLINE MATERIALS -EMI SUPPRESSION

One of the earliest uses of the amorphous material described in Chapter 18 was for a choke coil that Toshiba called the "Spike-killer". Presumably, only the cores are sold. The amorphous materials used are under license from Allied's Metglas® Division. Vacuumschmelze does market a Co-based amorphous material for EMI suppression applications. It is designated Vitrovac 6025 and is essentially a zero magnetostriction material. The characteristics are shown in Table 16.8. As an outgrowth of the amorphous materials, the iron-based nanocrystalline materials are the newest ones available and they have been used for EMI suppression. Their high permeabilities and low magnetostrictions made it very useful as a common-mode choke at relatively low frequencies. The earliest nanocrystalline soft magnetic alloy was made by Hitachi and names "Finemet". The properties of the Finemet nanocrystalline material are given in Table 16.9. The saturation is high but not in the Si-Fe class. The permeability is high but not in the 80 permalloy class. The magnetostriction is low and about the same as the Co-based amorphous material. The resistivity is the same as the amorphous alloys but many orders of magnitude lower than ferrite. It is really the combination of most of the good attributes that make it attractive. High permeability materials without the high resistivities are useful for the inductive or permeability portion of the impedance and are limited to the lower frequencies.

Vacuumschmelze has two nanocrystalline materials, Vitroperm 500F and 800F that they recommend for EMI suppression(common-mode chokes) along with their Co-based amorphous zero-magnetostriction material Vitrovac 6025. The permeability of the 800F is somewhat higher than the 500F. A comparison of the Co-based amorphous, the nanocrystalline material and a MnZn ferrite is given in Figure 9.31 The permeability is higher and the loss factor is lower for the metallic

Table 16.8 Properties of Amorphous and Nanocrystalline Materials for EMI Suppression Applications

				6025 F	500 F
Sättigungsinduktion /	saturation induction	B_s	[T]	0,55	1,2
Sättigungsmagnetostriktion /	saturation magnetostriction	λs	[ppm]	0,20	<0,5
Spez. elektr. Widerstand /	resistivity	ρ	[W mm²/m]	1,35	1,15
Curietemperatur /	curie temperature	T_c	[°C]	210	600
Banddicke /	tape width		[μm]	25	20
Stat. Koerzitivfeldstärke /	static coercive field strength	H_c	[A/cm]	0,003	0,005
Ummagnetisierungsverluste /	hysteresis losses	(P_{Fe} 100 kHz; 0,3 T)	[W/kg]	100	105
Dichte /	density	ρ	[g/cm³]	7,70	7,35

From Vacuumschmelze ,1994

Table 16.9- Properties of "FINEMET" Nanocrystalline Material

		FINEMET FT-1KM	Mn-Zn ferrite	Co-based amorphous	Fe-based amorphous
Initial magnetic permeability uri	10kHZ	>=50,000	5,300	90.000	4.500
	100kHz	16,300+/-30%	5,300	18,000	4,500
Maximumn flux density BMS*(T)		1.35	0.44	0.53	1.56
Coercive force Hc*(A/m)		1.3	8.0	0.32	5.0
Rectangular ratio Brms/Bms*		0.60	0.23	0.50	0.65
Core loss Pc**(kW/m^3)		350	1,200	300	2,200
Curie temperature Tc(C)		570	150	180	415
Saturation magnetostriction(x10^-6)		+2.3	-	-0	+27
Specific resistivity(ohm-m)		1.1x10^-6	0.20	1.3x10^-6	1.4x10^-6
Density ds(x10^3kg/m^3)		7.4	4.85	7.7	7.18

From Hitachi 1998

materials. The insertion damping curves (50 ohms) versus frequency for the Vitroperm 500F with different windings and that of a MnZn ferrite are in Figure 16.30

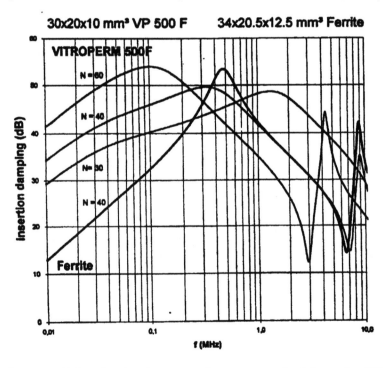

Figure 16.30-Insertion damping curves of a common-mode choke made of Vitroperm 500F for different numbers of turns compared to a choke made of ferrite. From Hilzinger, 1996

References

Fair-Rite (1996) Fair-Rite Soft Ferrites, 13th Ed. Fair-Rite Products Corp. One Commercial Row, Wallkill, NY 12589

Hill, L.(1994) MMPA Soft Ferrite Users Conference, Feb. 24-25, 1994, Rosemont, IL

Hilzinger, H.R. (1996)Soft Magnetic Materials '96, Feb.26-28, 1996, San Francisco, Gorham-Intertech Consulting, 411 U.S. Route One, Portland ME, 04105

Hitachi (1998) FINEMET FT-1KM-KN Series Core Page on Internet Product Guide

Magnetics (1998) Powder Cores MPP Cores for Filter and Inductor Applications, Magnetics, Div. of Spang and Co. Butler, PA 16001

Micrometals (1990) Micrometals Iron Powder Cores, EMI and Power Filters Micrometals, 1190 N. HawkCircle, Anaheim ,CA, 92807

Pyroferric(1984) Toroidal Cores for EMI and Power Filters, Pyroferric Internatiional, 200 Madison St., Toledo, IL 62468

Vacuumschmelze (1995) Vitrovac 500F-Vitroperm 6025, PK-004, Vacuunschmelze GMBH, Hanau, Germany

17 FERRITES FOR ENTERTAINMENT APPLICATIONS-RADIO AND TV

INTRODUCTION
The largest tonnage of soft ferrites goes into the radio and television applications. In particular, the television industry is the largest user since it is a popular consumer item. For this reason, the selling price of the ferrite component must be quite low, which in turn means that the ferrite cost material must also be low. Fortunately, the raw materials for the ferrites are inexpensive especially for the quality required which is fairly low. Consequently, the ferrites for entertainment uses have few competitors while the present system using the electron beam picture tube is in use. The components using ferrites for entertainment are;

1. Picture tube deflection yokes
2. Flyback Transformer
3. Audio and Intermediate Frequency Transformers
4. Antennae
5. Pin Cushion Transformer
6. Tuning Slug
7. EMI Suppression Cores
8. Power supply choke and transformers

The EMI cores were covered in Chapter 16 and the power materials will be covered in the next chapter.

FERRITE TV PICTURE TUBE DEFLECTION YOKES
In our discussions thus far, we have spoken of magnetic fluxes created by electron circulation in wires (currents) or in magnetic solids. These could be acted upon by external magnetic fields. Now we are about to discuss the control of an electron flow that is in the form of a beam in air or more appropriately in a vacuum. This is the type of beam found in cathode ray tubes such as a television picture tube or a display monitor. Other similar uses of electron beam control are found in radar, microwave tubes and satellite communications. Control or deflection of the beam is still subject to external fields. In this case the electron beam current is from the negative terminal to the positive or the opposite of the conventional current direction from positive to negative. Electric field deflection of an electron beam is possible but not as efficient as a magnetic deflection scheme. In magnetic deflection, the force between the beam and the magnetic field is given by the vector cross product;

$$F = v \times B \qquad [17.1]$$

Where F = force in dynes

426 HANDBOOK OF MODERN FERROMAGNETIC MATERIALS

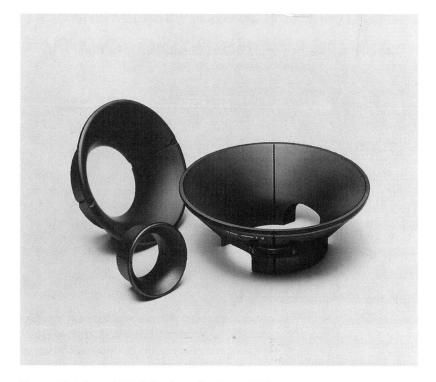

Figure 17. 1-Several TV deflection yokes From TDK

v = velocity of the electron
B = Flux density of coil

The direction of the vector cross product is perpendicular to the directions of the other two vectors. This means that the deflection will be in a plane perpendicular to the electron beam and that a vertical magnetic field will produce a horizontal force (and deflection) of the electron beam and vice versa. A simple air coil would produce a deflection but a coil wound on a magnetic material would increase B and therefore the deflection considerably. At the frequencies involved, a ferrite core would be the material of choice. The ease of molding the ferrite core to the shape of the tube neck where the yoke is placed is another cost-saving plus for ferrite. Since there are vertical and horizontal deflections used to complete the picture, two sets of coil windings are required.

The ferrite deflection yokes are funnel-shaped toroids to conform to the neck of the tube. Several typical shapes are shown in Figure 17.1. The yokes are pressed to the final shape but with two diametrically-opposed notches down the sides if the yoke. After firing, the yoke is split into two halves by placing a hot wire in the notches of the cold ferrite. The two halves are then wound with the two sets of deflection coils and the yoke reassembled. The two halves can then be held together

FERRITES FOR ENTERTAINMENT APPLICATIONS

Table 17.1

Comparison of Properties of MnZn and MgZn Ferrites for TV Deflection Yokes

Property	MnZn Ferrite	MgZn Ferrite
Permeability, μ	900-1100	350-500
Saturation, Bs (Gs)	3000-4000	2400-2700
Curie Temperature (°C)	180	150-160
Coercive Force, Hc (Oe)	0.25	0.35-0.5
Loss Factor	$5\text{-}12 \times 10^{-6}$	$15\text{-}50 \times 10^{-6}$
Density (g/cc)	4.8	4.5
Resistivity (ohm-cm)	10^2	$10^7\text{-}10^8$

with some tape. This procedure eliminates the costly procedure of toroidal winding of the yokes and with a good refit of the halves, produces only a slight drop in the permeability. There are many different shapes of these yokes for black and white or color TV's as well as for oscilloscopes and display monitors

MATERIALS FOR DEFLECTION YOKES

From the early days of ferrites when the TV deflection yoke was the first large commercial application of ferrites till just recently, manganese zinc ferrites was the material of choice. However, with the imminent introduction of HDTV (High Definition Television), the situation changed. Many suppliers have jumped the gun and changed over to magnesium zinc ferrites. The present day TV contains a raster or horizontal grid of 525 lines. The new HDTV is expected to increase this number to about 1200. This means that the horizontal sweep rate will have to increase proportionately. The ferrite material then must be effective at the higher frequency meaning higher resistivity. In addition, MnZn deflection yokes had to be wrapped with tape to protect against voltage breakdown between the wire and the low resistivity ferrite. Present day TV channels have a bandwidth of 6 MHz. HDTV will require a bandwidth of at least 18MHz. In addition the ratio of horizontal to vertical screen dimensions which is presently 12:6 will be raised to 16:9. With the wider screen in HDTV, the coils must wound directly on the yoke for wider deflection and better control and the tape would pose a problem. With the MgZn ferrites, the problem of higher frequency and elimination of the tape would both be solved. The windings could be applied directly on the ferrites. While the resistivity of the MnZn ferrites is only about 100 ohm-cm, that of the MgZn ferrite is about 105-6 ohm-cm. The properties of the MnZn ferrites are mostly superior to those of the MgZn ferrites (See Table 17.1) with the exception of the resistivity. In present day television sets, there are 525 horizontal lines to a raster. To the human eye, motion seems continuous if the successive frames are shown at a rate of 16 per second. This would make the minimum horizontal scanning frequency 8400 Hz. In actual fact, for good resolution this figure is considerable higher at 32-64 Khz. As we shall see, for high definition television and high resolution display monitors, the goal is 100 Khz. and higher.

At the ICF6 conference, Araki(1992) reported on a low-loss Ni-Zn-Cu ferrite for use in a high resolution display monitor at a frequency of 130 Khz. The chemistry was optimized at the following composition;
$Ni_{0.24}Zn_{0.57}Cu_{0.19}Mn_{0.01}Mg_{0.015}Ti_{0.015}Fe_{1.96(1+a/100)}O_4$.

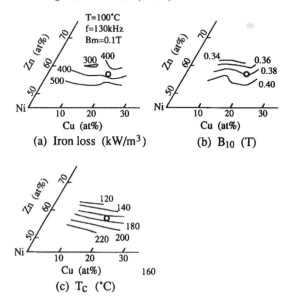

Figure 17.2- Variation of magnetic properties as a function of chemistry From Araki (1992)

In reviewing the major elements, if the sum of Cu + Ni +Zn equals 100%. The core losses were at a minimum with 10-15 atomic % Cu and 60-62% Zn. The results are shown on Figure 17.2. For additives, to reduce loss, Mg-Ti were found effective but B_{10} and T_c decreased. The optimum amount is shown in Figure 17.3. The decrease in core loss was found to be mainly due to a reduction in hysteresis loss which in turn was governed by H_c. The core loss was reduced to 270 KW/m3 at 100°C. With f=130 KHz. And B_m= 0.1 T. At the same conference, Kobyashi(1992) also reported on a ferrite deflection yoke for a high resolution display monitor but of the MgZn variety. Substitution of some of the iron with Mn gave a high flux density (B10) material with low core loss (30W/Kg.) at 100°C. With f= 32 KHz. and B =.1T., Bi_2O_3 was also used to reduce core loss. In the processing, a new granulation technique with lower PVA content also reduced variations in density in the fired yoke. Nomura and Ochiai (Nomura,1994)also reported on Mn_2O_3 additions to MgZn ferrite. The Mn increased the DC resistivity and retarded the permeability deterioration with frequency (Figure 17.4). The Mn_2O_3 addition prevented the formation of Fe^{+2}. The molar chemical composition of MgZn deflection yokes was about 49.-49.5% iron oxide with the sum of Mn_2O_3 and Fe_2O_3 at 51-53%. The effect of this sum on the degradation of perm with frequency is shown in Figure 17.5. Control of defect chemistry is important to suppress the degradation. At ICF7, Ikagami (1997) also reported on a MgMnZn ferrite deflection yoke material. The loss was lowest with 18% ZnO and increasing ratios of MnO to the sun of iron and manganese oxides (Figure 17.6). When the core loss was separated into hysteresis and

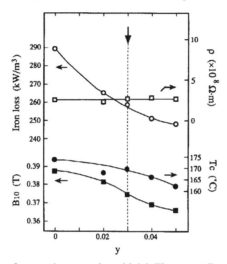

Figure 17.3- Variation of magnetic properties with MgTi content. From Araki (1992)

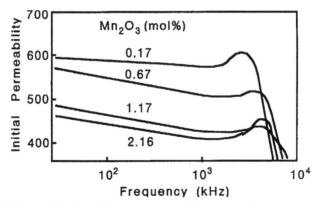

Figure 17.4-Effect of Mn_2O_3 additions on permeability versus frequency course. From Nomura (1994)

eddy current components, as a function of the above ratio, the hysteresis losses dropped with the ratio but the Eddy current losses remained flat. This confirms the earlier view that Mn affected the core losses through the hysteresis loss. The addition of Bi_2O_3 also increased density and lowered core loss up to a level of 0.6% Bi_2O_3. The reduction in core loss of the new material over conventional materials is shown in Figure 17.7.

A listing of typical deflection yoke materials commercially available at present is given in Table 17.2. The first two Materials 9H44 and H44F) with resistivities of 10^5 are MgZn materials. The other high resistivity material, D13 is a NiZn material. The one with the lower resistivity is presumably a MnZn material, Table 17.3 shows

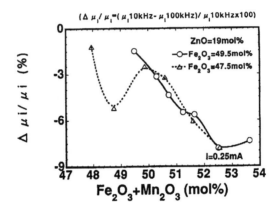

Figure 17.5- Variation of permeability change at higher frequency versus sum of Mn_2O_3 and Fe_2O_3 additions. From Nomura (1994)

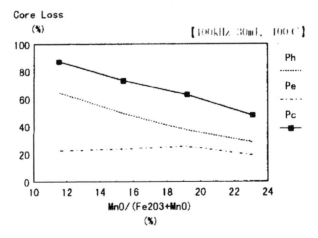

Figure 17.6- Variation of hysteresis, Eddy current and total Losses of a MgMnZn deflection yoke material as a function of ratio of Mn content From Ikegami (1997)

FERRITES FOR ENTERTAINMENT APPLICATIONS

applications of the various materials with the resolution required and the frequencies of use.

FLYBACK TRANSFORMERS

The flyback transformer in a television set works in conjunction with the deflection yoke in that it supplies the current for the horizontal deflection. The wave-form is a sawtooth type with a linearly increasing portion providing the steady deflection of the beam. At the end of the sweep, a rapid reversal of the beam is required to start

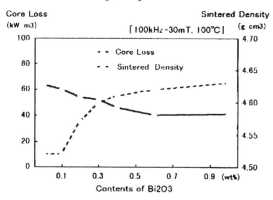

Figure 17.7-Variation of density and core loss vs Bi content in a deflection yoke material From Ikegami (1997)

Table 17.2
■ Standard material characteristics of ferrite cores for deflection yokes

Property	Symbol	Condition	Unit	H44	H44F	HD12	HD13
AC initial permeability	μ iac	0.1MHz	—	350	400	1800	500
Saturation magnetic flux density	Bs	23℃	mT	230	260	500	360
		60℃		190	220	450	325
		100℃		160	170	400	295
Residual magnetic flux density	Br	23℃	mT	160	190	150	210
Coercivity	Hc	23℃	A/m	40	24	20	36
Relative loss factor	tan δ / μ	0.1MHz	×10⁻⁶	<50	<120	<5	<30
Core loss 25kHz 100mT	Pc	23℃	kW/m³	420	230	90	280
		60℃		380	170	70	280
		100℃		430	170	60	260
Core loss 100kHz 100mT	Pc	23℃	kW/m³	1500	1000	270	960
		60℃		1470	850	230	950
		100℃		1640	860	180	940
Curie temperature	Tc	—	℃	>150	>150	>180	>200
Resistivity	ρ	—	Ω·m	10⁵	10⁵	1	10⁵
Apparent density	d	—	kg/m³×10³	4.5	4.6	4.8	5.0

Note: 1. The above values were obtained from FR25φ/15φ/5 troidal cores.
2. The values were obtained at 23±2°C unless otherwise specified.

432 HANDBOOK OF MODERN FERROMAGNETIC MATERIALS

the next line of the raster. This rapid dI/dt provides the high voltage to the secondary anodes(in the neck of the picture tube) which is used to accelerate the electron beam electrostatically. Therefore, a considerable proportion of the energy needed for the horizontal deflection is recovered during the flyback period. The flyback transformer is usually in the shape of a U-core and represents one of the few uses for this configuration that still exists. A typical flyback transformer core is shown in Figure 17.8. As expected, the saturation flux density is quite high with a moderate drop-off at 100°C. The permeability versus temperature and flux density curves are given in Figures 17.9 and 17.10. The core loss versus

Table 17.3
■ **Ferrite core materials for deflection yokes according to frequency**

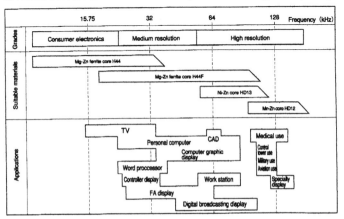

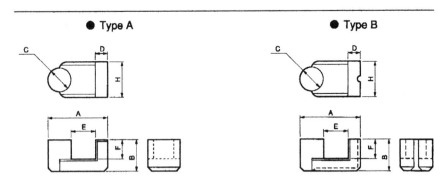

Figure 17.8 Schematic drawing of several TV flyback transformers From Fuji (1994)

frequency curves are given in Figures 17.11 and 17.12. These are obviously MnZn ferrites with compositions similar to those used for high frequency power supplies discussed in the next chapter.

GENERAL PURPOSE CORES FOR RADIO AND TELEVISION

There are numerous small ferrite cores in radio and television circuits for coupling and transformer functions of radio frequency signals. For example, in going from one amplification stage to another, it is desirable to pass signals within a certain frequency band. Figure 17.13 shows a circuit with an untuned primary of a first RF

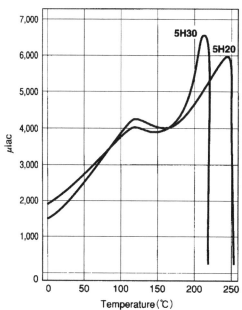

Figure 17.9- Permeability versus temperature for a TV flyback transformer material. From Fuji (1994)

amplifier coupled to a tuned secondary of the second RF amplifier. A tuned LC circuit produces a resonant condition that passes a desired frequency band. An intermediate frequency transformer is designed to operate in a definite narrow band of frequencies. For broadcast radio, it is 455 KHz. With a 5 %Hz. Bandwidth. Thus the IF transformer is designed to pass a band from 150 to 160 KHz. Such a double-tuned IF circuit is shown in Figure 17.14. Audio transformers are designed to pass parts of the audio frequency band from 20 Hz. To 20,000 Hz. Very good audio transformers can pass the principal part, i.e., 30-15,000Hz. Good but less expensive transformers pass 60-10,000 Hz. Those used only for voice transmission need only pass a band of frequencies from 150-3000 Hz. An audio output filter from a microphone to an audio input amplifier is shown in Figure 17.15. The ratio of primary to secondary turns is 1:44.6. This is an example of an impedance-matching function of a transformer. There is also the converse of the step-up transformer just mentioned in the audio output transformer, to a speaker, for example. This is shown in Figure 17.16. Here, the impedance of the audio output amplifier is 7,200 ohms and the impedance of the speaker is 8 ohms, so the ratio is 30:1. This is a step-down transformer and the turns-ratio is adjusted for impedance matching. Another example of a small ferrite core is used in the inductance tuner in a television set. This core is similar to the ones used in screw type tuners in pot cores. But less sophisticated.

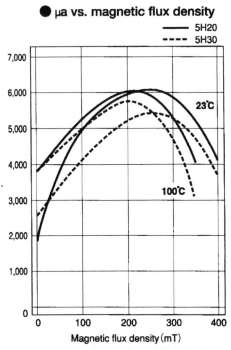

Figure 17.10- Permeability versus flux density for a TV flyback transformer material. From Fuji (1994)

core for these entertainment functions must be inexpensive so that the cores must be simple such as drum cores, balun cores, tread cores or very simple pot cores. Materials for these components range from higher permeability materials for the low frequency applications to the very low permeability cores for the higher frequencies.

FERRITE ANTENNAS FOR RADIOS
Early portable radios contained wire-wound loop antennas. When the radios became smaller in size, this was not possible. The problem was solved by the use of ferrite antennas. These could be made as small as a RF coil and could cover more than 100 times the are of the winding. All A.M. radios use this technique now. Most ferrite antennas are cylindrical. The height of the antenna is defined by the ratio of the voltage induced to that of the electric field;

$$h_c = E_s/E \qquad [17.2]$$

For a rod antenna, the induced voltage is

$E = \mu\omega EANFA/c_o$ [17.3]
where c_o = velocity of EM waves in vacuo
    = $1/\mu_o\varepsilon_o$ [17.4]

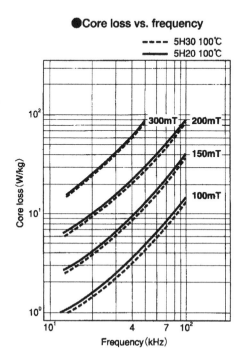

Figure 17.11- Core loss versus frequency for a TV flyback transformer material From Fuji (1994)

= 3×10^8 m/s
FA= emf averaging factor (depends largely on fraction of coil length to rod

The influence of the ferrite rod in concentrating the flux through the coil is shown in Figure 17.18. The effective permeability of the rod will change according to the *l/d* ratio as shown in Figure 17.19. For frequencies involved, nickel zinc ferrites with permeabilities from about 200 down to 12 are used. For very low permeabilities or high frequencies, cobalt is added to the nickel ferrite. The properties of a typical ferrite antenna rod material is given in Table 17.4.

436 HANDBOOK OF MODERN FERROMAGNETIC MATERIALS

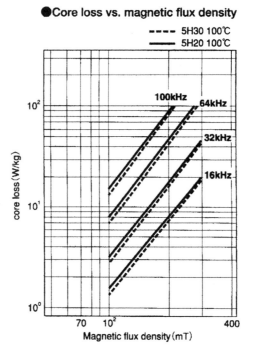

Figure 17.12- Core loss versus flux density for a TV flyback transformer material From Fuji (1994)

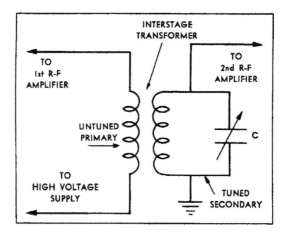

Figure 17.13- Schematic of a IF Transformer from an untuned input

FERRITES FOR ENTERTAINMENT APPLICATIONS

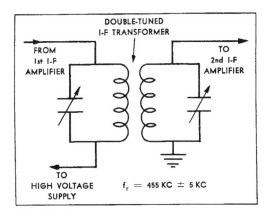

Figure 17.14- Schematic of an IF double-tuned transformer

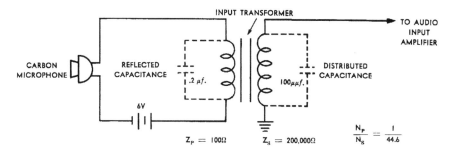

Figure 17.15-Schematic of an audio input transformer for a microphone

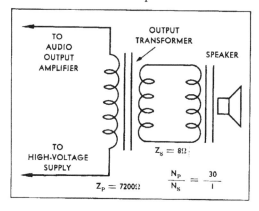

Figure 17.16 Schematic of an audio output transformer for a speaker

Table 17.4- A Typical Antenna Rod Material From MMG (1994)

Parameter	Symbol	Standard Conditions of test		Unit	F25
Initial Permeability (nominal)	μ_i	B<0.1mT 10kHz	25°C	-	50
Loss Factor (maximum)	$\dfrac{\tan \delta_{(r+e)}}{\mu_i}$	B<0.1mT 1MHz 2MHz 3MHz 5MHz 10MHz 15MHz 20MHz 40MHz	25°C	10^{-6}	50 50 55 65 75 100 125 300
Temperature Factor	$\dfrac{\Delta \mu}{\mu_i^2 . \Delta T}$	B<0.25mT +25°C to +55°C	10kHz	10^{-6} /°C	10 to 15
Curie Temperature (minimum)	Θ_c	B<0.10mT	10kHz	°C	450
Resistivity (typical)	ρ	1 V/cm	25°C	ohm-cm	10^5

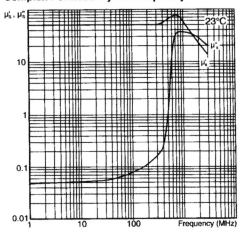

Figure 17.17-Real and imaginary permeabilities of an antenna rod material From MMG (1994)

FERRITES FOR ENTERTAINMENT APPLICATIONS

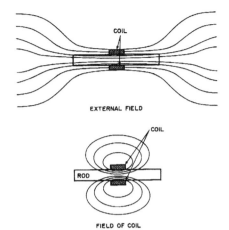

Figure 17.18- The influence of a ferrite rod in concentrating the flux in an antenna coil. From Philips.

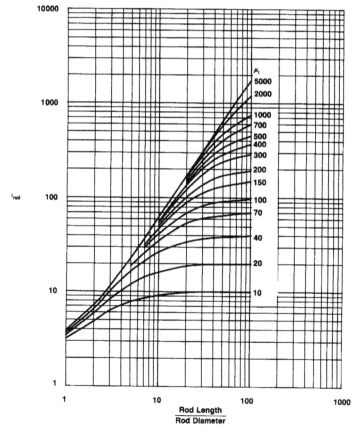

Figure 17.19- Variation of rod effective permeability in the l/d ratio. From Fair-Rite

SUMMARY

We have outlined some of the ferrites used for entertainment applications. We have confined these primarily to radio and television applications. Video cassette recorders (VCR's) which may be related to TV uses will be covered in the chapter on magnetic recording which is closer to the function of the ferrite. Some of the applications (such as IF transformers) in the present chapter are really related to the low-level uses described in Chapter 15. Still others such as TV flyback transformers are also related to the high power applications discussed in the next chapter. However, for the sake of keeping materials related to their functions in the devices, we have decided to both of these topics in the present chapter. The power magnetics discussed in the next chapter may be common to many of the other applications since they supply the power needed to drive them. It is also the area which is very timely now as the change from linear to switching power supplies accelerates.

References

Araki, T. (1992) Morinaga, H., Kobayashi, K.I.,Oomura, T. and Sato, K. Ferrites, Proc. ICF6, Jap. Soc. Powder and Powd. Met. Tokyo, 1185

FDK(1997) Ferrite Components for Audio and Visual Use, FE-11001-0907-010

MMG(1993)Product Catalog-Book 1-Soft Ferrite Components and Accessories

Ikegami, K. (1992) ,Masuda, Y.,Takei, H. and Maeda, T., Proc. ICF7, J. de Physique, IV, Vol.$\underline{7}$, C-1, 145

Nomura, T (1995)and Ochiai, T. Magnetic Ceramics, Ed. by B.B.Ghate and J.J.Simmins,,Ceramic Transactions Vol. 47, American Ceramic Society, Westerville OH. p.221,224

TDK(1990) TDK Deflection Yoke Ferrite Cores, BAE-032B

18 FERRITE TRANSFORMERS AND INDUCTORS AT HIGH POWER

INTRODUCTION
In the early days of ferrites, industry employed them at extremely low power levels because the major application at that time was in the telecommunications industry. Since the signal levels were very low, the ferrite core operated in the linear region of the magnetization curve (Rayleigh region) which was characterized by the initial permeability, μ_o. Chapter 13 describes the action of ferrite cores under these conditions. So ingrained was the electronic industry's preoccupation with the μ_o property of materials that even when higher power level ferrites were developed, the ferrite material was still characterized through the initial permeability. To this day, the principal specification given in ferrite vendors' catalogs is μ_o even through this term has only little relevance to the actual application.

THE EARLY POWER APPLICATIONS OF FERRITES
The first use of ferrite material in a power application was to provide the time-dependent magnetic deflection of the electron beam in a television receiver. The two-ferrite components used were the deflection yoke and the flyback transformer. This application was discussed in Chapter 16. This application remains the largest in tons of soft ferrite used.

Another early use of power ferrites was in matching line to load in ultrasonic generators and radio transmitters. Ferrites were not considered for line power inputs because at the lower frequencies (50-60 Hz.),they were economically unattractive (lower Bsat and higher cost than electrical steels). However, today's ferrites are employed as noise filters in power lines on the input to all types of electronic equipment. The potential for using ferrites at high frequencies was always there but the auxiliary circuit components (mostly semiconductors) were not yet developed. In addition, earlier there was no great market or stimulus for high frequency power supplies.

One envisaged use was in high frequency fluorescent lighting at about 3000 Hz. This idea was suggested in the early 1950's but the need for setting up line power at these frequencies was never fulfilled. (See Haver 1976)

In the 1970's, the rapid growth of ferrites for use at high power levels occurred shortly after the similar growth of power semiconductors that could switch at very high frequencies. This design specifically required moderate cost magnetic components with low losses at higher frequencies and elevated power levels. Thus the age of the switched mode power supply (SMPS) was born. Coincidentally, the rapid growth of computers and microprocessors has required small, efficient power supplies that could be constructed with power ferrite components. The computer and allied markets are certainly providing much of the present day impetus for today's power ferrite development.

Forrester (1994) has compiled a glossary of Power Conversion Technology to which the reader may find it useful to refer.

POWER TRANSFORMERS

Switching power supplies and ferrite expansion went hand in hand. To explain why ferrites were made to order for these applications, we must understand the implications of going to higher frequency operation. Ferrites have low saturation compared with most common metallic magnetic materials (such as iron) and also have much lower permeabilities than materials such as 80% NiFe. (See Table 4.2, Chapter 4). As we have said earlier, the low saturation of ferrites comes about from the fact that the large oxygen ions in the spinel lattice contribute no moment and so dilute the magnetic metal ions. This situation is compared to a metal such as iron where there is no such dilution. In addition, because of the antiferromagnetic interaction, not all the magnetic ions contribute to the net moment in ferrites but only those with uncompensated spins.

FREQUENCY-VOLTAGE CONSIDERATIONS

In Chapter 4, we described how an alternating electric current in a winding provides an alternating magnetizing field which creates a corresponding alternating magnetic induction in a magnetic material. This will induce an alternating voltage in another (secondary) winding. The general case of a voltage produced by a changing (not necessarily alternating) magnetic flux is given by Faraday's equation namely;

$$E = -N \, d\phi/dt = -N \, d(BA)/dt \qquad [18\text{-}1]$$

For a sine wave, the induced voltage is given by:

$$E = 4.4 BNAf \times 10^{-8} \qquad [18\text{-}2]$$

where: ϕ = Magnetic flux, maxwells or webers
E = induced voltage, volts
B = maximum induction, Gausses
N = number of turns in winding
A = cross section of magnetic material, cm^2
f = frequency in Hz.

For a square wave, the coefficient is 4.0 instead of 4.44.

If we, for the present, minimize the effects of complicating problems such as core losses and temperature rise (which we will discuss later), we can use this important induction equation to examine the use of the variables in the most preliminary de-

FERRITE TRANSFORMERS & INDUCTORS AT HIGH POWER

sign. To obtain a given voltage with the most efficient arrangement, the tradeoffs can be as follows;

1. Increasing B by using a material with high induction such as 50% Co-Fe. This material is used in aircraft and space application where space and weight are important. However, there is a material limitation on how high the B can go. Ferrites may have saturation of 4-5,000 gausses. The highest RT saturation of about 23,000 gausses is found in 50% Co-Fe, so that there is a possible 4:1 or 5:1 advantage here for metals.
2. N can be increased which leads to higher wire resistance losses. Also, there is a maximum number of turns that can be wound around a core with a window or bobbin area. Using small wire size allows more turns, but the increased resistance (due to the increased length to cross sectional area) limits the useable current through the wire.
3. A can be increased. In addition to higher core losses, the larger cross-section requires a longer length of wire per turn leading to higher copper losses and a larger and heavier device. The larger cross section in a poor thermal conductor such as a ferrite also creates the problem of how to remove the heat produced in a large core. If the heat isn't removed, the temperature rise lowers the saturation induction, B_s of the ferrite. Under these conditions, if the induction-swing, ΔB, is large, enough, the core may actually saturate and the current in the winding can become very large possibly causing catastrophic failure. This can damage the core, the winding and other components.
4. f can be increased. Here the effect can be quite dramatic depending on the frequency dependence of core losses. For instance, in going from a 60-Hz power supply to 100KHz supply, the factor is 1666. This coupled with a 4:1 reduction in going from high B metals to low B ferrite still leaves $400^+:1$ advantage. This permits a great reduction in the size & weight of the transformer, which reduces wire and core losses. In a high frequency power supply, increasing the frequency can exacerbate the thermal runaway problem if the exponent of frequency dependence is higher than that for flux density. We will deal with this subject with later in this chapter.

FREQUENCY-LOSS CONSIDERATIONS

We have shown that by increasing the frequency of a transformer, we can produce the desired voltage requirement at a greatly increased efficiency. However, we have neglected one consideration, that is, the increased losses that occur when we increase the frequency of operation. The additional losses incurred in the frequency increase are mainly eddy current losses caused by the internal circular current loops that are formed under AC excitation. The eddy current losses of a material can be represented by the equation:

$$P_e = KB_m 2f^2 d^2 / \rho \qquad [11\text{-}3]$$

where: P_e = Eddy Current losses, watts
K = a constant depending on the shape of the component

B_m = max induction, Gausses
f = frequency, Hz
d = thickness -narrowest dimension perpendicular to flux, cm
ρ = resistivity, ohm-cm

Again there is a trade-off for lower P_e. B can be lowered which means larger A to get the same voltage. Frequency, f, can be lowered which again means larger components. The thickness, d, can be made smaller, such as in thin metallic tapes, wire or powder. There are physical limitations to this variable, and also the high cost of rolling metal to very thin gauges.

The other measure we can take is to increase the resistivity. (See Table 4., Chapter 4) A comparison will demonstrate the advantage of ferrites. The resistivity for metals such as permalloy or Si-Fe is about 50×10^{-6} ohm-cm. The resistivity of even the lowest resistivity ferrite is about 100 ohm-cm. The difference then is about 2 million to 1. Since the effect of the frequency on the losses is a square dependence and that of resistance only a linear one, the net effect on frequency is about 1400 to 1. Thus, losses to the 60 Hz operation, for the same size core, extend to 84,000 Hz, close to the 100 KHz we postulated for the voltage calculation. Granted this calculation is simplified, having omitted wire losses and loss differences due to B variations, but the order of magnitude is probably reasonable. In actual cases, 60 Hz power supplies operate at efficiencies of about 50%, whereas the ferrite high frequency switching power supplies operate at 80-90%. Figure 18.1 shows typical ferrite cores for power applications along with the mounting and winding accessories.

We must include another consideration in the comparison. We have mentioned the poor thermal conductivity of ferrite and ceramics in general. Aside from the difficulty of firing very dense, large, ceramic parts without producing cracks, there is also the previously mentioned problem of heat transfer. Because of this limitation, ferrite switching power supplies have not been made larger than about 10 KW. This is in comparison to the over 100 KW supplies that are made of metallic materials. However, since the large markets in power supplies are for home computers or microprocessors, and since these are well within the operational size of ferrites, there is no real size problem here.

THE HYSTERESIS LOOP FOR POWER MATERIALS

In the applications of ferrites discussed in Chapter 10, the voltage input was sine wave and at very low power levels. In other power output stages in transformers for telecommunications, the input is still sine wave but the induction level is higher so as a result, the permeability is no longer constant but varies with excitation depending on the shape of the magnetization curve. In this case, the applicable permeability is no longer μ_o, the initial permeability, but the amplitude permeability, μ_a, which is the slope of the line from the origin to the point of maximum operating induction. This slope is always higher than the initial permeability and approaches the maximum permeability (the point of inflection on the magnetization curve). The maximum permeability is sometimes listed in vendor's catalogs and represents a

FERRITE TRANSFORMERS & INDUCTORS AT HIGH POWER

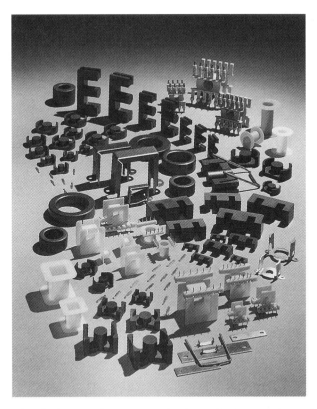

Figure 18.1- Typical ferrite cores for power applications with the mounting and winding accessories. Courtesy of Magnetics Division of Spang and Co.,Butler, Pa 16001

limit to the permeability obtainable with the material. The operation of a power transformer can be designed to have a bipolar drive as in the push-pull type or unipolar (forward or flyback mode). In the bipolar case, the course of the induction or the excursion is in both directions so that the magnetization is reversed. In the unipolar case, the induction is unidirectional and the magnetization is not reversed. In certain instances in power electronics, the limits of induction are from the remanent to the maximum inductions. The loop traversed is a minor loop in the first quadrant similar to the one shown in Figure 18-2. In this case, the B used in the induction equation is still $\Delta B/2$ even though the magnetization is not reversed. What produces the voltage is not the alternation but the rate of change of the flux. Examples of unipolar and bipolar inductions are shown in Figure 18-3a,3b and 3c.

Whereas the ΔB and thus the corresponding voltage are smaller in a unipolar than in the bipolar drive, the construction and operation are much simpler and more economical. As we said earlier, in the use of ferrites as flyback transformers for television receivers, the voltage and current waveforms are not sinusoidal but saw-

446 HANDBOOK OF MODERN FERROMAGNETIC MATERIALS

tooth or flyback shape. This permits the electron beam to sweep across the TV screen with its visual signal and when it comes to the end, it would rapidly return or

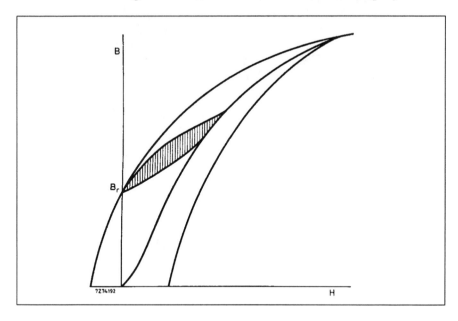

Figure 18.2-Minor hysteresis loop traversed when unipolar pulses are applied to a feed-through converter. The hatched area represents the portion of the hysteresis loop actually used. From Ferroxcube Application Note F602

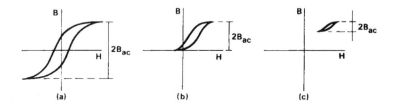

Figure 18.3-Unipolar(b) and bipolar(c) excursions in a power transformer. From Bracke (1983)

fly back to the starting horizontal sweep position to present the next lower line of information. In this case, the transformer operation is unipolar.

INVERTERS AND CONVERTERS
The basis of the modern electronic switching power supply is the action of the tran sistor as a switch. Early transistors were not built to carry much power and thus,

FERRITE TRANSFORMERS & INDUCTORS AT HIGH POWER

as was the case for early ferrite inductors and transformers, they were used mainly in telecommunication applications at low power levels. As power semiconductors

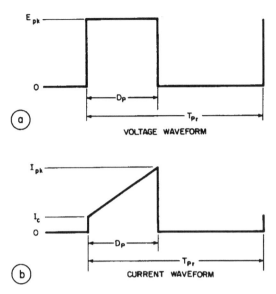

Figure 18.4-Voltage (a) and current(b) waveforms of a square wave. D_p ="on" time, T_{pr} = Pulse repetition rate.

became available, they were used as a switch to invert a D.C. voltage to produce voltage wave-forms (other than sine) such as the rectangular or square wave varieties. For a square wave, examples of both the voltage and current wave-forms are shown in Figures 18.4a and 18.4b. Note that the current waveform has a ramp shape that builds up gradually as the magnetic field in the core increases.

We should introduce two important terms now. The first is an inverter that is a device that takes a DC input and produces an ac output in a manner other than the usual rotary generator. A transformer may be incorporated in the device to give the required voltage. The device can be mechanical such as a vibrator or chopper or it can be of the solid state variety using a transistor. The word, oscillator, may sometimes be confused for an inverter but in the oscillator, the frequencies may be higher and the power levels lower. The second important tem, a converter, takes the DC of one voltage and converts it to DC at another voltage. One might call it a DC transformer. The intermediate step in a converter is that of an inverter namely the conversion of DC to AC. Of course, the additional step is rectification to D.C. The input to a converter can sometimes be a low frequency (50-60 Hz.) which is rectified, inverted, transformed, and then again rectified. The advantage over a conventional transformer is that the transformation is much more efficient at the higher frequency. The complete switching power supply may consist of several auxiliary sections in addition to the power transformer. A typical switching power supply is

448 HANDBOOK OF MODERN FERROMAGNETIC MATERIALS

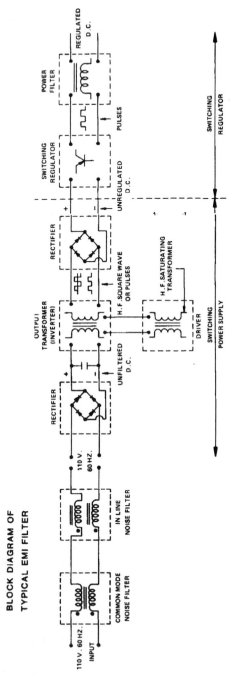

Figure 18.5a-Block diagram of a switching power supply From Magnetics PC-01(1982)

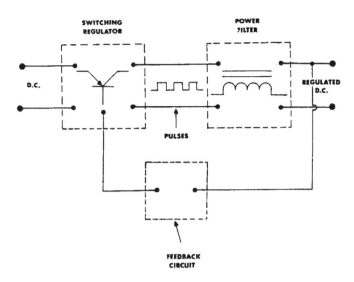

Figure 18.5b-An example of a switching regulator using feedback by pulse width modulation to control the output voltage. From Magnetics PS-01(1982)

shown schematically in Figure 18-5a (Magnetics 1984). If ac is the input, it must first go through a noise filter to keep out unwanted transients. It is then rectified before entering the power transformer where it is first inverted to a square wave of the higher frequency (or pulse repetition rate) and then transformed to the desired output voltage. The transistor is driven by an auxiliary timing transformer or a driver. After passing through the power transformer, the secondary voltage is again rectified. It then passes through a voltage regulator to maintain the voltage limits in the required range. Often this is done in a feedback circuit which controls the on-off ratio of the switching transistor. This technique is called "pulse width modulation" or PWM and is widely used. An example of such a circuit is shown in Figure 18-5b. Switching power supplies have efficiencies on the order of 80-90% compared to those of linear power supplies that may range from 30-50%. The switching supplies are therefore lighter and smaller than their counterparts.

Types of Converters
The type of converter circuit used will influence the operating conditions and therefore the choice of ferrite core to be used. There are three main types of converters: the flyback, the forward, and the push-pull.

Flyback Converter- This is one of the simplest designs for a converter. The basic circuit is shown in Figure 18-6a (Bracke 1983) with the associated hysteresis loop traversals. When the transistor switch is in the closed position, the transistor is in the

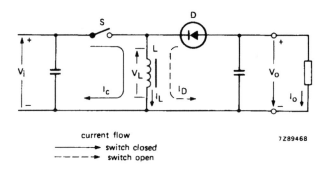

Figure 18.6a- A diagram of a flyback converter and the associated current and wave forms. From Bracke and Geerlings(1982).Philps HF Power Transformer and Inductor Design,Part 1.

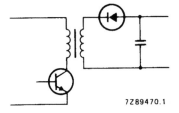

Figure 18.6b-A Flyback converter with associated transformer isolation. From Bracke and

Geerlings(1982).
conducting mode and the input voltage appears across the inductor, creating a magnetic field that stores the magnetic energy until the switch is opened. At that time, the current across the inductor reverses as the magnetic flux in the core decreases. The stored magnetic energy is transferred to the plates of the capacitor and the load. One way of varying the amount of energy stored and therefore the output voltage is by varying the ON time or duty cycle, D, of the transistor. Another variation of this circuit is shown in Figure 18-6b, which gives line isolation. The simplicity of design and fact that no output choke is needed make it a useful choice of design. Because it is unidirectional, a larger core may be needed than that used in other designs. It also has a higher "ripple" or residual ac component.

Forward Converter- Another type of converter, known as the forward converter is shown in Figure 18-7 (Bracke 1983). When the switch is closed and the transistor conducts, current again rises linearly in the inductor. However, since the inductor and load are in parallel to each other, part of the energy is stored in the inductor and part is transferred to the load. When the switch is opened, the current continues to flow to the load by way of the diode, known as the flywheel diode. The voltage can again be controlled by pulse width modulation. The forward converter is sometimes called the series converter and the flyback the parallel converter. The forward converter has lower ripple than the flyback as the combination of capacitor and inductor provide a better filter circuit.

Push-Pull Converter- This converter shown in Figure 18-8(Bracke 1983)s essentially a combination of two forward converters operating in opposite polarities giving it the name push-pull. It has two separate transistor switches, one for each forward converter with a split center-tapped winding. When S1(TR1) is open, diode D2 conducts and energy is partially stored in the inductor and partially supplied to the load. With both S1 and S2 (TR1 and TR2) open, the inductor will provide a load current by parallel diodes, D1 and D2. When S2(TR2) closes, D1 will still conduct and D2 will stop and the process repeats itself. A push-pull converter has twice the ripple frequency and therefore a lower ripple-voltage. Another advantage is its ΔB swing is twice as large as the unipolar variety and therefore, a smaller transformer can be used. Multiple outputs can be constructed using several secondary windings.
Flyback transformers are really more like power inductors than the other types of transformers since the energy is stored in the inductance during the current rise and discharged during the flyback period. The design, therefore, will be considered under the subject of power inductors. Variations of these circuits are possible. Table 18-1 gives the relative strengths and weaknesses of each type. For high power applications, the push-pull design is preferred. For less demanding versions, the forward converter is an alternative. In high voltage supplies, the flyback type is most suitable. Figures 18-9, 18-10, and 18-11 give the circuits and hysteresis loop traversals for the flyback, forward and push-pull types respectively.

452 HANDBOOK OF MODERN FERROMAGNETIC MATERIALS

In the design of transformers for inverters, the worst case scenario is used with regard to transient voltages that may increase the input voltage. Knowing the maximum and minimum voltages will help in the design process. Another operational

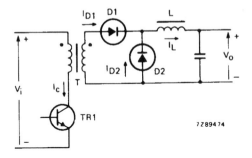

Figure 18.7-A Forward Converter with transformer isolation. From Bracke and Geerlings(1982)

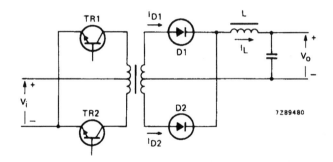

Figure 18.8-A push-pull converter and associated voltage and current waveforms. From Ferroxcube Design Manual

Table 18.1
Circuit Type Summary

Circuit	Advantages	Disadvantages
Push-pull	Medium to high power Efficient core use Ripple and noise low	More components
Feed forward	Medium power Low cost Ripple and noise low	Core use inefficient
Flyback	Lowest cost Few components	Ripple and noise high Regulation poor Output power limited (<100 Watts)

FERRITE TRANSFORMERS & INDUCTORS AT HIGH POWER

problem that must be considered in the design of push-pull converters is the possibility of D.C. imbalance in the two arms of the circuit. For this reason, full bridge converters are used for most high-power applications even though they have twice as many power semiconductor switches. The voltage stress is only on the DC Buss voltage, not twice the value. In addition, a DC blocking capacitor can be added to the full bridge where it cannot in a push-pull circuit.

CHOOSING THE RIGHT COMPONENT FOR A POWER TRANSFORMER

In choosing the best component for a power transformer, we use a process similar to that employed for a low level transformer inductor except that the necessary parameter are somewhat different. The choice will be determined by:

1. The type of circuit used.
2. Frequency of the circuit.
3. Power requirements.
4. The regulation needed (percentage variation of output voltage permitted)
5. Cost of the component.
6. The efficiency required.
7. Input and output voltages

From these considerations, the component requirements will be determined with respect to:

1. Ferrite material
2. Core configuration which includes associated hardware (tuning slugs, clamps, etc.)
3. Size of the core
4. Winding Parameters- (number of turns, wire size)

CHOOSING THE BEST FERRITE MATERIAL

We normally make this choice on the basis of frequency of operation. Vendors usually provide guidelines as to what materials are suitable for the various frequency ranges. The core losses are often given as a function of frequency. Although vendors generally list power materials separately, the user often has a choice of several available materials varying according to losses, frequency and sometimes, cost. Since power ferrites operate at the highest possible induction, we find, as we would expect, that they have the highest saturation of the ferrites consistent with maintenance of acceptable losses at the operating frequency. For frequencies up to about 1 MHz., Mn-Zn ferrites are the most widely used materials. Above this frequency,

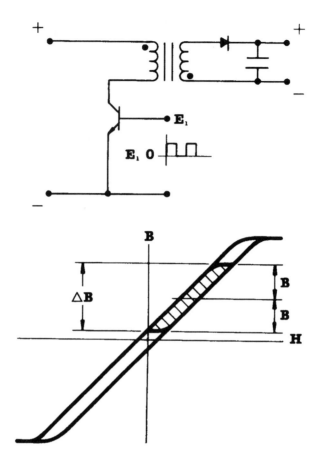

Figure 18.9-Block diagram of a flyback converter and the hysteresis loop traversal during operation. From W.A. Martin, Proc., Power Electronics Conf.(1987)

NiZn may be chosen because of its higher resistivity. For deflection yokes, Mg-Zn ferrites are finding favor.

Requirements for a Power Ferrite Material

A material slated for a power application must meet certain special requirements. Although ferrites in general have low saturations, we must, at least, provide the highest available consistent with loss considerations. This is mostly a matter of chemistry. Along with this consideration is the need for a high Curie point. This generally means maintaining a high saturation at some temperature above ambient which approaches actual operational temperature. In addition to the saturation re-

FERRITE TRANSFORMERS & INDUCTORS AT HIGH POWER

quirement, the material must possess low core losses at the operating frequency and temperature. The transformer losses, which include both the core loss and the into

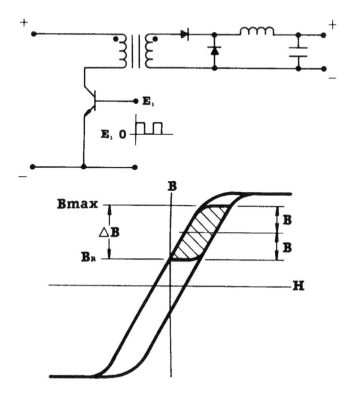

Figure 18.10- Block diagram of a forward converter and the hysteresis loop traversal during operation. From Martin (1987)

account, the core may saturate at the higher temperature with disastrous results. A runaway heating situation could develop leading to catastrophic failure. Many ferrite suppliers have redesigned their materials such that the core losses will actually minimize at higher operating temperatures preventing further heating of the cores. The negative temperature coefficient of core loss at temperatures approaching the operating temperature helps compensate for the positive temperature coefficient of the winding losses in the same region. Roess(1982) has shown that the minimum in the core loss versus temperature occurs at about the position of the secondary permeability maximum. Thus, if the chemistry of the ferrite can be designed to have the secondary maximum at the temperature of device operation of the transformer as described above, the core losses will also be low at that temperature. See Figure 18.12. However, we must consider that this is only a local minimum. Having the minimum at 75-100°C. is a tremendous aid to the designer in avoiding thermal runaway, but still requires careful design work as the core loss increases above this minimum and the capacity for thermal runaway is still very real.

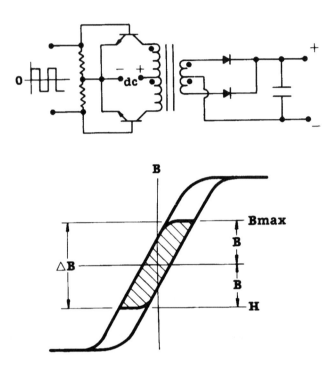

Figure 18.11- Block diagram of a push-pull converter and the hysteresis loop traversal during operation. From Martin(1987)

Flux Density. Probably the most important feature common to power transformer requirements is the relatively high flux density. Power ferrites are usually listed separately in vendors' catalogs and most show B_{sat} values of about 5,000 for materials to be used in the 25-100 KHz range. Earlier ferrite-cored power transformers operated at lower power levels and therefore, did not experience a large temperature rise. However, with present models, power levels are higher and temperature rises

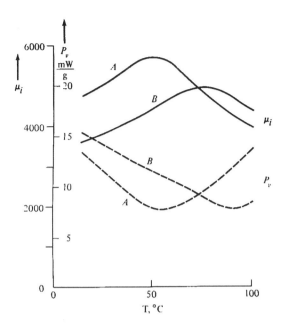

Figure 18.12- Temperature dependence of core losses for two different ferrite materials. The secondary maximum of the permeability and the minimum of the core losses occur at about the same temperatures for the respective materials. From Roess, Trans Mag. MAG 18 #6,Nov. 1982

of about 40° to 60°C are not uncommon. The ambient temperature of the core without excitation is not considered room temperature (20°C.) but one much higher,(on the order of 40-60°C.) because of the heat generated by other components. The temperature rise of the transformer is added to the elevated ambient. At these higher operating temperatures (80°-120°C) the saturation of the material will drop as shown in Figure 11.8. However, when the saturation is high, the Curie point will usually be correspondingly high so there is little likelihood of catastrophic failure with moderate caution in design. The reduced saturation must be considered in the choice of B_{max} (operating) and is the one used in the induced voltage calculation. Figure 18-13a shows the reduction in theΔ B in a power transformer as the operating temperature is raised.

Core Losses - Another property of the material that must be examined is the core losses at the frequency of operation. Here again, vendors' curves will show losses as a function of frequency and flux density. In recent years, the switch to higher operating temperatures has prompted the development of materials whose losses are lower at the elevated temperatures of operation. The difference between the older material & the newer one is shown in Figure 18-13b. If a higher frequency is chosen, the previously used flux density may have to be reduced to keep the losses from

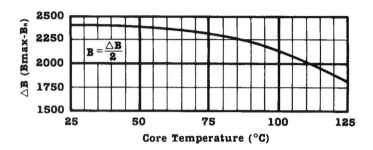

Figure 18.13a-Decrease of ΔB with temperature to maintain lower losses in a unipolar driven transformer. From Martin (1987)

becoming excessive. The power losses are either given in units of mW/gm(W/kg) or, more often, mW/cm^3. The units are given in terms of mW/cm^3/cycle, in which case the value must be multiplied by the frequency. A typical plot of losses versus f and B is shown in Figure 18-14.

McLyman(1982) has used an empirical equation similar to the Steinmetz equation to correlate losses with f and B. The applicable equation is

$$P_e = kf^m B^n \qquad [18\text{-}4]$$
where: P_e = core loss in watts/Kg
f = frequency, Hz
B = Flux density, Teslas (10^4 Gausses)

For the material shown in Figure 18.14, the equation would be;

$$P_e = f^{1.39} B^{2.19} \qquad [18\text{-}4a]$$

For power ferrites, the value of m is on the order of 1.5 while that of n is about 2.5. At higher frequencies, n will increase possibly due to the onset of ferromagnetic resonance. The higher value of the B exponent, n, over the f exponent, m, indicates that the losses are more sensitive to variations in flux density than in frequency. This equation is valid over a rather limited frequency range. Strictly speaking, this equation should only represent the hysteresis losses with the eddy current losses given by Equation 18-3. However, since the eddy current losses are low in ferrites,

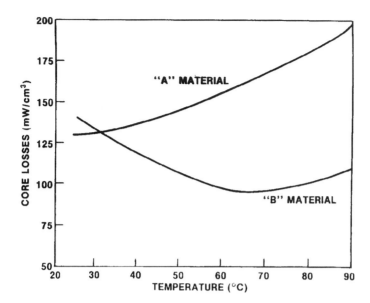

Figure 18.13b- Temperature dependence of core losses for two different ferrite materials. A material has lower losses at room temperature but the B material losses are lower at 60°C., the temperature that the transformer might be expected to operate.

this equation may be used over the limited frequency range. We will discuss the importance of these coefficients of f and B later in this chapter.

Snelling (1996) has shown the progression of ferrite materials with the year and frequency. See Figure 18.15 As the frequencies of operation for the SMPS have increased, new materials have been developed concurrently to operate under the new conditions.

PERMEABILITY CONSIDERATIONS

As mentioned earlier, the initial permeability is not a directly governing factor in power ferrite choice. However, it has been shown that the temperature where the secondary permeability maximum occurs is generally in the vicinity of the minimum in core losses at a particular frequency. See Figure 18.12. The permeability that should be considered in power applications is the amplitude permeability. This should be as high as possible. In Figure 18.16 is shown the variation of permeability with B level. Thus, when the flux density is very high, the permeability drops rapidly as we approach saturation, especially at 100°C. If the amplitude permeability is very low, then the magnetizing current or the number of turns must go up to provide the required H. This situation greatly increases the copper losses.

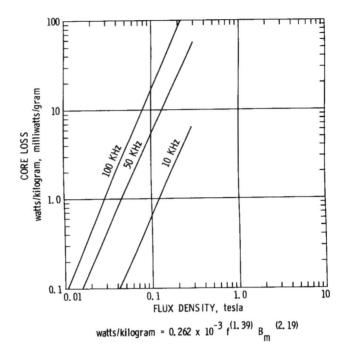

Figure 18.14- Core loss curves as a function of flux level and frequency. From Colonel W.T.McLyman, Magnetic Core Selection for Transformers and Inductors, Marcel Dekker, New York (1982)

LIMITATIONS IN FERRITE DESIGN

Depending on the operating conditions, the design of a power ferrite transformer will fall into one of three categories, namely;

1. Saturation-limited transformers
2. Core-loss-limited transformers
3. Regulation limited transformers

At low frequencies, the core losses are low and the advantage of frequency in the induced voltage equation is relatively small. The design is therefore saturation-limited. At higher frequencies, when the core losses increase, the design becomes core-loss-limited. Bracke (1982) has examined the transition from one state to another. Figure 18-17a shows the division of core to winding losses as a

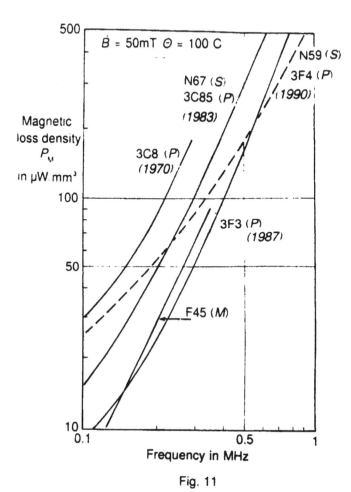

Fig. 11

Figure 18.15-Magnetic loss density versus frequency for successive developments of MnZn power ferrites

function of frequency. In region I (saturation-limited), throughput power is proportional to frequency. In region II, the output power increases as the square root of frequency and it still appears saturation limited. In region III, the core loss becomes a limiting factor and the design is based on the optimum ratio of winding to core losses which is taken to be 1. Bracke(1982) concludes that 44 percent is the ideal percentage of core to total losses. In region IV, less turns are needed and flux density decreases with frequency. As a result, the ideal ratio cannot be maintained and throughput power decreases with frequency faster than the core loss. In real cases, we find the curves corresponding to Figure 18-17b which are rounded but have the same general form.

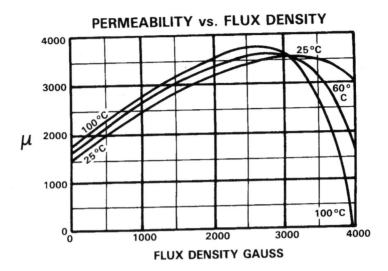

Figure 18.16- Amplitude permeability versus flux density for a typical ferrite power material at 25°C.,60°C. and 100°C. From Magnetics Catalog FC509(1989)

OUTPUT POWER CONSIDERATIONS

Stijntjes(1989) has proposed a new material criterion for power transformer ferrites. He cites an expression by Mulder(1989) for the throughput power;

$$P_{th} = W_d \times C_d \times f \times B \qquad [18\text{-}5]$$

Where P_{th} = Throughput power, Watts
W_d = Winding design parameter
C_d = Core design parameter

Using a reasonable figure of 200 KW/m^3 (mW/cm^3) proposed by Buthker(1986) for the allowable losses to maintain the required maximum temperature rise, Stijntjes (1989) develops a factor which he calls PF_{200}. This is the product of the material-dependent parameters in the equation (namely f and B) which will yield losses of 200 mW/cm^3. The other factors that determine the throughput power are related to winding and core design.

POWER FERRITES VS COMPETING MAGNETIC MATERIALS

There have been many comparisons of ferrites with other magnetic materials for power applications. The author (Goldman 1984) listed other metallic materials that were used for SMPS,s. Goldman (1995) compared metal strip, powder cores and ferrites for various applications including power. Bosley (1994) presented a rather extensive study of the different materials for transformers and inductors for frequency where the maximum flux was limited by saturation or core losses. For frequencies above 100 KHz., the ferrites, MnZn and NiZn had the highest values. The performance factor in Tesla-hertz vs frequency is shown in Figure 18.19. Here

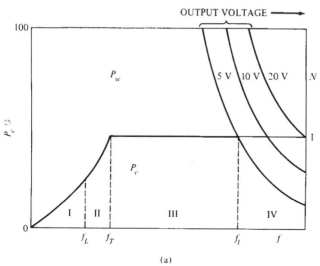

Figure 18.17(a)- Breakdown of percentage of total loss in a transformer into winding and core losses as a function of frequency. From Bracke(1983)

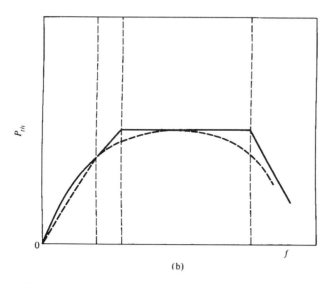

Figure 18.17(b)- Actual (dotted line) versus theoretical (solid line) breakdown of core losses as a function of frequency. From Bracke(1983)

again, above 100 KHz., the two ferrites were the highest. The economic trade-off for these materials is given in Figure 18.20 which charts the maxwells of flux per dollar as a function of frequency for the competing materials. Bosley (1994) also listed the advantages and disadvantages of the competing materials for SMPS transformers. Figure 18.21 show these for SiFe and Permendur, Figure 18.22 for the

464 HANDBOOK OF MODERN FERROMAGNETIC MATERIALS

NiFe alloys, Figure 18.23 for the amorphous alloys and Figure 18.24 for ferrites. Snelling (`1996) presents a plot of the power loss density of power ferrite, Co-Fe amorphous metal strip and the Vitrovac 6030 nanocrystalline material vs frequency in Figure 18.25. The relative advantages in core loss depend on the frequency and flux density. The previous comparisons did not include the nanocrystalline materials which gained recognition shortly after the Bosley article even though Yoshizawa (1988) reported on them earlier. The frequency dependence of the permeability and loss factor of a nanocrystalline material is given in Figure 18.26 (Herzer 1997)along with those for a Co-based amorphous material and a MnZn ferrite. The permeability is higher and the loss factor is lowest for the nanocrystalline material. In addition the saturation induction is shown for the same materials in Figure 18.27 (Herzer 1997). The nanocrystalline material has the advantage there. It remains to be seen whether the pricing of the nanocrystalline can be low enough to compete with the relatively inexpensive ferrite materials.

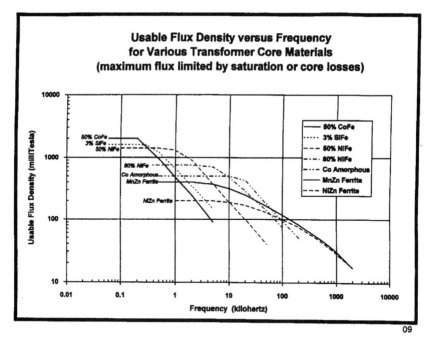

Figure 18.18 Usable flux density vs. frequency for various magnetic materials for SMPS transformer applications. From Bosley (1996)

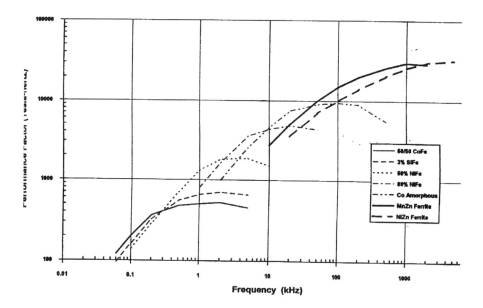

Figure 18.19- Performance factor in Tesla-hertz vs frequency for several magnetic materials for SMPS transformer applications From Bosley (1996)

POWER FERRITE CORE STRUCTURES

Power Ferrites come in a variety of shapes. Although pot-cores were the ferrite shapes of choice in telecommunication ferrites, several required or preferred features for this application are not as critical in power usage. These include:

1. Shielding
2. Adjustability

In addition, pot cores are more costly and power ferrite must compete with other alternative materials. Therefore, shapes such as E cores, U cores, and PQ cores are more applicable to power application. Other shapes including solid center post pot cores can be used. The following describes the types of shapes available.

The shape of the core has a bearing on the amplitude permeability since the inductance is given by;

$$L = .4\,\pi\mu N^2\,l/A \qquad [18\text{-}6]$$

where: l = length of the winding, cm.
A = Cross sectional area, cm^2

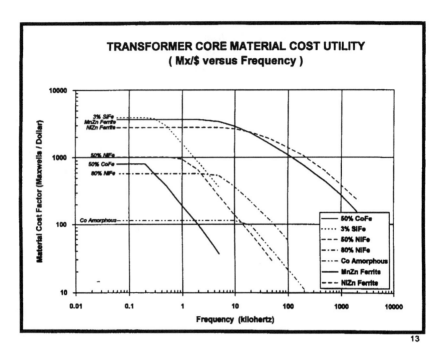

Figure 18.20 –The maxwells of flux per dollar versus frequency for the various magnetic materials for SMPS transformer applications

Therefore, the longer the section on which the winding is placed and the shorter the height of the winding, the higher the inductance.

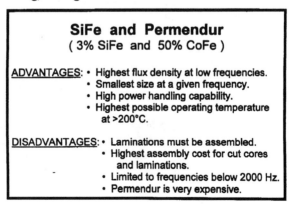

Figure 18.21- Advantages and disadvantages of SiFe as a SMPS transformer material From Bosley (1994)

FERRITE TRANSFORMERS & INDUCTORS AT HIGH POWER

Nickel Alloys
(50% NiFe and 80% NiFe)

ADVANTAGES:
- High flux density at lower frequencies (<10 kHz).
- Small size for a given frequency.
- High possible operating temperature (up to 200°C).
- High power handling capability.
- Unlimited size range for tape wound core.

DISADVANTAGES:
- Expensive materials.
- Need toroidal winding equipment.
- Normally used below 20 kHz.

Figure 18.22-Advantages and disadvantages of NiFe alloys for SMPS transformer applications. From Bosley(1994)

Low Profile Ferrite Power Cores

As mentioned in Chapter 15 on low power ferrite applications, the past 5-10 years have seen the introduction of low profile cores in several configurations. One reason for this change is explained in the last paragraph in which the permeability is maximizes by having the winding length large and the cross section small. This condition can be accomplished in a low profile or low height core. The other reason (also mentioned in Chapter 15) is the growing use of PC (printed circuit) boards on which to mount the magnetic cores. This method of attaching cores is even more important in the power ferrite area than in the low power telecommunications area

---

Amorphous Alloys
(Fe-Based and Co-Based)

ADVANTAGES:
- High flux density for Fe-based.
- Very low core losses for Co-based.
- Frequency range up to 100 kHz.
- Unlimited size available as tape cores.

DISADVANTAGES:
- Expensive materials.
- Low flux density for Co-based.
- Co-based limited to temperatures below 100°C.

Figure 18.23- Advantages and disadvantages of amorphous metal alloys for SMPS transformer applications. From Bosley(1994)

Ferrites
(MnZn and NiZn)

ADVANTAGES:
- Highest frequency range.
- Least expensive.
- Bobbins & hardware readily available.
- Allow operation above audible frequency.
- Round winding areas.
- Easliy adapted to mass manufacturing.

DISADVANTAGES:
- Lowest flux density.
- Largest core needed below 20 kHz.
- Poor heat transfer.
- Low temperature range.

Figure 18.24- Advantages and disadvantages of ferrites for SMPS transformer applications.From Bosley(1994)

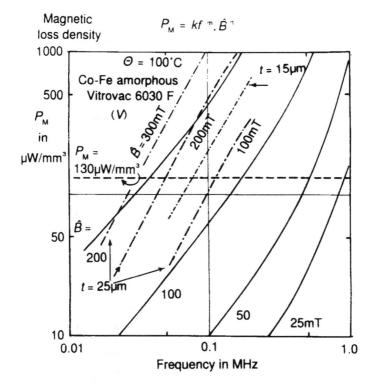

Figure 18.25-Power loss density of a power ferrite, a Co-based amorphous alloy and Vitrovac 6030 nanocrystalline material From Snelling 1997

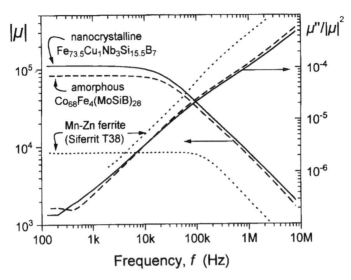

Figure 18.26- Real and imaginary parts of the permeability versus frequency for a MnZn power ferrite, a Co-based amorphous material and a nanocrystalline material. From Herzer (1997)

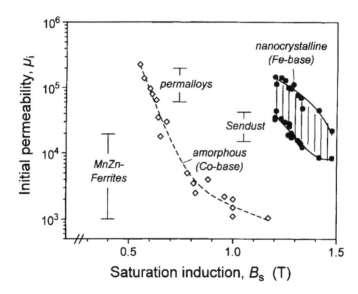

Figure 18.27-Saturation induction vs initial permeability for MnZn ferrites, permalloys, Sendust, a Co-based amorphous material and an Fe-based nanocrystalline material, From Herzer (1997)

since PC technology is increasingly placing the power supply for a circuit on the same PC board as the other circuit components. The space between the boards is one half inch so the power ferrite core must be designed to fit in that space with the bobbin and mounting hardware. The availability of low profile cores will be discussed under the following sections dealing with the various core shapes.

Surface-Mount Design in Power Ferrites
In Chapter 15, the use of surface mount design was described for low power ferrite applications. The motivation was the development of PC board technology surface-mount design (SMD. As with the low profile cores, the application has. The use of low-profile ferrite cores can be complemented to a large degree by been widespread mostly in the power ferrite application. The two terminal mounting types used for power ferrites are the gullwing and the J-type terminals shown in Figure 18.28. The gull wing form is used when thin wire up to .18 mm in diameter is used. The J-type design is used in wire sizes greater than .8 mm. Surface mount design lends itself to high speed automatic component placement on the PC board. A surface-mount bobbin with gullwing terminals is shown in Figure 18.29. The placement on the PC board is also shown.

Core Geometries
Pot Cores - Pot cores are sometimes used ungapped in power applications with a solid center post since there is no need for the adjustor found in telecommunication applications. The shielding to protect a low-level telecommunication signal in LC circuits is not necessary. There may be some advantage to the shielding in that it it does provide the lowest leakage inductance. Besides cost, another drawback to pot cores is the difficulty of bringing out heavy leads to carry the high currents. The closed structure also makes it difficult for heat from the windings to escape. Since pot core dimensions all follow IEC standards, there is interchangeability between manufacturers. See Figure 15.6a.
Double Slab Cores - In slab-sided solid center pot cores, a section of the core has been cut off on each side parallel to the axis of the center post. This opens the core considerably. These large spaces accommodate large wires and allow heat to be removed. In some respects, these cores resemble E-cores with rounded legs. See Figure 18.30

RM Cores and PM Cores- RM cores (See Figure 15.7) were originally developed for low power, telecommunications applications because of the improved packing density. They have since been made in larger sizes without the center hole. Their large wire slots are an advantage while still maintaining some shielding. PM cores are large RM-shaped cores specifically for power applications. Zenger(1984) feels that the geometry and self-shielding of RM cores make them useful at high
frequencies. Roess (1986) points out that the stray field from an E-42 core is 5 times higher than that of an RM core. With the trend towards increased operating frequencies, he feels that there may be a backswing to the RM cores in mains (line) appli

FERRITE TRANSFORMERS & INDUCTORS AT HIGH POWER

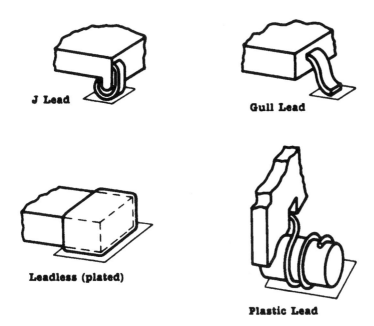

Figure 18.28-Various techniques of surface mounting of ferrite cores to printed circuit boards. From Huth, J.F III, Proc. Coil Winding Conf. Sept. 30-Oct. 3, 1986, 130

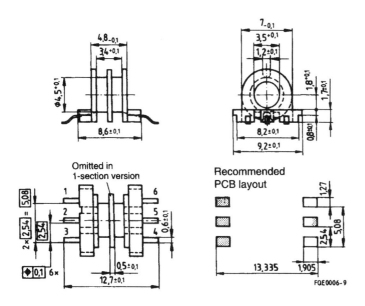

Figure 18.29- A surface-mount coil with gullwing terminals for an EP core. Also shown is the recommended PC layout. From Siemens (1997)

Figure 18.30- Double slab pot cores with and without a center hole.

cations. Since that time, the use of RM cores for power applications has grown significantly. Low-profile RM cores are available in the RM4, RM5, RM6, RM7, RM8, Rm10, RM12 and RM14 sizes. Surface mount bobbins are available in RM 4 Low Profile, RM5, RM6, and RM6LP. For power non-linear choke cores, Siemens offers special RM8 to RM 14 cores with tapered center posts. PM (Potcore Module) cores are used for transformers handling high powers, such as in pulse power transformers in radar transmitters, antenna matching networks, machine control systems, and energy-storage chokes in SMPS equipment. It offers a wide flux area with a minimum of turns, low leakage and stray capacitance. Because of the weight of these pot cores, they may not be suitable for mounting on PC boards.

E Cores - These cores are the most common variety used in power transformer applications. As such they are used ungapped. There are some variations that we shall discuss here. Their usefulness is based on their simplicity. Initially, E cores were made from metal laminations and the early ferrite E cores were made to the same dimensions and are called lamination sizes. However, as the ferrite industry matured, E core designs especially useful for power ferrite applications were developed.(Figure 18.31). Many standard E- cores have bobbins that permit horizontal

FERRITE TRANSFORMERS & INDUCTORS AT HIGH POWER

Figure 18.31- Ferrite E cores for power applications (Courtesy of Magnetics, Division of Spang and Co. Butler, Pa.)

mounting. Some of the smaller sizes also are available in surface mount design with gull-wing terminals.

E-C Cores-E-C cores are a modification of the simple E core. The center post is round similar to a pot core and since round center bobbins wind easier and are more compact than square center bobbins, this is an advantage. The length of a turn on the round bobbin is 11 percent shorter than the square bobbin that means lower winding losses. The legs of these cores have grooves to accommodate mounting bolts. (Figure 18.32)

ETD Cores- ETD cores are similar to E-C cores. They have a constant cross section for high output power per unit weight and simple snap-on clips for holding the two halves together. They also have a bobbins which provides for creepage for mains (line) isolation and have enough space for many terminals. Zenger(1984) suggests that the constant cross section of the ETD is an important attribute for high frequency and high drive levels. (Figure 18.33) These cores are available only in the large sizes and thus are not used with surface mount bobbins.

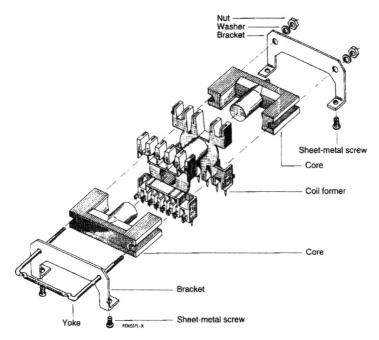

Figure 18.32- Ferrite EC cores for power applications (Courtesy of Siemens-Matsushita, Data Book, Ferrites and Accessories, 1997)

E-R Cores- These cores combine high inductance and low overall height. They have a round center post and surface mount bobbins available with the smaller sizes.

EP Cores- EP cores are a modification of a pot core but the overall shape is rectangular. A large mating surface allows better grinding and lapping, preserving more of the material's permeability. The EP core is usually mounted on its side with the bobbin below it facilitating printed circuit mounting. The best advantage of this core is in high permeability material. Shielding is very good. Some sizes of E-P cores su (EP7 and EP13) are available with surface-mount bobbins with gull-wing terminals.

PQ Cores- TDK says it stands for Power and Quality. These are one of the newest types of cores for power ferrites for switched mode power supplies. The lowest core losses in a transformer usually exist when the core losses equal the winding losses. The geometry in a PQ core is such as to best accomplish this requirement in a minimum volume. The clamp is also designed for a more efficient assembly. A more uniform cross sectional area is also achieved so that the flux density is uniform throughout the core so that the temperature will not vary much. See Figure 18.34

FERRITE TRANSFORMERS & INDUCTORS AT HIGH POWER

Figure 18.33- Ferrite ETD cores for power applications(Courtesy of Magnetics, Division of Spang and Co. Butler, Pa.)

Toroids- Toroids are sometimes used as power shapes because they take full advantage of the material permeability. Since there is no gap, leakage is very low. The toroid's main disadvantage is the high cost of winding as compared to an E or pot core. (Figure 18.35). Engelman (1989) constructed a multi-toroid power transformer that provides digital control. 40 toroids were used. Bates (1992) reported on a new SMP core technology combining new high frequency ferrite power materials as toroids in a matrix transformer that can deliver 2000 watts at 5V D.C. It has the advantage of being low profile, has low leakage inductance excellent winding isolation and higher thermal dissipation due to increased surface area.

EFD Cores- Probably the newest design in miniature power shapes is the EFD cores which stands for E- core with flat design. (See Figure 18.36) The center leg was flattened for the extra low profile needed for PC board mounting. Simple clips are available. As expected surface mount bobbins are available. Mulder (1990) has written an extensive application note on Design of Low-Profile High Frequency Transformers. He finds an empirical relation between effective volume and the thermal resistance of a magnetic device with which a CAD program can be constructed to develop the optimum range of EFD cores for the frequency band 100KHz to 1 MHz.

Figure 18.34- Ferrite PQ cores for power applications (Courtesy of Magnetics, Division of Spang and Co. Butler, Pa.)

Gapped Cores- In low power transformer or inductor applications, gapped cores were used to control the inductance and to raise the Q of the core. Although many ferrite power cores are used in the "ungapped" state, either as an E core or pot core without any intentional gap, in some situations, the intentional gap can be quite useful even in power applications. These often occur when there is a threat of saturation that would allow the current in the coil to build up and overheat the core catastrophically. The gap can either be ground into the center post or a non-magnetic spacer can be inserted in the space between the mating surfaces. The gapped core is extremely important in design of filter inductors or choke coils. We shall discuss this application later in this chapter. The basis of the gapped core is the shearing of the hysteresis loop shown in Figure 37a and 37b where 37a represents the ungapped and 37b the gapped core. The effective permeability, μ_e, of a gapped core can be expressed in terms of the material or ungapped permeability, μ, and the relative lengths of the gap, l_g, and magnetic path length, l_m :

$$\mu_e = \mu/[1+\mu_g \, l_g \, /l_m] \qquad [11\text{-}7]$$

With a very small or zero ratio of gap length to magnetic path length, the effective permeability is essentially the material permeability. However, when the

FERRITE TRANSFORMERS & INDUCTORS AT HIGH POWER 477

Figure 18.35- Ferrite toroids (Courtesy of Magnetics, Div. of Spang and Co., Butler,Pa.)

permeability is high(10,000), even a small gap may reduce the permeability considerably. For a power material with a permeability of 2,000 and a gap factor of .001, the effective permeability will drop to 1/3 of its ungapped value. When each point of the magnetization curve is examined this way, the result is the sheared curve shown in Figure 18.37b. Ito(1992) reported on the design of an ideal core that can decrease the eddy current loss in a coil by the use of the fringing flux in an air gap. The design includes a tapering of the core at the air gap. The reduction in temperature rise will depend on the operating frequency, the gap length and the wire diameter.

Prepolarized Cores- Another variation of the gapped core is one that is prepolarized with a permanent magnet. If the transformer operates in the unipolar mode and the polarity of the magnet is opposite to the direction of the initial ac drive, the starting point for this induction change will not be the remanent induction as is usually the case but a point much lower down on the hysteresis loop and in the opposite quadrant. The flux excursion will be much greater, possibly two or more times higher than the simple unipolar case. Magnetic biasing is old but the extension to this application has been described by Martin (1978) . To avoid eddy current losses, the magnet used may be a ferrite magnet often of the anisotropic variety.

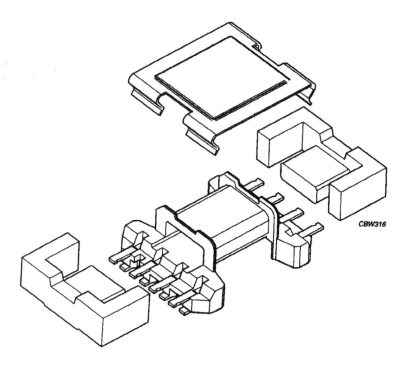

Figure 18.36 A low-profile EFD core with clip and coil former From Philips (1998)

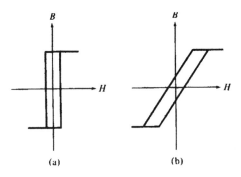

Figure 18.37-Shearing of a hysteresis loop by the application of an air gap in the magnetic circuit.

Shiraki (1978) reported a reverse-biased core for this purpose. Using a high-energy rare-earth cobalt magnet for the bias, he reduced the volume of the core 56% and the copper wire by a corresponding amount. Shiraki points out that the reverse-biased core has higher inductance near the normal saturation than the unbiased core. With this device, he more than doubled the volt-amp rating of the transformer. Nakamura (1982) reported a 70% increase in the figure of merit namely the LI^2. Thus, size and weight was reduced. The losses were not significantly higher under these conditions. Sibille (1982) also reported on several different geometries to implement the prepolarized core. Prepolarized cores are especially useful in flyback and inductor applications with high DC components. Huth (1986) has described a clever way of biasing a core using orthogonal winding techniques. (See Figure 18.38).

PLANAR TECHNOLOGY

Continuing with the low-profile design tendency particularly with PC board mounting has led to a completely new generation of cores called planar cores. Huth(1986) reported on this earlier and now, most ferrite companies offer planar cores in several varieties. Some of the arrangements are shown in Figure 18.39. Either the E-E or E-I configuration is used. The I core is actually a plate completing the magnetic circuit. In many cases the windings are fabricated using printed circuit tracks or copper stampings separated by insulating sheets or constructed from multilayer circuit boards.(See Figure 18.40) In some cases, the windings are on the PC boards with the two sections of the core sandwiching the board. Philips (1998) claims the advantages of this approach as;

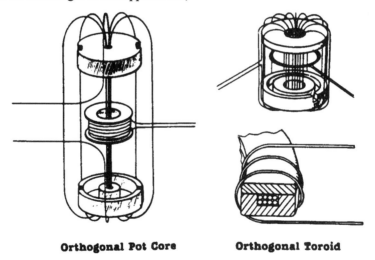

Orthogonal Pot Core   **Orthogonal Toroid**

Figure 18.38-Orthogonal windings on ferrite pot cores From Huth, J.F. III, Proc. Coil Winding Conf. Sept. 30-Oct. 2, 1986

1. Low profile construction
2. Low leakage inductance and inter-winding capacitance.
3. Excellent repeatability of parasitic properties.
4. Ease of construction and assembly

480   HANDBOOK OF MODERN FERROMAGNETIC MATERIALS

5. Cost effective
6. Greater reliability
7. Excellent thermal characteristics-easy to heat sink.

Yamaguchi(1992) performed a numerical analysis of power losses and inductance of planar inductors. A rectangular conductor was sandwiched with magnetic substrates. He suggested that the air gap between two magnetic substrates is an important factor governing the trade off between inductance and iron losses. Sasad (1992) examined the characteristics of planar indutors using NiZn ferrite substrates. A planar coil of meander type is embedded in one of the NiZn ferrite substrates and covered with another with a specified air gap. A buck converter of the 10 Watt class was constructed using the inductors with an efficiency as high as 85 percent and a switching frequency of 2 MHz. Varshney (1997) has described a monolithic module integrating all of the magnetic components of a 100 Watt 1 MHz. forward converter using a plasma-spray process for deposition of the ferrite which serves as the core.

Mohandes (1994) used integrated PC boards and planar technology to improve high frequency PWM (Pulse Width Modulated Converter) performance. Estrov (1986) has described a 1 MHz resonant converter power transformer using a new spiral winding with flat cores which solved eddy current losses, leakage inductance

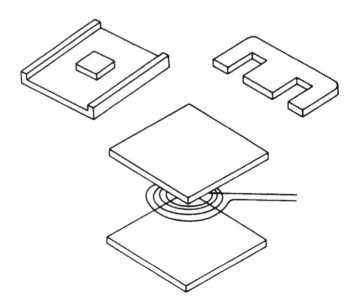

Figure 18.39-Planar ferrite cores representing a new generation of low profile cores From Huth, J.F. III, Proc. Coil Winding Conf. Sept. 30-Oct. 2, 1986

FERRITE TRANSFORMERS & INDUCTORS AT HIGH POWER   481

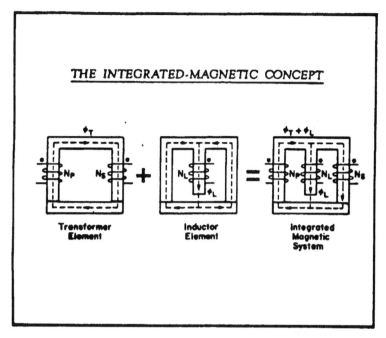

Figure 18.39a-Example of Integrated Magnetic From Bloom (1994)

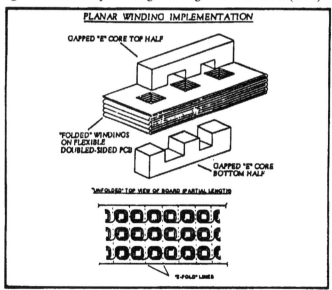

Figure 18.39b- Use of 3-bobbin integrated magnetic design. The need for 3 separate bobbins is eliminated by use of folded PC board design. From Bloom(1994)

and other problems. He also used planar magnetics and low-profile cores to cut the height and improve converter efficiency from 20 KHz. to 1 MHz.(See Figure 15.11) Brown (1992) replaced the traditional copper wire with a winding from the PC board or stamped copper sheet and using a low-profile ferrite core improved the performance and manufacturability of HF power supplies. Huang (1995) described design techniques for planar windings with low resistance. Three representative pattern types were explored; circular, rectangular and spiral. Gregory (1989) has described the use of flexible circuits to work with new planar magnetic structures. He claims that printed circuit inductors reduce losses and increase packing density making them an excellent choice for high-frequency magnetics.

Bloom (1994) has shown the application of planar-type "integrated" magnetics wherein the transformer and inductor element can be combined on the same core with separate wing. An example of this technique is shown in Figure 39a. The use of folded windings on printed circuit boards with flexible fold lines is shown in Figure 39b.

HIGH FREQUENCY APPLICATIONS

Special attention must be paid if the frequencies of power supplies extend past 100 KHz and even to the 1 MHz region. First, the size of the core may be reduced significantly. Second, the core material must be modified to lower the core losses at these frequencies. The maximum flux density or B level used which, at lower frequencies, may have extended to 2000-2500 gausses may have to be reduced to something on the order of 500-600 gausses to attain the lower losses. The increase in frequency with smaller size and better efficiency may more than offset the lower saturation used. We will discuss designs at these higher frequencies at a later point in this chapter. Figure 18-41 shows how the flux density must drop to lower values at higher frequencies in order to keep the core loss constant at 100 mW/cm^3.

DESIGN OF A FERRITE COMPONENT FOR POWER TRANSFORMERS

A large number of good books have been published on analysis and design of Switched-Mode Power Supplies. Among them are those by Smith(1983 a), Severn (1985), Hnatek (1981), Grossner (1983), Roddam (1963), Wood(1981), and Pressman (1977). The actual design of the supply will be left to these books and will not be treated here. In addition, a veritable storehouse of mathematical analyses of various electrical design factors in power applications is contained in Snelling(1988).
The reader is strongly advised to consult this book for detailed supplementary information. Recent articles in technical and trade periodicals dealing with design of ferrites in power applications includes ones by Smith(1983), Martin (1982), Sum(1986),Badsch(1971),Bledsoe(1984),Middlebrook(1983),Turnbull(1977), Hew (1982), Fluke(1986), Bloom(1986), Chen(1978), Mullett(1986), Hill(1975), Cuk(1981), Triner(1981), Brown(1986), Kitagawa(1985), Margolin(1983), Konopinski (1979), Harada(1985), Stratford(1983), Dull(1975), and Ciarcia (1981). These publications are listed in Appendix A at the end of this chapter.

FERRITE TRANSFORMERS & INDUCTORS AT HIGH POWER

The practical design given by ferrite vendors of the power transformer may vary considerably from company to company and from engineer to engineer. Usually, previous experience in the field will provide a good starting point. In addition, vendors often supply appropriate tables, graphs and figures to assist in the design. In this chapter, we will discuss some of the practical methods suggested. Recent safety requirements entering into the design will also be covered. As is the case in many

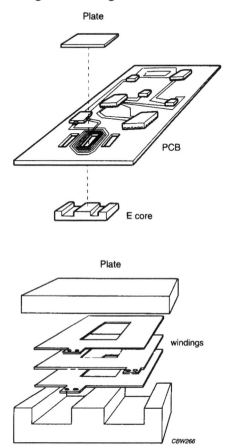

Figure 18-40 Illustrations of Planar technology for mounting cores on PC boards From Philips (1998)

higher-frequency digital devices, consideration must be given to the electromagnetic interference, both on the transmission and reception sides. Filters are therefore necessary additions to the design. Incidentally, these also use ferrite components.

CREEPAGE ALLOWANCE

The design of a transformer that must satisfy the safety requirements of line or mains isolation provides for an 8 mm. gap between the primary and secondary winding. This is most often accomplished by leaving a space of 4mm on each end of the bobbin winding space of both the primary and secondary windings. This type of construction is shown in Figure 18.42a and 18.42b taken from Snelling(1988). Many of the core selection curves including some of Snelling's (1988) take this into

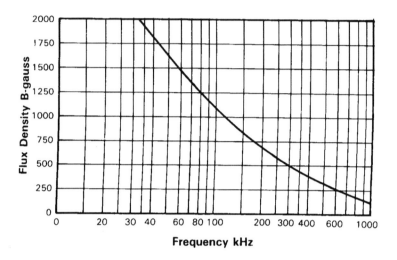

Figure 18.41-Decrease in the flux density of a ferrite as a function of frequency in order to keep the losses to 100 mW/cm^3. From Snelling(1988)

account in the transformer design. The specification is either spelled out in the German VDE 0806 or the International IEC 435 documents. The increase in output power, P_0 without this allowance is given by;

$$P_{0n} = P_{0c} [\text{Full } W_w/(\text{Full } W_w - 8 \text{ mm.})] \qquad [11\text{-}8]$$

Where; W_w = Winding width (mm.)
and c = core with creepage allowance
n = core with no creepage allowance

This amounts to about 25% increase for small cores and about 10% for large cores. Therefore, this is an important design consideration, especially if the line or mains voltage in the country of use is 230 V. or more.

THERMAL CHARACTERIZATION OF POWER FERRITE CORES

Since the temperature rise, θ, is an important factor in the operation of the transformer in a switching power supply, it is useful to have a figure which relates the

FERRITE TRANSFORMERS & INDUCTORS AT HIGH POWER

temperature rise to the output power. This parameter is known as the thermal resistance, R_{th} (Bracke 1982) and is given by;

$$R_{th} = \theta / P_o \quad\quad [18\text{-}9]$$
Where; θ = Temperature Rise in °C.
P_o = Output Power, Watts

The R_{th} is a function of the material as well as the shape and size of the core, the mounting method and the insulation used. It does not depend on the ratio of winding to core losses and so can be measured by noting the temperature rise as a function of the DC power loss dissipated in a winding. This figure can then be used at high frequency operation as done by Bracke (1982). He measured the temperature rise at D.C. and also at 50 and 100 KHz. The thermal resistance was the same at the varying frequencies. An inverse relationship between R_{th} and core size was also shown. As expected, the cores in which creepage allowance was included in the design had larger R_{th} than the same cores without the creepage allowance.

Hess (1985) found a correlation between the thermal resistance and the thermal conductivity. He found that the special power materials with the presence of CaO at the grain boundaries increased the thermal conductivity and reduced the thermal resistance. Heavier cores here also had lower thermal resistance and the ETD was particularly good.

Snelling (1988) suggests a safe value for heat loss from the surface of the core to be $300\mu W/mm^2$ or $.2W/in^2$. He, therefore, recommends the cores to be broad and thin.

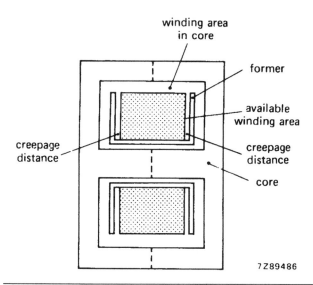

Figure 18.42a- Cross section of a ferrite core and winding showing the provision for "creepage" distance. From Snelling(1988).

DETERMINING THE SIZE OF THE CORE

Years ago, transformers were designed by using cut-and-try methods involving many modifications and final optimization. Such techniques are time-consuming and ineffective procedure and although some use of them remains, many design aids have been established to assist the designer in at least a close fit to the required

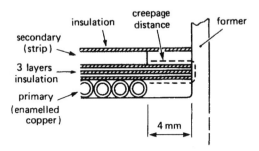

Figure 18.42(b)-The arrangement of a winding and insulation to provide the required 8mm. "creepage" distance. From Snelling(1988)

circuit with only some minor adjustment needed. Several schemes of sizing the core and completing the circuit design are presented in this chapter.

Initial Considerations in Designing a Transformer Core

In the design of a core for a power transformer used in SMPS converters, we must take into account the input current requirements to provide the AC field to drive the core to the proper B level. This will be determined by the following equation;

$$H = .4\pi NI / l \qquad [18\text{-}10]$$

In strict operational terms, the NI of the primary winding will provide the flux variation to induce the necessary secondary voltage. This voltage is related to the operating conditions by the following equation;

$$E = 4.44 \, BNAf \times 10^{-8} \qquad [18\text{-}11]$$

for sine wave with the coefficient changing to 4 for square wave.

Although part of the dimensions (cross sectional area) of the magnetic core is related directly to the flux requirements imposed by the second equation, all the windings in a power core are contained inside the core. This includes the primary turns, N_p, determined by the magnetizing current equation and the secondary turns, N_s, given by the induction equation. These windings are contained either inside the window of the toroid or a U or E core or are on a bobbin surrounding the center post in a pot core. Consequently, the size of the window or bobbin winding space does affect the overall size of the core. Therefore, it is these two requirements that are related in the design determining the shape and size of the core.

In other words, the flux equation contains the cross sectional area of the core. The NI requirements must be met by a certain number of turns each having a certain capacity to carry a current I. Achieving a higher current may allow only a few turns with a larger cross sectional area per turn as opposed to a design carrying a larger number of turns with a smaller cross sectional area per turn. It is the product of the NI which is a measure of the total copper cross sectional area and which will determine the window area.

Therefore, there are two areas that will at first determine the size of the core. One criterion used for years by design engineers is the product of these areas which is called the Area Product, A_p, (Magnetics 1987) described by;

$$A_p = W_a A_c \quad (cm^4) \quad [18\text{-}12]$$

Where; A_p = Area Product
W_a = Area of the window (cm^2)
A_c = Area of the window (cm^2)

Of course, the A_c is the area transverse to the flux and the W_a is the area transverse to the current flow. The area of the window is not completely usable because of the space between the wires and also the insulation thickness. Therefore, we introduce a copper-filling factor, K, which is the fraction of the window containing the copper. The total cross sectional area of the copper is given by;

$$A_{cu} = NA_w$$

where; A_{cu} = area of copper (cm^2) [18-13]
A_w = Cross sectional area of the copper wire, cm^2.

Therefore, the copper filling factor, K, is;

$$K = A_{cu}/W_a = NA_w/W_a \quad [18\text{-}14]$$

$$N = KW_a/A_w \quad [18\text{-}15]$$

If we multiply by A_c, we get;

$$NA_c = W_a K A_c/A_w \quad [18\text{-}16]$$

Now, from Equation 11.2 for a square wave;

$$NA_c = E \times 10^8/4Bf \quad [18\text{-}17]$$

Setting the two equations equal and rearranging;

$$W_a A_c = EA_w \times 10^8/4BfK \quad [18\text{-}18]$$

The area of the wire is related to its current-carrying capacity by one of several analogous factors. Traditionally, electrical engineers have spoken of wire sizes in circular mils instead of cm^2 (possibly because the number is quite small in cm^2). A circular mil is the cross-sectional area of a wire whose diameter is 1mil or .001 inches. The area is then .7854 square mils or 5.0671×10^{-5} cm^2. Therefore to convert from a A_w or possibly a W_a in cm^2 to circular mils, divide the circular mils by this last number. If, as is a common practice, the current carrying capacity of the wire is given in terms C, in circular mils/ampere, the relevant equations are;

$$C = A_w/I \qquad \text{Circular mils/ampere} \qquad [18\text{-}19]$$

Then;
$$A_w = IC \qquad [18\text{-}20]$$

The input power, P_i is

$$P_i = EI \qquad [18\text{-}21]$$

If we further define the efficiency;

$$e = P_o/P_i \qquad [18\text{-}22]$$

Where;

P_o = Output power

we can then relate the output power, P_o to the Area Product, A_p;

$$W_a A_c = P_o C \times 10^8 / 4 B e f K \qquad [18\text{-}23]$$

If some assumptions are made about C (800-1000 circular mils/amp), e at about 80-90%, and K (about .2-.3) we can simpify the equation. Note that the K value is only the copper-filling factor only for the primary which normally occupies about 50% of the winding space. The rest is occupied by the secondary winding. Based on these assumptions, we arrive at an equation relating P_o to operating conditions with a single constant, k_1;

$$W_a A_c = k_1 P_0 \times 10^{11} / Bf \qquad [18\text{-}24]$$

If the B level is set at 2000 Gausses, families of curves relating the output power, P_o to the Area Product, $W_a A_c$, can be generated as shown in Figure 18.43 The various cores having the corresponding $W_a A_c$ values are also shown. The P_o calculated from this equation is compared to the measured values in Table 18.2. The agreement is quite good. The $W_a A_c$ ranges can also be correlated to the temperature rise of the core in operation. The table below from Magnetics Catalog(1987) gives an approximation of the temperature rise that can be expected.

FERRITE TRANSFORMERS & INDUCTORS AT HIGH POWER

Approximate Expected Temperature Rise from the Core W_aA_c Values

W_aA_c RANGE	T RANGE
$< .2 \times 10^{-6}$	$<30°$
.2 to 0.9×10^{-6}	30°C to 60°C
.9 to 3.0×10^{-6}	60°C to 90°C

Table 18.2-
Calculated and Measured Output Levels for Power Ferrite Components

		CORE VOLUME CM^3	CORE WEIGHT GMS	SURFACE[1,2] AREA, A_s CM^2	CORE LOSS[3] @ 100°C Watts	W_aA_c CIR MILS CM^2 ($\times 10^4$)	P_o FROM P_o/W_aA_c GRAPH	MEASURED[4] OUTPUT POWER ($\Delta T \approx 50°C$)[5] WATTS
Pot cores	42616-UG	3.52	20	24	.35	.08	30	40
	43019-UG	6.12	34	33	.6	.14	55	92
	43622-UG	10.6	57	45	1.0	.30	110	184[6]
	44229-UG	18.2	104	66	1.8	.73	280	260
E cores	44020-EE	17.7	87	88	1.8	.68	260	220[7]
	45528-EE	42.3	212	142	4.2	1.78	630	530
	47228-EE	52.7	264	190	5.3	2.88	1,000	760
Toroids	42206-TC	1.47	6.9	25	.15	.08	25	50
	43813-TC	10.3	52	68	1.0	.68	210	230
	44925-TC	18.8	91	136	1.9	2.3	700	680

Other Area Product Relationships

McLyman (1982) points out a similar relationship between the area product, A_p, to the power handling capability as well as to several other important parameters used in transformer design. For current carrying capacity, a new constant, K_j is introduced making his equation;

$$A_p = (P_t \times 10^4/K_fB_mfK_uK_j)^x \qquad [18\text{-}25]$$

where: K_f = wave form coefficient
   = 4.00 for square wave
   = 4.44 for sine wave
K_u = window utilization factor as previous K
K_j = current density coefficient which is related to copper losses
x = exponent related to geometry (for pot cores, x = 1.2)
B_m = Maximum induction in <u>Teslas</u> (Note the change from Gausses. (1 Tesla = 10^4 Gausses)

Both K_j and x values are listed by McLyman (1982) for different core configurations

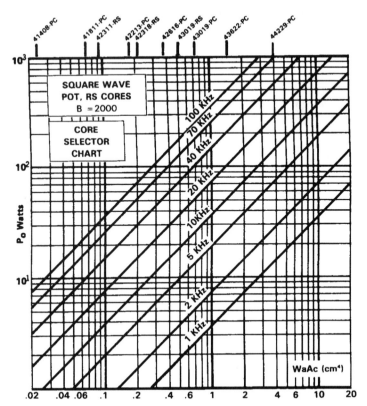

Figure 18.43- Families of curves showing output power as a function of W_aA_c at several different frequencies. At the top are listed the cores which possess the corresponding W_aA_c values. From Magnetics Catalog FC509(1997).

VOLTAGE REGULATION IN TRANSFORMERS

McLyman(1982) has also developed a new criterion and design method for transformers and inductors where the so called "regulation" is an important consideration. Smith (1985) describes regulation as "the variation in voltage from no load to full load expressed as a percentage". Thus, regulation of 5% means that 5% of the input voltage is dropped across the series resistances and reactances and the balance is transmitted to the load.

McLyman)1982) combines the power handling ability and the regulation by relating them to two constants, one a function of geometry and the other related to magnetic and electrical operating conditions. The equation is:

$$\alpha = P_t/2K_gK_e \qquad [18\text{-}26]$$

where P_t = apparent power
α = regulation in percentage

FERRITE TRANSFORMERS & INDUCTORS AT HIGH POWER

K_g = geometry coefficient
K_e = electrical coeffecient

The apparent power, P_t, is the sum of the input power, P_i, and the output power, P_0;

$$P_t = P_i + P_0 \quad [18\text{-}27]$$

With a given efficiency, η, and in the typical D.C.-D.C Converter, the equation becomes;

$$P_t = P_o\{2/\eta + 2\} \quad [18\text{-}28]$$

The geometry constant, K_g is given as;

$$K_g = W_a A_c 2 K_u / MLT \quad [18\text{-}29]$$

where MLT = Mean length per turn

Rather than using circular mils for the wire size, A_w and window area, W_a, these units are each given in cm^2.

The constant K_e or the electrical constant is given by;

$$K_e = 0.145 \, K_f^2 \, f^2 B_m^2 \times 10^{-4} \quad [18\text{-}30]$$
(B is in Teslas or Wb/m^2)

The current density, J, in A/cm^2 is given by;

$$J = P_t \times 10^4 / K_f K_u f B_m A_p \quad [18\text{-}31]$$

McLyman gives an example of the K_g approach in the design of a transformer for a single-ended forward converter. It is shown in Appendix 2. This approach may appear long and and requires data which may not be in the manufacturers catalog. Fortunately, McClyman has supplied a compilation of the needed data in his books(1982,1988).

Grossner (1983) has also called attention to the dependence of P_o on the area product. He uses the same approach as previously discussed with several geometrical coefficients to approximate the output power, P_o. Grossner is more concerned with the temperature rise in the calculation that is expressed as;

$$P_0 = C_2 f B g_6 (h\theta)^{1/2} \quad [18\text{-}32]$$

where;

C_2 = a constant involving core and winding fractions and wire resistance

h = Coefficient of heat transfer
θ = Temperature rise
g_6 = Geometrical constant involving surface area, magnetic path length and wire length per turn

Thus, Grossner concludes that the power level is more responsive to increases in frequency and flux density than to an increase in the temperature rise. In practice, f is defined by the circuit and B is limited by the core material. With B and f fixed, keeping a small size and a high power level are aided by operating at the highest possible temperature rise. Because circuits may be designed to optimize different requirements, Grossner develops the parameters, g_1 - g_8, which in some combination will lead to optimization of power, inductance, and optimum power.

Smith (1985) uses the area product as a design criteria, but reduces it to Normalized core dimensions so that any size core can be calculated.

Another author using the W_aA_c approach is Pressman (1977). Here the tables of the supplier are used to approximate the core for the power level required.

DeMaw (1981) approximates the temperature rise in a core as

$$T_{rise} = 50P_t/P_o \qquad [18-47]$$

Where P_o is not the output power as previously used but the power dissipation level within a specified core that will cause a 50°C temperature rise. P_t is the total power dissipated in an inductor including core loss and winding loss. The core loss data can be obtained from the manufacturing catalog while the winding losses can be estimated by formulae.

Watson (1986) also uses the Area product and McLyman's K_g method but derives equations based on two kinds of current density, rms and instantaneous. This distinction is important for flyback transformers.

Although the W_aA_c approach is merely a starting point in the design, many other factors may have to be considered in finalizing the design. With ferrites in power supplies some of these factors are as follows;

1. Max temperature of the core (<100°C.)
2. DC imbalance
3. Magnitude and linearity of magnetizing current
4. Magnitude of transient current
5. Under transient loading, need to limit Bm to avoid saturation

OTHER TRANSFORMER DESIGN TECHNIQUES

Snelling(1988) has divided the design of power transformers into several different categories. They are:

1. Winding loss limited
2. Saturation limited
3. Regulation limited
4. Core loss limited
5. High frequency limited

Items numbered 2-5 are discussed previously. Snelling has added the other two. He discusses them each separately, stating that, in general, they come into play as the frequency of operation increases.

Winding-Loss-limited Design

This is almost the same situation we have been discussing using the $W_a A_c$ approach or the wire current density approach. As in the previous cases, there is also some limit on the B level. However, here the only real source of dissipation is the winding loss and, depending on the size of the transformer, the treatment is only applicable to a frequency of about 5-10 KHz. The equation for the input power, P_i Snelling(1988) gives as;

$$P_i^2 = P_w f^2 B_e^2 / m k_1 \qquad [18\text{-}33]$$

Where; m = Fractional increase in the resistivity of the copper over that at 25°C

and; $k_1 = 2\rho_c l_w / \pi^2 A_e^2 A_w F_w$
l_w = mean length of a turn of the winding
F_w = Winding factor of the copper

Regulation-Limited Design

In this category, the regulation can be a constraint on the winding loss listed above. Again, the winding loss is really the only source of dissipation, that is, the core loss, P_c is much smaller than P_w at these frequencies. The voltage regulation, , is given as

$$\alpha/100 = P_w / P_i \qquad [18\text{-}34$$

Then; $P_i = (f^2 B_e^2 \alpha / m k_1) \times 100 \qquad [18\text{-}35]$

Since the core loss is negligible, we can ignore it and the output power is given by;

$$P_0 = P_i - P_w = P_i [1 - \alpha/100] \qquad [18\text{-}36$$
$$= f^2 B_e 2(\alpha - .01\alpha^2) / m k_1 \times 100 \qquad [18\text{-}37]$$

B_e^2 is limited by the saturation flux density of the material. The other parameters are given so that k_1 can be calculated. Snelling (1988) presents a table of the values of k_1 in his book for many power core sizes in so that the choice can be made. Most ferrite transformers are not regulation limited.

Saturation-Limited-Design

As the flux density increases, the hysteresis curve will flatten out as saturation is approached. When this happens, the incremental permeability drops sharply. In this case, the impedance (or inductive reactance) becomes quite small and the current, therefore, increases. See Figure 18.44. The manufacturer will often give limit values

494 HANDBOOK OF MODERN FERROMAGNETIC MATERIALS

for the maximum flux density that the designer should not exceed. Since the saturation drops at higher temperatures, the saturation value at the operating temperature should be examined in this regard. For operation at 100°C., the value of 3200 gausses would be a real maximum.

The constant k_1 is still a valid constant for saturation-limited designs. Since the efficiencies at these lower frequencies are close to 100 percent, the output power, P_o, can be considered the same as the input power, P_i. Therefore;

$$P_o^2 = P_w f^2 B_e^2 / m k_1 \qquad [11\text{-}38]$$

If the flux, ϕ^2, is used instead of B_e^2, the factor k_2 is used where $k_2 = k_1 A_c^2$. If core loss is included in the saturation limited case, the equation for the square wave drive becomes;

$$P_0 = (P_t - P_c)^{1/2} k_3 F_w^{1/2} f \quad \text{Watts} \quad [18\text{-}39]$$
$$\text{where;} \quad P_t = \text{Total Losses} = P_w + P_c \qquad [18\text{-}40]$$

P_t is also listed in Snelling's (1988) Table 9.3. At low frequencies, the core loss is less than half the total loss and so may be set to 0 especially because of the square root dependence.

$$k_3 = (A_w / m_c l_w)^{1/2} \phi \qquad [18\text{-}41]$$

Again, Snelling lists the values of k_2 and k_3 in Table 9.3 of his book (1988). In addition the value of P_t which is the sum of all the losse (winding and core) is also given in the same table. The permitted temperature rise in this table is 40°C. If the proposed temperature rise is different, the new value of B_e or can be recalculated from the thermal resistance, R_{th}.

At low frequencies, where core loss is a small fraction of the total loss, the output power is proportional to f. If the operating variables such as frequency, temperature rise and copper factor are assumed, the power handling values for a given core can be given. Most manufacturers provide such information. The values of P_o are similar to the ones derived earlier from the area product technique.

Core-Loss-Limited Design
Traditionally, design of a transformer is optimized by making the winding losses equal to the core losses. It has generally been taken as a device Other calculations place the division as;

$$P_c = (2/n) P_w \qquad [18\text{-}42$$

When n=2, the two losses are indeed equal. However at higher frequencies, the value of n is between 2 and 3. For core loss limited designs assuming the output power, P_0, is equal to the input power, P_i, the equation for the core loss, P_0, for a square wave is;

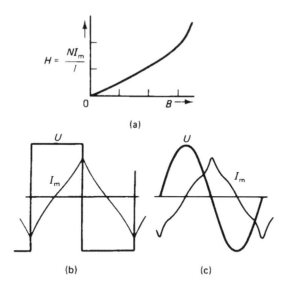

Figure 18.44-The effect of increasing induction on the field and thus the current associated with it as a ferrite core approaches saturation-(a) The increase in slope of H near saturation;(b)The voltage and current waveforms(Note the sharp increase in the slope of current at the end of the pulse. The drive is square wave. (c) The same with sine wave drive.

$$P_o = [P_t/(1+2/n)]^{1/2} \times k_3 F_w^{1/2} f \phi_{p-p} \quad [18\text{-}43]$$

where K_3 is defined as before.

With P_t and k_3 given in the table, the output power can be given in the case of a core-loss limited design. To check the flux density, the manufacturers' graphs showing core loss as a function of frequency and flux density can be consulted. When P_t is known, the core loss can be estimated from the previous equation or just set to $1/2$ P_t. From this core loss and the operating frequency, the B value can be read off the graph. If there is a varying cross sectional area of the core, the equation is modified as such for square wave;

$$P_o(1+2/n)^{1/2}]/2fBF_w^{1/2} = P_tA_w/m \; l_w]^{1/2} \times A_{min} = k_4 \quad [18\text{-}44]$$

The values for k_4 are also tabulated in Snellings Table 9.3.

Based on the input design specifications, k_4 and B can be calculated from the minimum area and a core can be chosen. For the division of losses, a value of 2.5 is typical. For a fully-wound transformer, F_w can be set at .5. With these assumptions, k_4 can be written as;

$$k_4 = 0.845P_o/fB \quad \text{for sine wave} \quad [18\text{-}45]$$
and
$$k_4 = 0.949P_o/fB \quad \text{for square wave} \quad [18\text{-}46]$$

The manufacturers' data is certainly a good way to check the core loss assumptions. These design methods are useful in initially picking a core and modifications must be made if one or more condition is not met.

MAGNETIC AMPLIFIER-MULTI-OUTPUT DESIGN

In a multi-output power converter, it is often important to control one or more of the outputs independently. One method of doing this is by using a core having a square-loop material as a magnetic amplifier. As pointed out by Snelling (1997б), this delays the leading edge of the secondary circuit 'on' pulse by an amount depending on the re-set condition. The re-set condition determines the amount of volt-seconds needed to drive the core to saturation. When saturation occurs, the inductance falls to a low value and the energy transfer can commence. Bosley (1994) has compared three available materials for this type of design. They are Permalloy

Square Permalloy

ADVANTAGES:
- Moderate price.
- Highest flux density.
- Low core loss below 100 kHz.
- Low saturated inductance

DISADVANTAGES:
- Stress sensitive, must use a core box.
- High core losses above 100 kHz.

Figure 18.45-Advantages and disadvantages of square Permalloy as a magnetic amplifier for SMPS applications

Cobalt-based Amorphous

ADVANTAGES:
- Lowest core losses.
- Moderate flux density.
- Highest squareness.
- Lowest saturated inductance.
- Can be used encapsulated, without a core box.

DISADVANTAGES:
- Highest price.
- Encapsulated version compromises the high squareness and the low saturated inductance.
- Must be used below 100°C.

Figure 18.46-Advantages and disadvantages of Co-based amorphous material as a magnetic amplifier for SMPS applications

FERRITE TRANSFORMERS & INDUCTORS AT HIGH POWER

(NiFe) alloy, amorphous alloy and square loop ferrites. We have considered the use of square loop ferrites previously in conjunction with the ferroresonant transformer design. The advantages and disadvantages of the square permalloy are shown in Figure 18.45. Those of the cobalt-based amorphous metal material are given in Figure 18.46 and the corresponding ones for the square-loop ferrite are given in Figure 18.47. The 100 KHz. B-H.loops for these three materials are given in Figure 18.47. The Philips 3R1 material is used for this purpose. The square loop amorphous material has a higher squareness and give better performance. They are, however, more expensive but require fewer turns.

POWER FERRITE DESIGN FROM VENDORS' CATALOGS

The vendors of ferrite cores have proposed several different design schemes. The one used in the Magnetics Catalog(1987) has been discussed. Design methods described by other ferrite vendors are:

Square-Loop Ferrite

ADVANTAGES:
- Lowest price.
- Stress resistant.
- Lowest losses above 250 kHz.

DISADVANTAGES:
- Limited sources.
- Highest saturated inductance.
- Lowest flux density.
- Flux density varies with temperature.

Figure 18.47-Advantages and disadvantages of square loop ferrite material as a magnetic amplifier for SMPS applications

Philips

Philips- In the case of Philips catalog, the throughput power, P_o, information is supplied in the form of graphs of the P_o and output voltage, V_o for each type of converter. Such a graph is shown in Figure 18.49. In addition, the performance factor (f x B_{max}) is graphed as a function of frequency for their power ferrite materials. A graph of this type is given in Figure 18.50.

Siemens

Siemens lists the output powers for each power shape in several power materials and for each converter type. The output powers are given at a typical frequency and a cut-off frequency for each material. A portion of this table is given in Table 18.3 The material specific values on which the table values are based were taken from the maximum temperature rise for each material (given in Table 18.4) and the ther-

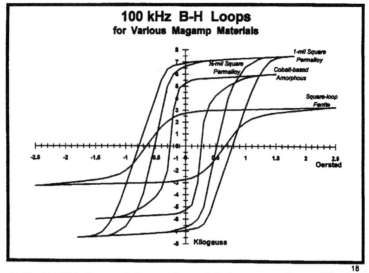

Figure 18.48- 100 KHz hysteresis loops of materials for a magnetic amplifier in SMPS applications[18]

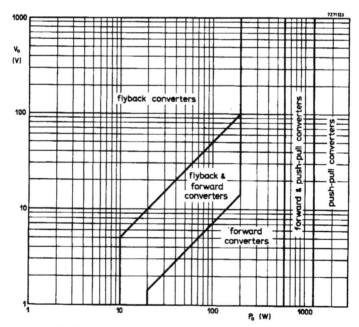

Figure 18.49-Suggested out power, P_o and output voltage for the different types of converters. From Philips (1998)

FERRITE TRANSFORMERS & INDUCTORS AT HIGH POWER

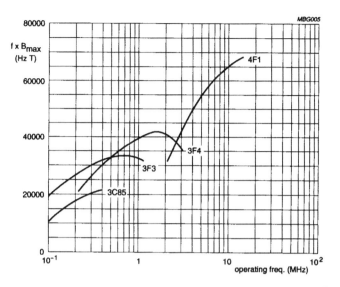

Figure 18.50-Performance factor (f x B) versus frequency for several different power ferrite materials. From Philips (1998)

Table 18.3-Power handling capabilities of various shape cores and materials

Core shapes	Material	Version (LP=Low profile)	f_{typ} kHz	f_{cutoff} kHz	Power capacities					
					Push-pull converter		Single-ended converter		Flyback converter	
					$C = 1$ [1]		$C = 0{,}71$ [1]		$C = 0{,}62$ [1]	
					P_{trans} (f_{typ}) W	P_{trans} (f_{cutoff}) W	P_{trans} (f_{typ}) W	P_{trans} (f_{cutoff}) W	P_{trans} (f_{typ}) W	P_{trans} (f_{cutoff}) W
EFD25/13/9	N59	Normal	750	1500	311	417	221	296	193	258
	N49		500	1000	196	263	139	187	122	163
	N67		100	300	175	280	124	199	109	173
	N87		100	500	242	482	172	342	150	299
EFD30/15/9	N59	Normal	750	1500	401	343	285	244	249	213
	N49		500	1000	253	544	180	386	157	337
	N67		100	300	226	365	160	259	140	227
	N87		100	500	312	630	221	447	193	390
U cores										
U15/11/6	N27	Normal	25	150	31	81	22	58	20	50
U17/12/7	N27	Normal	25	150	37	97	26	69	23	60
U20/16/7	N27	Normal	25	150	74	161	52	114	46	100
U25/20/13	N27	Normal	25	150	198	432	141	306	123	268
UU93/152/30	N27	Normal	25	150	2527	5508	1794	3910	1567	3415

[1] Numerical data are stated in accordance with the publication "Effect of the magnetic material on the shape and dimensions of transformers and chokes in switched-mode power supplies", G. Roespel, Siemens AG München, J. of Magn. and Magn. Materials 9 (1978) 145-49

From Siemens (1998)

Table 18.4- Maximum Temperature Rise and Typical and Cut-off frequencies for various ferrite power materials

	ΔT_{max} K	f_{typ} kHz	f_{cutoff} kHz
N59	30	750	1500
N49	20	500	1000
N62	40	25	150
N27	30	25	100
N67	40	100	300
N87	50	100	500
N72	40	25	150
N41	30	25	100

Table 18.5- Thermal Resistances for Main Power Transformer Shapes

Core shapes	R_{th} (K/W)	Core shapes	R_{th} (K/W)	Core shapes	R_{th} (K/W)
E 20/6	50	ETD 29	28	PM 50/39	15
E 25	40	ETD 34	20	PM 62/49	12
E 30/7	23	ETD 39	16	PM 74/59	9,5
E 32	22	ETD 44	11	PM 87/70	8
E 40	20	ETD 49	8	PM 114/93	6
E 42/15	19	ETD 54	6	U 11	46
E 42/20	15	ETD 59	4	U 15	35
E 47	13	ER 42	12	U 17	30
E 55/21	11	ER 49	9	U 20	24
E 55/25	8	ER 54	11	U 21	22
E 65/27	6	RM 4	120	U 25	15
EC 35	18	RM 5	100	U 26	13
EC 41	15	RM 6	80	U 30	4
EC 52	11	RM 7	68	U 93/20	1,7
EC 70	7	RM 8	57	U 93/30	1,2
EFD 10	120	RM 10	40	UI 93	5
EFD 15	75	RM 12	25	UU 93	4
EFD 20	45	RM 14	18		
EFD 25	30				
EFD 30	25				

mal resistance for each size and shape of core which is listed in Table 18.5. The Total core losses are related to these factors by ;

$$P_{V, tot} = \Delta T/R_{th} \quad [18.47]$$

The assumption is made that the temperature rise and the losses in the core are evenly distributed. The application area for flyback transformers were restricted to , 150 KHz. . The overtemperature, ΔT is the sum of the temperature rises resulting from the core and winding losses. The maximum flux densities were <200mT for flyback converters and <400mT for push-pull converters

Thomson
Thomson(1988) presents charts of average wattage for the various size and shape power cores listed according to inverter type. The frequencies are 25KHz(2000 Gausses), 100 KHz(1000 and 1200 Gausses).

FERRITE TRANSFORMERS & INDUCTORS AT HIGH POWER

TDK
TDK lists the calculated output power under the specifications of each type of power core. These power levels are given at 50 and 100 KHz for their standard power materials. These power levels are given for the forward converter mode. TDK also gives the power l. The conditions are 25 KHz(2000 Gausses) and 100 KHz(2000 Gausses). The temperature rise is also plotted against the power loss for each core. An example of the data provided for one core is shown in Figure 11-26.

CORE DATA ON CD ROM OR FLOPPY DISKS
Many ferrite core manufacturers offer their core data in the form of tables and graphs on either floppy or CD ROM disks. These include

1. Siemens
2. Magnetics
3. Philips
4. Kaschke

INTERNET WEB SITES
Just about all the manufacturers of ferrites have Web sites on the Internet. In most cases, the core data can be down loaded and printed by the user. Where large catalogs are involved, the use of the Acrobat reader is needed but the vendor can often download this as well. Some of the vendors that maintain Web sites are;

1. Magnetics
2. Philips
3. Siemens
4. Fair-Rite Products
5. Steward
6. MMG North America

COMPUTER-AIDED POWER TRANSFORMER DESIGN
As we would expect, several different schemes are available for the use of computer- aided design in the choice of the appropriate ferrite core and the windings. Some are user-generated and some are available commercially. One such method is based on the paper by Martinelli (1988). The software program involved in the design is coupled with various data disks that list the properties of all the presently available cores in the different materials. These include ferrite as well as many materials of the metallic and powder core varieties. A typical transformer spread sheet(Analytic Artistry 1988) is shown in Figure 18.52. The designer enters theconditions such as whether the design is core-loss or saturation limited, and input and output voltages and currents. Using certain assumptions on current density, flux density and ratio of core to wire losses, the designer chooses a certain core and

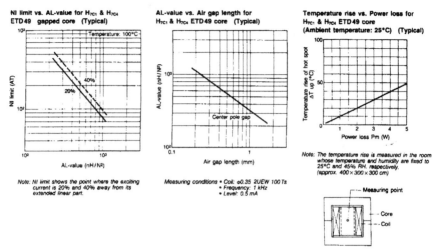

Figure 18.51-Listing of calculated output power and power loss for several different ETD cores. The temperature rise associated with that power loss is also listed. From TDK Ferrite Cores Catalog BLE873-001C(1987)

checks for conformance to specifications. If one or more design specifications are off, another core is chosen and the process reiterated. When the same core is picked twice, the design is complete. The basis of initially picking the core is based on the $W_a A_c$ approach. P_c, B_{ac}, P_{cu}, J, and N_p are obtained from the core dimensions in the data file

McLyman (1990) has offered Magnetic Component Design Software Programs for the IBM PC or the MacIntosh computers. The programs available are;

1. Computer-Aided Designs for Inductors and Transformers Mark III
2. Flyback Converter Magnetics Design Mark III
3. Specialty Design Magnetics Mark III
4. Magnetic Core Data Conversion Mark III
5. Computer Aided Transformer and Inductor Analysis Mark III

WINDING LOSSES

Although copper losses are not a magnetic phenomenon, the windings that produce them also carry the currents which create the fields in the primary and sense the induced voltage in the secondary. They must therefore be considered in the design. The number of turns is discussed under core losses and completion of the design.

Smith (1983) has written an algorithm for the optimum copper losses, the core losses and the sum of the losses which provides the dimensions of the cross sectional area and window area. The core loss is estimated from the equation:

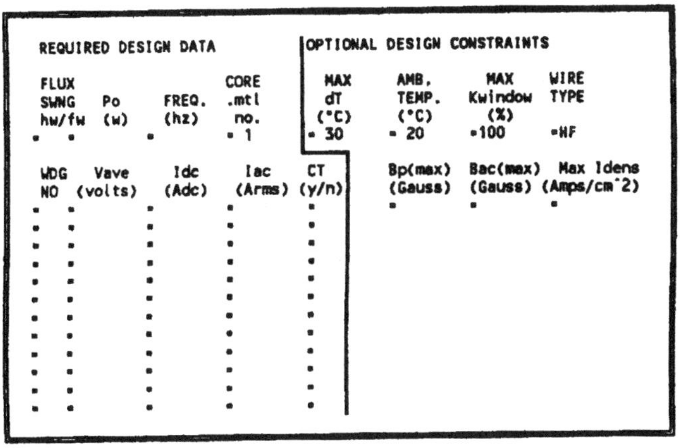

REQUIRED DESIGN DATA				OPTIONAL DESIGN CONSTRAINTS			
FLUX SWNG hw/fw	Po (w)	FREQ. (hz)	CORE .mtl no. • 1	MAX dT (°C) • 30	AMB. TEMP. (°C) • 20	MAX Kwindow (%) •100	WIRE TYPE •HF
•	•	•					
WDG NO	Vave (volts)	Idc (Adc)	Iac (Arms)	CT (y/n)	Bp(max) (Gauss)	Bac(max) (Gauss)	Max Idens (Amps/cm^2)
•	•	•	•	•	•	•	•
•	•	•	•	•			
•	•	•	•	•			
•	•	•	•	•			
•	•	•	•	•			
•	•	•	•	•			
•	•	•	•	•			
•	•	•	•	•			

Figure 18.52-Transformer spread sheet for a transformer design by Computer-Aided Design(CAD).From Analytic Artistry(1988)

$$P/V = C_1 B^{2.5} \quad \text{(loss/volume)} \quad [18\text{-}48]$$

Both the simplification by removing the frequency dependence and concentration on toroidal shapes makes the procedure more specific to tape wound cores than ferrites. Nevertheless, the approach can be used for ferrites.

More to the point of ferrites is an article by Carsten (1986) which contains a very complete discussion on high frequency conductors in switched-mode power supplies. This article fully discusses the origin of skin, proximity and related effects. The high frequency design is especially applicable for forward and flyback converters where the inputs are below 50V. Carsten also gives a method of calculating the losses in a particular winding. The article is aimed at pulse currents with more attention than duty-cycle modulated square waves. Other wave-forms can be calculated if the harmonic content is known.

Snelling's(1988) book also has a very complete section on the design of windings.

COMPLETING THE DESIGN OF THE TRANSFORMER-WINDING DATA
Once the appropriate core is picked based on the limitations of saturation or core loss, the number of turns and wire size is chosen for the primary and secondary windings;

$$N_p = E_p \times 10^8 / 4BNA_c f \quad [18\text{-}49]$$
$$N_s = E_s N_p / E_p \quad [18\text{-}50]$$

and

$$I_p = P_{in}/E_{in} = P_{out}/eE_{in} \quad [18\text{-}51]$$
$$I_s = P_o/E_{out} \quad [18\text{-}52]$$

The wire size can be obtained by referring to the winding area of the bobbin, the number of turns, the copper fill factor and a table of the wire sizes.

$$F_w W_a = N_p A_{wp} + N_s A_{ws} \qquad [18\text{-}53]$$

For a first approximation, the available window area is assumed to be divided equally between the primary and the secondary so that each winding size can be evaluated separately. The copper fill factor can be assumed to be .4 for toroids and .6 for pot cores and E cores. The number of turns fitting in the winding space of individual cores can by found from the winding tables published by the vendor and the window area published under each core. From the wire tables, the resistance per unit length of winding can be found for a particular wire size. This unit resistance, when multiplied by the length of winding (number of turns times the average length per turn given in the core data), yields the total resistance of each winding. Then the winding loss per winding can be calculated by;

$$P_{wp} = I_p 2 R_p \qquad [18\text{-}54]$$
$$P_{ws} = I_s 2 R_s \qquad [18\text{-}55]$$

The total winding loss,
$$P_w = P_{wp} + P_{ws} \qquad [18\text{-}56]$$

This can be compared with the optimum ratio of winding loss to total loss;
$$P_w = P_t/[1+2/n] \qquad [18\text{-}57]$$

If the calculated loss is close to the optimum, the design is probably pretty good. Otherwise further modifications in the winding or core size may be necessary.

VERY-HIGH-FREQUENCY POWER FERRITE OPERATION

In the last five years, the frequency of switching power supplies has increased dramatically. While initially it was 25 KHz., the present state of the art is 100 KHz., and new designs for the 200-500 KHz. range extending upwards to 1 MHz. are being developed and as a result, there has been a large reduction in the size and weight of the transformers. Engineers working with ferrite materials have responded with materials capable of operating at these frequencies. This has been done as described in a previous chapter by a combination of chemistry and microstructural improvements. We may be approaching the limit of operation of MnZn ferrites and new chemistries will be forthcoming in the NiZn materials. Snelling(1989) points out that problems with the conventional switching supply using pulse width modulation techniques is also reaching its limit because of circuit problems such as switching transients and the radio interference caused by harmonics. Snelling (1989) predicts a limit of 500 KHz. due to these effects. He also predicts that resonant power conversion may take over at the higher frequencies.

For materials up to this limit, he advises the minimization of m in the equation

$$P_m = kf^mB^n \qquad [18\text{-}58]$$

This would include materials of lower permeabilities. There should be an optimum composition for minimum power loss at a given frequency.

As previously mentioned, the design of ferrites for very high frequency application should also involve lowering of the B_{max} of the material because the flux density dependence (n) is of higher order than the frequency dependence. Therefore for the same power level, decreasing the losses by using lower flux density has more leverage than increasing the losses by the higher frequency. The ratio of eddy Current to magnetic losses is shown in Figure 18-53. Even at 250 KHz, the ratio of Eddy Current to magnetic losses, while increasing is still only about 20%. Another design factor in the higher frequencies is the advisability of operating at as high a temperature as is feasible. Better design of cores such as those of the planar type is also being used as this increases surface area for removal of heat from the ferrite.

In a recent paper, Buethker(1986) has considered the breakdown of losses in power ferrites. He lists them as

Hysteresis Losses, $P_{hyst} = C \times f^x B^y$ [18-59]
Eddy Current Losses, $P_{e.c.} = .8f^2 B^2 A_e / \rho$ [18-60]
Residual Losses, $P_{res} = 2.5 \times 10^{-3} f B^2 \tan \delta / u$ [11-61]

At 100 KHz., the hysteresis loss is predominant (Figure 18-54a) and a large reduction in these losses in 3C85 over 3C8 accounts for a significant overall loss reduction. However, at 400 KHz., the 3C85 (Figure 18.54b) shows a greater increase in eddy current and residual losses which now can be lowered by rather drastic changes in microstructure in the new material 3F3 (Figure 18-54b) so that now again, the hysteresis losses are predominant. Historically, it seems that when the state of the art of power supply design requires power ferrite material for higher

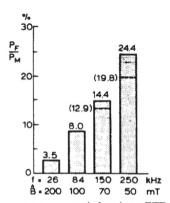

Figure 18.53-Ratio of Eddy current to magnetic loss in an ETD core. From Snelling(1989)

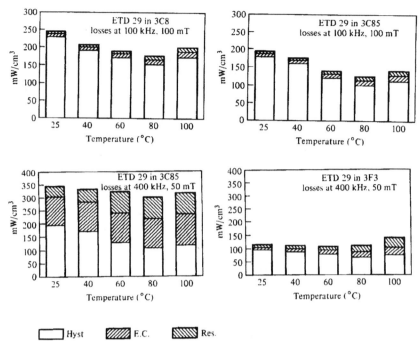

Figure 18.54(a)-Breakdown of core losses into Eddy current, Hysteresis and Residual Losses at 100 KHz.for two ferrite materials. From Buthker,C.,and Harper,D.J., Transactions HFPC, May,1988,p.186.

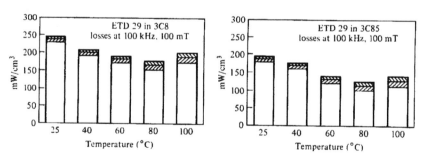

Figure 18.54(b)- Breakdown of core losses similar to Figure 18.54(a) but at 400 KHz. From Buthker(1988)

frequencies, the ferrite designer produces materials with lower eddy current losses to meet the challenge.

Earlier, we discussed the performance factor PF_{200} proposed by Stijntjes(1989) which is the product ($B_m f$) of the frequency, f, and the B level which in a given material will give losses of 200 mW/cm^3. Figure 18.55 shows the PF_{200} of 4 different materials as a function of frequency. The optimum operating conditions for a material with regard to power occurs where the curve is a maximum. For material

A(MnZn Ferrite with Ti and Co additions), the maximum PF_{200} is 35,000 at .5 MHz(500 KHz). The corresponding figure for material C(NiZn+Co[fine]) is 110,000 at 30 MHz. It would appear that operation at 30 MHz with material C would be more desirable but other considerations such as availability of semiconductors and the cost of the NiZn ferrite appear to be more important. Thus, for the present, improved MnZn ferrites are the major power materials.

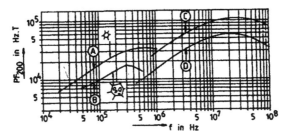

Figure 18.55-The Performance Factor, PF_{200}, for several types of MnZn and NiZn ferrite materials as a function of frequency. From Stijntjes (1989)

Recent papers on high frequency transformers or materials include; Sano(1988), Finger(1986),Kamada(1985),Carlisle(1985),Hiramatsu(1983),Martin(1984), Cattermole(1984), Shiraki(1890), Kepco(1986), Mochizuki (1985),Schlotterbeck), and Zenger(1984).

COMPETITIVE POWER MATERIALS FOR HIGH FREQUENCIES
Roess(1987) has recently emphasized that a great virtue of ferrite power material is their adaptability, and even at higher power frequencies. He compares the losses of several competing power materials for the higher frequency operation. Trafoperm is a NiFe strip material. Vitrovac is an amorphous metal strip material and Siferrit is of course, a ferrite. The results are given in Figure 18.56 . Up to 100 KHZ., the amorphous metal materials have lower losses than the ferrite especially the thinner gage type which remains lowest even at the higher frequencies. Roess points out that despite this disadvantage, a ferrite core is still the magnetic component of choice because of its much lower cost and its adaptability to be produced in many different shapes. The strip, on the other hand, has limitations on the shapes in which it can formed as shown in Figure 18.57. The new nanocrystalline materials were developed after this study.

A new fine-grained rapidly solidified (not amorphous) strip material has just been introduced by Hitachi Ltd. It has much higher saturation (13,500 Gausses), higher permeability (16,000 at 100 KHz.) and very low losses at 100KHz. While having the same lack of versatility as ferrites, it remains to be seen if the price and performance will allow it to compete with ferrites.

FERRORESONANT TRANSFORMERS
We have spoken of resonance as it relates to low level linear ferrite components. Here a series or parallel combination of an inductor and capacitor acted as an LC circuit for frequency control in low level filters. The term resonance (more properly,

ferroresonance) here has more of a connotation of resistance to changes in the input voltage and current by storing energy in the resonant circuit. As a matter of fact the first uses of ferroresonance was in the construction of a constant-voltage 60 Hz. transformer by Sola. In power supplies, an important use of the ferroresonant transformer is as a regulator. The early 60Hz transformers have given rise to the high-frequency type which, as noted earlier, may be even more useful at the highest frequencies than the conventional switching transformer design.

As a high-frequency power inductor, the ferroresonant transformer has a quite different function. For one thing, the magnetic circuit is non-linear and because of the high currents and fields, operation is close to saturation. Most often when used as a power inductor, it is necessary to insert an air gap or spacer to avoid saturation. Figure 18.58 shows a simple ferroresonant regulator that consists of a linear inductor, L_1, a non-linear saturating inductor, L_2 and a capacitor, C_1 in parallel with L_2. It is the latter two that form the ferroresonant circuit that controls the input voltage.

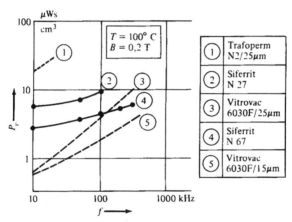

Figure 18.56-Core losses as a function of frequency for several different types of magnetic materials ;(1) is a NiFe metal strip material; (3) and (5) are amorphous metal strip materials; (2) and (4) are ferrites. From Roess(1987)

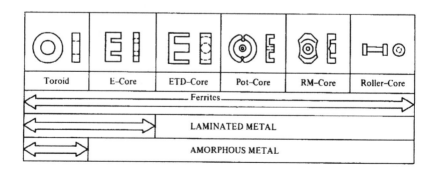

Figure 18.57-Formability of ferrites versus NiFe metal strip and amorphous metal strip. From Roess(1986)

FERRITE TRANSFORMERS & INDUCTORS AT HIGH POWER

The input energy is stored in L_1 and the resonant circuit acts to pass a uniform voltage to the load. Although the linear transformer may be of the typical power ferrite found in transformers, the saturating transformer is quite different. In addition to the usual attributes of power ferrites, it should possess a rather square hysteresis loop. The squareness ratio, B_r/B_s should be over 85%. The permeability over the linear portion of the loop should be as high as possible with the saturation permeability quit low ($\mu = 20\text{-}30$)

By combining the ferroresonant regulator with a high frequency inverter, a ferroresonant converter can be constructed as shown in Figure 18.59 Then, with the addition of a rectifier in front, a switching power supply can be made. See Figure 18.60.

McLyman(1969)has shown how a high frequency ferroresonant transformer, tuned to about 20 KHz. can be used to stabilize high frequency inverters.

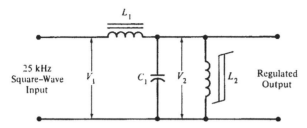

Figure 18.57-Formability of ferrites versus NiFe metal strip and amorphous metal strip. From Roess(1986)

POWER INDUCTORS

Power Inductors differ from the low-level inductors that we have dealt with in Chapter 10. They are not used in LC circuits for frequency control. In power inductors, use is made of their ability to store large amounts of power in their magnetic field. As such, they can limit the amount of ac voltage and current. When this is done in the presence of a high D.C. current, the inductor, usually in combination with a capacitor, serves as a smoothing choke to remove the ac ripple in a D.C. supply. This is often done in the output circuit of the supply after rectification. Since there are large D.C. and superimposed a.c. currents, they usually need gaps to prevent saturation. In addition to the increase in current and possible catastrophic failure at saturation, the incremental permeability drops close to zero and therefore, the required inductance specification is not met. With the gap, the magnetization curve is skewed to avoid saturation (See Figure 18.37).

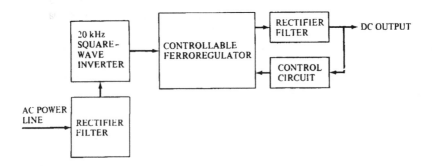

Figure 18.59 A Ferroresonant inverter. From Fletcher (1977)

With regard to the ac component, the permeability of the gapped core is larger than one operating at saturation. The amount of gap depends on the maximum D.C. current, the shape and size of the core and the inductance needed for energy storage. The a.c. ripple is usually on the order of 10% of the D.C. signal. To estimate the maximum current, I_m, an extra 10% or more safety factor for transients is inserted in the design making I_m on the order of 1.2-1.3 I_0.

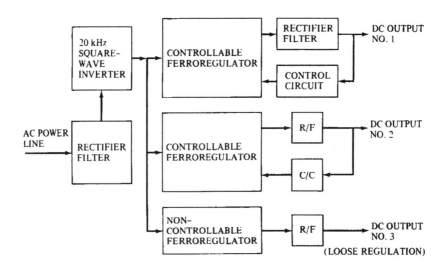

Figure 18.60 - A Ferroresonant switching power supply From Fletcher (1977)

FERRITE VS METALLIC POWER INDUCTOR MATERIALS

Earlier in this chapter, we compared ferrite power transformer materials with their counterparts in metallic materials. Whereas for ferrite power inductors, the materials are mostly the same as the transformer materials, the for power inductors, many metallic materials are different. Whereas gapped cores are used for many power inductor applications, in the case of the metallic cores, the gap is a distributed

FERRITE TRANSFORMERS & INDUCTORS AT HIGH POWER

one as found in powder cores. Bosley (1994) who did the analysis on SMPS transformer materials compared the materials for SMPS inductors in the same paper. The materials evaluated were;

1. NiFe Powder Cores- Molypermalloy and Hi_Flux Cores
2. Sendust Powder Cores- Kool-Mu and MSS cores
3. Amorphous Choke Cores
4. Powdered Iron Cores
5. Gapped Ferrite Cores

Some of these materials were discussed in the chapter on powder cores for low level telecommunications applications. For power applications at high power levels, the materials may be somewhat different. Bosley (1994) listed the advantages and disadvantages of the above mentioned materials for power inductor applications. Figure 18.61 lists these for NiFe powder cores, Figure 18.62 for Sendust(FeAlSi) powder cores. Figure 18.63 for amorphous metal choke cores, Figure 18.64 for powdered iron cores and Figure 18.65 for gapped ferrite cores. Since power inductor often must operate under high D.C. bias conditions, the effective permeability for these materials are given. The DC bias curves for several of these materials are shown in Figure 18.65. Again, the nanocrystalline materials were not considered. Bosley also listed the core losses of the various inductor materials compared to ferrites. They are listed in Table 18.3. Although the ferrites are lower in losses than the others listed, Bosley notes that, with a medium to large gap, there may be increased losses due to fringing flux This may increase ac copper losses near the gap. Nanocrystalline materials were not considered here as well.

Design of Power Inductors

Some of the same factors of concern in transformers are applicable in inductors. The amount of power that can be handled and thus, the temperature rise are important. Regulation needs may also be present. To specify the inductor, we must have a minimum inductance and a maximum current. The stored energy of the inductor is given as;

$$E_i = 1/2 \, LI^2 \qquad [18\text{-}62]$$

In the design of an inductor, we must specify the minimum inductance, L_{min} and the maximum current, I_m. This permits us to calculate the LI^2 product. The conventional manner and time tested design technique is through the use of Hanna Curves (Hanna 1927). An example of a Hanna curve is given in Figures 18.67 or 18.68 for a specific material. Here, $LI^2/Volume$ is plotted against H. Also shown is the ratio of the gap that must be inserted in the core to the magnetic path length. Although the volume is not known at the start of the design, a trial number can be used and then by iteration of the calculation, the optimum core chosen. A point on the center of the scale can be a starting point. From the chosen core, the core volume is used to recalculate LI^2/V and the H and with the l_w of the core and the given I, the number of turns is found. The size of wire will be dictated by the current carrying capacity. This figure can be checked by comparing the NA_wF_w of the core checked against the winding volume of the core. From the ratio l_g/l_m and the l_m of the core, the gap can be calculated. An example using this approach is described in the Magnetics Catalog and is given in Figure 18.68.

Si-Al Powder Cores
(Sendust, Kool Mµ, MSS)

ADVANTAGES:
- Moderate core losses at high frequency.
- Moderate price.
- Distributed gap, no fringing flux.
- Good temperature and frequency stability.
- Very low magnetostriction.

DISADVANTAGES:
- Only available as toroids.
- Winding costs higher than ferrites.

Figure 18.61-Advantages and disadvantages of FeSiAl (Sendust) cores as power inductor materials for SMPS applications

Amorphous Choke Cores

ADVANTAGES:
- Very high DC bias capability.
- Flat permeability vs. DC bias curve.
- Moderate core loss at high frequency.
- Low magnetostriction.

DISADVANTAGES:
- Only available as toroids.
- Higher price than ferrites.
- Core must be boxed, can limit LI^2.

Figure 18.62-Advantages and disadvantages of amorphous metal cores as power inductor materials for SMPS applications

Another example from Snelling (1988)(Figure 18.69) shows a family of curves for different sizes of a particular type of power ferrite core. With the LI^2 known, a horizontal line is drawn until it reaches the first curve and the vertical line from that point gives the NI of the core. The smallest core can be chosen unless other consideration require the use of a larger one. The NI_o is matched against Snellings table of the NI for the various power cores. Also shown on the Hanna curve is the spacer thickness where the total gap is about $2l_s$. The value of l_g is the gap in the center leg giving the same inductance as $2l_s$.

FERRITE TRANSFORMERS & INDUCTORS AT HIGH POWER

Powdered Iron Cores

ADVANTAGES:
- Most are low cost.
- Distributed gap, no fringing flux.
- Available in various shapes.
- Available in large sizes.
- Most economical for DC and low frequency applications.

DISADVANTAGES:
- High magnetostriction.
- May need larger size when high frequency ripple is present.

Figure 18.63-Advantages and disadvantages of iron powder cores as power inductor materials for SMPS applications

Cut Cores

ADVANTAGES:
- Highest DC bias capability in CoFe.
- High temperature capability.
- Low core losses with NiFe and Amorphous alloys.
- Economical winding.
- Unlimited size range.

DISADVANTAGES:
- Expensive.
- Difficult assembly process.
- Large gaps may cause high copper loss due to AC fringing flux.

Figure 18.64-Advantages and Disadvantages of metallic strip cut-cores as power inductor materials for SMPS applications

Smith (1983) shows how to use optimization theory to design inductors for minimum resistive loss per unit volume. He has published tables of core geometries which maximize the efficiency of inductors as well as saturable reactors.

Bloom (1989) has described a design method called *Integrated Magnetics* in which the transformer and inductor are combined in a single converter core.

Design of an Inductor for a Switching Regulator

A switching regulator takes an unregulated D.C. output and produces a regulated DC output. In the design of the switching regulator shown in Figure 18.5, an LC filter is used to smooth the ripple in the D.C. output. A typical regulator circuit consists of three parts: transistor switch, diode clamp, and an LC filter. An un-regulated

Gapped Ferrite Cores

ADVANTAGES:
- Wide variety of shapes.
- Economical winding.
- Flat permeability vs. DC bias curve.
- Lowest core losses at high frequency.
- Widest choice of gapping.
- Low price.

DISADVANTAGES:
- Requires assembly.
- Lowest flux density.
- Large gaps may cause high copper loss due to AC fringing flux.

Figure 18.65-Advantages and disadvantages of gapped ferrite cores as power inductor materials for SMPS applications

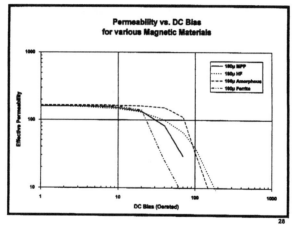

Figure 18.66-Effect of DC bias conditions on the effective permeability of several materials as power inductors for SMPS applications

DC voltage is applied to the transistor switch which usually operates at a frequency of 1 to 50 Kilohertz. When the switch is on, the input voltage, E_{in} is applied to the LC filter, thus causing current through the inductor to increase. Excess energy is stored in the inductor and capacitor to maintain output power during the off time of the switch. Regulation is obtained by adjusting the on time, t_{on}, of the transistor switch, using a feedback system from the output. The result is a regulated DC output, expressed as:

$$E_{out} = E_{in} t_{on} f \qquad [18-63]$$

The off time of the transistor switch is related to the voltages by

$$t_{off} = (1-E_{out}/E_{inmax})/f \qquad [18-64]$$

For E_{inmin}: $f_{min} = (1-E_{out}/E_{inmin})/t_{off} \qquad [18-65]$

If we assume the ripple current, i, through the indictor to be equal to $2I_{0min}$, the inductance is;

$$L = E_{out}t_{off}/\Delta i \quad [18\text{-}66]$$

The ferrite core to supply this inductance can be obtained by again calculating the LI^2 product and using the charts such as the one shown in Figure 18.70. From the intersection of the LI^2 with one of the core lines, the appropriate A_L can be read. In this case it is convenient to refer to the standard gapped cores available under each core's description. An example of this listing is shown in Table 18-3. The number of turns can be calculated from;

$$N = \sqrt{L/A_L} \quad [11\text{-}67.]$$

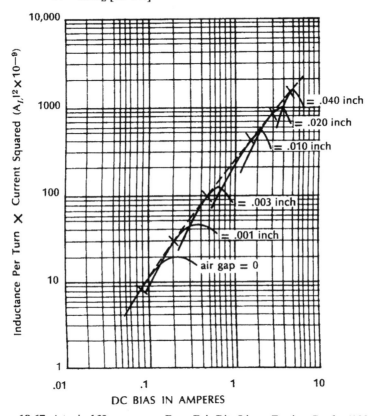

Figure 18.67- A typical Hanna curve . From Fair-Rite Linear Ferrites Catalog(1996)

The wire size is chosen from the wire tables using a current density of 500 circular mils/amp. An example of this method from the Magnetics Catalog is shown in Appendix D.

An approach given by Jongsma and contained in the Philips Catalog is shown in Figure 11-71. The LI^2 is plotted against the spacer thickness or center leg gapwidth for a series of different core shapes and sizes. When a core supplying the required LI^2 is chosen, reference is made to the data for the individual core chosen. The specific graph of LI^2 versus spacer thickness for that core is given for various choke designs (depending on the I_{ac}/I_0 ratio). A graph of this type is shown in Figure 18.72. On the same graph, the curve of LI^2 versus A_L for that particular core is given. For the particular LI^2 chosen, the intersection with the line for the converter is found. The working point must be below this line. A vertical can be dropped to the spacer thickness axis and from the tolerance on the spacer thickness, s_{min} and

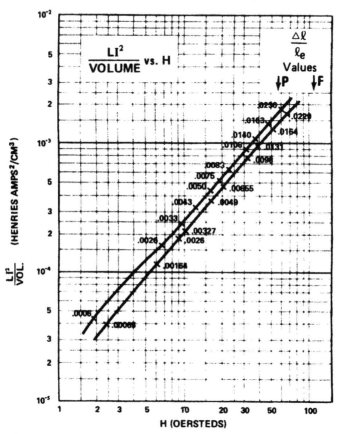

Figure 18.68- A Hanna curve of LI^2/V vs H giving the ratios of gaps to magnetic path lengths to achieve them. From Magnetics Catalog(1989)

s_{max} can be chosen on the axis. These lines can be extended to the A_L curve for the converter type. The two intersections when read across to A_L (to right hand scale) will give the limits of A_L. To avoid saturation N_{max} is given by;

$$N_{max} = (I^2L)_{max1}/I_m 2A_{L1} \qquad [18\text{-}68]$$

To achieve L_{min}, the N_{min} is given by;

$$N_{min} = L_{min}/A_{L2} \qquad [18\text{-}69]$$

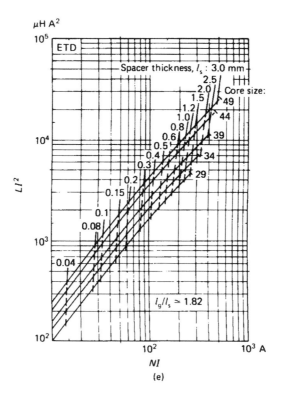

Figure 18.69- A Hanna curve for several ETD core sizes with the spacer thickness required to produce the LI^2 at a particular NI. From Snelling(1988)

An integral number of turns is chosen. The winding procedure can be completed as outlined under transformers or if special considerations are needed, the design by Jongsma is recommended.

MCLYMAN TREATMENT OF INDUCTOR DESIGN
Following a treatment similar to the one used for transformers, McLyman(1982) employs the K_g constant. The applicable expression is:

$$\alpha = (\text{Energy})^2/K_g K_e \qquad [18\text{-}70]$$

Ferrite DC Bias Core Selector Charts

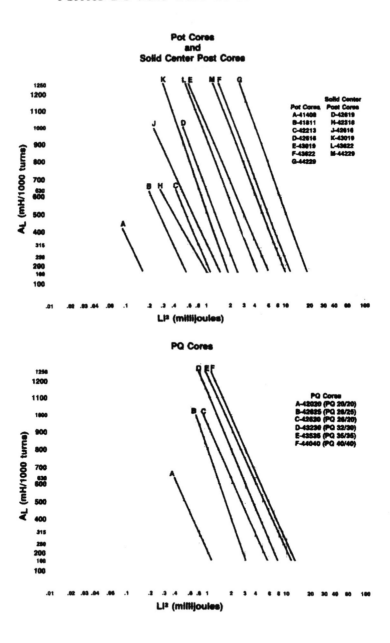

Figure 18.70- A Graph showing the A_L needed in a particular core to furnish a specific LI^2. From Magnetics Catalog (1989)

FERRITE TRANSFORMERS & INDUCTORS AT HIGH POWER

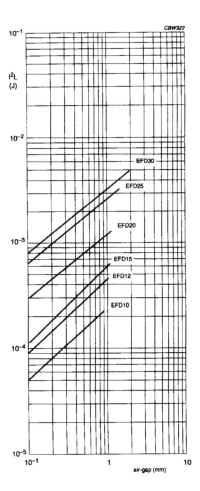

Figure 18.71 - The LI^2 for several EFD cores as a function of the center leg gap lengths. From Philip Catalog(1998).

where α and K_g have been defined under the transformer calculation. The energy in an inductor is given by

$$\text{Energy} = 1/2 \, LI^2 \qquad [18\text{-}71]$$

The K_e constant is varied somewhat from the transformer equation. It is represented by:

$$K_e = W_a A_c 2 K_u / l_t \qquad [18\text{-}72]$$

The area product approach can also be used for inductors:

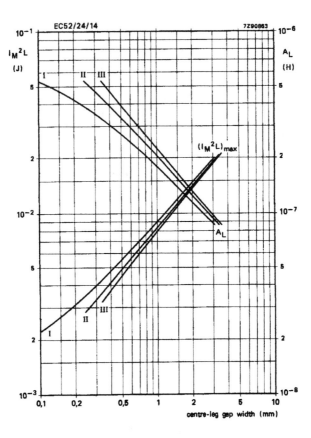

Figure 18.72- For a specific core, the LI^2 and A_L are given for different gap lengths for different converter types. From Philips Catalog(1998)

$$A_p = [2 \, (\text{Energy}) \times 10^4 / B_m K_u K_j]^n \quad cm^4 \quad [18\text{-}73]$$

The fraction, K_u, of the available winding space that will be occupied by the copper is given by

$$K_u = S_1 \times S_2 \times S_3 \times S_4 \quad [18\text{-}74]$$

where S_1 = conductor area/ wire area
S_2 = wound area/usable window area
S_3 = usable window area/window area
S_4 = usable window area/ usable window area + insulation

The design of an inductor using McLyman's approach is given in Appendix 5.

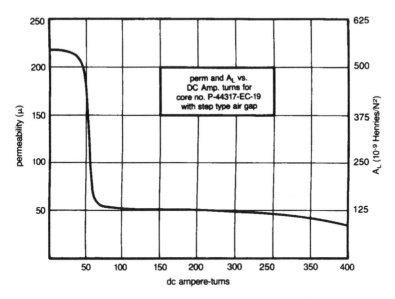

Figure 18.73-The Construction of a swinging-gap choke. From Martin, W.A. Powertechnics Magazine, Feb. (1986) p.19

FLYBACK CONVERTER DESIGN

Previously, we stated that the design of a flyback converter is similar to that of a power choke or inductor because in both cases, the energy is stored in the inductor during the current rise period and released when the current is turned off. If the converter is a simple non-isolating type (no transformer coupling as shown in Figure 18-6), the design (Jongsma 1982) is treated as a power inductor where;

$$L_{min} = 9\, \delta_{min}\, V_{imax})^2 \text{ and:} \quad [18\text{-}75]$$
$$I_m = I_{dcmax} + 2I_{ac} \quad [18\text{-}76]$$
$$= (P_o/\delta_{max}V_{imin}) + (\delta_{max}V_{imin}/2fL) \quad [18\text{-}77]$$

With the calculated values of I_{max} and L_{min}, the design can then be completed using the power inductor methods. If however, there is transformer coupling as shown in Figure 18-7, the turns ratio must be controlled to avoid damage to the semiconductor switches. Jongsma gives the limiting equations in this case as;

For $V_{imax}/V_{imin} < 2$
$$r = 3/7\{V_{imax}/V_o + V_F + V_R)\} \quad [18\text{-}78]$$

Where, V_F = Voltage drop across the output choke
and V_R = Voltage drop across the rectifier

We see then that not only the characteristics of the magnetic devices must be considered but also the voltage drop and current distribution in many of the auxiliary circuit elements. Because of this, when completing the design of these and other magnetic components, the reader is advised to consult the many books, vendors' literature and various periodicals that deal with this subject.

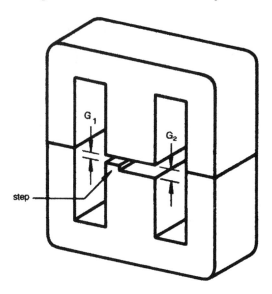

Figure 18.74-The permeability versus D.C. bias characteristics of a swinging-gap choke. From Martin(1986)

SWINGING CHOKE

We have spoken of the use of the air gap and the prepolarized cores as techniques in the design of power inductors. Another such design variation that is used to improve the regulation and efficiencies of choke is called the swinging choke or divided gap choke. This is described by Keroes(1969), Martin(1982) and by Snelling(1988). The action is non-linear as shown in Figure 18-73. The use of the stepped gap (Figure 11-43) allows for a wide swing of D.C. currents or magnetic fields. At low D.C. levels, the ripple current is a large part of the total current so a high inductance is needed and is provided by the small gap. However as the D.C. level increases, the ferrite at the small gap will saturate and the large gap will take over, protecting the circuit and main core from saturation and overheating. Thus, a dual action is accomplished.

APPENDIX 1
RECENT ARTICLES ON DESIGN OF FERRITES FOR POWER APPLICATIONS

Baasch, T.L., Electronic Products, Oct. 1971, 25

Bledsoe, C., Electronic Business, June 1,1984, 128
Bloom,E., IEEE Transactions on Magnetics,(1986), 141
Bosley, L.M. (1994) Magnetics Brochure.
Brown, B.(1992) PCIM, July 1992) 46
Brown, J.F., Powetechnics. Dec.1986, 17
Chen, D.Y., Solid State Power Conversion, Nov/Dec.,1978,50
Ciarcia, S.A.,Byte,Nov.1981, 36
Cuk,S., Power Conversion International, 1981, 22
Dull,W., Kusko, A.& Knutrud,T., EDN, Mar.5,1975, 47
Engelman, R.(1989)PCIM, 15, #7, 14
Estrov, A.(1989) PCIM, May, 1989 16
Estrov, A.(1986) PCIM, August 1986, 14
Fluke,J.C.,Proc. Power Electronics Show, 1986, 128
Gatres B.(1992) PCIM, 18, #7 July 1992, 28
Harada, H. and Sakamoto, K., IEEE Translation Journal of Magnetics in Japan, #7, Oct.1985
Hew, E., Power Conversion International, Jul./Aug. 1982, 14
Hill, P.C., Proc. Powercon 2, 1975, 243
Kitagawa, T. and Mitsui, T.,IEEE Translation Journal of Magnetics in Japan, Sept, 1985
Konopinski , T. and Szuba, S. Electronic Design, 12,June 7, 1979, 86
Margolin, B., Electronic Products, Mar.28,1983, 53
Martin, W.A. Proc. Powercon 9, 1982,
Middlebrook, R.D., Power Conversion International, Sept. 1983,20
Mohandes B.E.(1994) PCIM, July 1994, 8
Mullett, C.E.,Proc. Power Electronics Show,1986,36
Smith, S., Power Conversion International, May 1983, 22
Stratford, J.M., EDN, Oct. 13,1983, 140
Sum, K.K.,Power Electronics,1986, 153
Triner, J.E.,Power Conversion International, Jan. 1981, 69
Turnbull, J., Electronic Products, May,15,1972, 53
Ying X. and Zhi,Z.,IEEE Transactions on   Magnetics,Feb.1985,148

Appendix 2
Design Example of McLyman K_g Approach

The following section is abstracted from Magnetic Core Selection for Transformers and Inductors by Colonel Wm. T. McLyman, Marcel Dekker, New York,1982

Single-Ended Forward Converter Design
The following parameters are given:
Input voltage (V_{in}) = 140 V min
Output voltage(V_{out})= 10 V
Output current (I_o) = 5.0 A
Frequency(f)   = 20 KHz
Switching efficiency= 90%

Regulation = 1.0%
Ferrite toroid matl.= 5000µ (Magnetics)
The design steps used can be summarized as follows

1. Calculate output power which is equal to $(V_{out} + V_{(diode\ drop)}) \times (I_o) = (10 +1)5 = 55$ Watts
2. Calculate the apparent power using Equation 11-28
   $$P_t = P_0\{\sqrt{2/\eta} + \sqrt{2}\}$$
   $= 55\{1.41/.9 +1.41\} = 164$
   10 % is added to the apparent power for the demagnetizing winding;
   $P_t(1.1) = 180$ Watts
3. Calculate the electrical conditions assuming $B_m = .2T$ and square wave $(K_f=4.0)$ using Equation 11-30.
   $K_e = 3712$, $K_g = 180/2(3712)$
4. Calculate $K_g = P_t/2K_e$, using Equation 11-29
   $K_g = .0242$
   K_g is then recalculated for additional insulation because of the high voltage between primary & secondary windings.
   $K_g = .03025$
5. Select a toroid from McLyman's table (Table 11.5) with the comparable K_g and record the data regarding the toroid.
   Magnetics 52507, $K_g = .0352$
6. Calculate primary turns, N_p, using Equation 11-2 using coefficient 4 for square wave in place of 4.44 for sine and using Teslas for the units for B_m ($1T = 10^4$ Gausses).
   $N_p = V_p \times 10^4/ K_f B_m A_c$
   $= 140 \times 10^4/ 4 \times .2 \times 2 \times 10^4 \times .393 = 222$
   $N_p = N_m$ (Demagnetizing winding)
7. Calculate primary current, I_p using a duty cycle, D, of .5 and a switching efficiency, η, of .9.

   $I_p = P_0/DV_p\eta = 55/(.5 \times 140 \times .9) = .873$ A.
   $I_m = I_p \times .1 = .0873$ A.

8. Calculate current density, J, from Equation 11-31 Use K_u (Window Utilization Factor) = .4
   $J = 380$
9. Calculate bare wire size $A_w(B)$. For forward converter, I_p and I_m must be multiplied by .707

   $A_{w(B)} = I_p(.707)/J = (.873 \times .707)/380 = .00162$
   $A_{w(B)} = I_m(.707)/J = (.0873 \times .707)/380 = .000162$
10. Select wire size from table

    AWG #25 has bare area of .00162

FERRITE TRANSFORMERS & INDUCTORS AT HIGH POWER

11. Calculate primary winding resistance, R_p

    $R_p = (MLT)(N) \times u\ /cm. = 3.3 \times 1062 \times 223 \times 10^{-6}$
    $= .781\ \Omega$

12. Calculate primary copper loss. P_p

    $P_p = (I_p \times .707)^2 R_p$
    $= (.873 \times .707)^2 \times .781 = .297$ Watts

13. Calculate secondary turns, N_s

    $V_s = (V_0 + V_d)/D$
    $= (10 + 1)/.9 = 22$
    $N_s = N_p V_s/ V_p$
    $= (223)(22)/140 = 35$ Turns

14. Calculate bare wire size, $A_w(B)$ for secondary

    $A_{w(B)} = I_0(.707)/J$
    $= (5)(.707)/380 = .00930\ cm^2$

15. Select wire size from table

    AWG Wire with area of $.00823\ cm^2$
    $\mu\Omega/cm = 209.5$

16. Select secondary winding resistance

    $R_s = (MLT)(N)(\mu\Omega/cm)$
    $= (3.3)(35)(209) \times 10^{-6} = .0242\ \Omega$

17. Calculate secondary copper loss, P_s

    $P_s = (I_0 \times .707)^2 R_s$
    $= (5 \times .707)^2 .0242 = .302$ Watts

18. Calculate transformer regulation

    $\alpha = P_{cu} \times 100/(P_0 + P_{cu})$
    where P_{cu} = sum of primary and secondary copper losses
    $= (.297 + .302) = .599$ Watts
    $= (.599 \times 100)/(55 + .599) = 1.08\ \%$

19. Calculate core loss, P_e from core loss curves and core weight.

    $P_{fe} = (Milliwatts/gm)\ W_{fe} \times 10^{-3}$
    $= (20)(12.1) \times 10^{-3} = .242$ Watts

20 Calculate Total Losses

$P_\Sigma = P_{cu} + P_{fe}$
$= (.599) + (.242) = .841$ Watts

21. Calculate efficiency for transformer

$e = (P_o \times 100)/(P_o + P_\Sigma)$
$= [(55) \times 100]/(55 + .599) = 98.9\%$

22. Calculate Watts/Unit Area from surface area of core.

$\Psi = P_\Sigma / A_t$
$= (.841)/33.4 = .025$ Watts/cm^2

A value of .03 Watts/cm^2 normally gives a 25°C. rise.

APPENDIX 3

The following section is abstracted from the Magnetics Catalog FC405, published by Magnetic, Division of Spang and Co., Butler PA 16001, 1987

Magnetics Inductor Design Method using Hanna Curves

Example - The following example illustrates the use of a Hanna curve to find the core for a particular power inductor.

Let L = .1 mH and I_{DC} = 10 amperes. Find the core, the air gap, and number of turns required.

1. Calculate LI^2
   $LI^2 = (.1 \times 10^{-3}) \times (10)^2 = 10 \times 10^{-3}$

2. Refer to Hanna Curve in Figure 11-37. Assume $(LI)^2/V = 5 \times 10^{-4}$ (from center of vertical scale).

3. Core Selection- Choose a core geometry, for example an E core, and select a size from Table 11-4 with the volume nearest to 20 cm^3. Use P45021-EC.
   Volume = 21.6 cm^3, l_e=9.58cm, W_a=.351 $\times$ 10^6 circ.mils

4. Recalculate

   $LI^2/V = 10 \times 10^{-3}/21.4 = 4.6 \times 10^{-4}$

5. Determine H and l_g/l_e from the Hanna curve (P material), using recalculated value of LI^2/V.

   H=18 and l_g/l_e = .006

FERRITE TRANSFORMERS & INDUCTORS AT HIGH POWER

6. Calculate N

   $H = .4\, NI/l$, $N = Hl/.4$ $I = 13.7$ Turns-Use N=14

7. Calculate Wa needed. For $I_{DC} = 10$ amperes, use AWG #11 wire.

   Aw = 9×10^3 cir. mils per turn
   Wa needed = $A_w \times N/K$

   where Wa = core or bobbin window area
   Aw = cross sectional area of the wire
   N = number of turns
   K = winding (or space utilization) factor (K varies with the designer and operating conditions of the inductor. Typically, this factor is 0.4).

   Wa needed = $(9 \times 10^3) \times (14/0.4) = 315 \times 10^3$ circ. mils

8. Compare Wa values

   Wa needed = 315×10^3 circ. mils
   Wa available in 45021-EC = 351×10^3 circ. mils
   At this point, the designer can use the core selected or repeat this process to select a smaller (or larger) core.

9. Gap calculation. If the P-45021-EC core is chosen, the air gap is calculated as follows.

   $l_g/l_e = .006$, $l_e = 9.58$ cm.
   $l_g = .006 \times 9.58 = .057$ cm.(.023 in.)

APPENDIX 4

Magnetics Inductor Design for Switching Regulators

The following is abstracted from Magnetics Catalog FC405, published by Magnetics, Division of Spang & Co., Butler, PA 16001, 1987

Only two parameters of the design application must be known:
(a) Inductance required with DC bias
(b) DC current

1. Compute the product of LI^2 where:
   L = inductance required with DC bias (millihenries)
   I = maximum DC output current = $I_{omax} + i$

2. Locate the LI^2 value on the Ferrite Core Selector charts such as the one shown in Figures 11-39. Follow the LI^2 coordinate to the intersection with the first core size curve. Read the maximum nominal inductance, A_L, on the Y axis. This represents the smallest core size and maximum A_L at which saturation will be avoided.

3. Any core size line that intersects the LI^2 coordinate represents a workable core for the inductor if the core's A_L value is less than the maximum value obtained on the chart. If possible, it is advisable to use the standard gapped cores because of their availability. These are indicated by dotted lines on the charts and can be found in the catalog.

4. Required inductance L, core size, and core nominal inductance (A_L) are known. Calculate the number of turns using

$$N = 10^3 \sqrt{L/A_L}$$

where L is in millihenries.

5. Choose the wire size from the wire tables using 500 circular mils per amp.

Example - Choose a core for a switching regulator with the following requirements:

E_0 = 5 Volts
e_o = .5 Volts
I_{omax} = 6 amp
I_{omin} = 1 amp
E_{inmin} = 25 Volts
E_{inmax} = 35 Volts
f = 20 KHz.

1. Calculate the off-time and minimum switching, f_{min}, of the transistor switch using equations 11-66 and 11-67.

$t_{off} = (1-E_{out}/E_{inmax})/f$
$t_{off} = (1-5/35)/(20,000) = 4.3 \times 10^{-5}$ sec.
$f_{min} = (1-E_{out}/E_{inmin})t_{off}$
$f_{min} = (1-5/25)/(4.3 \times 10^{-5}) = 18,700$ Hz.

2. Let the maximum ripple current, i, through the inductor be

$\Delta i = 2I_{omin}$
$\Delta i = 2(1) = 2$ Amps

3. Calculate L using Equation 11-68.

$L = (E_{out} \times t_{off})/\Delta i$
$L = 5(4.3 \times 10^{-5})/2 = .107$ millihenries

4. Calculate the value of the capacitance, C and and maximum equivalent series resistance, ESR max

$C = i/8f_{min} \Delta e_o$
$C = 2/8(18700)(.5) = 26.7$ u farads

FERRITE TRANSFORMERS & INDUCTORS AT HIGH POWER

$ESR_{max} = \Delta e_o / \Delta i$
$ESR_{max} = .5/2 = .25$ ohms

5. The product of $LI^2 = (.107)(8)^2 = 6.9$ millijoules.

6. Due to the many shapes available in ferrites, there can be several choices for the selection. Any core size that the LI^2 coordinate intersects can be used if the maximum A_L is not exceeded. Following the LI^2 coordinate, the choices are:

(a) 45224 EC 52 core, A_L315
(b) 45015 E core, A_L250
(c) 44229 solid center post core, A_L315
(d) 43622 pot core, A_L400
(e) 43230 PQ core, A_L250

7. Given the A_L, the number of turns the required inductance can be found for each core using Equation 11-69.

A_L	Turns
250	21
315	19
400	17

7. Use #14 wire.

APPENDIX 5

MCLYMAN DESIGN OF SWITCHING INDUCTOR USING K_G APPROACH

This section is abstracted from Magnetic Core Selection for Transformers and Inductors by Colonel Wm. T. McLyman, Marcel Dekker, New York, 1982

Design of a Buck Switching Inductor
Given;
Input Voltage, V_i = 28 +/- 6 V.
Output Voltage, V_o = 20 V.
Output Current Range, I_0 = 5 - 0.5 A
Frequency, f, = 20 KHz.
Switching Efficiency, = 98%
Regulation = 1.0%
Ferrite Pot Core

Step 1. Calculate time period, t, of operation

$T = 1/f = 1/20 \times 10^3 = 50 \times 10^{-6}$ s.

Step 2. Calculate Minimum duty cycle
$D_{min} = V_0/V_{in(max)} = 20/34(0.98) = .60$

Step 3 Calculate maximum duty cycle
$D_{max} = V_0/V_{in(min)} = 20/22(0.98) = .927$

Step 4. Calculate Load Resistance at Minimum Load Current
$R_0 = V_0/I_{0(min)} = 20/0.5 = 40$

Step 5. Calculate Minimum Required Inductance
For a Buck Converter
$L_{min} = R_{0(min)}t(1-D)_{min})/2 = (40)(50 \times 10^{-6})(1-0.6)/2$
$= 400 \times 10^{-6}$ H.

Step 6. Calculate I in the Inductor
$\Delta I = tV_{in(max)}D_{(min}(1-D_{(min)})/L$
$= (50 \times 10^{-6})(34)(0.6)(1-0.6)/400 \times 10^{-6}$
$= 1.0 = 2I_{0(min)}$ A.

Step 7. Calculate $LI^2/2$ using Equation 11-73
$I = I_{0(max)} + I/2 = (5.0) + 1.0/2 = 5.5$ A.
$LI^2/2 = (400 \times 10^{-6})(5.5.)^2/2 = .00605$ W-s.

Step 8. Calculate K_e using Equation 11-74.
$P_0 = V_0I_0 = (20)(5) = 100$ W.
Assume $B_m = .35$ T (3500 Gausses)
$K_e = 0.145 P_0B^2 \times 10^{-4} = 0.000178$

Step 9. Calculate K_G
$K_G = (Energy)^2/K_e$ using Equation 11-72
$= (0.00605)^2/(0.0001781) = 0.213$

Step 10. Select a comparable core geometry K_g from listing of pot cores (McLyman, 1982). Record all pertinent dimensional data.
Pot core = B65611, 36x22(Siemens)
For this pot core, $K_g = 0.221$
G(window height) =1.46cm.
W_{tfe}(weight ferrite) = 26 gm.
MLT = 7.3 cm., $A_c = 2.01$ cm^2, $W_a = 1.00$ cm^2
A_t(Surface area) = 45.24 cm^2

Step 11. Calculate current density (Correlation with K_g is derived in book(McLyman 1982)
$J = 2(Energy) \times 10^4/B_mK_uA_p$
Use K = 0.4 (window utilization factor)

FERRITE TRANSFORMERS & INDUCTORS AT HIGH POWER

$J = 2(0.00605) \times 10R4F/(0.35)(0.4)(2.01) = 430 \text{A/cm}^2$

Step 12. Calculate bare wire size $A_{w(B)}$
$A_{w(B)} = I_0/J = 5/430 = .0116$

Step 13. Select a wire size from the wire table. If area is not within 10%, take smaller size. Record wire data.
AWG #17 with 0.01039 cm^2
  $\mu\Omega/\text{cm} = 166$
  $A_w = 0.0117 \text{ cm}^2$ (with insulation)

Step 14. Calculate the effective window area, $W_{a(eff)}$
  $W_{a(eff)} = W_a S_3$
For single section bobbin on pot core, $S_3 = 0.75$
  $W_{a(eff)} = (1.00)(0.75) = 0.75$

Step 15. Calculate number of turns
  $N = W_{a(eff)} S_2/A_w$
Typical value of $S_2 = 0.6$
  $N = (0.75)(0.6)/(0.0117) = 38$ turns

Step 16. Calculate gap required for inductance
  $l_g = 0.4 \, N^2 A_c \times 10^{-8}/L$
  $= (1.26)(38)^2(2.01) \times 10^{-8}/412 \times 10^{-6} = 0.089$ cm.
For fishpaper spacer, thickness is given in mils.
  $l_g = 0.089$ cm $\times 393.7$ mils/cm $= 35$ mils
Paper comes in 10 and 7 mils. One of each across entire pot core mating surface doubles gap. (a gap each for skirt and centerpost) Therefore the total gap is 17 mils or $0.034 \times 2.34 = 0.0864$ cm.

Step 17-18. Recalculate new turns correcting for fringing flux (not shown here)

  $N = 34$

Step 19. Calculate Winding Resistance
  $R = (MLT)(N) \, /\text{cm} \times 10^{-6}$
  $= (7.3)(34)(166) \times 10^{-6}$
  $= 0.041 \Omega$

Step 20. Calculate Copper Loss, P_{cu}
  $P_{cu} = I^2 R = (5.5)^2 (0.041) = 1.24$ W.

Step 21. Calculate Regulation, .
  $\alpha = P_{cu} \times 100/(P_0 + P_{cu}) = (1.24)(100)/(100+1.24)$
  $= 1.22\%$

Step 22. Calculate total a.c. + d.c. flux density

$B_m = 0.4\pi N(I_{dc} + \Delta I/2) \times 10^{-4}/l_g$
$= (1.26)(34)(5.5) \times 10^{-4}/0.0864 = 0.273 T (2730 G.)$

Step 23. Calculate a.c. flux density
$B_{mac} = 0.4\pi N(\Delta I/2) \times 10^{-4}/l_g$
$= (1.26)(34)(0.5) \times 10^{-4}/0.0864 = 0.0248 T$

Step 24. Calculate Core Loss P_{fe}. Use core loss curves for 0.0248 T. or 248 Gausses. Use the ferrite weight given before.
$P_{fe} = (mW/gm)(W_{tfe}) \times 10^{-3}$
$= (0.06)(57) \times 10^{-3} = 0.0034 W.$

Step 25. Calculate total loss
$P_t = P_{cu} + P_{fe} = (1.24) + (0.0034) = 1.2434 W.$

Step 26. Calculate the efficiency
$e = (P_0)(100)/(P_0 + P_t) = (100)(100)/(100+1.2434)$
$= 98.8\%$

Step 27 Calculate the Watts/unit area
$\psi = P/A_t = (1.2434)/45.24 = 0.0275 W/cm^2$
The value of .03 W/cm^2 corresponds to a temperature rise of 25°C.

References

Analytic Artistry (1988), Inductor Spread Sheet, Analytic Artistry, Torrance CA,
Bloom, G.(1989) Powertechnics April 1989, 19
Bracke, L.P.M.,(1983) Electronic Components and Applications, Vol.5, #3 June 1983, p171
Bracke L.P.M.(1982) and Geerlings, F.C., High Frequency Power Transformer and Choke Design, Part 1, NV Philips Gloeilampenfabrieken, Eindhoven, Netherlands
Buthker,C.(1986) and Harper, D.J., Transactions HFPC, 1986,186
Carlisle, B.H.(1953),Machine Design,Sept.12,1985,53
Carsten, B.(1986), PCIM,Nov.1986,34
Cattermole, P.(1988) and Cohn, Z., Proc. HFPC, May 1988, 111
De Maw,M.F.(1981),Ferromagnetic core Design and Applications Handbook, Prentice Hall, Englewood Cliffs, NJ, 1981
Dixon L.(1974), and Pale, R., EDN, Oct.20,1974,53 and Nov. 5,1974, 37
Estrov, A.(1989) and Scott,I., PCIM, May 1989
Finger, C.W.(1986), Power Conversion International,1986
Forrester, S (1994) PCIM Dec. 1994 , 6
Grossner, N.R.(1983), Transformers for Electronic Circuits, McGraw-Hill Book Co., New York

FERRITE TRANSFORMERS & INDUCTORS AT HIGH POWER

Hanna,C.R.,J.Am.(1927) I.E.E., 46,128,
Haver, R.J.(1976), EDN,Nov.5,1976, 65
Hess, J.(1985) and Zenger, M., Advances in Ceramics, Vol.16 501
Hiramatsu, R.(1983) and Mullett, C.E.,Proc. Powercon 10,F2, 1
Hnatek,E.R.(1981), Design of Solid State Power Supplies, Van Nostrand Reinhold,New York
IEC ( ) Document 435,International Electrotechnical Commission
Jongsma, J.(1982),High Frequency Ferrite Power Transformer and Choke Design, Part 3, Pilips Gloeilampenfabrieken, Eindhoven Netherlands
Jongsma, J.(1982a) and Bracke, L.P.M. ibid Part 4
Kamada, A. (1985) and Suzuki, K. Advances in Ceramics Vol.16, 507
Kepco (1986), Kepco Currents,Vol1#2
Magnetics (1987) Ferrite Core Catalog, Magnetic Div.,Spang and Co, Butler, PA 16001
Magnetics (1984) Bulletin on Materials for SMPS
Martin,H.,(1984), Proc. Powercon 11, B1, 1
Martin, W.A.(1978), Electronic Design, April 12,1978, 94
Martin, W.A.(1986), Powertechnics Magazine,Feb.1986,p.19
Martin, W.A.(1982), Proc. Powercon 9
Martin, W.A.(1987), Proceedings,Power Electronics Conference (1987)
Martinelli, R.(1988), Powertechnics,Jan.1988
McLyman, Col. W.T.(1969) JPL, Cal Inst. Tech. Report 2688-2
McLyman, Col.W.T.,(1982), Transformer and Inductor Design Handbook, Marcel Dekker, New York
McLyman, Col.W.T.(1982), Magnetic Core Selection for Transformers and Inductors, Marcel Dekker, New York
McLyman, Col. W.T. (1990) KG Magnetics Magnetic Component Design Software Program
Mochizuki, T.(1985),Sasaki,I. and Torii,M.,Advances in Ceramics Vol. 16, 487
Nakamura, A.(1982) and Ohta, J.,Proc.Powercon 9,C5, 1
Ochiai, T and Okutani, K, ibid ,447
Pressman, A.(1977) Switching and Linear Power Supply Converter Design, Hayden Book Co., Rochelle Park, N.J.
Philip Catalog,(1986) Book C5, Philips Components and Materials Div., 5600Md, Eindhoven, Netherlands
Roddam, T.(1963), Transistor Inverters and Converters, Iliffe, London and Van Nostrand Reinhold, New York
Roess, E.(1982), Transactions Magnetics MAG18,#6,Nov.1982
Roess, E.(1986), Proc. 3rd Conf. on Phys Mag Mat. Sept.9-14,1986,Szczyrk-Bita, Poland, World Scientific
Roess, E.(1987) ERA Report-0285
Sano, T.(1988),Morita, A. and Matsukawa, A., Proc. PCIM,July 1988, 19
Sano,T.(1988a) Morita, A. and Matsukawa, A. Proc. HFPC
Schlotterbeck, M.(1981), and Zenger, M. Proc. PCI 1981,37

Severn, R.P. (1985) and Bloom, G.E.,Modern D.C. to A,C. Switchmode Power Converter Circuits, Van Nostrand Reinhold, N.Y.

Siemens (1986-7) Ferrites Data Book, Siemens AG, Bereich Bauelemente, Balanstrasse 73, 8000 Munich 80 Germany

Shiraki, S.F.(1978), Electronic Design, 15, 86

Shiraki, S.F.(198) Proc. Powercon 7, J4, 1

Sibille,R. (1981), IEEE Trans. Magnetics, Mag.22 #5,Nov. 1981, 3274

Sibille, R.(1982), and Beuzelin, P., Power Conversion International,1982, 46

Smith, S.(1983), Magnetic Components, Van Nostrand Reinhold, New York

Smith, S. (1983a) Power Conversion International, May 1983, 22

Snelling E.(1988) Soft Ferrites, Properties and Applications Butterworths, London

Snelling, E(1989) presented at ICF5

Stijntjes,T.G.W.(1985), and Roelofsma, J.J., Advances in Ceramics, Vol 16, 493

Stijntjes, T.G.W.(1989) Presented at ICF5, Paper C1-01

TDK (1988) Catalog BLE-001F, June 1988, TDK, 13-1 Nihonbashi, Chuo-ku, Tokyo, 103, Japan

Thomson (1988) Soft Ferrites Catalog, Thomson LCC, Courbeville, Cedex, France

VDE ( ) Document 0806

Watson, J.K.(1980) Applications of Magnetism, John Wiley and Sons, New York

Watson, J.K.(1986) IEEE Trans Magnetics

Wood, P.(1981) Switching Power Converters, Van Nostrand Reinhold, New York

Zenger,M. (1984), Proceedings, Powercon 11(1984)

19 MATERIALS FOR MAGNETIC RECORDING

INTRODUCTION

Magnetic Recording has become the most important method of storing large amounts of information of all types for computers as well as audio and video entertainment systems. At the present time, there are three main techniques used in coding and decoding the magnetic system:

1. Impressing a magnetic state on particles or films of magnetic media and reading these states inductively in playback. This method includes tapes, drums and disks.
2. Using the magnetic state in a material to interact with the optical properties of the material (usually in the form of a thin film) which is then read back in play/back. This is known as magneto-optical recording.
3. Using the magnetic state to vary the resistance of a magnetoresistive material. Here, the technology is mostly used in thin films or most recently in multilayer films. This technique is the newest and shows great promise.

Of these techniques, at the present time, the first is by far the most widely used and as such, ferrites play a most important part as the preferred media for magnetic tapes and disks and also as a material for heads. However, with the computer and information explosion, we can expect the usage of magnetoresistive heads to increase in dramatically in the near future primarily in hard disks.

History of Magnetic Recording

The earliest magnetic recorders were ones in which a moving wire was magnetized by a magnetic head and played back by detecting the magnetic state of the wire. Metallic tape was then used as the media but this was too costly and bulky. Since 1947, practically all magnetic material used in recording tape, disks or drums has been fine-particle γ- Fe_2O_3 and its successors. We have said that γ- Fe_2O_3 is a spinel with a defect structure but instead of having divalent ions, there are vacancies that cause the imbalance of Fe^{+++} ions on the two different types of sites leading to a net magnetic moment. Ferrites (including γ-Fe_2O_3) have been used in digital as well as audio and video recordings. Although most of the present day computer applications use digital recording, most of the entertainment (audio and video) applications are use analog recording. However, there is a trend developing to use digital recording for the latter purposes as well. More recently, there have been inroads by plated metal films and fine-particle metallic materials.

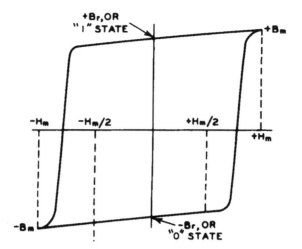

Figure 19.1- B-H Loop of a Square Loop Ferrite showing the two states of Remanent Induction, From RCA Guide to Memory Products, 1965

Digital Recording Applications

Square Loop Memory Cores

At the end of World War II, work started on the construction of digital computers. Albers-Schoenberg (1954) developed some of the first square loop Mg-Mn ferrite cores for digital applications. In 1953, the first core memory was used in the main memory of a computer. Later an upgraded model was used in the first SAGE prototype in January, 1955. The computers were driven and controlled by circuits with vacuum tubes. Later, when transistors replaced vacuum tubes, new circuits were designed using much less power. In addition, automated core winding was developed. With these innovations and others, the ferrite core memory was the basis of the IBM 360 computer that later became the standard of the industry for some time.

Since digital computation uses binary logic, there is a need for bistable elements,(that is, those which would have two stable states) and a method of switching rapidly from one to the other. Now, ferrites have 2 states of saturation but when the magnetizing current removed, the induction drops to B_r. In round loop materials, such as the those typical of the early ferrites, the two B_r's were not sufficiently stable to function as the bistable element. Also, the flux excursion in switching (and the voltage) was small. However, when square loop ferrites were developed, they became a natural for this type of system. If we look at Figure 19.1 showing a loop for a typical square loop ferrites the two states become obvious. In binary logic these two states are known as "0" and "1". Since it is necessary to keep the circuit at Br without demagnetization leading to shearing of the loop, a closed circuit toroid is the only choice of component. The toroids were usually made of square loop Mg-Mn ferrites. By applying a pulse of the right polarity the core can reset in an initial state "0". In the binary system if a core had to be set to

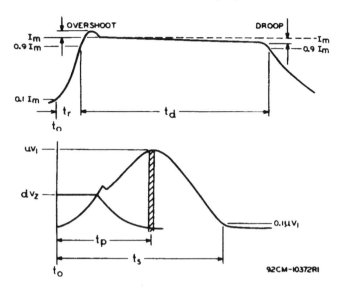

Figure 19.2-Upper curve- Current waveform for driving a ferrite memory core. Bottom curve- Voltage waveforms of Undisturbed "1" (uV_1) and Disturbed "0"(dV_z) states of a ferrite memory core. From RCA Applications Guide to Memory Products, 1965

"1" a pulse in the opposite direction set the core in the upper state of saturation and to Br after the pulse was removed. If a "0" was meant for that digit- no pulse was used. To read the contents of that core, a reset pulse in the original direction was used. If a "1" was set in the core, there will be a flux swing from the upper B_r to the Bs in the opposite direction. The change in flux or the ΔB (roughly 2 B_r) is detected in a "read" wire also threading the core. A voltage would be produced in the read wire if the core was set at "1" and almost no voltage would be produced if the core was in the "0" state. It would only have a small voltage since the induction would go from B_r to B_{sat} in the same direction which should be a small ΔB. The difference in the voltage waveforms for the "1" state or large ΔB and that for the "0" or small ΔB is shown in Figure 19.2. The upper curve shows the driving current waveform while the lower one shows the voltage waveforms of the "uV_1 " (undisturbed "one") which is the "full read " output compared to the "dV_z" (disturbed "zero"). There is normally enough difference between these outputs to discriminate between a "1" and a "0".

We have spoken about an isolated core but in a large array, there must be means of accessing that core. The method used is the "coincident current" method. Let us examine a two-dimensional array of cores as shown in Figure19.3. The magnetizing wires are threaded through the cores in horizontal row as and vertical columns. The position of a single core can be described by its row and column. Now, if half the current needed to set a core, ($+I_m/2$), is passed through one row and coincidentally the same current is passed through the appropriate column only one core in the array will have the necessary field to set the core in the "1" position. To

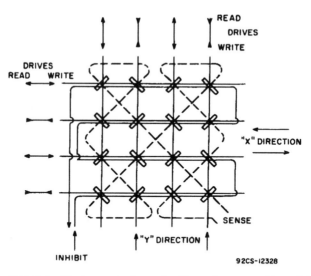

Figure 19.3. Coincident-current plane of a 16-word memory. From RCA Applications Guide to Memory Products, 1965

permit a zero to be written in the other cores, another wire threading all the cores in either the horizontal or vertical direction is subjected to a reverse pulse of $-I_m/2$ canceling the positive half pulse in that line. This winding is called the "inhibit" line. To read the core, the same technique is followed except the pulse will be one of $-I_m/2$ in the row and column of the desired core. A sense wire threading all the cores will detect the "uV_1" voltage if a "1" had been set in that core. Note that in reading the state of magnetization of a core, the memory is removed and no future use can be made of that core information. To overcome this, provision for reinserting the information must be provided. This read operation is called "destructive" read out. To store words of more than 1 bit requires an additional plane for each bit. The drive wires for all the planes in a stack are linked together and driven by a common source.

As previously stated, the ferrite core memory system is discussed mostly on a historical basis since they are not used commercially today. The advent of semiconductor memories and oxide coated disks and drums dealt a deathblow to ferrite core memories in mainframe memory systems.

Oxide-Coated Tape, Disks and Drums

In the early days of computers, although oxide-coated reel to reel tape was used in audio electronics in an analog fashion, the access time was very slow for digital uses and therefore it had limited use in mainframe memory systems. The problem of the slow access time in digital tape as a memory storage system was solved by the use of oxide coated disks and drums. In the oxide-coated material, the toroid of a core memory is replaced by a magnetic particle or rather the region containing an assembly of the particles of the magnetic media. Here, the individual particles comprising the region have properties reminiscent of the toroids. However, since a closed magnetic path is not available to keep the remanence high, another scheme must be used. This new method uses that fact that the particles are made of a semi-hard material with magnetic properties somewhere between the soft magnetic ferrite and hard ferrite. In addition to the crystal anisotropy of the material, advantage is also taken of the shape anisotropy of Fe_2O_3 particles. The particles themselves are on the order of 1 micron or single domain to allow them to act as permanent magnets. Otherwise the domain walls would demagnetize them. Since the particle are so small they obviously could not be detected by any type of reasonably sized physical arrangement. Instead, regions of tape containing many particles that are magnetized similarly are used as the individual memory elements. In most of the present day uses, the regions are oriented longitudinally along the tape and are magnetized or written to record and later read by detecting the voltage as the magnetized region passes over a gap. A new technique uses perpendicular orientation, which increases the bit density.

The magnetizing or "write" process involves passing the section of the tape over an air gap in a wound magnetic core called the recording head. The tape is magnetized by the fringing flux of the head gap. By changing the polarity of the current in the head winding, the oxide regions can take on one of two conditions of magnetization representing "0" and "1" states. As in the case of the ferrite core memory, the magnetic induction of the particle and therefore, the region will drop back to remanence. There will be some interaction between the poles of the particles and this must be accounted for in the design. The core is then read by passing the appropriately magnetized sequence of "1" and "0" magnetized sections over a "read" recording head, which now becomes the sensing device. As each section passes the head, the presence of a voltage output or the lack of one indicates a "1" or "0" output. The difference between the core case and the tape case lies in the fact that instead of using a reset or read pulse to read a core voltage, in the tape case, the linear passage of the sections of different polarity will cause Δ B excursions in the magnetic circuit of different polarities. The output voltage is given by:

$$E = -N \, d\phi/dt = -NA \, dB/dx \cdot dx/dt$$

Th dB/dx is the change in B with distance on the tape and the dx/dt is just the tape speed.

The γ-Fe2O3 material for practical tape recording usually has a specific magnetization, σ, of about 74-76 emu/gm. If we assume a density of 5.074, the theoretical saturation would be 4800 Gausses. As a tape material, the B_r is about 1300-1500 Gausses and the squareness ratio, B_r/B_s, is about 90%. One of the most important properties for a recording medium is the coercive force, H_c. In lower coercivity gamma iron oxides, as used in audio cassettes, the coercive force of the powder is usually about 350-500 Oe. For the Cobalt-treated gamma iron oxides, the coercive force of the powder is about 550-650 Oe. and that of the tape from 650-750

Oe. Chromium dioxide has a coercive force of about 450-600 Oe. One drawback of CrO_2 is the low Curie temperature of only 126°C. Some of the early problems of temperature instability seem to have been solved as CrO_2 audio-cassettes are marketed successfully.

The preparation of the elongated particles and the coating and alignment process are quite important in the manufacture of the tape. The particles should have a high aspect ratio (length to cross-section). A value of 6:1 is found in commercial materials. The distance between the tape and head (gap) and the density or thickness of the oxide coating(usually about 15 microns) is quite critical. Therefore, the slurry preparation and coating process must be controlled carefully. The Fe_2O_3 particles are packed on the tape at about 40-50% by volume. Orientation of the particles is accomplished after the coating but before the solvent dries. The particles must be shaped properly and also be in the right particle size range. If they are too small, they will be superparamagnetic rather than ferrimagnetic and thus be ineffective. The tape is usually of a Mylar type polymer. The tape thickness cannot be too thin (usually about 35 microns) or interactions between neighboring wraps of tape may occur. In ½ inch digital tapes, there are usually nine tracks.

Floppy and Hard Disks

For Floppy disks, the same coating procedure as in the tape is used except that there is no orienting step. The remanence of a γ- Fe2O3 disk material is about 1650 Gausses with a squareness ratio of about .8.

Coating hard disks is done somewhat differently. The oxide dispersion (usually at about 20-25% pigment volume concentration) is sprayed onto a rotating aluminum disk, then oriented, dried and polished. Hard disks have remanences of only about 800-900 Gausses.

Other Digital Magnetic Recording Systems

While all these developments were taking place, two other milestones were reached in memory and magnetic recording. First, bubble domains were used in memory storage devices and while no large-scale use of bubbles has been achieved, there were some commercial applications where bubbles are used. However, their usage is had been replaced by other methods. The other type of recording that was investigated for some time is known as magneto-optic recording. In this case, light is used to detect the states of magnetization. In an earlier book I remarked about magneto-optic recording that "no large scale commercial usage has developed and it remains a laboratory phenomena whose time has not come". As of this publication, the time has indeed come and many magneto-optical disks are being offered commercially. It appears that the future of magnetic recording for commercial digital applications is increasingly being based on thin film or multilayers. for cost considerations, entertainment video and audio recording should stay with oxide particles for some time to come.

Bubble Domain Devices

In magneto optical work with single crystal orthoferrites in the 1960's, cylindrical domain in an otherwise large single domain area were found to be generated with a magnetization opposite to the surrounding area. Later, it was discovered that with additional accessory deposited structures, the domains could be generated, moved and detected. This discovery became the basis of the bubble domain memory system. Presence of a domain in a position could mean a "1" and absence could

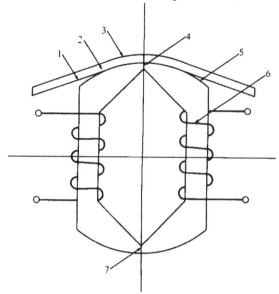

Figure 19.4- Diagram of Construction of a Recording Head; 1- Magnetic Coating, 2-Carrier Tape, 3- Magnetic Tape, 4-Gap, 5-Winding, 6-Gap. From Heck, C.H., Magnetic Materials and their Applications, Crane Russack & Co., New York (1974)

mean a zero. The pathways in which the bubbles moved were permalloy films. The domains could be generated by a current loop of deposited permalloy, moved by passing a current through another such loop in the neighboring cell and detected by a Hall effect probe. The main use of bubble domain memories is in auxiliary memory circuits such as ring counters. However much has been done to miniaturize the domain size and increase bit density so that some main memories have been constructed.

Magnetic Recording Head Properties

Magnetic recording heads as described above are used to write and read magnetic information stored on the magnetic media. The overall construction of both types of heads is similar with only the associated electronics varying. A typical magnetic recording head is illustrated in Figure 19.4. Magnetically, it would be more desirable if there were only one gap instead of the two shown. One, of course is the recording gap while the other is there simply for case of construction. It would be very different to produce a single-gap head with the exact gap width such a small small gap. Although the older tape heads actually made contact with the tape, the new ones actually ride on a cushion of air and are called "flying heads". However, with dirt and attracted oxide particles getting between the tape and the head, abrasion takes place. Therefore, one of the requirements for the head material is that it be relatively hard and resist pull out. Fine-grain, dense material is ideal. For construction of the head with regard to stability of dimensions during the

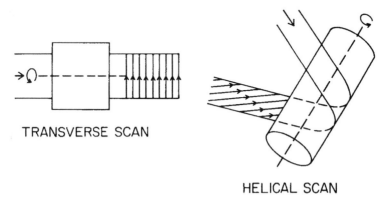

Figure 19.5- An example of Transverse and Helical Scan Arrangements in Video Tape Recorder Heads. Note the tracks in the Transverse Mode are short while those of the helical mode are longer. From Mallinson, J.C., The Foundations of Magnetic Recording, Academic Press, San Diego (1987), 139

construction, glass bonding is used. A low melting glass is often used since a high temperature would change some of the electrical properties. This is especially true for manganese-zinc ferrites that are unstable to higher oxygen pressures at elevated temperatures. Nickel ferrites were used for some time but low melting glass and Mn-Zn ferrites are the choice because of lower losses and lower cost.

The front gap in a recording head should be quite small to generate as strong a fringing flux as possible in the gap. For digital recording, the particle alignment is in the same direction as the velocity of travel and the gap must be equal to or smaller than the ratio of the tape speed to maximum for frequency.

Digital recording writing heads magnetize the recording media to saturation so that the saturation flux density of the head must be high to avoid saturation. However, the remanence of the core should be low so that when the write current is off to avoid erasure.

Audio and Video Magnetic Recording
While magnetic recording for computer memory applications naturally use digital recording, most audio and video recording is done by analog methods. To obtain good linearity characteristics, an ac bias frequency of about 100 KHz is used. The

recording of digital information was accomplished by saturating the media particles. Consequently, the flux excursion was between two saturations or between two remanences. In the case of audio recording, since different sounds require different voltage wave forms, the situation is quite different. For faithful recording and playback, the voltages and therefore the flux excursions must be continuously variable, linear and reproducible. To accomplish this requires a linear section of the magnetization curve rather than extending up to saturation. To assure the reproducibility, the audio signal is superimposed on an ac signal leading to what is known as an anhysteritic B-H loop. This removes much on the non-linearity about the origin (Westmiize 1953).

For audio heads, which must be low cost and do not require high speed recording, laminations of permalloy or amorphous metal alloys may be used instead of ferrites. For video recording, tape speeds of 1500 inches/sec. were used in the first video recorder. This is much too high a speed for fixed head machines. As a result, a new technology was developed using a rotary head. The axis of rotation of the head was parallel to the direction of the tape motion. In this case, the tracks that the head scanned were quite short (See Figure 19.5) To make the tracks longer so more video information could be inserted, the rotary was placed at at an angle to the direction of tape motion.(see Figure 19.5)This is known as a helical scan technique. For consumer video recorders, the tape speed is 220 inches/sec and the track density is 1400 tracks/inch. The drum speed is 200 revolutions/sec.

MAGNETIC RECORDING MEDIA
Magnetic recording media consists of several types
1. Oxide particulate, magnetite, γ-iron oxide, chromium dioxide, barium ferrite
2. Metal particulate media
3. Metal thin film media
4. Oxide thin film media

At ICF6 in 1992, Hirotaka (1992) conducted a panel discussion to assess the merits of each of these media as well as recording heads to predict the future course for materials for both functions. A synopsis of the talks by experts follows;

For Magnetic Recording Tapes
*Metal particulate Tape-(N. Nakahara)*Introduced in 1984for video and broadcast tapes. Consist of very thin (100-500 nm) upper later and thick TiO_2 pigment underlayer. Has high signal output at high density region and expect it to be higher tham thin film metal tape in future.

Oxide Particulate media-(A.E.Berkowitz)-Except for barium ferrite(below) improvements are expected to be minimal. Still has the largest volume of any media.

*Barium Ferrite Particulate Media-(T. Suzuki)*Has unique characteristic of being oriented longitudinally or perpendicular. Coercivity (650-1300 Oersteds) can be controlled, has high density performance. Used in high density floppy disks, VCR and data tape. Expect coercivity to reach 1800-2000 Oersteds. Very promising

*Obliquely Evaporated Metal Thin Film Tape-(K.Sato)*Excellent short-wave characteristics. Corosion resistance improved. Signal output now 5 dB higher than metal particulate tape. Most suitable for future high density recording.

Thin film Perpendicular recording tape-(R.Sugita)-Consist of Co-Cr, Co-O, Fe-Co-O, Co-Ni- Mn-RE-P and have been proposed for magnetic layers of perpendicular recording media. Co-Cr have best performance. Considered to yield recording density of 1 bit/$(\mu m)^2$ for digital VCR system.

High-Density Rigid-Disk Technology-(M. Futamoto)-Recording Density for rigid disk has increased by factor of 10 every 10 years. Recording density of 1- 2 GBits/in^2 have been reached in laboratory. New techniques are available to reach 10 GBits/in^2 for longitudinal recording and 20-30 GBits/in^2 is forecast with the factor of 10 fold increase for each following 10 year period.

Ba Ferrite Rigid Disk-(D.E. Speliotis)-Ba ferrite rigid disk has excellent high density recording characteristics, large signal output and low noise. They are also corrosion free.

Perpendicular recording Media-(Y.Miura)- For ultra-high recording density(10 GBits/in^2, grain size for crystalline media must be reduced and remove exchange interactions between grains of sputtered metal thin film media. In the case of longitudinal recording, very thin media of 10 nm. is needed. Therefore, perpendicular recording is indispensable for ultra high densities. Areal densities of 2 GBits/in^2 were obtained in CoCrTa/NiFe double layered film.

MAGNETIC RECORDING HEADS

Inductive Recording Heads
Heads for VCR Recording-(H.Hayakawa)-(These are primarily ferrite heads, thin metal films or MIG (metal-in gap) heads.

Laminated Bulk Head-K.Takahashi)-Track width must be reduced for recording density of 1 GBits/in^2 . When track width is narrower than 10 μm or less, reproduction efficiency drops and deterioration occurs. Permeabilities of 2000 or greater are necessary in the material for reprocibilitty in narrow track recording. Laminated bulk yoke structure are candidates for most practical for higher frequency higher track density VCR systems. Higher saturation and higher permeability materials are needed.

Thin Film Head for Video Recording-(Y. Noro)-Thin film heads are suitable for wide-band signal system such as HD VCR because of ease of getting narrow track, low inductance and no abrasive noise.

MATERIALS FOR MAGNETIC RECORDING 545

Heads for Rigid Disks

MIG and Composite Bulk Heads-(M. Kakizaki)-Three approaches are suggested for improvement;
1) Substitute Sendust with High saturation material
2) Adopt double-sided MIG
3) Adopt enhanced dual gap (EDG). For MIG heads, maximum density is 190 Mbits/in^2.

Thin film head for Rigid Disk-(M.Aihara)-Thin film heads have several advantages over conventional bulk heads. However, increases in linear recording density and track density are necessary to get much higher recording density. Coercivity and saturation must be increased. Material such as CoTaZr has a saturation of 1.3 Teslas. High saturation gives a coercivity of up to 3000 Oersteds. Reduction of head noise is also needed. Also by adopting a multi-layered head, we can decrease head noise and increase areal density to 1 GBits/in^2.

MAGNETORESISTIVE HEADS

The magnetic recording heads discussed above are all considered inductive, that is, the signal written or read was based on changing the magnetic state of the material(write) or getting a voltage from the flux change in the material (read). In other words, the effect was purely magnetic.

Recently, a completely new concept in magnetic heads has been intr4oduced and by most accounts will replace a large part of the inductive- head application. We are, therefore, giving it a great deal of attention although the main effect operative is only coincidentally magnetic. Several new terms will be introduced namely;

1. Magnetoresistance
2. Giant Magnetoresistance
3. Colossal Magnetoresistance
4. Magnetic Multilayers
5. Spin Valves

All of these terms are related in the new technology and we will develop them chronologically.

Magnetoresistance-This effect refers to the change in resistance of a material in the field direction compared to that perpendicular to it when a magnetic field is applied. For many metals, the effect is small. The first use of a magnetoresistive recording head was made by in 1970 Hunt (1970, 1971).(See Shelledy (1992) in Figures 19.6 and 19.7) The first commercial tape product in the IBM 3480 in 1985. The first hard disk file with an MR head was introduced in 1991. The original Hunt head was unshielded thin film and had unsatisfactory resolution compared with inductive heads. Shielding was inserted and a transverse bias scheme using another magnetic layer and a non-magnetic layer separating the MR element and the shield. These and other improvements made for better heads but the change was still small (about 5%). An improved version with shielding and another magnetic layer for biasing is added. Another non-magnetic film(insulator or conductor) separates the two magnetic layers. The MR scheme used by Hunt and his followers is called Anisotropic Magnetoresistance or AMR. It has been used for many years in a number of improved variations. However, in 1988 a new development revolutionized the MR head technology. It was called Giant Magnetoresistance or GMR.

Giant Magnetoresistance (GMR)- Magnetic Multilayers-In 1988, Babich (1988) reported changes in magnetoresistance of as much as 50% at low temperatures in multilayer ultrathin films. This huge effect originally found in $(FeCr)_n$ multilayer films was found to occur in a number of different multilayer films. The effect was labelled Giant Magnetoresistance. This effect is quite different than the bulk

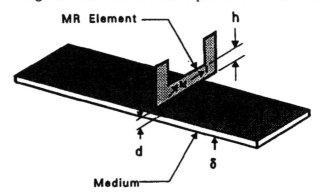

Figure 19.6- A Magnetoresistive Head as Invented by Hunt (From Shelledy (1992)

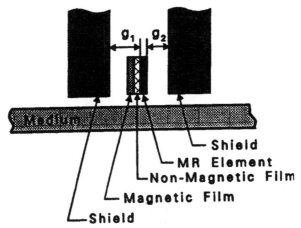

Figure 19.7- An Anisotropic Magnetoresistive Head with Shielding and Biasing From Shelledy (1992)

MATERIALS FOR MAGNETIC RECORDING

magnetoresistive effect described earlier. A figure showing the GMR effect reported by Babich is shown in Figure 19.8 (White (1992)).

The magnetoresistive effect occurs in ultrathin multilayer arrays of alternate layers of magnetic metal separated by layers of non-magnetic material. To obtain the MR effect, the following conditions must be met.

1. There must be a way to change the relative orientations of the magnetizations in adjacent layers.

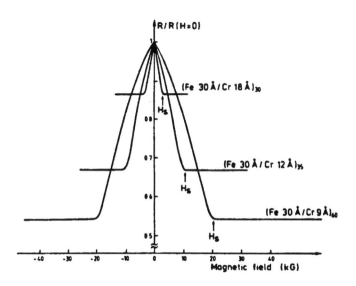

Figure 19.8- Giant Magnetoresistive Effect reported by Babich From White (1992)

2. The thickness of the film must be only a fraction of the mean free path of the array.

In the Babich multilayer film, an antiferromagnetic coupling between adjacent FeCr layers and through the non-magnetic Cr layer kept successive FeCr layers in antiparallel coupling. When a large in-plane magnetic field was applied the exchange interaction could be overcome and the magnetization in all layers brought into parallel orientation. Other non-magnetic layers were found to give the same effect. Also other combinations of metal non-metal combinations also gave the GMR effect. The thickness of the non-metal layer was also critical in determining the type of coupling it produced in the magnetic layers. The explanation of the electron scattering theory for explaining the GMR effect is handled excellently by

548   **HANDBOOK OF MODERN FERROMAGNETIC MATERIALS**

White (1992) and the reader is recommended to refer to the paper for further review. The primary application for GMR heads is in the read head of a magnetic recording system.

Spin Valves-For use in magnetic recording head, in addition to a large $\Delta R/R$ ratio, the material for a GMR head must have a large resistance change for a modest magnetic field, The Babich GMR material $(FeCr)_n$ which required up to 20,000 Oersteds to switch from a parallel to antiparallel orientation is not attractive as a head material. Any thin film system held antiparallel by exchange interaction through the spacer material has the same problem. However there are other schemes to switch uncoupled films from antiparallel to parallel configurations from. Figure 19.9 shows a three-layer system of two different magnetic films separated by a pinned in orientation by an antisymmetric coupling to an antiferromagnet, the lower the same effect is through the use of a higher coercive field material for the upper

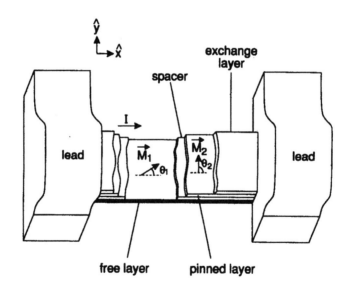

Figure 19.9- A 3-layer Spin Valve Configuration showing the free and pinned layers From Chang (1994)

material than the lower to switch the relative orientations. The spin-valve technique is widely used in MR read heads. An excellent review of GMR is found in White(1992)

MATERIALS FOR MAGNETIC RECORDING

Materials for the Different Layers in Spin Valves

One promising feature of the spin-valve approach is the large variety of configurations, film thicknesses, and material choices compared to the AMR system which in most cases are single Permalloy films. Kools (1996) has listed the various materials that have been used for the different layers in a spin-valve.

1. Ferromagnetic Layer-These materials are chosen in the fcc (face-centered – cubic) range of the FeBiCo ternary alloy system.The most widely used are Permalloy 80 and Co materials but others have also been used.
2. Non-magnetic Layer- In this case a lattice match must be made between the Ferromagnetic layer material and the the non-magnetic layer. If the fcc NiFeCo alloys are the FM materials, then only using the noble metals, Cu, Ag and Au give interesting MR values. Since the Cu gives the least mismatch of the three, it is most widely(universally) used.
3. Antiferromagnetic Layer- This layer has the largest choice of materials available. Three types of materials have been used. First the fcc alloys such as γ- FeMn,; second the amorphous transition metal-rare earth alloys such as TbCo and third, the oxides, such as NiO, NiCoO and NiO/CoO multilayers. NiO can be deposited by reactive sputtering in an Ar/ O_2 atmosphere. FeMn and TbCo are very sensitive to oxidation in contrast to NiO. Since some groups have used these two materials indicates that this problem can be solved. Figure 19.11 shows the configuration of a spin-valve head and the differences in the signal amplitude for both the MR and spin-valve head. Note that the spin-valve head does not require the very thin films of the MR head. Figure 19.12 shows a conceptual cutaway of a magnetoresistive head.

Colossal Magnetoresistance

The discovery of colossal magnetoresistance (CMR) has not been utilized in devices yet but since this development shows such great potential, we have included it in our discussion of magnetic recording.

The giant magnetorestive materials we have discussed thus far have all involved the use of metallic films or multilayers for the ferromagnetic element of the system. Subsequent to the discovery of GMR in metal films, the effect was also found, first, in single crystals and finally in ceramics. It is in the ceramic or oxide materials that we find the colossal magnetoresistance.

The materials in which the CMR effects were found are the perovskites mentioned at the end of Chapter 11. In this case, it is the manganites having the formula $Ln_{(1-x)}A_xMnO_3$ where A= Ca, Ba or Sr) and Ln is usually a rare earth ion. The reason they are called colossal is that their magnetoresistance ratios are many orders of magnitude larger than those of the GMR materials. Unfortunately, the temperatures at which the "colossal " MR ratios occur are well below room temperature(on the order of 77 K or about $-200°$ C). In addition, the magnetic fields necessary to accomplish the CMR effect are on the order of 6-8 T. or about 60,000-80,000 Oersteds. Several papers on the subject of CMR were presented at ICF7 (Seventh International Conference on Ferrites) which was held Sept. 3-6, 1996 in Bordeaux France.

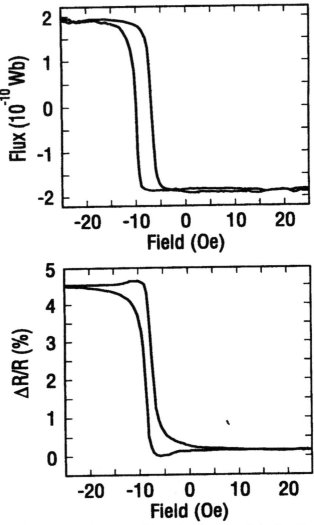

Figure 19.10- (Upper)Hysteresis Loop and (Lower) Magnetoresistive Response in a Spin Wave Head (From Chang 1994)

Raveaux (1997) described the recent trends in the exploration of CMR in the manganites whose general formula is given above. There are two types of CMR manganites, Type I in which $0.2 < x < 0.5$ and the type II in which $x = 0.5$. The two factors that determine the CMR properties are the average size of the interpolated ion and the hole carrier density [Mn(III).Mn(IV) mixed valence. Doping of the Mn

MATERIALS FOR MAGNETIC RECORDING 551

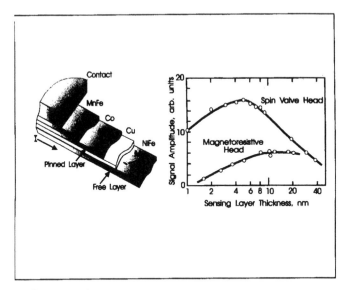

Figure 19.11- Configuration of spin valve head and and differences in signal amplitude in MR and spin valve heads.

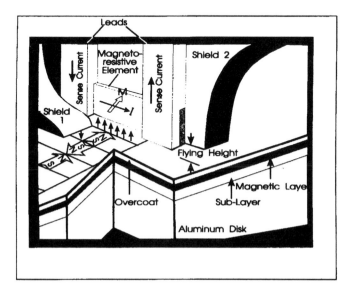

Figure 19.12- Conceptual Cutaway of a Magnetostrictive Read Head. From Grochowski (1994)

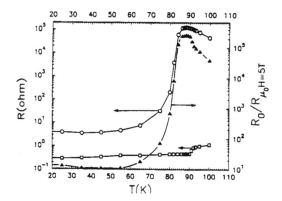

Figure 19.13- Temperature dependence of the resistance and magnetoresistance ratio of a colossal magnetoresistance material. From Raveau (1997)

sites by various cations was studied with spectacular results with Mg and Fe. The manganite, $Pr_{0.7}Sr_{0.05}Ca_{0.25}MnO_3$ exhibited a very large CMR effect with a RR (Resistance Ratio) of 2.5×10^5 at 85K in a magnetic field of 5 T. (50,000 Oe.) In these Type I materials, the peak in resistance ratio occurs at the Curie point. Figure 19.13 shows the temperature dependence of the resistance and resistance ratio of this material and Figure 19.14 shows the magnetization versus temperature curve. The aim of course is to fins a material with high ratios around room temperature and

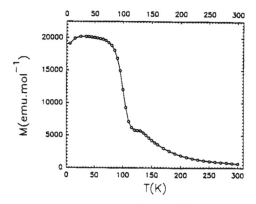

Figure 19.14 Magnetization versus temperature curve of a colossal magnetoresistance material. From Raveau (1997)

at lower magnetic fields. The general belief is that ordering of the Mn^{+3}-and Mn^{+4} species should play an important role in the transition. Electron diffraction and high resolution electron microscopy seem to support this view. Subramanian (1997) looked at the structure, magnetic properties and CMR of the pyrochlores, $A_2Mn_2O_7$ materials where A = Dy-Lu Y, Sc, In, or Tl. Ichinose (1997) examined the MR effect in $La_{(1-x)}Sr_xMnO_3$ ceramics. The materials were prepared by conventional ceramic techniques, calcined at 1273 K in air, pulverized, , granulated with binder and pressed into disks. The MR ratios were as high as 12% at room temperature at a fiels of 1000 KA/m. Holzapfel (1997) prepared thin films of $La_{(1-x)}Pb_xMnO_3$ by pulsed lazer deposition. The ferromagnetic Curie points were between 200-300 K. A film with a T_c of 193 K had a resistance change of 25% at 300 K. For T_c of 220, the 180 K change was 70%. Kitagawa investigated the magnetoresistance and magnetization of thin films of $LaCa_{0.25}Mn_{1.2}O_{(3+\delta)}$ grown by rf-sputtering. An MR ratio of 29%. One sample had an MR shoulder of 20% at 240K. A larger ratio of 175% at 108K was obtained for a film deposited on MgO. Chen (1996) examined the MR behavior of $La_{0.06}Y_{0.07}CaMnO_x$ very large MR values of 10^8 % were obtained at 60 K. and 7 T. This is the largest MR value to date. Figure 19.15 shows Chen's magnetoresistance percentage as a function of temperature.

Although the MR ratios are high at low temperatures and at high fields, there is much more improvement needed before the CMR materials can be used in commercial devices. However, the potential is also quite impressive.

Magneto-optic Recording
Magneto-optic recording is somewhat similar to CD-ROM recording(which not magnetic-based). Most CD's use a factory-installed memory while the magneto-optic memory can be written, erased and re-recorded with the same equipment. (Equipment to record on CD's are now being made available.) Kryder(1992) thinks that magneto-optic recording will assume a portion of the 50 billion US dollar magnetic storage market. Magneto-optic recording media consists of a thin film made in a way that the easy magnetization direction is perpendicular to the film plane. Writing is done by the combined action of a submicron-sized beam of light from a laser and a magnetic field perpendicular to the film plane. Heat from the absorption of light raises the temperature of the region and lowers the coercive force of the film material. When the external magnetic field exceeds the coercive force, the magnetization can be oriented by the direction of the magnetic field. The pattern of up and down areas represents digital information. This type of writing is similar to the Curie point method previously used. Reading the information is done magneto-optically using Faraday Rotation and the Kerr (reflection) effect. The equipment is shown in Figure 19.16. Here the laser intensity is reduced so that magnetization state is not altered. The beam is plane polarized and on interaction with the two states of magnetization will be rotated clockwise or counter-clockwise. The analyzer will distinguish the rotation and give two different light intensities that is sensed by the photodetector. The materials used for the magneto-optic media are;

554 HANDBOOK OF MODERN FERROMAGNETIC MATERIALS

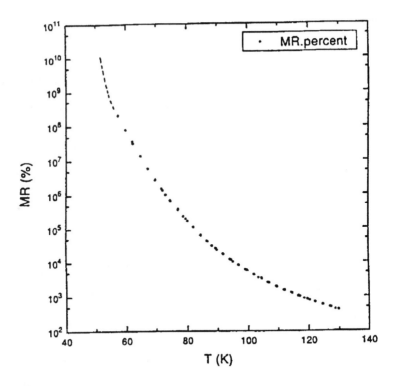

Figure 19.15-Magnetoresistance percentage of a colossal magnetoresistance material as a function of temperature. From Chen(1996)

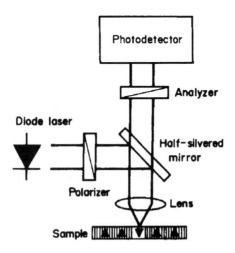

Figure 19.16-Equipment for magneto-optical recording using the Faraday rotation of a thin film with perpendicular magnetization. From Kryder (1994)

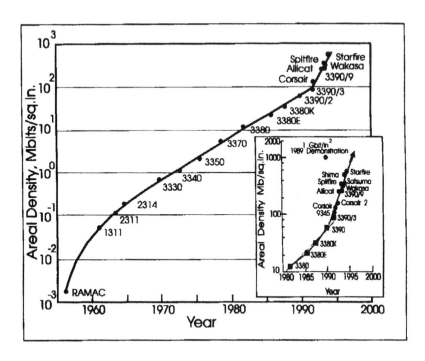

Figure 19.17- Chronology of Areal Density in Recording Heads and blow-up of period from 1990. From Grochowski 1994

1. Amorphous Rare-Earth-Transition Metal Thin Films.
2. Magnetic Oxides- Bi-doped garnets,
3. Co-Ti doped Ba ferrite
4. Co-Pt and Co-Pd Multi-layers-very promising

Outlook for Areal Densities in Magnetic Recording
Although spin-valve technology has gotten most of the attention on magnetic recording research, several groups have obtained up to 5 Gbits/in^2 in laboratory tests and 1-2 Gbits/in^2 appear possible in production. In addition to advances in the materials, physical and configurational improvements have also been made. Figure 19.17 shows the chronological increase in recording densities from about 1960. Figure 19.18 shows the more recent advances and the projections for future improvements.

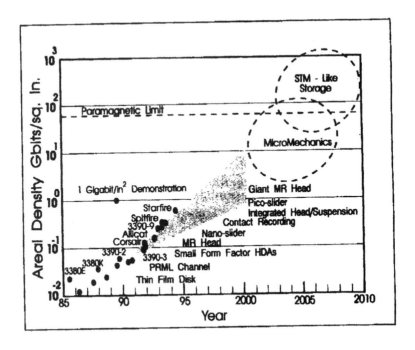

Figure 19.18- Recent Areal Density for Recording Heads History and Future Projections From Grochowski (1994)

References

Albers-Schoenberg, E. (1954), J. Appl. Phys., 25, 152
Babich, M.N.(1988), Broto, J.M.,Fert, A.,Nguyen van Dau, F.,Creuzet, G.,Friedich, A. and Chazelas, J., Phys. Rev. Lett.,61, 2472
Bertram, W.H. (1994), Theory of Magnetic Recording, Cambridge Press, Cambridge
Bertram, W.H. (1995) IEEE Trans. Mag., 31 #6 2573
Hunt, R.P. (1970)U.S. Patent 3,493,694,Feb 3, 1970
Hunt, R.P. (19710 IEEE Trans. Mag., 7,150
Khizroev, S.K.(1997) Bain, J.A.,and Kryder, M.H.,Trans Mag.,33, #5,2893
Kools, J.C.S.(1996)ibid, 32 #4, 3165
Kryder, M.H. (1992) in Conscise Encyclopedia of Magnetic and Superconducting Materials, Ed by J. Evetts, Pergamon Press, Oxford, 275
Shelledy, F.B.(1992) and Nix, J.L. IEEE Trans. Mag, 28, #5, 2283

Tsang, C. (1994) Fontana R.E.,Lin, T.,Heim, D.E.,Speriosu, V.S.,Gurney, B.A.,and Williams, B.L., Trans. Mag., 30 , #6 3801
Tsang, C. (1997), Lin, T.,Mac Donald, S., Pinbarsi, M.,Robertson, N.,Santini, H., Doerner, M., Reith, T.,Vo, L.,Diola, T.,and Arnett, P.,ibid 33, #5, 2866
Westmijze, W.K.(1953), Philips Res. rep., 8, 148
White, R.L.(1992) IEEE Trans Mag. 28 #5, 2482

20 FERRITES FOR MICROWAVE APPLICATIONS

INTRODUCTION
We have spoken in Chapter 4 of the gyromagnetic effect of ferrites. For many applications of soft ferrites, the absorption of energy due to the onset of ferromagnetic resonance may be detrimental. This is true at higher frequencies where the tail of the ferromagnetic resonance curve overlaps the permeability versus frequency curves of the soft ferrites. It is partially the cause of the limitation of Snoek's µf limit as given in Equation 4.1. It also contributes to the anomalous loss coefficient, a, in Legg's equation (Equation 4.12). In this chapter, we will examine the applications where the absorption of energy by ferrites at microwave frequencies can be an extremely valuable tool. These properties form the basis for the technologies of space telecommunication and radar. In many instances, they are the only materials available for these applications.

THE NEED FOR FERRITE MICROWAVE COMPONENTS
Our previous discussion in Chapter 4 dealt with the ferromagnetic resonance or in general, the interactions of microwave energy with ferrites. We have also said that at microwave frequencies, the conventional magnetic phenomena involving domain wall motion and rotation do not apply as the whole domain structure breaks down. Therefore, the use of wound components are not available at frequencies such as 100- 500 MHz. Use may be made of coaxial and strip circuits with ferrites but at higher frequencies, even these cannot be used. The transition frequency between conventional and microwave usage is not distinct but at frequencies of 1000 MHz and above, we can consider only in the microwave realm. In this region, the electrical energy is not transmitted through wires, but through electromagnetic waves usually propagated or contained in wave-guides and transmitted through space. The dimensions of the wave guide are directly related to the frequency so that if d is the broad dimension of the waveguide and λ is the wavelength then:

$$d < \lambda < 2d \quad [20.1]$$

The other dimension is d/2. Circular wave-guides can also be used. The wavelength, λ, is related to the frequency by the velocity of light or electromagnetic radiation in the following equation;

$$\lambda = c/f \quad [20.2]$$

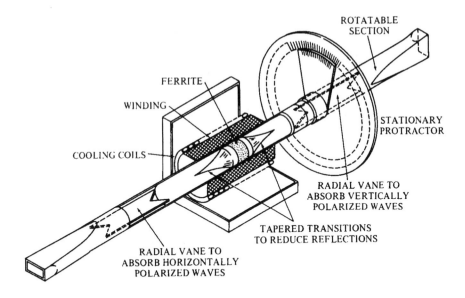

Figure 20.1- An example of a longitudinal field microwave device, an early Faraday Rotator. From Hogan, C.L., Bell System Telephone Journal, 31, #1, 1 (1952)

where; λ = wavelength in cm.
c = velocity of light = 3×10^{10} cm/sec
f = frequency in Hertz (cycles/sec)

Thus; for a typical microwave frequency of 9.5 GHz (10^9Hz.)
$\lambda = (3 \times 10^{10}) / (9.5 \times 10^9) = 3.16$ cm

With electromagnetic fields, the classical means of switching, voltage dividing, current directional control (diodes) are no long useful. We must therefore develop new methods of controlling the electromagnetic fields to perform useful operations. For this purpose, specialized microwave components are necessary. Some of the functions designed around gyromagnetic effects are:

1. Isolator - This is a device to isolate the transmitted and reflected waves.
2. Circulation - This is a device which directs the various waves entering the device from different channels into other specified channels.
3. Phase Shifters - These devices change the phase of the input electromagnetic wave.

FERRITE MICROWAVE COMPONENTS

Several different arrangements for the use of ferrite components have been advanced. One arrangement is known as longitudinal field device in which the ferrite material is biased by a D.C. magnetic field oriented in the direction of

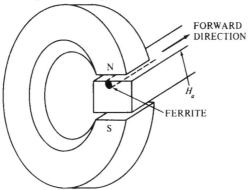

Figure 20.2- An example of a transverse field microwave device. From Anderson, Magnetism and Magnetic Materials, 201

propagation. See Fig. 20.1. The other arrangement is the transverse field device in which the material is magnetized transverse to the direction of propagation. See Figure 20.2. Since the treatment of microwave device design is quite complex and would require much exposition, we will limit are discussion to the action of the ferrite in the system.

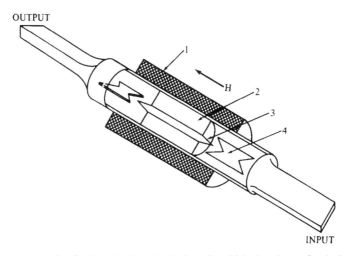

Figure 20.3- An example of a Faraday Rotator Isolator in which the plane of polarization is rotated by 45° in each direction so that the returning wave is rotated 90° from the original input plane and thus cannot enter and is absorbed by the vanes. From Heck, C.H., Magnetic Materials and their Applications, Crane Russack, & Co.,New York,(1974)

Longitudinal Field Devices

A simple longitudinal field device is a Faraday rotator as shown in Figure 20.3. The wave guide is round so that, if other sections are rectangular, a round section must be inserted for this purpose. The ferrite is a pencil shaped rod magnetized by a solenoidal field. The AC magnetic field is parallel to the applied D.C.field. Depending on the direction of the field the moments in the material will precess around the DC field in a particular direction. If a plane polarized alternating microwave field is propagated down the wave guide, the magnetic field component interacts with the spin system in the ferrites. As described in Chapter 4, The plane of polarization will be rotated by a certain angle after it interacts with the ferrite. The effect is similar to the rotation of the plane of polarization of light as the electromagnetic wave. In most cases the rotation, governed by the rod length and DC field, is designed to be 45°. Low field operation is preferred since the losses are lower. The frequency is usually chosen to be somewhat removed from the resonance frequency.

Microwave Isolators

If the microwave field is propagated from the direction opposite to the previous case with the ferrite magnetized as before, one would expect the rotation of the plane of polarization to be opposite to the original case (Reciprocal action). However, while the sense of the circular rotation of the ac wave is reversed, the sense of precession will also be reversed so that the absolute direction of rotation will be the same giving a total rotation of twice the original one.(This action is called non-reciprocal and is encountered in other gyroscopic phenomena.) The net difference between the two rotations is 90°, so that the second wave cannot exit through the port of the first input port. By use of an appropriate absorber vane (carbon), the returning wave can be completely absorbed. This allows for isolation of the input and the reflected wave or basically a one-way transmission device called a Rotation Isolator (See Figure 20.3)

Rotation Phase Shifters

In the device described above, if the angle of rotation for each direction had been 90° instead of 45°, the returning signal would have been 180° out of phase with the input signal. This then would make the device an ideal phase shifter

Rotation Circulators

By arranging a series of different input and output ports having different angles relative to the various propagation directions, the various waves can be directed to specific ports. This type of device is called a rotation circulator. An example of such a device is shown in Figure 20.4.   This type of device is used to route power between generators and antennas, and between antennas and receivers Use of this type of circulation device in radar. The input signal that is very large can be directed to the antenna while the reflected wave that is very weak can be directed from the antenna to the receiver.

FERRITES FOR MICROWAVE APPLICATIONS

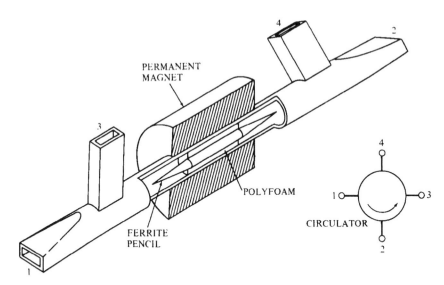

Figure 20.4- An example of a Faraday circulator. The arrow shows the direction in which the signal is circulated, ie.,1→2, 2→3, 3→4, 4→1.From Fox, A.G., Miller, S.E. and Weiss, M.T., Bell System Telephone J., 34, 5,(1955)

If a coil is used to provide the magnetic field, the attenuation can be modulated by variation of the DC field. These devices are called gyrators or modulators.

Transverse Field Devices
In transverse field devices, the ferrite is placed in the wave-guide as shown in Figure 20.5 with the biasing field transverse to the propagation direction. For transverse field devices, some of the same interactions occur here as did with the longitudinal field case. In the case of the transverse field devices, a rectangular wave-guide (or strip-line) is used with the biasing field often provided by a permanent magnet with poles at designed positions on the broad section of the wave-guide. See Figure 20.2.

Resonance Isolators
Most of the transverse field devices are designed to operate at or near resonance. The resonance frequency of the spin precession (Larmor Frequency) is given by;

$$\omega = \gamma H_{eff} \quad [20.3]$$
where $\quad \gamma$ = Gyromagnetic ratio
ω = Angular frequency = $2\pi f$
H_{eff} = Effective D.C. in the ferrite

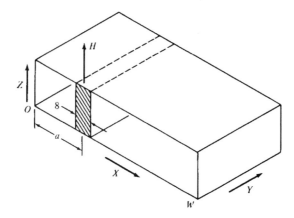

Figure 20.5- A transverse field device with a ferrite slab positioned at an optimum point. Y is the direction of propagation. From Bradley, F.N., Materials for Magnetic Functions, Hayden Book Co., New York, 1971, p.277

In terms of frequency;

$$f = 2.8 \text{ MHz/Oe} \times H_{eff} \quad [20.4]$$

H_{eff} depends on the shape of the sample and may be calculated from the Kittel equation and a knowledge of the demagnetizing factors for that shape. The magnetic field arrangement when viewed across the broad section of the waveguide is shown in Figure 20.6. The field arrangement will move in the direction of propagation progressing as shown in the succeeding frames of the figure. If we examine the field direction or vector at point A as shown by the arrows, it rotates counterclockwise as the wave pattern moves from left to right and clockwise as it moves from right to left. A ferrite slab placed at position A would have its precessional spin system (biased by the magnet interact with the rotating magnetic field vector in a non-reciprocal manner. If the direction of the circularly rotating field at A is the same as the spin precession in the ferrite, resonance absorption as described in Chapter 4 will occur. A rf field propagated in the reverse direction would produce rotation of the field vector in the opposite sense of the precession with the same DC field direction. Hence, the reciprocal action. Typical positions for resonance isolators for high frequency operation is shown in Figure 20.7. The reverse to forward attenuation is shown in Figure 20.8 as a function of position of the ferrite.

Another type of transverse field non-reciprocal isolator is called the field-displacement type. It operates on the distortion of the field when the ferrite is placed against one of the side- walls. This type is only rarely used.

FERRITES FOR MICROWAVE APPLICATIONS

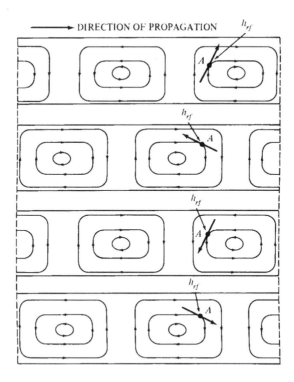

Figure 20.6- Magnetic field Contours of an electromagnetic wave as viewed through the broad side of the guide. The succeeding frames from top to bottom show the magnetic field vector at point A rotating as the wave propagates from left to right. From Anderson, Magnetism and Magnetic Materials, p.203.

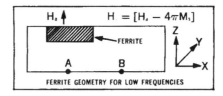

Figure 20.7- Position of a ferrite in the wave guide in a resonance circulator at high frequencies. Courtesy of Trans-Tech Inc.P.O.Box 69, Adamstown, MD 21710.

Junction Circulators

Junction Circulators are devices consisting of three or four wave-guides radiating out from the central junction at angles of 120° or 90° from each other with a ferrite cylinder situated at the center of the junction. The 3 junction circulator is called a Y circulator(Figure 20.9) and the 4 junction type is known as an X

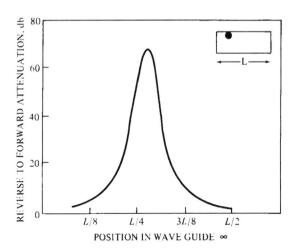

Figure 20.8-The reverse to forward attenuation ratio in a resonance isolator as a function of position of the ferrite in the wave-guide. From Anderson, Magnetism and Magnetic Materials, p. 204

circulator (Figure 20.10). The ferrite cylinder is magnetized axially. The circulation usually is of the type; 1→2, 2→3, 3→1 in the Y circulator. Junction Circulators are compact and can operate over a broad frequency band. The circulation is controlled by the disk diameter and magnetic field or by saturation magnetization and magnetic field. The operation of the junction circulator is not completely understood from a theoretical point of view but it would appear to be based on the interaction of the magnetic field at the two positions in the wave-guide where circular polarization occurs(See Figure 4.6). When these rotating fields interact with the spin precession of the ferrite rod, they produce a bending of the beam that can be directed into the next wave-guide.

Digital Phase Shifters
By placing a ferrite slab is placed in a rectangular wave-guide in a position shown in Figure 20.7 and it is D.C. magnetically biased, it will produce a phase shift of the rf wave propagating down the wave-guide. This phase shift can be very useful in modifying the direction of propagation. A variation of this device is obtained by using a rectangular ferrite toroid in the wave-guide in place of the slab. If the toroid is magnetized by a wire through the center, it will fall back to remanence. If the ferrite is a square loop material, the remanence can be moderately high. The remanence can then provide the biasing field for the phase shifter, eliminating the need for the external (permanent magnet). The amount of phase shift depends on the thickness of the ferrite toroid. If a number of the toroids of different thickness and with separate magnetizing wires are combined, a digital phase-shifter can be

FERRITES FOR MICROWAVE APPLICATIONS

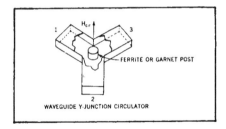

Figure 20.9- An example of a Y circulator, Courtesy of Trans-Tech Inc., P.O. Box 69, Adamstown, MD 21710

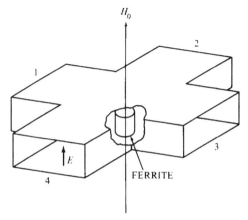

Figure 20.10- An example of an X circulator. From Heck,C.H., Magnetic Materials and their Applications, Crane, Russack &Co., New York, (1974) p.572

constructed to give varying degrees of phase shift. An array of these phase shifters form the basis of the so called Phased Array Radar, which when hooked up to a computer, can electrically scan a large quadrant of the horizon. This system is sometimes referred to as Electronically Steerable Radar. It eliminates the bulky mechanical means of rotating the radar antenna. Figure 20.11 shows an example of a digital phase shifter.

Radar Absorbing Ferrites
In some cases, it is necessary for the ferrite to act as a microwave field attenuator or absorb an incoming microwave signal. Often a ferrite and a dielectric-absorbing medium are combined to produce this material. The ferrite is made rather lossy with high resistivity and a permeability of 10-15. An important application of this absorbing medium is the case in which ferrite coatings are used to make an object invisible to radar by absorbing the microwave energy that is used to detect the object. The task is a difficult one as the thickness of the coating must be uniform but not

too heavy that it adds excess weight as in aircraft or space applications. This technique is a basis for the radar invisibility of the Stealth bomber.

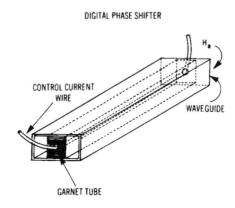

Figure 20.11 An example of a digital phase shifter, Courtesy of Trans-Tech Inc., P.O. Box 69, Adamstown, MD 21710

COMMERCIALLY AVAILABLE MICROWAVE MATERIALS

As might be expected the quantity of microwave ferrite materials marketed is relatively small and might be considered a special or premium material. As a result and also due to the high cost of some raw materials, the price per pound is also relatively high. In addition, because of the unusual shapes of microwave components, many have to be specially machined by grinding or by ultrasonic techniques. There are very few companies marketing microwave ferrites and garnets. For informational purposes we have reprinted the catalog list of Trans-Tech microwave garnets in Table 20.1 and that of their microwave ferrites in Table 20.2.

SUMMARY

Microwave garnets complete our discussion of materials according to frequencies since the operate at the highest frequencies. Having discussed most of the common magnetic application groups, we find that some others do not fit into any one special classification. Therefore, the next chapter will discuss the materials that may best be listed as miscellaneous magnetic applications.

References

Anderson, J.C.(1968) Magnetism and Magnetic Materials. London, Chapman and Hall, Ltd.
Bradley, F.N.(1971)Materials for Magnetic Functions, New York,Hayden Book Co.
Fox, A.G.(1955 Miller, S.E. and Weiss, M.T.,Bell System Tel. J., 34, 5
Hogan, C.L.(1952) Bell Syst. Tel. J.,31, #1,1
Trans-Tech (1973) Tech-Briefs, TransTech Inc. P.O.Box 69 Adamstown, MD 21710

FERRITES FOR MICROWAVE APPLICATIONS

Table 20.1 – Listing of Trans-Tech Microwave Garnet Materials

COMPOSITION AND TYPE NUMBER	PAGE NUMBER	SATURATION MAGNETIZATION 4πM$_s$ (Gauss)	LANDÉ' g-FACTOR g-eff	LINE WIDTH ΔH oe @ -3dB	CURIE TEMPERATURE T$_c$ (°C)	SPIN WAVE LINE WIDTH ΔH$_k$ oe	REMANENT INDUCTOR B$_r$ (Gauss)	COERCIVE FORCE H$_c$ (oe)	INITIAL PERMEABILITY μ$_0$
		(Nominal Value)	(Nominal Value)		(Nominal Value)	(Nominal Value)	(Nominal Value)	(Nominal Value)	(Nominal Value)
ALUMINUM DOPED									
G-1009	6	175 ± 25g	2.03	≤50	85	1.5	40	0.90	11
G-250	6	250 ± 25g	2.02	≤45	105	1.4	123	0.62	34
G-300	7	300 ± 25g	2.02	≤45	120	2.0	182	0.62	46
G-350	7	350 ± 25g	2.01	≤45	130	1.4	213	0.66	31
G-400	7	400 ± 25g	2.01	≤45	135	1.4	224	0.69	41
G-475	8	475 ± 25g	2.01	≤45	140	1.4	310	0.60	40
G-510	8	550 ± 5%	2.00	≤48	155	1.3	398	0.55	37
G-610	8	680 ± 5%	2.00	≤48	185	1.5	515	0.70	50
GADOLINIUM DOPED									
G-1005	10	725 ± 5%	2.02	≤300	280	7.6	357	1.51	26
G-1003	10	870 ± 5%	2.00	≤188	280	6.4	543	1.10	36
G-1002	10	1000 ± 5%	1.99	≤132	280	5.8	672	0.93	48
G-1001	11	1200 ± 5%	1.99	≤96	280	4.3	717	1.00	72
G-1600	11	1600 ± 5%	1.96	≤68	280	3.8	986	0.83	115
GADOLINIUM ALUMINUM DOPED									
G-1006	11	400 ± 25g	2.01	≤78	150	4.2	186	1.00	23
G-500	12	550 ± 5%	2.00	≤78	180	3.5	280	0.80	28
G-600	12	680 ± 5%	2.00	≤72	200	4.0	375	0.89	34
G-1004	12	800 ± 5%	2.00	≤90	240	5.2	493	0.83	38
G-800	13	800 ± 5%	2.00	≤66	230	4.3	504	0.69	60
G-1000	13	1000 ± 5%	1.99	≤66	250	3.6	641	0.97	56
G-1021	13	1100 ± 5%	1.99	≤108	260	5.4	722	0.76	54
G-1200	14	1200 ± 5%	1.96	≤60	260	3.2	795	0.83	65
G-1400	14	1400 ± 5%	1.96	≤60	265	3.1	918	0.69	88
HOLMIUM DOPED									
G-4250	14	550 ± 5%	2.00	≤120	180	8.5	280	0.80	28
G-4259	15	800 ± 5%	2.00	≤132	240	8.1	493	0.93	38
G-4258	15	1000 ± 5%	1.99	≤156	260	8.9	672	0.93	48
G-4257	15	1200 ± 5%	1.99	≤120	280	8.1	717	1.00	72
G-4256	16	1600 ± 5%	1.96	≤84	280	5.4	986	0.83	115
NARROW LINE WIDTH SERIES									
G-113	6	1780 ± 5%	1.97	≤30	280	1.4	1277	0.45	134
G-610	9	900 ± 5%	1.99	≤30	200	1.5	543	0.62	46
G-1010	9	1000 ± 5%	1.99	≤30	210	1.4	694	0.55	66
G-1210	9	1200 ± 5%	1.98	≤30	220	1.3	784	0.69	87
CALCIUM VANADIUM DOPED									
TTVG-800	16	800 ± 5%	2.00	≤15	192	2.0	560	0.60	129
TTVG-930	16	930 ± 5%	2.00	≤10	198	2.0	380	0.40	225
TTVG-1000	17	1000 ± 5%	2.00	≤10	199	2.0	320	0.30	210
TTVG-1100	17	1100 ± 5%	2.00	≤10	205	2.0	600	0.60	209
TTVG-1200	17	1200 ± 5%	2.00	≤10	208	2.0	635	0.30	221
TTVG-1400	18	1400 ± 5%	2.00	≤10	215	2.0	825	0.30	263
TTVG-1600	18	1600 ± 5%	2.00	≤15	220	2.0	—	0.60	46
TTVG-1850	18	1850 ± 5%	2.00	≤15	214	2.0	1000	0.60	227
TTVG1950	—	1950 ± 5%	2.00	≤15	235	2.0	1232	0.50	388

Table 20.2-Listing of Trans-Tech Microwave Ferrite Materials

	COMPOSITION AND TYPE NUMBER	PAGE NUMBER	SATURATION MAGNETIZATION 4πM$_s$ (Gauss)	LANDE' g-FACTOR g-eff (Nominal Value)	LINE WIDTH ΔH oe @ -3dB	CURIE TEMPERATURE T$_c$ (°C) (Nominal Value)	SPIN WAVE' LINE WIDTH ΔH$_k$ oe (Nominal Value)	REMANENT INDUCTION" B$_r$ (Gauss) (Nominal Value)	COERCIVE FORCE" H$_c$ (oe) (Nominal Value)	INITIAL PERMEABILITY† μ$_0$ (Nominal Value)	AVAILABLE AS SUBSTRATES GRADES 1,2
MAGNESIUM FERRITES	TT1-414	22	750 ± 5%	1.98	≤114	90	5.1	544	0.48	120	■
	TT1-1000	22	1000 ± 5%	1.98	≤120	100	3.1	627	0.82	93	■
	TT1-109	22	1300 ± 5%	1.98	≤182	140	2.5	940	0.87	30	■
	TT1-1500	23	1500 ± 5%	1.98	≤216	180	2.3	968	0.99	51	■
	TT1-105	23	1750 ± 5%	1.98	≤270	225	2.2	1220	1.20	55	■
	TT1-2000	23	2000 ± 5%	1.98	≤300	290	2.1	1385	1.60	52	■
	TT1-390	24	2150 ± 5%	2.04	≤648	320	2.5	1288	1.80	50	■
	TT1-2500	24	2500 ± 5%	2.03	≤624	275	3.0	1410	1.33	57	■
	TT1-2650	24	2650 ± 5%	2.02	≤636	245	2.8	1511	1.33	85	■
	TT1-2800	25	2800 ± 5%	2.01	≤648	225	2.2	1477	0.83	140	■
	TT1-3000	25	3000 ± 5%	1.99	≤228	240	3.2	2100	0.85	54	■
NICKEL FERRITES	TT2-113	25	500 ± 10%	1.54	≤190	120	-	140	2.00	23	■
	TT2-125	26	2100 ± 10%	2.30	≤575	560	6.1	1426	4.42	26	■
	TT2-102	26	2500 ± 10%	2.25	≤610	570	6.9	1485	4.42	23	■
	TT2-2750	26	2750 ± 10%	2.20	≤540	580	9.0	1130	3.00	20	■
	TT2-101	27	3000 ± 10%	2.19	≤575	585	12.4	1853	5.70	17	■
	TT2-3250	27	3250 ± 10%	2.10	≤440	550	10.5	1200	2.20	36	■
	TT2-3500	—	3500 ± 10%	2.10	≤500	540	9.0	1260	2.40	50	
	TT2-4000	27	4000 ± 10%	2.22	≤425	470	7.0	1800	3.00	93	
MILLIMETER E FERRITES	TT2-111	28	5000 ± 5%	2.11	≤200	375	6.0	1956	0.96	317	■
	TT1-4800	28	4800 ± 5%	2.01	≤240	400	-	3360	0.89	-	
	TT86-6000**	28	5000 ± 5%	2.11	≤200	363	6.0	3600	1.50	317	■

LITHIUM FERRITES Please consult the factory for information on Trans-Tech Lithium ferrite materials.

21 MISCELLANEOUS MAGNETIC MATERIAL APPLICATIONS

INTRODUCTION
In addition to the applications discussed in previous chapters, there are additional ones for ferrites that don't conveniently fit into any of the previously described categories. However, these miscellaneous uses still take advantage of a combination of properties somewhat unique to ferrites. These properties might include high resistivity, chemical inertness, temperature-coefficient of permeability, magnetostriction and economy of materials and manufacture. All most cases, it is the magnetic properties of the ferrite that are used to best advantage. In some instances, such in magnetomechanical uses, they compete with other electronic ceramic such as the piezoelectrics typified by the titanates.

Magnetostrictive Transducers
In designing ferrite materials for most of the previously discussed applications, we were looking for low or practically zero magnetostriction because this led to high permeability and low loss. However, there are cases when we may take advantage of magnetostriction. One such case occurs when we require transducers to convert electrical energy to oscillatory mechanical or acoustic motion. When a magnetic energy is the one being converted, these are called magnetostrictive devices or magneto mechanical devices. Some applications using these are ultrasonic cleaners, ultrasonic machining, ultrasonic delay lines, and devices for generating and detecting underwater sound (sonar) The latter may be used for detecting submarines, fish etc.

A high magnetostriction is a necessary material requirement for the component. In addition it should also possess high resistivity, high Curie point, and a high magneto-mechanical coupling factor. The latter refers to the coupling between the mechanical resonance of the component (related to the dimensions) and the electrical resonance. The quality factor for each of these resonances is given by the Q of the system. Both of these should be as high as possible.

Cobalt alloys & oxides have high magnetostrictions but require high power to saturate. Nickel and nickel alloys are often used because of there is lower power requirements. Ferrites are very useful at high frequencies (over MHz) but their

Table 21.1-Properties of Plastic-bonded Ferrite and other competing materials

Material Comparison			
	K 1	Carbonyl iron	Plastic ferrites *)
μ_i	80	15	11
$f_{max.}$ (MHz)	12	100	> 10
$H_{max.}$ (A/m)	500	> 10,000	5,000
B (mT)	360	1,600	—

ferrites are very stable in this respect. Pure nickel ferrite and nickel ferrite with a small amount of cobalt ($Ni_{.972}Co_{.028}Fe_2O_4$) have high magnetostrictions. Another material used for this purpose is a Cu substituted Ni ferrite ($Ni_{.42}Cu_{.49}Co_{.01}Fe_2O_4$).

Sensors
There are a wide variety of magnetic effects that can be used to detect changes in force, torque displacement, magnetic fields, acoustic, and displacement. Some of the more recent uses of magnetic materials as sensor have been to detect proximity, temperature and for electronic article surveillance (EAS). For instance, altering the position of a movable ferrite component may change the magnetic circuit (as changing the variable air gap) so that the inductance may be monitored in an LC circuit and denote proximity. Rotational changes can be accomplished in a similar manner with displacment being circular rather than linear. This type of application is used to detect the closing of water-tight doors on ships and doors in high-security areas. Strasser (1991) stated that proximity switches are a main application for ferrites in Europe. There is a need to increase the distance in the response. Siemens has designed appropriate shapes of cores which must also function in the presence of high DC currents. They have come out with a line of plastic-bonded ferrites. The properties of the plastic-bonded ferrites are given in Table 21.1 compared to other competing materials. K1 is a sintered ferrite material. TDK markets a product called Sensing Door Latches to sense magnetically if a door is open or shut. Here, the main element is a reed switch. A plate of ferromagnetic material is mounted on the stationary surface while a magnet in the housing attracts the plate holding the door closed. In addition the reed switch is activated to sense the closing.

The use of ferrites in determining when a certain temperature has been reached is an interesting application. While the actual temperature is not measured as an object is heated, a maximum predetermined temperature for either control, safety or equipment protection can be sensed by a ferrite device. Commercially, such devices are used in electric irons, soldering irons, heaters, motors, etc. Use is made of the ability to vary the Curie temperature of a ferrite by accurately varying the

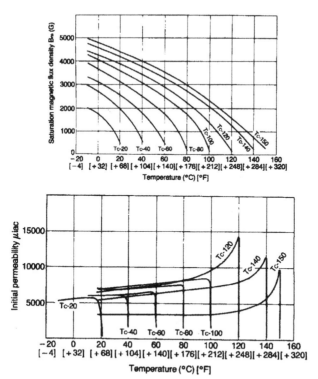

Figure 21.1-Magnetic characteristics of some thermal ferrites. The upper curve is the saturation magnetization versus temperature plots whereas the lower curve is that of the permeability versus temperature properties of the same thermal ferrites. From TDK

composition. In this manner, a whole array of different ferrites with moderately reproducible Curie points can be produced. Now, the permeability in the region of the Curie point drops sharply to 0 from a high value. Again, if the sensor ferrite is used in an inductor in an LC circuit, the inductive reactance of that circuit will decrease abruptly and this can be detected and the heat turned off. TDK markets a series of temperature responsive reed switches. These consist of temperature responsive ferrites, ferrite magnets and reed switches. The flux density of the temperature sensitive ferrite changes and the flux from the magnet can be controlled and the reed switch contact can be switched ON and OFF. The saturation magnetizations and permeabilities of thermal ferrites are shown in Figure 21.1

For the electronic surveillance application (also known as Theft Deterrence), this use has grown dramatically in recent with the increase in shoplifting. It consists of magnetic of tagging items such as clothing, books and other merchandise in stores and libraries and scanning the person leaving the premises through a magnetic detector. Large magnetic antenna dipoles near the exits are used. Magnetic markers perturb the spatial distribution of the magnetic field of the antenna but this is difficult to detect. More useful is the change of frequency of the exciting field by the marker. This is a result of harmonic generation by the magnetic tag so these tags are harmonic tags. Time perturbations of the exciting field can also be effected by a tag that returns a delayed signal, echo or response after the excitation field has been turned off. These tags are called resonant or magnetoelastic tags. The delay caused by the mixing of electric and acoustic waves is also discussed in the section on delay line following. A diagram of the harmonic and magnetoelestic tags is shown in Figure 21.2a. If a square loop material such as Permalloy 80 is used as the marker, the interaction of the excitation frequency and the square hysteresis loop will produce a induced voltage rich in harmonic content. Use of a frequency sensitive detector set for the harmonics produced will indicate the presence of the magnetic tag. In the case of the magnetoelastic tag using pulse excitation The presence of a signal after removal of excitation indicates the presence of the tag. The current and voltage wave forms for both cases are shown in Figure 21.2b. The comparison of the field dependence of the square loop and rounded loop materials is shown in Figure 21.3. The great non-linearity of the square loop material produces the high harmonic content needed in the harmonic tag. Permalloy 80 has been traditionally used for harmonic tags. Unfortunately it is quite sensitive to processing and handling and can be degraded by plastic deformation. Thus the tags (thin narrow strips) must be packaged securely and protected from handling stresses. Amorphous Metglas^{TM} alloys 2705M and 2826MB have higher yield strengths than the Permalloy and thus are less susceptible to degradation. Harmonic tags as described are always ON and thus the legitimate purchased article or checked book must bypass the scanning system. One method of removing the ON after legitimate acquisition is by biasing the soft material with a adjacent strip of permanent magnet material such as Vicalloy. Demagnetizing the Vicalloy(and Permalloy leaves the system ON and magnetizing both at the counter turns it OFF. The use of the bias strip and the resultant hysteresis loops are shown in Figure 21.3. Another family of harmonic tags have grown up around the Perminvar phenomenon described by Bozorth (1951) involving materials with wasp-waist hysteresis loops. If the alloy is annealed in a certain manner, the domain walls are pinned under weak fields until subjected to a threshold field, H_p, when they move abruptly causing a large flux change rich in harmonic content. In a large sample with many domain walls, there may be a number of H_p's. In a thin sample only one domain wall may be present so a single flux jump will occur.

In contrast to the harmonic tags, the magnetoelastic tags use the response that is delayed in time. The ring down of a ferroresonant tag is after the drive antenna is switches off. With coupling of the magnetic and elastic properties by magnetostriction changing the direction of magnetization produces a strain. The converse is also true. Since the acoustic wave travels slower, there is a delay of the signal in time. The energy stored in the system leaks back during the ring down period. The materials therefore must possess a high magnetostriction. Typical magnetostrictive tags measure a few centimeters in length and have resonant frequencies from 30-100 KHz. The frequency of the acoustic wave is directly related to the dimensions by ;

$$f = \sqrt{Y/\rho}/2\pi l$$

where l = length of sample
Y = Young's modulus
f = frequency
ρ = mass density

This is the frequency of a standing longitudinal wave of wavelength $2l$ in length of material. Atomic displacements are 0 in center and maximum at ends.

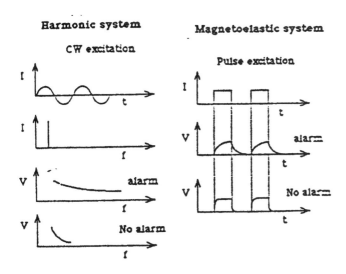

Figure 21.2a Current and voltage waveforms of the harmonic and magnetoelastic EAS detection systems with "alarm" and "no alarm" conditions. From O'Handley (1993)

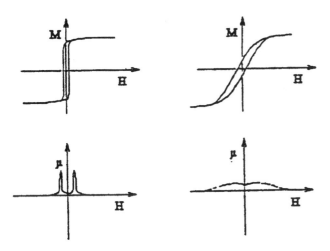

Figure 21.2b- Hysteresis loops and field dependences of permeability for a square and rounded loop material. Permeability is highly non-linear in the left case and nearly constant at right. From O'Handley 1994.

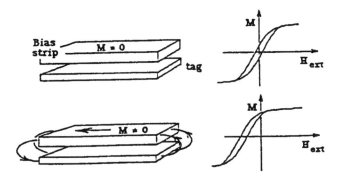

Figure 21.3-Schematic of a bias strip used in some tags. Right-M-H loops of tags with bias material or in its remanent state.

Copier Powders

In copier powders, the magnetic properties of the ferrite are necessary but not critical. Here it is just the ability to magnetically remove the carrier powder from the rotating drum that is required. However, the powder can also be the pigment and many simple copying powders use magnetite or variations of such to act as the toner (pigment) and the carrier. The materials for the single-component variety are thought to be substituted magnetites.

Ferrofluids

Ferrofluids are colloidal suspensions of magnetic particles which often are ferrites. Metal particles of that size might oxidize easily, which is not a problem in ferrites. The particles are so small that they never settle and therefore the suspension behaves as a liquid, Strictly speaking the particles are not ferromagnetic but super paramagnetic. They show no hysteresis but are attracted by a magnetic field. Ferrofluids are used in mechanical devices where the position of the "liquid" can be controlled by a magnetic field. Applications include gears, clutches, vents, etc.

Electrodes

In electroplating, the action of the electrolyte often corrodes the electrodes. Although high resistivity is usually required in a ferrite, for this application, the resistivity should be low or the conductivity high. Then if the ferrite electrode which has a large enough cross sectional area, it can carry enough current to perform the electrolysis. The resistance of the ferrite to corrosion makes it attractive in some applications.

MISCELLANEOUS MAGNETIC MATERIAL APPLICATIONS

Delay Lines

A variation of the magnetostrictive transducer is the delay line. This device is used to place a time delay in the transmission of an electrical signal or pulse. This is done by passing the signal into a coil that couples it to the magnetostrictive ferrite converting it into an acoustic wave. The acoustic wave is passed down the length of the rod and is then reconverted to an electrical signal by the reverse process. The velocity of the acoustic wave is many about 5 orders of magnitude slower than the electomagnetic wave so there is a time delay essentially equal to the time for the acoustic wave to traverse the length of the rod. This delay is given by;

$$t = l_m / V_a$$

Where
t = Time delay, sec.
V_A = Velocity of acoustic wave in ferrite, cm/s
l_m = length of rod, cm.

Thus, if the acoustic velocity is 5×10^5 cm/s and the length is 5 cm. the time delay is:

$$t = 5 \text{ cm} / (5 \times 10^5 \text{cm/s})$$
$$= 10 \times 10^{-6} \text{ or } 10 \text{ microsec.}$$

Ferrite Tiles for Anechoic Chambers

For proper testing of EMI characteristics (radiated or immunity) of a device or electronic equipment, an anechoic chamber is essential to prevent reflections that would invalidate the results. In the past traditional foam-type absorbers have been used that are quite bulky. Ferrite tile absorbers are relatively new and have come into use wherever high absorption (-15 to –25 dB at ,100 MHz.) and compact size (6mm versus 2400 mm for foam absorbers. Thereare now hundreds of installations worldwide in FCC certified chambers. Ferrites are immune to fire, humidity and chemicals providing a compact solution for attenuating plane wave reflections in shielded enclosures.

When an electromagnetic wave travelling through free space encounters a different medium, the wave will be relected, transmitted or absorbed. The relected wave is the one of interest. The thickness of the tile is tuned so that the phases of the reflected and exit wave cancel to form a reresonant condition. The resonant condition appears as a deep "nul" in the return loss response. The resonance is also a function of the frequency dependent electrical properties of the ferrite material such as the relative permeability (μ_r) and permittivity, (ε_r), which determine the reflection coefficient (Γ), impedance (Z), and return loss (RL. Ferrite tiles for anechoic chambers are usually nickel-zinc ferrites that have been optimized to produce consistent broadband absorption at frequencies down to 26 MHz. The properties of a ferrite material for tiles for anechoic chambers are given in Table 21.2. The permeability and permittivity for the material is shown in Figure 21.4. The return loss is given in Figure 21.5. The wide angle return loss is given in Figure 21.6 and the effect of gap between tiles on the reflectivity. Shown is the importance of precise machining of the tiles.

Physical Characteristics of 42 Material

Specific Gravity	5.2	
Young's Modulus	1.8×10^4	kgf/mm²
Tensile Strength	4.9	kgf/mm²
Compressive Strength	42	kgf/mm²
Flexural Strength	6	kgf/mm²
Vickers Hardness	740	
Coeff. of Thermal Expansion	9	10^{-6}/°C
Initial Permeability (relative)	2100	μ_r
Relative Permittivity	14	ε_r
Resistivity	5×10^6	ohm-cm
Curie Temperature	> 95	°C
Composition	Nickel-Zinc Ferrite	
Power Handling (CW)	400	V/m

Table 21.2- Properties of a ferrite anechoic chamber tile absorber material

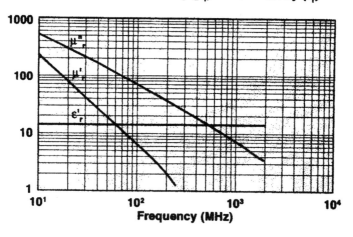

Figure 21.4- Permeability and Permittivity of a ferrite anechoic chamber tile absorber material. From Fair-Rite

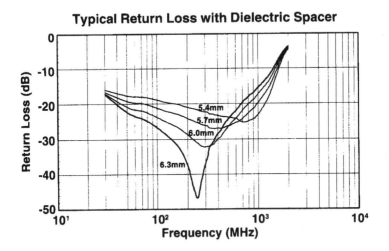

Figure 21.5-Typical return loss with dielectric spacer for a ferrite absorber tile material. Spacer thickness = 13mm. From Fair-Rite

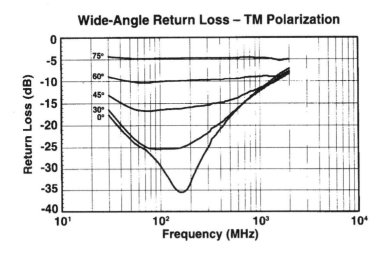

Figure 21.6 Wide angle return loss with TM Polarization for a ferrite tile absorber material. From Fair-Rite

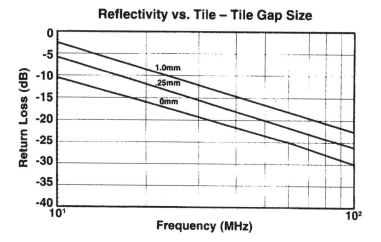

Figure 21.7 Effect of gap on the reflectivity in a ferrite tile absorber material. From Fair-Rite

Reed Switches

We have spoken of the soft and hard magnetic materials and the medium coercive force materials for inductive magnetic recording. There is another class called semi-hard materials that are mostly used in reed switches. These operate by the application of a magnetic field to close the contacts that are two rhodium plated metal reeds in an evacuated sealed glass tube. For their action they must be ductile. Some require one coil to produce the magnetic field to close the contact with the reverse current to release. Others have two coils, one for opening and one for closing. Others are self-latching. Many early materials for this application contained equal amounts of iron and cobalt with small additions of vanadium. Coercive forces for this material were about 29 oersteds. Fujitsu developed a new material for the purpose using about 85% Co 12% iron and 3% niobium called Nibcolloy. The coercive forces were about 20 oersteds. This material is self latching eliminating the need for maintaining the magnetic field. Special glass with matching thermal expansion similar to the metal must be used.

SUMMARY

We have listed some of the miscellaneous applications of magnetic materials. The next chapter relates to the physical and thermal characteristics primarily for ferrites. The final chapter will deal with the magnetic measurement techniques.

References

Fair-Rite Products (1996) Fair-Rite Soft Ferrites, 13th Edition, p.62
O'Handley, R.C. (1993) J.Mat. Eng. and Perf. 2(2) 211

22 PHYSICAL, MECHANICAL AND THERMAL ASPECTS OF FERRITES

INTRODUCTION

The physical and thermal characteristics of some of the metallic materials discussed were included in the relevant chapter especially in the case of permanent magnets. This chapter will deal with these properties in ferrites. While the magnetic and electrical properties are most important in ferrites, the applications will also require that certain other properties relating to their physical, mechanical or thermal condition be satisfied. Sometimes this means degrading the magnetic properties in a compromise between magnetic and other parameters. The ceramic nature of a ferrite makes it vulnerable to impact, thermal shock and tensile failure. A designer of ferrites must, therefore, be aware of these limitations as well as virtues of the material.

Densities of Ferrites

The densities of ferrites are significantly lower than thin metal counterparts so that a component of the same size would be lighter in a ferrite. However, because of the low saturation of ferrites, we have seen this advantage disappear. The densities of ferrites are given in Table 22.1 and compared with other magnetic materials. The densities listed are the X-ray densities, that is those calculated for single crystals or from the X-ray diffraction data assuming no porosity. In fact, these are not attained in polycrystalline materials where porosities from about 5-25% can be present. In actual practice, the materials with the highest permeabilities are the ones with the highest densities second, of course, to single crystals. Hot pressed materials also produce high densities. On the other hand, some of the high frequency, low-permeability materials have lower densities. While there are some differences in X-ray densities due to the difference in divalent ions present, a major contribution to effective density of a ferrite part is its porosity. A listing of densities of commercial materials with their densities is also given in Table 22.1.

Mechanical Properties of Ferrites

A previously-mentioned disadvantage of ferrites that must be considered is their low mechanical strength, particularly the tensile strength. They are, however, high in compressive strength. The strength is generally related the porosity with the lower strength present in the more porous materials. A listing of the various mechanical properties is given in Table 22.2. In some cases, the variation with porosity is given.

Table 22.1-Densities of Some Ferrites

Spinels

Ferrite	X-Ray Density, gm/cm^3
Zinc Ferrite	5.4
Cadmium	5.76
Copper	5.28
Cobalt	5.27
Magnesium	4.53
Manganese	4.87
Nickel	5.24
Lithium	4.75
Ferrous	5.24

Hexagonal

Ferrite	X-Ray Density, gm/cm^3
Barium	5.3
Strontium	5.12
Lead	5.62

Commercial	Measured Density, g/cm^3
MnZn(High perm)	4.9
MnZn(low perm)	4.5-4.6
MnZn(power)	4.8
NiZn(Recordindg Head)	5.3
MnZn (Recording Head)	4.7-4.75

Workability and Hardness of Ferrites

While metallic magnetic materials can be rolled to thin sheet, coiled and oher wise worked, ferrites cannot be worked in this manner. A great virtue of ferrites is that they can be cast into complex shapes and with control of firing shrinkage, can be made close to final size. Another convenient feature is that they can be ground and lapped easily and will take a fine finish. This becomes important in their application as a recording heads.

Another important feature of ferrites for recording head application is the high hardness, improving their wear resistance. The hardness values of some ferrites are shown in Table 22.2. Test procedures for measuring the hardness as it affects head wear have been proposed.

Break Strength

Many operation such as tumbling, grinding, winding, clamping, and handling put severe stresses on ferrites causing them to break or chip. A study by Johnson (1978) investigated the cause of lowered strength in some ferrites. He relates the weakness to an oxidized layer on the surface produced during the firing operation.

Table 22.2- Mechanical Properties of Ferrites

Mechanical Property	Value
Tensile Strength (general ferrites)	20 N/mmm^2
NiZn (5% porosity)	5 Kg/mm^2
NiZn (40% porosity)	2 Kg/mm^2
Compressive Strength (general ferrites)	100 N/mm^2
NiZn (5% porosity)	200 Kg/mm^2
NiZn (40% porosity)	10 Kg/mm^2
Modulus of Elasticity	15 x 10^4 N/mm^2
Youngs Modulus	80-150 N/mm^2
Hardness	6 (Moh's)
Vickers (HV)	8000 N/mm^2
Knoop	650
Recording Head, H_V-.06Kg,30s.	560-750

Thermal Properties

In common with other ceramics, the thermal conductivity of ferrites is rather low. This feature becomes quite important in its application in power transformers where the considerable heat generated is not lost easily and thus the center of a core will accumulate the heat and lead to lowering of saturation and possibly exceeding the Curie point. The thermal conductivities of ferrites are listed in Table 22.3. Attempts have been made to alter the thermal conductivities of ferrites. A recent paper (Hess 1985) suggests the use of CaO and NaO to raise the thermal conductivity.

Coefficient of Expansion

Another thermal factor to be considered is the coefficient of thermal expansion. The thermal expansions are similar to those of other ceramics.

In the manufacture of recording heads, the ferrite core is often assembled to form the gap using a glass bonding technique. In this case it is quite important to match the coefficient of expansion of the glass to that of the ferrite. The coefficients of expansion for several ferrites are given in Table 22.3.

Specific Heat

Ferromagnetic materials have greater specific heats than non- ferromagnetic materials. Part of the energy that is needed to align the electron spins is an additional contribution the normal vibrational and valence electron specific heats.

The specific heat increases greatly at the Curie point where the orientation against thermal energy is greatest. This gives further credence to the theory of spin

Table 22.3-Thermal Properties of Ferrites

Thermal Property	Value
Thermal Conductivity	$10\text{-}15 \times 10^{-3}$ cal/sec/cm/°C.
	4 W/m/°C.
	4.7×10^{-3} J/mm^2/s/°C.
Coefficient of Expansion	$7\text{-}10 \times 10^{-6}$/°C.
Recording Head	
MnZn RT to 200°C.	11 to 13×10^{-6}(+ slope)
" 200-600°C.	13 to 10×10^{-6}(- slope)
NiZn (RT)	8×10^{-6}/°C.
NiZn (800°C.)	10×10^{-6}/°C.
Specific Heat	700-1100 J/kg/°K
Melting Points	
Barium Ferrite	1390°C.
Cadmium "	"1540°C.
Cobalt "	1570°C.
Copper "	1560°C.
Magnesium "	1760°C.
Manganese "	1570°C.
Nickel "	1660°C.
Lead "	1530°C.
Zinc "	1590°C.

clusters between ferromagnetic and paramagnetic behavior. The specific heats of some ferrites are given in Table 22.3.

Thermal Shock Resistance

In common with most other ceramics, ferrites have poor thermal shock resistance. Unfortunately, the higher the density, the more prone ferrites are to cracking by this mechanism. During the sintering process, fast cooling is conducive to thermal shock cracking, which may not always be visible from the exterior. One of the factor that determine thermal shock resistance is thermal conductivity which, as we have already seen, is low in ferrites.

Melting Points

The melting points of ferrites are difficult to measure because of their loss of oxygen at high temperatures. However, Van Arkel (1936) measured them with an oxy-hydrogen flame. Table 22.3 lists them.

Pressure Effects on Ferrites

Tanaka (1975) measured the permeability of MnZn ferrites at pressure up to 2000 kg/cm^2. At low pressures, u increases with pressure when anisotropy constant $K_1 < O$ or decreases when $K_1 > 0$. Above 1000 kg/cm^3, the rate of decrease goes down with decreasing values of the magnetostriction or with decreasing oxygen content. LeFloc'h (1981) measured the effect of pressure on the magnetization mechanism in ferrites in which the hydrostatic pressure was lateral and perpendicular to the plane surface. He explains the differences based on changes in domain topography. They are due to the unbalanced stresses induced in these materials at the grain boundaries by the existence of a closed porosity. Loaec (1975) found that in nickel ferrite, the susceptibility decreased with pressure. There is evidently a pressure induced hysteresis.

In addition to the applied hydrostatic pressure introduced externally, there are also pressure variations which can be introduced by such things as polymer encapsulation. The shape of the hysteresis loop can be drastically changed by such procedures. These effects can be related to the stresses set up by pressure variations.

Effects of Machining and Grinding of Ferrites

Snelling (1974) performed studies on the effect of tensile and compressive stresses or Mn-Zn ferrites. The measurements were made on a manganese-zinc ferrites of the composition Mn_{52}, Zn_{40}, $Fe_{2.08}O_4$, and one with Ti addtion (Stijntjes 1971) of compositon $Mn_{.64}Zn_{.30}Ti_{.05}Fe_{2.01}O_4$. The values of the permeabilities at 5Hz. and residual loss factors(including Eddy current losses) at 100 KHz and 2.5 gausses are given in Figure 22.1. The differences were explained by the variation in K_1. There is a temperature, at which K_1 goes through zero known as the compensation temperature, θ_o. Below this, K_1 is negative,(<111> is the preferred direction). Above the temperature, K_1 is small and positive and <100> is the preferred direction. The position of this peak is dependent on the compostion and the sintering conditions. The magnetostriction λ_{111} is positive and that of λ_{100} is negative. Therefore, above θ_o, a small tensile stress will reduce the permeability and a small compressive stress will increase it. The residual loss factors that usually have minima in the positions where μ is a maximum show this pattern in the stress free sample. In the Ti-substituted ferrites, the residual loss factors are stress-dependent. In general, stress of either kind increases the loss factor. If we want to look at losses of the Ti-substituted case separated from the permeability, we may look at the imaginary part of the complex permeability u" which relate to the losses. Here, the pattern shown is the same as that of the real permeability. At higher temperatures, tension decreases $\mu"$ and compression increases it. As low temperatures, the situation is the reverse.

Knowles (1975) examined the increase in magnetic loss due to machining of ferrites such as surface grinding or center post reaming. He divided the situation into longitudinal effects, those in which the ground surface is parallel to the flux path and those in which the are perpendicular. He had previously shown that grinding the longitudinal surfaces caused a change in the μ vs T curve but did not

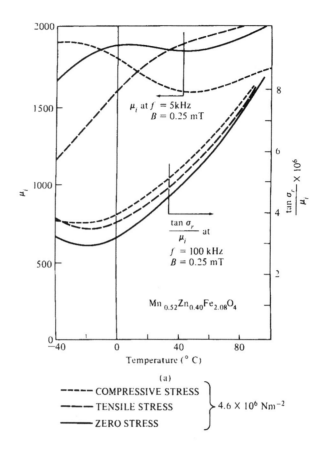

Figure 22.1 - The initial permeabilities at 5 KHz and the Loss Factors at 100 KHz of a MnZn ferrous ferrite showing the variation with the application of tensile or compressive stresses of 4.6 x 10^6 N/m². From Snelling, E., IEEE Trans. Mag. MAG 10, Sept.,1974, p.616.

change the loss factor. However, grinding the transverse surface increased the loss factor by as much as 33%.

The ground surface causes permanent damage to the surface layer several microns deep causing compressive stresses in the layer. This is the primary stress but smaller secondary stresses are generated in the interior and may depend on the shape and size of sample.

To examine stress profile a flat plate 15 x 15 x 2 mm was made from a MnZn ferrite of $Mn_{.64}Zn_{.30}Ti_{.05}O_4$ and both sides of the plate was ground with a diamond wheel. One side was polished with Syton that gives a stress-free surface. When released from the substrate, the slice took on a curvature showing that the ground side was in compression. The slice was the progressively etched with concentrated hydrochloric acid. The original curvature was removed. Even though the damaged layer is small, there is a disproportionate increase in the material loss factor which, in turn, causes an increase in the loss factor of the component. In layers parallel to the flux, the flux density is decreased so that the decrease in loss factor is small.

PHYSICAL-THERMAL ASPECTS OF MAGNETIC MATERIALS

Encapsulation of Ferrite Cores

From the above considerations one would certainly predict that putting a ferrite core in compression by coating it would change the magnetic properties. Dramatic changes in the B-H loops are produced when a ferrite is surrounded with a glass ring fitted around the toroid and heated to a high temperature. On cooling, the glass will put the tangential compressive stress. When this is done,the loop becomes quite rectangular. Van de Poel (1981) examined the effect of various varnishes used in vacuum impregnation of wound ferrite components from the point of view of stress effects. Parts with small thicknesses and porous ferrite materials showed the largest effects. Prevention of penetration of the varnish into the ferrite material reduces the effect. Where feasible, a wound winding is preferable to a wound ferrite core.

The author of this book has found that using a "pillow" of soft foam between a core (toroid) and the potting material is useful in reducing the adverse effects.

Shock and Impact

As might be expected the properties relative to shock and impact in ferrites are inferior to those of metals. The Charpy impact strengths are only about .032 ft lb. but obviously, this will vary greatly according to the type of ferrites. The brittleness of ferrites is certainly a factor where stability is required after a large impact or vibration is experienced. This holds true for military aircraft and space applications.

Moisture and Corrosion Resistance

Unlike some metals, ferrites are resistant to moisture and salt water corrosion. They are, however, attacked by stong acids. In the more porous low-permeability ferrites used for high-frequency inductors, moisture may be absorbed into the ferrite and cause increased losses.

Radiation Resistance

Siemens Catalog (1986-7) list the exposures to the following radiation which can be encountered without significant variation in inductance in ungapped ferrite cores. ($\Delta L/L < 1\%$)

Gamma quanta	10^9 rad
Quick neutrons	2×10^{20} neutrons/m^2
Thermal neutrons	2×10^{22} neutrons/m^2

SUMMARY

This chapter has dealt with the physical and thermal characteristics of ferrites. The next and last chapter will deal with methods of making making magnetic measurements primarily on ferrites.

References

Hess, J.,(1978) and Zenger, M., Advances in Ceramics, Vol. 16, 501

Johnson. D.W.,(1978), Processing of Crystalline Ceramics, Plenum Press, New York, 381

Knowles, J.E.,(1975), IEEE Trans Mag. MAG 11,(#1),Jan.,1974, p. 44

Le Floc'h, M.(1981), Loaec, J. Pascard, H., and Globus, A., IEEE Trans Mag.,MAG 17, (#6), 3129

Loaec, J, (1975), Globus, A., Le Floc'h, M. and Johannin, P IEEE Trans. Mag., MAG 11, (#5), Sept.1975, 1320

Siemens Catalog,(1986/7)Siemens AG, Munich 80, Germany

Snelling, E. (1974), IEEE Trans. Mag. MAG 10,Sept,1974, 616

Stijntjes, T.G.W.,(1971), Ferrites, Proc. ICF1, University Press, Tokyo, 191

Takada, T.,(1975) Jap. J. Appl. Phys. 14, 1169

Van Arkel, A.E.,(1936), Rec. trav. chim., 55,331

23 MAGNETIC MEASUREMENTS ON MATERIALS AND COMPONENTS

INTRODUCTION

Although this chapter deals mainly with magnetic measurements on ferrites, the same techniques with some adjustment for frequency may be used for other materials. Proper evaluation of ferrite materials and components requires the use of many different magnetic and electrical measurement techniques. In some cases, D.C. or low frequency methods are used but since ferrites are primarily high frequency materials, measurements at the higher frequencies are more common. While some of the requirements for ferrites are similar to those used for other magnetic materials, many such as disaccomodation are unique to ferrites. In addition, the users of ferrites have put together a combination of what would seem to the ferrite producers a very large number of requirements which must be met in the same material or on the same component. For example, for soft magnetic ferrite, Table 23.1 gives a listing of some of the requirements that may exist for a particular core. To add to the number of measurements to be made is the fact that, due to variations, known or unknown, in raw materials, processing conditions and firing conditions, ferrite cores depend on frequent magnetic in-process testing to provide process feedback. There are trade associations and standards groups that provide assistance in standardization of ferrite measurements. Foremost is the TC51 Committee of the IEC (International Electrotechnical Commission) (IEC,1989) headquartered in Switzerland. Lists of documents of the IEC and ASTM for metallic and ferrite materials appear in Table 32.2 at the end of this chapter. In addition, there is a very helpful User's Guide to Soft Ferrites, MMPA-SFG96 (MMPA 1996) published by the Magnetic Materials Producers Association.

MEASUREMENT OF MAGNETIC FIELD STRENGTH

Before undertaking the measurement of many of the magnetic parameters of ferrites, it is essential to be able to measure the applied magnetic field. This can be subdivided into the D.C. and ac fields. Some of the measurement techniques for the two are different.

Table 23.1 - Soft Ferrite Specifications
1. Initial Permeability, μ_o
2. Incremental Permeability, μ_Δ
3. Loss Factor, $1/\mu Q$
4. Temperature Factor, TF
5. Disaccomodation Factor, DF
6. Watt Loss, P_e vs T and f
7. μ vs B
8. μ vs DC bias
9. B_s at RT and high T
10. Pulse Permeability, μ_P
11. Quality Factor, Q
12. Hysteresis Constant, h_{10}
13. Coercive Force, H_c
14. Squareness, B_r/B_s

Measurement of D.C. Fields

The traditional method for measuring magnetic fields until recent times involved the measurement of the magnetic flux change and the corresponding induced voltage in a search coil as the magnetic field was applied. As we will show later, this is the same technique used to measure the magnetization since it, too, will produce a similar flux change. This method is called the Fluxmeter Method or in the case of the magnetic field, the instrument is a Gaussmeter. The old method of changing the magnetic field seen by the search coil was to mechanically pull it out of the field or to flip the coil 180° to get twice the effect. The emf produces a charge (current-time integral) depending on the resistance in the system. This is detected in a ballistic galvanometer circuit. This ballistic method is shown in Figure 23.1. It is quite simple and can often serve as a referee method. Another method using the fluxmeter technique is called the Rotating Coil Gaussmeter in which the rotating coil transverse to the field cuts lines of flux which produces a voltage proportional to the field. The mechanical method of flux change has been replaced with the Electronic Integrating Fluxmeter which is much faster. This method will be discussed under the section on Measurement of Magnetization.

There are several new methods of measuring magnetic field strength which do not involve the flux method. One of these is the Hall Probe method and is probably the most widely used for field measurement at present (Figure 23.2). A Hall probe is a semiconductor plate that, when a current is flowing along one axis, produces a voltage tranverse to the current when a magnetic field is applied perpendicular to the plane of voltage and current .It is convenient in that no coil movement is necessary and it is a constant-field direct reading device. Another very accurate method of measuring a D.C. field is with the use of an N.M.R. (Nuclear Magnetic Resonance) Gaussmeter.

The measurement of a.c. fields is normally done by measuring the induced a.c. voltage in a coil with a knowledge of the frequency and cross-sectional area and number of turns of the coil

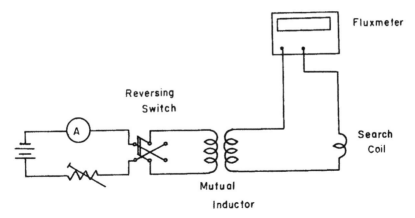

Figure 23.1- Ballistic method of measuring flux. Figure from Magnetic Measurement, J.M.Janicke, IEEE Magnetics Workshop, Marquette University, June,1975

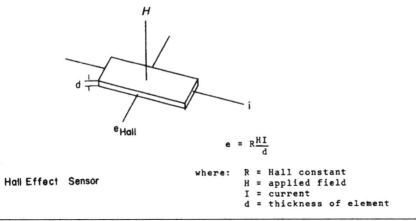

$$e = R\frac{HI}{d}$$

where: R = Hall constant
H = applied field
I = current
d = thickness of element

Figure 23.2- Hall Probe for measuring magnetic field. Figure from Magnetic Measurement, J.M. Janicke, IEEE Magnetics Workshop, Marquette University, June, 1975.

Measurement of Magnetization

There are many diffent methods of measuring the magnetization. For materials evaluation, the quantity of interest is the saturation magnetization, M_s. Usually, this is done on powders, very small samples or thin films. In the case of powders or irregularly shaped samples, the magnetization or rather the magnetic moment is given per weight or per gm. The commonly used is called σ, or emu/gm which then makes it a fundamental unit independent of the actual density of the sample. As

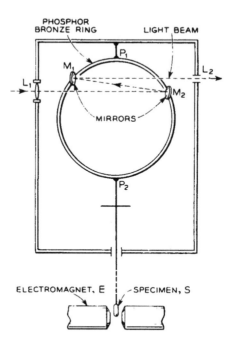

Figure 23.3- Sucksmith Magnetic Balance- from Bozorth,R.M., Ferromagnetism, D.Van Nostrand Co., New York, (1951) 859

shown in Chapter 3, the σ value can be converted to the B value from the theoretical density of the material.

One of the earliest methods of measuring M_s was by a Magnetic Balance Method. This method had been used for measuring the susceptibilities of paramagnetic and diamagnetic substances and can measure small samples of ferromagnetic substances. One such instrument developed by Sucksmith (1938-1939) is shown in Figure 23.3. Magnetic balances measure the force exerted on a sample under the influence of the field gradient of a permanent or electromagnet. If the gradient of field to vertical distance, dH/dy, can be made constant, the downward force is;

$$F = -M_s V \, dH/dy \qquad [23.1]$$

The difference in weight before and after application of the saturating field when multiplied by the acceleration of gravity(980 dynes/cm) allows determination of M_s. Another magnetic balance method of measuring M_s is through the use of a pendulum magnetometer (Rathenau and Snoek, 1946). It is shown in Figure 23.4. Here, the period of oscillation of a pendulum holding the sample at its end

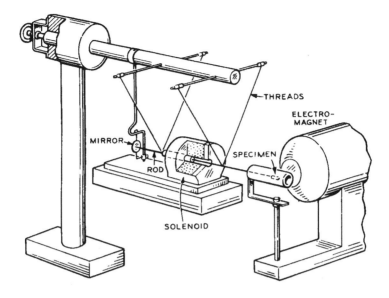

Figure 23.4- Pendulum Magnetometer- From Bozorth, R.M.,Ferromagnetism, D.Van Nostrand Co., New York,(1951), 860

in the shaped magnetic field of an electromagnet is altered by the interaction of the magnetic moment of the sample and the field. From the change in period , the magnetization can be calculated. The magnetization can also be measured using microwave resonance techniques.

We have spoken of the measurement of fields by the use of fluxmeter techniques. The measurement of the magnetization can be made in a similar manner. In this case, the flux is a combination of H and M lines. To calculate the magnetization, the H value must be subtracted from the total induction, B or ϕ/A.

$$M = (B-H)/4\pi \qquad [23.2]$$

In soft ferrites, the 4π M is practically equal to B but in hard ferrites, the subtraction must be made by measuring the H field in the same or similar coil but without the ferrite sample. The sample is preferably in the form of a toroid so that there are no demagnetizing fields which would require higher magnetizing currents to saturate if a winding was used to produce the field. However, it is also possible to use bars or cylinders of constant cross section magnetized between the pole tips of an electromagnet. The applied field in either case should be can be increased point by point until it is high enough to saturate the sample. The measurement can be made ballistically as discussed in the section on field measurement. The integrated induced voltage can also be measured by increasing the field manually until the magnetization curve flattens out. As previously mentioned, the preferred method is

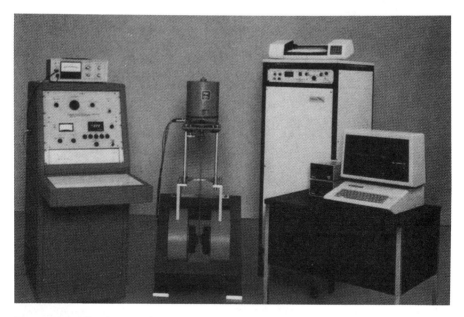

Figure 23.5- Vibrating sample magnetometer, Photo Courtesy of EG&G PARC

with the use of a electronic integrating fluxmeter. A solenoidal field can also be used for longer specimens.

Another widely used method which uses the fluxmeter approach to measure magnetization is through the use of the Vibrating Sample Magnetometer developed by Foner(1959). This instrument is capable of measuring the magnetization of powder, small samples and thin films. Although the sample is magnetized by a D.C. field, the pickup is accomplished through the action of the sample vibrating inside of the coil, thus its lines of flux cut the windings of the coil producing a voltage. It is really mainly used for small samples. An example of such a device is shown in Figure 23.5.

Magnetization Curves and Hysteresis Loops

The magnetization curves of a material can be measured by either D.C. or a.c. means. The D.C. methods use fluxmeter techniques, either ballistically point by point or by electronic integrators in a continuous mode. The H or field strength is usually measured by a Hall Probe Gaussmeter and connected to the X axis of an X-Y plotter or to the X axis of an oscilloscope. The B value is obtained by a search coil wound on the toroid or solenoidally around a bar or cylinder and sent through an electronic fluxmeter. The induced voltage is connected to the Y axis of the plotter or ocilloscope. When the instrument is arranged (usually by computer) to increase the field automatically, reverse at saturation and continue to complete the B-H loop, the device is called an automatic hysteresigraph. An example of such an

MAGNETIC MEASUREMENTS- MATERIALS- COMPONENTS

instrument is shown in Figure 23.6. Another means of displaying the B-H loop is through the use of a.c. driveof a preset voltage preferably on a toroid. The input a.c.

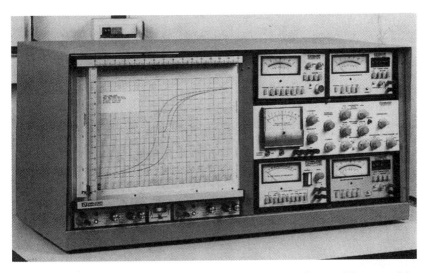

Figure 23.6-Magnetic hysteresigraph- Photo Courtesy of O.S.Walker Co.,Worcester, Mass.

current determining the applied H is sent through a 1 ohm resistor so it reads directly in volts and is connected to the X- axis of the oscilloscope. The induced secondary is sent through an RC integrator and the integrated voltage which is proportional to B is connected to the Y-axis of the oscilloscope. If phase relationships are considered, the output will be the B-H loop displayed on the oscilloscope screen. With calibration, the B_s and B_r can be determined. Such an oscilloscope-based hysteresis loop tracer is shown in Figure 23.7. The RC time constant or 1/f is;

$$\tau = 1/f = RCE_p \quad [23.3]$$

The H value in Oersteds can be determined for each point from the input current, number of turns in the coil and the effective magnetic path length. The B value can be calibrated from the frequency, induced secondary voltage, number of turns and the cross sectional area.

Measurements on Hard Ferrites

As we have mentioned previously in Chapter 8, the permanent magnet properties of hard ferrites are concerned with the second quadrant or demagnetization curve of the hysteresis loop. Because high fields are normally needed to saturate these hard ferrites, the samples which are rods or cylinders are usually placed between the pole tips of an electromagnet. There are now automatic B-H measuring instruments that use flux integration for B and Hall probes for H to display the demagnetization curve. For special shapes, such as arc segments for motors, there are special yokes to accomodate the shapes. Figure 23.8 shows such a B-H plotter for hard ferrites.

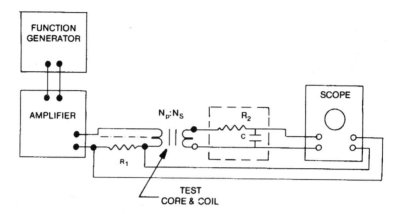

Figure 23.7- Schematic circuit for displaying hysteresis loop on oscilloscope, From MMPA SFG-96, Soft Ferrites, User's Guide (1996),33

Magnetocrystalline Anisotropy

There are several way in which this measurement can be made. The measurements are usually made on single crystals of on a grain-oriented or textured materials which are not common in ferrites. The measurement is made with a knowledge of the direction of the crystallographic axes. This is usually determined by X-Ray Diffraction techniques. Once determined, there are several different method of anisotropy measurements.

1. Measurement of the Magnetization Curves in different crystallographic directions, that is the flux direction is made to coincide with one of the directions. The results are curves similar to that in Figure 23.9. The difference in energy or the area between the curves for the easy and hard directions represents the anisotropy energy.

2. Torque Curves. Here, use is made of an instrument called a torque magnetometer. The sample is aligned with a specific crystallographic direction in a plane in the middle of and parallel to the D.C. field of an elelectromagnet. The sample is rotated on this plane so that during the revolution, the easy or hard directions will be alternately parallel and perpendicular to this direction. The interaction of the magnetization and the field will create differences in the amount of torque encountered. The shaft rotating the sample can detect and measure the value of this torque. A typical torque curve of a CoZn single crystal ferrite is shown in Figure 23.10. By analysis of these torque curves, the magnetocrystalline anisotropy can be determined.

Figure 23.8- B-H Plotter for hard and soft magnetic materials, Photo, Courtesy of LDJ Inc., Troy, Michigan.

3. Anisotropy can also be measure by microwave techniques. The method is beyond the scope of this book

Magnetostriction
Most Magnetic measurements on Magnetostriction are made today through the use of strain gauges, which are sensitive devices for measuring strain by the increase in resistance. This technique was developed by Goldman(1947) and is most effectively done on single crystals to determine the anisotropy of magnetostriction, in other words, how the magnetostriction depends on crystallographic direction. There had been attempts to measure the Joule magnetostriction as it is called, involving the use of levers and mirrors to amplify the displacement on magnetization. In one case, the displacement changed the capacitance of a capacitor or changed the mutual inductance of two coils. By changing resonant frequencies, this method was made quantitative.In the strain gage technique, a wire is folded back on itself many times and the resistance change is measured on a Wheatstone bridge. Neither high or low temperature properties can be measured. A magnetostriction measurement circuit is shown in Figure 23.11.

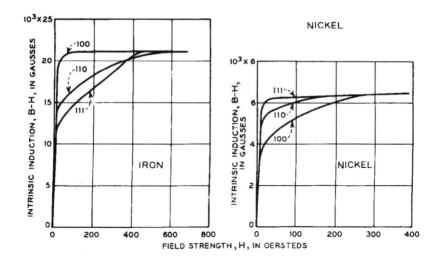

Figure 23.9- Magnetization curves for different directions in a single crystal. From Bozorth, R.M., Ferromagnetism, D.Van Nostrand Co.,New York(1951), 478

Curie Point

The Curie point of a material can be measured very simply and conveniently by noting at which temperature a peice of magnetic material will drop off of a permanent magnet. The magnet must maintain sufficient field at the Curie point of the test material. The method is not too exact especially for weakly magnetic materials. A more accurate method is to note the temperature at which the inductance (See later measurement techniques) drops to zero.The last portion of this curve is non-linear so normally, the linear portion is extrapolated to zero permeability and the temperature noted. Thermogravimetric Analysis or TGA has recently become a popular method of detecting Curie points as thermal changes also occur at the Curie point. Perkin-Elmer has a commercial model designed specifically for this application.

Structure Sensitive Properties

The measurements discussed thus far have been primarily to determine the intrinsic or microstructure-insensitive properties depending only on chemistry and crystal structure. There are many other important factors such as those listed below that depend additionally on things such as microstructure. In most cases, the material properties are measured on components where the dimensions are used in the calculations. Therefore, the component measurement methods will be considered concurrently.

MAGNETIC MEASUREMENTS- MATERIALS- COMPONENTS

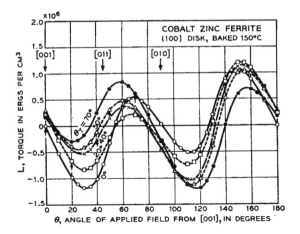

Figure 23.10 - Torque curves for measuring magnetic anisotropy, from Bozorth, R.M., Tilden, E.F., and Williams, A.J., Bell Telephone System Monograph 2513, (1954)

Inductance and Permeability Measurements

Permeability may be measured several ways but in most cases it is the a.c. permeability that is of interest. D.C. permeability can be measured from the B and H values from ballistic or fluxmeter methods. For a.c. methods a useful method which is useful at low frequencies(not too useful for ferrites) is the simple "lumped parameters" E-I method. This is an a.c. method where the H is measured from the primary current and the B is measured from the secondary voltage. Here, the inductance is considered a pure inductance with no resistive losses and therefore no impedance separation is made. Bridge methods are the methods of choice in measuring the permeability and loss factor, which are material properties, since the high frequencies normally used with ferrites increase the signal voltages. An a.c. impedance bridge has been the method of choice for high frequency inductance measurements on ferrites, which is the usual way of measuring permeability. In any type of bridge, the two measurements obtained are L_s and R_s. Note that in this case, in contrast to the E-I method, the resistive and inductive values are separated. The complex permeability can be derived from these two quantities.

$$Z = R_s + j\omega L_s \qquad [23.4]$$
$$R_s = (N^2 A/D) \times 10^{-8} \text{ Ohms} \qquad [23.5]$$
$$L_s = (N^2 A/D) \times 10^{-8} \text{ Henries} \qquad [23.6]$$

The equations can be solved simultaneously for the permeability of a toroid;

$$L = .4\pi N^2 \, A/l_e \qquad [23.7]$$

where A = Cross sectional area of the toroid
and l_e = Effective magnetic path length of the toroid

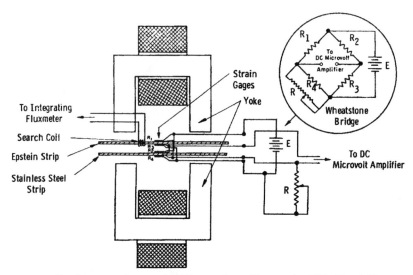

−Circuit for measuring magnetostriction showing exciting magnet and Wheatstone bridge

Figure 23.11- Circuit for measuring magnetostriction using the strain gage technique.

while tan δ can be derived from;

$$\tan \delta = \mu''/\mu' = R_s/\omega L_s = 1/Q = R_s/X_L \quad [23.8]$$
$$\text{Loss factor} = \tan \delta /\mu = R_s/\mu\omega L_s = 1/\mu Q = R_s/\mu X_L \quad [23.9]$$

The method of calculation of the effective length of a toroid as well as the other effective parameters when the sample is not a toroid will be discussed later on the section of inductance measurements on components.

The Maxwell Bridge-(Figure 23.12) employs a capacitor as the standard which has advantages in that it is small, easy to shield and has no external field. The Maxwell Bridge may experience difficulty when the Q is high.

The Hay Bridge- (Figure 23.13) -High Q's can be measured conveniently. There is overlap but if the coil Q is higher, the Hay Bridge is preferred.

The Owen Bridge-(Figure 23.14)-is another type of bridge used in inductance measurement. It has both adjustable elements on the same arm. This makes the reactive adjustment independent of the resistive adjustment, thus avoiding the interlocking action.

Most of the bridges described above are quite old and of course are of the analog variety using manual operation. With the advent of digital electronics, new digital bridges have been developed which lend themselves to computerized programming, data aquisition, calculation and plotting. Several versions of the

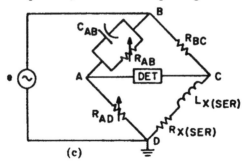

Figure 23.12-Schematic Diagram of a Maxwell inductance bridge, from IEEE Standard 393-1977, 26

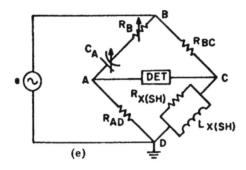

Figure 23.13-Schematic Diagram of a Hay inductance bridge, from IEEE Standard 393-1977, 26

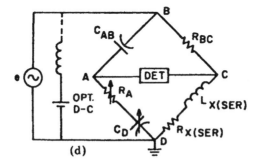

Figure 23.14- Schematic Diagram of an Owen inductance bridge, from IEEE Standard 393-1977, 26

digital LCR meters, impedance analyzers and network analyzers are available. Although the manufacturers of these instruments do not feel that these are bridges and don't usually call them bridges, many people concerned with these measurements still call them "digital bridges". Balancing the bridges has been taken over by very accurate digital techniques and the impedance standards are usually very precise resistors. The schematic circuit of one such instrument is shown in Figure 23.15. Several are shown in Figures 23.16 and 23.17. General Radio makes a Mini-bridge which is microprocessor based and features automatic balancing thereby eliminates manual operation. They can be interfaced with a computer to program the various sequences of flux densities and frequencies as well as temperatures. The information can then be stored on floppy disks, recovered later in the appropriate order and plotted. They all have BCD outputs for IEEE busses and can be combined with other accessory devices such as temperature cabinets, D.C. biases and such. The time-saving features of these instruments are dramatic and only in special cases, would manual bridges be required.

We have not previously mentioned the possible need for magnetic conditioning (demagnetization) of cores before measurement. This is often required since the permeability among other things may be sensitive to past magnetic history. For accurate μ_o readings, the core should be demagnetized to start out at the origin (zero H and zero induction). This can be done by several means, the preferred method being that of damped oscillations (alternating feld gradually reduced to zero amplitude). Heating the core above the Curie point will also demagnetize. In this case, disaccomodation must be considered. (See later)

Loss Factor

Much of the same equipment and procedures used for permeability can be used for measuring low field losses. On the bridges described which measure R_s, series resistance, the data can be expressed as the permeability and loss factor, tan δ/u or the two parts of the complex permeability, μ' and μ''. Still another method of measuring loss at low levels is the measurement of Q which is tied in with the loss factor (L.F. = $1/\mu Q$)

Q Factor

Q can be determined from the permeability and loss factor as shown above but frequently, if only Q is needed, there are Q meters specifically for this purpose. See Figure 23.18. In addition to the manually operated Q meters, there are also digital Q meters. External oscillators are necessary if frequencies outside of the range are required. Usually, in the ferrite suppliers' catalog, Q values are plotted versus frequency.

High Frequency Measurements

When high frequencies are involved, special precautions must be followed to have meaningful results. For example, capacitive effects may cloud the inductive effects. To minimize this difficulty, very short leads are suggested.

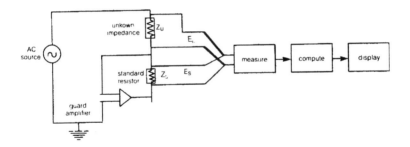

Figure 23.15 - Schematic circuit of a Wayne -Kerr inductance analyzer-From Wayne Kerr Instrument Catalog 002 (1989), Wayne Kerr, Inc. Woburn, MA.

In addition, to minimize the wire losses at these high frequencies braided Litz wire should be used.

Permeability Measurements on Ferrite Components

Publication 367-1(1982) from the IEC(IEC 1982) gives many of the precautions and procedures needed for accurate permeability measurements. This is especially needed in non-toroidal cores such as those composed of several sections; pot cores, E-cores etc. In the generalized calculation of permeability from the inductance, the equation becomes

$$L = .4\pi N^2 \mu A_e/l_e \quad [23.10]$$

This means that in a magnetic circuit composed of different sections, the individual $\mu A/l$ for each section must be considered. If the μ is considered constant, the summation is over the A/l's. The IEC approved method does this by defining three constants related to these summations. There is defined;

Effective area, $A_e = C_1/C_2$ (cm^2)   [23.11]

Effective Length, $l_e = C_1 2/C_2$ (cm)   [23.12]

Effective Volume, $V_e = C_1 3/C_2 2$ (cm^3) [23.13]

For a toroid;

$$C_1 = 2\pi / h \log_e(r_2/r_1) \quad [23.14]$$

$$C_2 = 2\pi /h^2 \log_e 3(r_2/r_1) \quad [23.15]$$

The calculations involved in determining the constants C_1 and C_2 for the whole series of IEC cores can be found in the documents relating to these cores. A listing

604 HANDBOOK OF MODERN FERROMAGNETIC MATERIALS

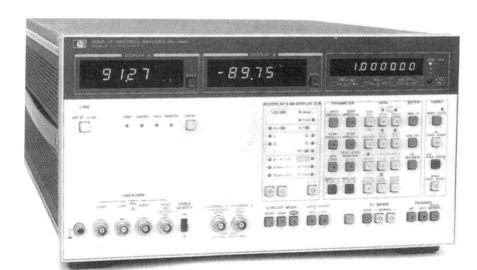

Figure 23.16- Hewlett-Packard impedance analyzer, from Hewlett-Packard Test and Measurement Catalog(1989) Hewlett-Packard Co.,Palo Alto, CA

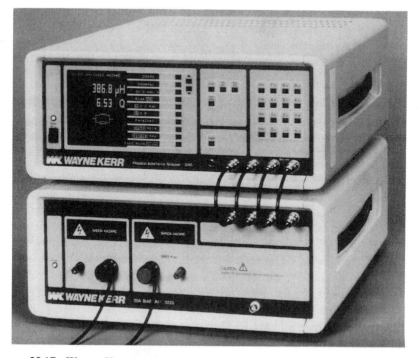

Figure 23.17- Wayne Kerr precision inductance analyzer and 20A. Bias Unit. Photo Courtesy of Wayne Kerr Inc., Woburn, MA

MAGNETIC MEASUREMENTS- MATERIALS- COMPONENTS 605

Figure 23.18- Boonton Model 260A Q Meter, Boonton Radio Corp. ,Boonton, NJ

of these documents is found in Table 23.2 at the end of this chapter. While Standard 367-1 and its attachments do not go into detail on the equipment or specific procedure, conditions for the test are standardized. One important example is the design of measuring coils for new types of cores such as RM cores.

Pot Core Adjuster Variation

To achieve a very precise inductance in an a.c. circuit, a screw-type adjuster is used. The component design engineer must know the range of the particular adjustor as well as the linearity. This can be done by measuring the inductance variation as the screw is successively rotated into the pot core gap. Again IEC standard 367-1(1982) gives guidelines for this measurement.

Permeability Variations with Temperature

The techniques of measuring permeability described above can be used in studying the variation of permeability with temperature. There are especially designed temperature cabinets for this purpose where the temperature can be adjusted manually or automatically by computer and thermostat. IEC Publication 367-1(1982) discusses the general techniques that can be used. With the computerized measurement of permeability described above, the temperature is one of the variables which can be set by the computer. When only the percent variation of permeability with temperature is needed, a simple method can be used. One such method involves a circuit using a very stable capacitor in a resonant LC circuit. The circuit is called a free-running oscillator which will automatically oscillate at the resonant frequency determined by the L and C of the circuit. The relative change in inductance as determined by the relative change in frequency is given by ;

$$\Delta L/L_0 = -2\Delta f/f_0 \quad [23.16]$$

where ΔL = Change in inductance indicated by
Δf = change in resonant frequency
and L_0 = Reference inductance associated with
f_0 = Reference frequency

so that by noting the change in frequency with temperature, the corresponding percent change in inductance can be measured.(See Magnetics Catalog,1989.) To get the Temperature factor. T.F., the following equation is used;

$$T.F. = \Delta L/(L_0\, \mu \Delta T) \quad [23.17]$$

Disaccomodation

As we had mentioned in Chapter 4, disaccomodation is a property that is unique to ferrites. It was defined as the time decrease in permeability. and the Disaccomodation Coefficient is the relative decrease per decade of time;

$$D.A. = (\mu_2 - \mu_1)/(\mu_1 \log_{10}[t_2/t_1]) \quad [23.18]$$

where μ_1 and μ_2 are two permeabilities measured at times t_1 and t_2 after demagnetization. To make the mathematics easier, the time is taken in successive decades such as 10, 100, and 1000 minutes. As in the case of the

$$D.F. = D.A./\mu_1 = (\mu_2-\mu_1)/\mu_1 2\log_{10} t_2/t_1 \quad [23.19]$$

Since the changes in permeability measured may be quite small in some cases, it is important to avoid sample temperature changes that might mask the disaccomodation effect. A good constant-temperature cabinet is needed.

Effect of D.C. Bias on Permeability

Since a D.C. bias will shift the point of measurement on the magnetization curve, the permeability will change. In gapped parts, the permeability may drop dramatically with D.C. bias.(See Figure 23.19). Provision should be made for application of bias. Several digital briges provide this internally but the Wayne-Kerr model allows for a large bias (1.0 ampere internal). However, accessory external bias modules of 20 A. per unit can be cascaded for a maximum of 100A.

Loss Separation at Low Flux Density

The individual contributions of Eddy current, and hysteresis effects to tan δ can be determined by some sort of loss separation. Publication 367-1(1982) describes the method of measuring the $(\tan \delta)_{e+r}$ which is indicative of both Eddy current and residual losses. At low flux densities, the hysteresis loss is quite low. At low frequencies, the Eddy current losses are low. To measure the hysteresis loss tangent, the measurement is made at two flux levels and the $\tan \delta_h$ is calculated from;

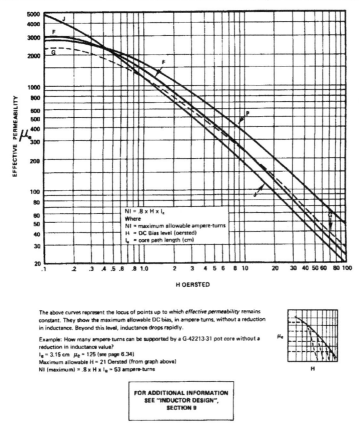

Figure 23.19- Effect of D.C. bias on permeability of gapped cores. From Magnetics Catalog FC509, Magnetics Div, Spang and Co., Butler, PA 16003

$$\tan \delta_h = E\Delta R/\omega L_s \Delta E \quad [23.20]$$

where E = Max voltage
ΔR_s = change in resistance
L_s = Inductance at low flux level
ΔE = Change in voltage

Another expression involving the loss breakdown involved the Legg equation;

$$R_s/\mu fL = hB + ef + r \quad [23.21]$$
where h = hysteresis coefficient
e = Eddy current coefficient
r = residual coefficient

If the left side is plotted against B, the slope of the curve is h. If the same is plotted against f, the slope is e. The intercept is r.

Losses at High Power Levels-Core Losses

Here, we are interested in measuring the total core loss. There are methods using bridges that will convert to core losses but these methods are not often used. One very time-consuming method that gives very accurate results is the calorimetric method that measures the heat generated during the excitation of the core. This method may be a good referee method but is not recommended for repetitive or routine measurements. There are several other electronic methods that can be used. These are;

1. Electronic Wattmeter Method
2. Multiplying Voltmeter Method
3. Bridge Methods (See above)
4. Digitizing Oscilloscope Method

Electronic Voltmeter Method

The instrument using this method is commonly called a VAW meter. It is a direct reading Volts-Amps-Watts meter hence the name. For the early days of power loss measurements, the Fluke VAW meter was a household name. The company has since stopped marketing this instrument. The new VAW meters are aimed at the higher frequency range typified by ferrites. The schematic diagrams of such a unit (Clarke-Hess Model 258) are shown in Figures 23-20(a), (b) and (c). This instrument is extremely useful in repetitive and routine production measurements on power ferrite cores. The normal current range is from 50-2500 mA., and the voltage range is from 1-100 V. The power ranges are determined by the product of all the voltage and current combinations.

Multiplying Voltmeter Method

For many years, the heart of this system was the Norma Mu-Function Meter which was a two channel rms voltmeter with a very large range of voltages. Any voltage waveform(sine, square, triangular,etc.) could be used. The voltage wave form is transformed into an equivalent volt-second area composed of a train of identical pulses totaling the original area. Knowing the ET of each pulse, the digital representation of the rms voltage is made. The magnetizing current is coincidentally measured across a standard resistor (1Ω) so that V = I. This voltage is then digitized in a manner similar to the previous one. The two digitized voltages are the multiplied and the product read on an analog meter. The Norma Co. has recently introduced a new model called a power analyzer which presumably works on the same principle but is computerized and handled digitally on an IEEE bus. The schematic of their power analyzer is shown in Figure 23-21.

Schlotterbeck (1981) describes the use of the Mu-function meter to measure the power losses. He notes that the method is applicable to 2 Mz. down to some milliwatts. Figure 23.22 is a diagram of his measurement set-up. Gaudry (1985) used the mutiplying voltmeter to get the instantaneous product and cause the peak voltage to coincide with peak flux density. Mochizuki(1986) also used the

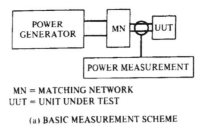

MN = MATCHING NETWORK
UUT = UNIT UNDER TEST

(a) BASIC MEASUREMENT SCHEME

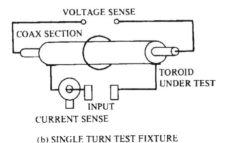

(b) SINGLE TURN TEST FIXTURE

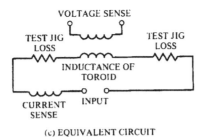

(c) EQUIVALENT CIRCUIT

Figure 23.20- Schematic Diagrams of a Clarke Hess VAW Meter, Clarke Hess Communications Research Co., New York, NY

Mu-function meter, but stored the instantaneous values in a Waveform Memory Device which was processed by a microcomputer. Evidently this is in anticipation to the next method, namely the digitizing oscilloscope method.

Digitizing Oscilloscope Method
This is the newest method available to date and although it may be somewhat more time-consuming than the previously mentioned ones, it has many virtues;

1. The actual measurement can be made with a very short operating time of the transformer core(several cycles of the B-H loop). This is desirable because prolonged excitation will heat the sample and actually change the temperature of the measurement. The drive generators are external but can be programmed

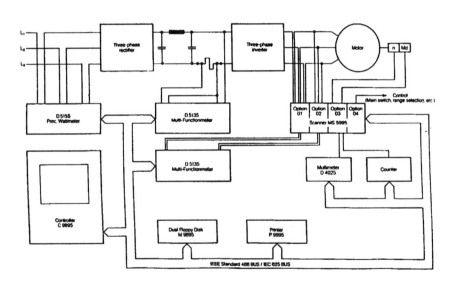

Figure 23.21- Schematic diagram of a Norma power analyzer, Norma Messtechnik, Gmbh, Vienna,Austria

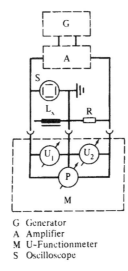

G Generator
A Amplifier
M U-Functionmeter
S Oscilloscope

Figure 23.22-Schematic Diagram of Core Loss Measurement with Multiplying Voltmeter. From Schlotterbeck, M. and Zenger M., Proc. PCI (1981), 37

by the computer as can most of the other conditions. By skillful programming, this type of core loss measurement can be automated
2. The measurement can be made over a wide frequency range.
3. The voltage waveform can be monitored during the measurement to assure no distortion. This can be a serious problem in loss measurements.
4. Families of hysteresis loops at various frequencies can be generated. The method depends on the digitizing of each point of the hysteresis loop, storing the values in a computer and either displaying or plotting the loop.

Thottuvelil(1985) states that to calculate the core losses, the values representing the instantaneous induced secondary voltage and those representing the instantaneous input currents (using a current probe) or across a current-sensing resistor are sampled at a high rate of speed and stored in the computer. The instantaneus power can then be calculated. If the primary and secondary turns are equal, the instantaneous power per cycle is just the product. If the turns ratio is not one, then the result must be multiplied by the turns ratio. The average core loss is obtained by computing the mean of the instantaneous power over once cycle. Multiplying this by the frequency and dividing by the volume of the core gives the core loss.

Thottuvelil found difficulties in measuring low-loss, low permeability cores because of phase shift. The method has indeed posed a challenge because of wave distortion, high frequency problems and phase shift. Several papers on the use of this method have appeared recently. Figure 24-23 shows the setup proposed by Sato

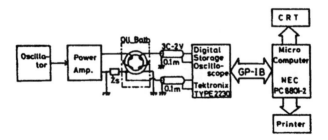

Figure 23.23- Schematic diagram of Core Loss Measurement Set-up using a Digitizing Oscilloscope From Sato, T. and Sakaki, Y., IEEE Trans Mag. MAG 23, Sept. 1987, 2593 (1987).

To overcome the problem of phase shift, the voltage wave forms are analyzed by Fourier Analysis to compensate for the harmonic frequencies. The flow chart of his measurement is shown in Figure 23.24.

The heart of the new system is the digitizing Oscilloscope. Several such units are shown in Figures 23.25 and 23.26.

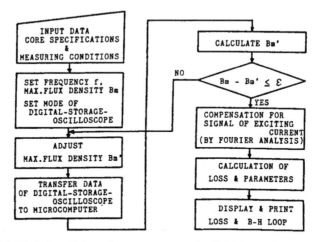

Figure 23.24- Flow sheet for core loss measurement by digitizing oscilloscope. From Sato, T and Sakaki, Y., Trans. Mag. MAG 23, Sept 1987, 2593

A schematic of equipment for measuring core losses by this technique at Magnetics, Division of Spang and Co, is shown in Figure 23.27. A picture of the equipment is shown in Figure 23.28

MAGNETIC MEASUREMENTS- MATERIALS- COMPONENTS

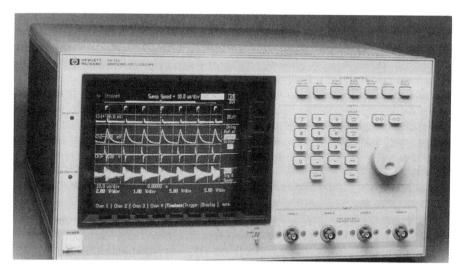

Figure 23.25- Hewlett-Packard Digitizing Oscilloscope From Hewlett Packard Test and Measurement Catalog(1989)

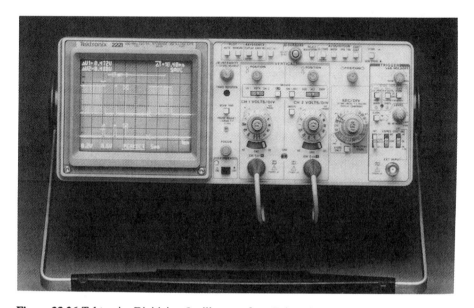

Figure 23.26-Tektronics Digitizing Oscilloscope, from Tektronix, Inc., Beaverton OR 97075

614   HANDBOOK OF MODERN FERROMAGNETIC MATERIALS

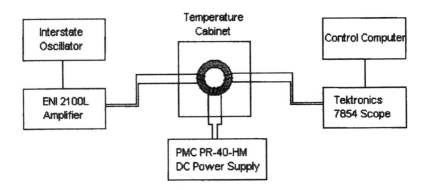

Figure 23.27- Schematic of power ferrite core loss measurement at Magnetics, Div. of Spang and Co. using the digitizing oscilloscope technique. Courtesy of Magnetics.

Figure 23.28-Photo of core loss measurement by digitizing oscilloscope technique at Magnetics, Div. of Spang and Co.

Pulse Measurements

Ferrites for pulse applications must be measured under the conditions of actual usage therefore each user of these cores will generally have his own measurement setup. The core is driven with a square voltage pulse from remanence to some higher level and the permeability is measured by observing the voltage and current wave forms. A typical pulse wave form is shown in Figure 23.29. The magnetizing current that appears as a ramp function is monitored by a current probe or through a precision resistor. The voltage is read off the oscilloscope plot and must be corrected for overshoot. The pulse inductance is;

$$L_p = E_m/(dI/dt) \quad [23.22]$$

where E_m = Maximum voltage
dI/dt = Rate of change of current with time, (I/t_d)

Amplitude Permeability

This is the permeability at high flux levels and is usually made on a specially designed bridge to take the higher drive levels. This measurement is useful for power applications and it often measured near the μ_{max} point. The measurement is done with a sinusoidal wave form and is quite simple. The peak magnetizing field, H, is measured across a series resistor and the voltage across the measuring coil. The B level is calculated from;

$$B = E_{av}/(4.44fNA \times 10^{-8}) \quad [23.23]$$

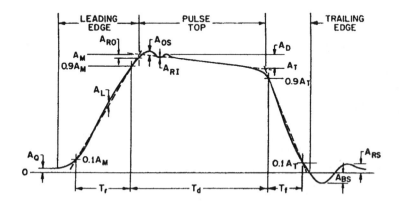

Figure 23.29- A typical voltage wave form found in measuring pulse permeability. From IEEE Standard 393-1977

The amplitude permeability is calculated from the B/H ratio. Schlotterbeck (1981) has measured the amplitude permeability of Siemens N27 with the setup shown in Figure 23.30.

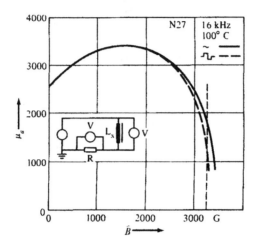

Figure 23.30 Equipment set-up and results of measurement of Siemens ferrite N27 for amplitude permeability. From Schlotterbeck, M. and Zenger,M., Proc. PCI (1981), 37

References

Foner, S.(1959)- Rev. Sci. Inst. 30, 548
Gaudry P.(1985)- Proceedings PCIM, Nov. 1985, 28
IEC (1989)- Documents published by Bureau Centrale de la Commission Electrotechnique Internationale, 3, rue de Varembe, Geneva, Switzerland
IEC (1982)- Publication 367-1 (1982) Cores for Inductors and Transformers, Part 1.-Measuring Methods
Magnetics (1989)- Catalog FC509, Magnetics Div. , Spang and Co., P.O. Box 391, Butler PA 16003
MMPA (1989)- Soft Ferrites, A Users Guide, Magnetic Materials Producers Association, 800 Custer Ave.,Evanston, IL 60202
Mochizuki,T.(1986), Sasaki, T., and Murakawa, K., IEEE Trans. Mag., MAG-22,#5,Sept. 1986,668
Rathenau, G.W.(1946) and Snoek, J.L., Philips Res. Rept. 1, 239
Sato, T.(1987) and Sakaki, Y., IEEE Trans. Mag., MAG 23,#5,Sept.,1987, p.2593
Schlotterbeck,M.(1981),and Zenger, M.,Proc. PCI 1987, 37
Sucksmith, W.(1938) and Pierce,R.R., Proc, Roy. Soc.(London)167A, 189
Sucksmith,W.(1939), ibid, 170A, 551
Thottuvelil, V.J.(1985),Proc. PESC, 412

Table 23.2-IEC Publications on Soft Ferrites

133	(1985)	Dimensions of pot-cores made of magnetic oxides and associated parts. (Third Edition).
205	(1966)	Calculation of the effective parameters of magnetic piece parts. Amendment No. 1 (1976). Amendment No. 2 (1981).
205A	(1968)	First supplement.
205B	(1974)	Second supplement.
220	(1966)	Dimensions of tubes, pins, and rods of ferromagnetic oxides.
221	(1966)	Dimensions of screw cores made of ferromagnetic oxides. Amendment No. 2 (1976).
221A	(1972)	First supplement.
223	(1966)	Dimensions of aerial rods and slabs of ferromagnetic oxides.
223A	(1972)	First supplement.
223B	(1977)	Second supplement.
226	(1967)	Dimensions of cross cores (X-cores) made of ferromagnetic oxides and associated parts. Amendment No. 1 (1982).
226A	(1970)	First supplement.
367:	–	Cores for inductors and transformers for telecommunications.
367-1	(1982)	Part 1: Measuring methods. (Second Edition). Amendment No. 1 (1984). Amendment No. 2 (1992).
367-2	(1974)	Part 2: Guides for the drafting of performance specifications. Amendment No. 1 (1983).
367-2A	(1976)	First supplement.
401	(1993)	Ferrite materials - Guide on the format of data appearing in manufacturers' catalogues of transformer and inductor cores. (Second Edition).
424	(1973)	Guide to the specification of limits for physical imperfections of parts made from magnetic oxides.
431	(1993)	Dimensions of square cores (RM cores) made of magnetic oxides and associated parts. (Second edition). Amendment No. 1 (1995).
492	(1974)	Measuring methods for aerial rods.

Table 23.2 (Continued)

525	(1976)	Dimensions of toroids made of magnetic oxides or iron powder. Amendment No. 1 (1980).
647	(1979)	Dimensions for magnetic oxide cores intended for use in power supplies (EC cores).
701	(1981)	Axial lead cores made of magnetic oxides or iron powder.
723:	—	Inductor and transformer cores for telecommunications.
723-1	(1982)	Part 1: Generic specification.
723-2	(1983)	Part 2: Sectional specification: Magnetic oxide cores for inductor applications.
723-2-1	(1983)	Part 2: Blank detail specification: Magnetic oxide cores for inductor applications. Assessment level A.
723-3	(1985)	Part 3: Sectional specification: Magnetic oxide cores for broadband transformers.
723-3-1	(1985)	Part 3: Blank detail specification: Magnetic oxide cores for broadband transformers. Assessment levels A and B.
723-4	(1987)	Part 4: Sectional specification: Magnetic oxide cores for transformers and chokes for power applications.
723-4-1	(1987)	Part 4: Blank detail specification: Magnetic oxide cores for transformers and chokes for power applications - Assessment level A.
723-5	(1993)	Part 5: Sectional specification: Adjusters used with magnetic oxide cores for use in adjustable inductors and transformers.
723-5-1	(1993)	Part 5: Sectional specification: Adjusters used with magnetic oxide cores for use in adjustable inductors and transformers. Section 1: Blank detail specification - Assessment level A.
732	(1982)	Measuring methods for cylinder cores, tube cores and screw cores of magnetic oxides.
1007	(1994)	Transformers and inductors for use in electronic and telecommunication equipment -Measuring methods and test procedures. (Second Edition).
1185	(1992)	Magnetic oxide cores (ETD-cores) intended for use in power supply applications - Dimensions. Amendment No. 1 (1995).
1246	(1994)	Magnetic oxide cores (E-cores) of rectangular cross-section and associated parts - Dimensions.

MAGNETIC MEASUREMENTS- MATERIALS- COMPONENTS

IEC Standards for Metallic Materials

Under the Responsibility of IEC/TC68 *MAGNETIC ALLOYS AND STEELS*

IEC 60404-1: 1979	Magnetic materials. Part 1: Classification. (A new version will be published in 1999)
IEC 60404-2: 1996	Magnetic materials. Part 2: Methods of measurement of the magnetic properties of electrical steel sheet and strip by means of an Epstein frame.
IEC 60404-3: 1992	Magnetic materials. Part 3: Methods of measurement of the magnetic properties of magnetic sheet and strip by means of a single sheet tester.
IEC 60404-4: 1995	Magnetic materials. Part 4: Methods of measurement of the d.c. magnetic properties of iron and steel.
IEC 60404-5: 1993	Magnetic materials. Part 5: Permanent magnet (magnetically hard) materials - Methods of measurement of magnetic properties.
IEC 60404-6: 1986	Magnetic materials. Part 6: Methods of measurement of the magnetic properties of isotropic nickel-iron soft magnetic alloys, types E1, E3 and E4.
IEC 60404-7: 1982	Magnetic materials. Part 7: Methods of measurement of the coercivity of magnetic materials in an open magnetic circuit.
IEC 60404-8-1: 1986	Magnetic materials. Part 8: Specifications for individual materials. Section One - Standard specifications for magnetically hard materials. Amendment 1: 1991 Amendment 2: 1992, (A new version will be published in 1999)
IEC 60404-8-2: 1985	Magnetic materials. Part 8: Specifications for individual materials. Section Two - Specification for cold-rolled magnetic alloyed steel strip delivered in the semi-processed state. (A new version will be published in 1998)
IEC 60404-8-3: 1985	Magnetic materials. Part 8: Specifications for individual materials. Section Three - Specification for cold-rolled magnetic non-alloyed steel strip delivered in the semi-processed state. (A new version will be published in 1998)
IEC 60404-8-4: 1986	Magnetic materials. Part 8: Specifications for individual materials. Section Four - Specification for cold-rolled non-oriented magnetic steel sheet and strip. (A new version will be published in 1998)
IEC 60404-8-5: 1989	Magnetic materials. Part 8: Specifications for individual materials. Section Five - Specification for steel sheet and strip with specified mechanical properties and magnetic permeability.
IEC 60404-8-6: 1986	Magnetic materials. Part 8: Specifications for individual materials. Section Six - Soft magnetic metallic materials. Amendment 1: 1992, (A new version will be published in 1999)

IEC Standards for Metallic Materials (Continued)

IEC 60404-8-7: 1988	Magnetic materials. Part 8: Specifications for individual materials. Section Seven - Specification for grain-oriented magnetic steel sheet and strip. Amendment 1: 1991, (A new version will be published in 1998).
IEC 60404-8-8: 1991	Magnetic materials. Part 8: Specifications for individual materials. Section Eight - Specification for thin magnetic steel strip for use at medium frequencies.
IEC 60404-8-9: 1994	Magnetic materials. Part 8: Specifications for individual materials. Section Nine - Standard specification for sintered soft magnetic materials.
IEC 60404-8-10: 1994	Magnetic materials. Part 8: Specifications for individual materials. Section Ten - Specification for magnetic materials (iron and steel) for use in relays.
IEC 60404-9: 1987	Magnetic materials. Part 9: Methods of determination of the geometrical characteristics of magnetic steel sheet and strip.
IEC 60404-10: 1988	Magnetic materials. Part 10: Methods of measurement of magnetic properties of magnetic sheet and strip at medium frequencies.
IEC 60404-11: 1991	Magnetic materials. Part 11: Method of test for the determination of surface insulation resistance of magnetic sheet and strip.
IEC 60404-12: 1992	Magnetic materials. Part 11: Guide to methods of assessment of temperature capability of interlaminar insulation coatings.
IEC 60404-13: 1995	Magnetic materials. Part 12: Methods of measurement of density, resistivity and stacking factor of electrical steel sheet and strip.

ASTM Standards for Magnetic Materials

Alternating Current

Test Methods for:

A 346 – 74 (1993)	Alternating Current Magnetic Performance of Laminated Core Specimens Using the Dieterly Bridge Method
A 932 – 95	Alternating-Current Magnetic Properties of Amorphous Materials at Power Frequencies Using Wattmeter-Ammeter-Voltmeter Method with Sheet Specimens
A 927/A 927M – 94	Alternating-Current Magnetic Properties of Torodial Core Specimens Using the Volmeter-Ammeter-Wattmeter Method
A 772/A 772M – 95	ac Magnetic Permeability of Materials Using Sine Current
A 912 – 93	Alternating Current Magnetic Properties of Amorphous Materials at Power Frequencies Using Wattmeter-Ammeter-Voltmeter Method With Torodial Specimens
A 697 – 91 (1996)	Alternating Current Magnetic Properties of Laminated Core Specimen Using the Voltmeter-Ammeter-Wattmeter Methods
A 804/A 804M – 94	Alternating-Current Magnetic Properties at Power Frequencies Using Sheet-Type Test Specimens
A 889/A 889M – 93	Alternating-Current Magnetic Properties at Low Inductions Using Wattmeter-Varmeter-Ammeter-Voltmeter Method and 25-cm Epstein Frame
A 343 – 93a	Alternating Current Magnetic Properties of Materials at Power Frequencies Using Wattmeter-Ammeter-Voltmeter Method and 25-cm Epstein Test Frame
A 348 – 95a	Alternating-Current Magnetic Properties of Materials Using the Wattmeter-Ammeter-Voltmeter Method, 100 to 10 000 Hz and 25-cm Epstein Frame

Amorphous Materials

Specification for:

A 901 – 90	Amorphous Magnetic Core Alloys Semi-Processed Types

Direct Current

Test Methods for:

A 341/A 341M – 95	Direct-Current Magnetic Properties of Materials Using D-C Permeameters and the Ballistic Test Methods
A 773/A 773M – 96	dc Magnetic Properties of Materials Using Ring and Permeameter Procedures with dc Electronic Hysteresigraphs
A 596/A 596M – 95	Direct-Current Magnetic Properties of Materials Using the Ballistic Method and Ring Specimens

MAGNETIC MEASUREMENTS- MATERIALS- COMPONENTS

ASTM Standards for Magnetic Materials (Continued)

Electrical Steel

Specifications for:

A 726 – 92	Cold-Rolled Magnetic Lamination Quality Steel, Semiprocessed Types
A 726M – 92	Cold-Rolled Magnetic Lamination Quality Steel, Semiprocessed Types (Metric)
A 345 – 90 (1995)	Flat-Rolled Electrical Steels for Magnetic Applications
A 876 – 92	Flat-Rolled, Grain-Oriented, Silicon-Iron, Electrical Steel, Fully Processed Types
A 876M – 92	Flat-Rolled, Grain-Oriented, Silicon-Iron, Electrical Steel, Fully Processed Types (Metric)
A 840 – 91 (1996)	Magnetic Lamination Steel, Fully Processed
A 840M – 91 (1996)	Magnetic Lamination Steel, Fully Processed (Metric)
A 677 – 96	Nonoriented Electrical Steel, Fully Processed Types
A 677M – 96	Nonoriented Electrical Steel, Fully Processed Types (Metric)
A 683 – 91	Nonoriented Electrical Steel, Semiprocessed Types
A 683M – 91	Nonoriented Electrical Steel, Semiprocessed Types (Metric)

Test Methods for:

A 720 – 91	Ductility of Nonoriented Electrical Sheet Steel
A 721 – 92	Ductility of Oriented Electrical Sheet Steel
A 712 – 75 (1991)	Electrical Resistivity of Soft Magnetic Alloys
A 937 – 95	Interlaminar Resistance of Insulating Coatings Using Two Adjacent Test Surfaces, Determining
A 900 – 91 (1996)	Lamination Factor of Amorphous Magnetic Strip
A 718 – 75 (1991)	Surface Insulation Resistivity of Multi-Strip Specimens (*Discontinued 1996†*)
A 717/A 717M – 95	Surface Insulation Resistivity of Single-Strip Specimens

Practice for:

A 664 – 93	Identification of Standard Electrical- and Laminations-Steel Grades in ASTM Specifications

Magnetic Amplifier Cores

Test Method for:

A 598 – 92	Magnetic Properties of Magnetic Amplifier Cores

Magnetic Shields

Test Method for:

A 698/A 698M – 92	Magnetic Shield Efficiency in Attenuating Alternating Magnetic Fields

Metallic Materials

Specifications for:

A 838 – 90a	Free-Machining Ferritic Stainless Soft Magnetic Alloys for Relay Applications
A 848/A 848M – 96	Low-Carbon Magnetic Iron
A 801/A 801M – 92	Iron-Cobalt High Magnetic Saturation Alloys
A 867/A 867M – 94	Iron-Silicon Relay Steels
A 753 – 85 (1990)	Nickel-Iron Soft Magnetic Alloys

Test Methods for:

A 900 – 91	Lamination Factor of Amorphous Magnetic Strip
A 719 – 90	Lamination Factor of Magnetic Materials
A 342 – 95	Permeability of Feebly Magnetic Materials

Practice for:

A 34 – 96	Sampling and Procurement Testing of Magnetic Materials

Nonmetallic Materials

Test Methods for:

A 893 – 86 (1991)	Complex Dielectric Constant of Nonmetallic Magnetic Materials at Microwave Frequencies
A 883 – 96	Ferrimagnetic Resonance Linewidth and Gyromagnetic Ratio of Nonmetallic Magnetic Materials
A 894/A 894M – 95	Saturation Magnetization or Induction of Nonmetallic Magnetic Materials

Sintered Powder Materials

Specifications for:

A 839/A 839M – 96	Iron-Phosphorous Powder Metallurgy (P/M) Parts for Soft Magnetic Applications
A 811 – 90	Soft Magnetic Iron Fabricated by Powder Metallurgy Techniques
A 904 – 90	50 Nickel - 50 Iron Powder Metallurgy Soft Magnetic Alloys

Terminology

Terminology Relating to:

A 340 – 96	Magnetic Testing

Bibliography

Bozorth, R. M. 1951. *Ferromagnetism*. Princeton, N. J.: D. Van Nostrand Co., Ltd.

Bradley, F. N. 1971. *Materials for Magnetic Functions*. New York: Hayden Book Co., Inc.

Brailsford, F. 1951. *Magnetic Materials*. London: Methuen and Co., Ltd.

Brailsford, F. 1966. *Physical Principles of Magnetism*. London: D. Van Nostrand Co., Ltd.

Chikazumi, S. 1964. *Physics of Magnetism*. New York: John Wiley and Sons.

DeMaw, M. F. 1981. *Core Design and Application Handbook*. Englewood Cliffs, N.J.: Prentice–Hall, Inc.,

Goldman, A. 1988. "Magnetic Ceramics." In *Electronic Ceramics*, ed. Lionel M. Levinson, pp. 147–189. New York: Marcel Dekker, Inc.

Grossner, N. 1983. *Transformers for Electronic Circuits*. New York: McGraw–Hill Book Co.

Heck, C. 1974. *Magnetic Materials and Their Applications*. New York: Crane–Russack and Co.

Hoshino, Y., Iida, S., and Sugimoto, M. 1971. *Ferrites: Proceedings of the International Conference*. Tokyo: University of Tokyo Press.

Hnatek, E. R. 1974. *Design of Solid State Power Supplies*. New York: Van Nostrand Reinhold.

Kampczyk, V. W., and Roess, E. 1978. *Ferrite Cores (Ferritkerne)*. Berlin: Siemens AG.

Kittel, C. 1971. *Introduction to Solid State Physics*. New York: John Wiley and Sons.

McLyman, C. T. W. 1982. *Magnetic Core Selection for Transformers and Inductors*. New York: Marcel Dekker, Inc.

Magnetic Materials Producers' Association. 1989. *Soft Ferrites: A User's Guide*. Evanston, Ill.: Magnetic Materials Producers Association.

Magnetics, Division of Spang and Co. 1989. *Ferrite Cores: Catalog FC509*. Butler, Penn.: Magnetics, Division of Spang and Co.

Mallinson, J. C. 1989. *The Foundations of Magnetic Recording*. San Diego: Academic Press.

Morrish, A. H. 1965. *The Physical Principles of Magnetism*. New York: John Wiley and Sons.

Olsen, F. 1966. *Applied Magnetism*. Eindhoven: Philips Technical Library.

Onoda, G. Y., and Hench, L. L. 1978. *Ceramic Processes Before Firing*. New York: John Wiley and Sons.

Palmer, H., Davis, R. F., and Hare, T. M. 1977. *Processing of Crystalline Ceramics*. Material Science Research, Vol 2. New York: Plenum Press.

Parker, R. J., and Studders, R. J. 1962. *Permanent Magnets and Their Application*. New York: John Wiley and Sons.

Philips Electronics Co. 1986. *Components and Materials; Book C-5*. Eindhoven: Philips Electronics and Materials Division.

Polydorf, W. J. 1960. *High Frequency Magnetic Materials*. New York: John Wiley and Sons.

Quartly, C. J. *Square Loop Ferrite Circuitry*. London: Iliffe Books, Ltd.

Rado, G., and Suhl, H. 1963. *Magnetism*, Vol. III. New York: Academic Press.

Reed, J. S. 1988. *Introduction to the Principles of Ceramic Processing*. New York: John Wiley and Sons.

Richards, C. E., and Lynch, A. C. 1953. *Soft Magnetic Materials for Telecommunications*. London: Pergamon Press, Ltd.

Roddam, T. 1963. *Transistor Inverters and Converters*. New York: Van Nostrand Reinhold.

Schieber, M. M. 1967. *Experimental Magnetochemistry*. Amsterdam: North-Holland Publishing Co.

Siemens AG. 1990–1991. *Ferrites: Data Book*. Munich: Siemens AG.

Smit, J. and Wijn, H. P. J. 1959. *Ferrites*. New York: John Wiley and Sons.

Smith, S. 1985. *Magnetic Components, Design and Applications*. New York: Van Nostrand Reinhold.

Snelling, E. C. 1988. *Soft Ferrites, Properties and Applications*. 2nd ed. London: Butterworths.

Snelling, E. C., and Giles. 1983. *Ferrites for Inductors and Transformers*. Letchworth: Research Studies Press, Ltd.

Soohoo, R. 1960. *Theory and Application of Ferrites*. Englewood Cliffs: Prentice–Hall, Inc.

Srivastava, M., and Patni, M. J. 1989. *Advances in Ferrites: Proceedings of the Fifth International Conference*, Vol. 1. New Delhi: Oxford and IBH Publishing Co. PVT.

Srivastava, M., and Patni, M. J. 1989. *Advances in Ferrites: Proceedings of the Fifth International Conference*, Vol. 2. Switzerland: Trans-Tech Publications.

Standley, K. J. 1962. *Oxide Magnetic Materials*. London: Oxford University Press.

TDK Corporation. 1987. *TDK Ferrite Cores for Power Supplies and EMI/RFI Filter*. Tokyo: TDK Corporation.

Tebble, R. S., and Craik, D. J. 1969. *Magnetic Materials*. London: Wiley Interscience.

Thompson, J. E. 1968. *The Magnetic Properties of Materials*. Cleveland: CRC Press.

Thomson LCC. 1988. *Soft Ferrites*. Book 1. Courbevoie: Thomson LCC.

Von Aulock, W. H. 1965. *Handbook of Microwave Ferrites*. New York: Academic Press.

Wang, F. F. Y. 1985. *Advances in Ceramics*. Vol. 15: Fourth International Conference on Ferrites—Part 1. Columbus, Ohio: American Ceramic Society.

Wang, F. F. Y. 1985. *Advances in Ceramics*. Vol. 16: Fourth International Conference on Ferrites—Part 2. Columbus, Ohio: American Ceramic Society.

Watanabe, H., Iida, S., and Sugimoto, M. 1981. *Ferrites, Proceedings of the ICF3*. Tokyo: Center for Academic Publications.
Watson, J. K. 1980. *Applications of Magnetism*. New York: John Wiley and Sons.
Wohlfarth, E. P. 1980. *Ferromagnetic Materials*, Vol. 2. Amsterdam: North-Holland Publishing Co.
Wood, P. 1981. *Switching Power Converters*. New York: Van Nostrand Reinhold.

Supplementary Bibliography

APPENDIX 1
ABBREVIATIONS AND SYMBOLS

A	Cross-sectional area, cm^2
A	Attenuation or insertion loss, dB
A	Atomic weight
A	Angstrom units (10^{-8} m)
A_L	Inductance factor (inductance per 1000 turns), mH
A_c	Area of core, cm^2
A_W	Area of wire, cm^2 or circ. mils
A_p	Area product, cm^4
A_e	Effective area, cm^2
A_g	Cross-sectional area of gap
A_m	Cross-sectional area of magnet
a_0	Unit cell length, A
a	Distance between atoms, A
a	Anomalous loss coefficient
AC	Alternating current
B	Magnetic induction, gausses or teslas (webers/m^2)
B_s	Saturation induction
B_r	Remanent induction
B_m	Maximum induction or induction in magnet
B_g	Induction in gap
ΔB	Change in induction
B_e	Effective flux density
C	Centigrade temperature
C	Capacitance (farads)
C	Curie constant
C	Current-carrying capacity
c	Speed of light (3×10^{10} cm/s)
DA	Disaccommodation
DF	Disaccommodation factor
DC	Direct current

d	Thickness, cm
d	Density (g/cm^3)
d_o	Outer diameter
d_i	Inner diamter
d	Grain diameter, microns
E	Voltage, volts
E	Energy, ergs
E_p	Magnetostatic energy, ergs/cm^3
E_k	Anisotropy energy, ergs/cm^3
E_w	Wall energy, ergs/cm^2
E_l	Energy stored in an inductor, ergs/cm^3
E_{out}	Output voltage
E_{in}	Input voltage
e	Charge on an electron, coulombs
e	Efficiency
e	Eddy current coefficient (Legg)
F	Force, dynes
F	Farads (capacitance)
F_W	Winding factor
F_{cu}	Copper factor
f	Frequency, hertz
f_r	Resonant frequency, Hz
Δf	Bandwidth, Hz
f and F	Leakage factors, magnets
G	Gravitational constant
g	Spectroscopic splitting factor
H	Magnetic field strength, oersted or A/m
H_d	Demagnetizing field
H_m	Magnetic field in magnet
H_g	Magnetic field in gap
H_c	Coercive force, oersteds or A/m
H_{ci}	Intrinsic coercive force
Hz	Hertz, frequency
ΔH	Linewidth, oersted
ΔH_k	Spinwave linewidth, oersted
h	Planck's Constant
h_{crit}	Critical current (microwave power)
h	Hysteresis coefficient (Legg)
I	Current, amperes
I_p	Pulse current
J	Current density, A/cm^2
j	Unit imaginary vector

ABBREVIATIONS AND SYMBOLS

K	Absolute temperature, Kelvin
K	Constant
K	Copper-fill factor
K_1, K_2	First and second anisotropy constants
K_w	Winding utilization factor (McLyman)
K_j	Current density factor
K_g	Geometrical factor
K_e	Electrical factor
k	Boltzman constant
L	Torque
L	Inductance, henries
L_N	Inductance for N turns
L_p	Parallel inductance
L_p	Pulse inductance
L_S	Series inductance
LF	Loss factor
l	Distance between poles
l_m	Magnetic path length, cm
l_e	Effective path length, cm
l_g	Length of gap
M	Intensity of magnetization, emu/cm^3
M_s	Saturation magnetization
MGO	Mega gauss-oersteds
m	Pole strength
m	Mass of an electron
m	Exponent of frequency in modified Steinmetz equation
N	Number of turns
N	Demagnetizing factor
N_p	Turns in primary winding
N_s	Turns in secondary winding
n	Number of unpaired electrons in an atom or ion
n	Exponent of B in modified Steinmetz equation
Oe	Oersteds, field strength
Pa	Pascals
P	Power, watts
P_O	Output power
P_i	Input power
P_t	Apparent power
P_t	Total loss, watts
P_c	Core loss, watts
P_w	Winding loss, watts
P_{hyst}	Hysteresis losses

P_e	Eddy current losses
P_r	Residual losses
p_{O_2}	Oxygen partial pressure
p	Total angular momentum of an electron
p	Porosity
Q	Quality factor—in a series inductance, $Q = X_L/R_s$
q	Electric charge in coulombs
R	Resistance, ohms
R_s	Series resistance
R_p	Parallel resistance
R	Reluctance, magnets
R_{th}	Thermal resistance
r	Distance between masses or charges
s	Skin depth
T	Temperature, degrees Kelvin or centigrade
T_C	Curie temperature
T_N	Néel temperature
TC	Temperature coefficient
TF	Temperature factor
T	Pulse width
t	Time, seconds
t_1, t_2	Specific times
t_{on}	On time of a transistor
t_{off}	Off time of a transistor
V	Volume, cm^3
V_m	Volume of a magnet
V_e	Effective volume
W_a	Window area, cm^3
X_L	Inductive reactance, ohms
X_C	Capacitive reactance, ohms
Z	Impedance, ohms
α	Regulation, %
$\alpha_1, \alpha_2...$	Direction cosines
δ	Angle between H and M
γ	Gyromagnetic ratio
δ	Thickness of domain wall
δ	Oxygen parameter (Tanaka)
ε	Phase angle between the direction of magnetization on the surface and at a depth in material
Δ	Change in a unit
η	Efficiency

ABBREVIATIONS AND SYMBOLS

θ	Angle in degrees
θ	Angle between a magnetic field and the axis of a magnet
θ	Angle between the magnetization and the easy direction of magnetization
θ	Temperature rise in a power material, degrees
θ_C	Curie temperature
θ_N	Neel temperature
λ	Magnetostriction constant
λ_S	Saturation magnetostriction
λ_{100}	Magnetostriction in a cube edge direction
λ_{111}	Magnetostriction in a cube body diagonal direction
λ	Wavelength, meters
μ	Microns, 10^{-6} m
μ	Magnetic moment
μ_B	Bohr magneton
μ_0	Permeability of free space
μ	Permeability
μ_0	Initial permeability
μ_e	Effective permeability
μ_a	Amplitude permeability
μ_p	Pulse permeability
μ_{DC}	DC permeability
μ'	Real permeability
μ''	Imaginary part of permeability
μ^+	Clockwise rotating part of permeability of a circularly polarized electronmagnetic wave
μ^-	Counterclockwise part of permeability of above
μ_{max}	Maximum permeability
ρ	Resistivity, ohm-cm
σ	Applied stress
σ	Magnetic moment per unit weight, emu/g
τ	Pulse width, seconds
ϕ	Magnetic flux, maxwells or webers
χ	Magnetic susceptibility
ω	Angular velocity, radians/s

APPENDIX 2

LIST OF THE WORLD'S MAJOR FERRITE CORE SUPPLIERS

North America

United States

Arnold Engineering Co. P.O. Box G Marengo, Ill. 660152	H*
Ceramic Magnetics 16 Law Drive Fairfield, N.J. 07006	S**
CMI Technology 3401 Leonard Court Santa Clara, Calif. 95054	S
Colt Industries Crucible Magnetics Div. P.O. Box 100 Elizabethtown, Ky. 42701	H
Fair-Rite Products Corp. Box J Walkill, N.Y. 12589	S
Ferrite International Co. 15280 Wadsworth Road Wadsworth, Ill. 60083	S
Ferronics Inc. 45 O'Connor Road Fairport, N.Y. 14450	S

*H = Hard ferrites
**S = Soft ferrites

Hitachi Metals America   S,H
2400 Westchester Ave.
Purchase, N.Y. 10577

MMG/Krystinel Corp.   S
126 Pennsylvania Ave.
Paterson, N.J. 07509

National Magnetics Group, Inc.   S
250 South Street
Newark, N.J. 07114

Discrete Products Division   S
Materials Group
Philips Components
5083 Kings Highway
Saugerties, N.Y. 12477

Siemens Corp., Special Products Div.   S
186 Wood Street, South
Iselin, N.J.

Magnetics Div., Spang and Co.   S
P.O. Box 391
Butler, Penn. 16003

D. M. Steward Mfg. Co.   S
East 36th Street
P.O. Box 510
Chattanooga, Tenn. 37401

Kane Magnetics International   H
700 Elk Avenue
Kane, PA 16735

TDK Corp. of America   S,H
1600 Feehanville Drive
Mount Prospect, ILL. 60056

Mexico

TDK de Mexico S.A. de C.V.   S
Carr, Juarez-Porvenir,
Parque Ind. A.J. Bermudez, iCd
Juarez, Chin. Mexico

LIST OF WORLD"S MAJOR FERRITE CORE SUPPLIERS

South America
Brazil

TDK Brasil Ind. E. Com. LTDA Avenida Brigadeiro Luis Antonio, 2367 Conjunto 301 Bela Vista CEP 01401 Sao Paulo, Brazil	S

Asian Countries
Korea

Isu Ceramics Co. C.P.O. Box 5680 Seoul, Korea	S
Korea Ferrite Corp. Rm. 301 Dongjin Bldg. 218 2-ka, Hankang-ro, Yongsan-ku Seoul, Korea	S
Korea TDK Co., Ltd. 642-3 Doksan Dong Yeongdeungpo-ku Seoul Korea	S
Sam-wha Electronic Co., Ltd. 38, Ojeon-ri, Euiwang-myeon Siheung-kun, Gyonggi-do, Korea	S

India

Cosmo Ferrites Ltd. 30 Community Center, Saket New Delhi, 11017, India	S
Morris Electronics Ltd. Poona-26, India	S
Permanent Magnets Ltd. Sylvester Bldg, 20, S. Bagatsingh Road Bombay, 400023, India	H

Japan

Daido Steel Co. 11-18, Nishiki 1-chome, Nakagu Nagoya, 460, Japan	H
Fuji Electrochemical Co. 5-36-11, Shinbashi, Minato-ku Tokyo, 105, Japan	S,H
Hitachi Metals Ltd. 1-2, Maunouchi 2-chome, Chiyoda-ku Tokyo, 100, Japan	S,H
Hokko Denshi Co. Ltd. 26-41, Oiwake-Nishi, Tenno-cho Minami-Akita-gun Akita-ken, 010, Japan	S,H
Matsushita Electric Co., Ltd Kadoma-shi Osaka, Japan	S
Mitsubishi Electric Co., Ltd 2-3, Marunouchi 2-chome, Chiyoda-ku Tokyo, 100, Japan	S
Murata Mfg. Co., Ltd. 26-10, Tenjin 2-chome, Nagaoka-kyo-shi Kyoto, 617, Japan	S
Nippon Ferrite Ltd 1-25-1, Hyakunincho, Shinjuku-ku Tokyo, 160, Japan	S
Sony Corp. 7-35, Kitashinagawa 6-chome Tokyo, 141, Japan	S
Sumitomo Special Metals Co., Ltd. 22, Kitahama 5-chome, Higashi-ku Osaka, 541, Japan	S,H
Taiyo Yuden Co., Ltd. 2-12, Ueno 1-chome, Taitoh-ku Tokyo, 110, Japan	S,H

LIST OF WORLD"S MAJOR FERRITE CORE SUPPLIERS   637

TDK Corporation 13-1, Nihonbashi 1-chome, Chuo-ku Tokyo, 103, Japan	S,H
Tohoku Metal Industries Ltd. 5-8, Aoyama 2-chome, Minato-ku Tokyo, 107, Japan	S
Tokyo Ferrite Mfg. Co. 1-14, Tabat-Shinmachi 1-chome, Kita-ku Tokyo, 114, Japan	H
Tomita Electric Co., Ltd. 123, Saiwai-cho, Tottori-shi, Tottori-ken, 680, Japan	S

People's Republic of China

Shanghai Magnetic Materials Factory Shanghai Instrumentation and Electronics Import and Export Corporation 68 Guizhou Road Shanghai, China	S,H
Southwest Institute of Appl. Magnetics P.O. Box 105 Sichuan, China	S

Singapore

Taiyo Yuden (Singapore) PTE, Ltd. 9 Joo Koon Rd. Jurong Town Singapore	S

Taiwan

Philips Electronics Industries Ltd. San Ming Bldg., Fourth Floor 57-1 Chung Shan North Rd. Sec. 2 Taipei, Taiwan	S

Super Electronics Co., Ltd 726, Chung Shan North Rd. Sec. 5 Taipei, Taiwan	S,H
Tatung Co. 22, Chung Shan North Rd. Sec. 3 Taipei, Taiwan	H
TDK Electronics (Taiwan) Corp. 159, Sec. 1, Chung Shan Rd. Tatung Li, Yangmei Taoyuan, Taiwan	S
Yeng Tat Electronics Co., Ltd. P.O. Box 2-30, Shu-Lin (238) Taipei Hsien, Taiwan	S

Europe

Austria

Gebrudder Boehler and Co. AG Vienna, Austria	H

Czechoslavakia

Pramet Unecowska, Sumperk, Czechoslavakia	S

England

Mullard Ltd. Mullard House Torrington Place London WC1E 7HD, England	S,H
Neosid, Ltd. Edward House, Brownfields, Welwyn Garden City, Herfordshire, AL7 1AN, England	S
Salford Electrical Instuments Ltd. Peel Works, Barton Lane, Eccles Manchester, M30 OHL England	S

LIST OF WORLD"S MAJOR FERRITE CORE SUPPLIERS

France

Thomson LCC   S
50 Rue J-P Timbaud
BP13/92403
Courbevoie, Cedex, France

East Germany

Kombinat VEB Keramische Werk   S
Hemsdorf
Friedrich-Engels-Str. 79
Postfach 2,
653 Hemsdorf/Thuringen, German
Democratic Republic

West Germany

Freid. Krupp Gmbh.   S,H
Munchener Strasse 90
Postfach 102161
4300 Essen, 1, Federal Republic of
Germany

Kaschke KG Gmbh and Co.   S
Rudolf-Winkel Str. 6
Postfach 771
3400 Gottingen, Federal Republic of
Germany

Siemens AG   S
Bereich Bauelemente
Balanstrasse 73
D-8000, Munich, 80, Federal
Republic of Germany

Valvo Gmbh   H
Burchardstrasse 19
Postfach 106323
2000 Hamburg, 1, Federal Republic
of Germany

Vogt Gmbh and Co. KG   S
D-8391 Erlan uber Passau, Federal
Republic of Germany

Italy

Industria Ossidi Sinterizzati   H
21 023 Malagesso/Varese, Italy

Netherlands

N. V. Philips Gloeilampenfabriken   H
Afd. Elonco, Boschdiijk 525
5600 PD Eindhoven, The Netherlands

N. V. Philips Gloeilampenfabriken   S
Commercial Department Materials
Bldg. BE-4
Eindhoven, The Netherlands

Poland

Polfer Zaklad Materialow   S
Magtycznych
Warsaw-47, Poland

Yugoslavia

Iskra, Nabavne Organizacije,   S,H
Kranj, Yugoslavia

APPENDIX 3
UNITS CONVERSION FROM CGS TO MKS (SI) SYSTEM

Symbol	Quantity	To convert from CGS Units	MKS Units	Multiply by Factor
l	Length	cm	m	10^{-2}
m	Mass	g	Kg	10^{-3}
F	Force	dyne	N, Newton	10^{-5}
E or W	Energy, Work	erg	Joule	10^{-7}
H	Magnetic Field Strength	Oersteds	A/m	79.58
B	Magnetic Induction or Flux Density	Gausses (Maxwell/cm^2)	Teslas(Wb/m^2) (volt-s)	10^{-4}
Φ	Magnetic Flux	Maxwells	Webers(Wb)	10^{-8}
μ	Permeability	Unitless	Henries/m	4×10^{-7}
F	Magnetomotive Force	Gilberts	Amp-turns	.7958
$(BH)_{max}$	Maximum Energy Product	Gauss-Oersteds (ergs)	Joules/m^3	7.96×10^{-3}

Other quantities which are the same in both systems and their common units include: Power, P (Watts, W); Charge, q (Coulombs); Potential, E or V (Volts); Time (Seconds); Resistance, R (ohms); Capacitance, C (Farads); Current, I (Amperes); Inductance, L (Henries).

Index

Adjustor, pot core, 370
Air gap, 478
Alnico, 72, 82-86
Aluminum oxide, substitution, 261
Amorphous metal alloy, 120, 122, 124-5
Anechoic chamber tiles, 571-574
Anisotropic permanent magnet, 76, 82, 85-86
Anisotropy, see Magnetocrystalline anisotropy
Anomalous losses, 66, 251
Antennas, 434
Antiferromagnetism, 34-36
Area product, 488-491
Atomic magnetism, 21
Audio frequency applications, 8
Auger microscopy, 256, 281-283, 289

Baluns, 189, 194
Bandwidth, 367, 369
Barium ferrite, 260
Beads, ferrite, 405
Bias, D.C., 519
Binary logic applications, 73
Bipolar drive, 446
Bohr magneton, 23, 49
Bohr theory of magnetism, 22
Bridges, inductance, 600-602
Broadband transformers, 283-284
Bubble domains, 262, 538

Cadmium ferrite, 215
Calcium oxide, 246-247, 249, 253, 256, 293
Carbonlyl iron, 184-185, 188
Chalcogenides, 226
Channel filters, 366
Chemical analysis, 309
Choke, swinging, 423
Chromium oxide, 249
Circulator, microwave, 558
Cobalt ferrite, 211, 214-215, 236, 249
Cobalt-iron alloy, 135-142
Cobalt-zinc ferrite, 217-218
Coercive force, 53
Common-mode filters, 403-404
Compensation point, 218, 220, 233
Competitive materials, 461-468, 498-499, 508-512
Compressive strength, 583
Computer-aided design, 501
Converters, 447-455, 503, 524-533
Copier powders
Copper-zinc ferrite, 428
Copper ferrite 214-215, 236
Coprecipitation, 305, 317-320, 325

Core losses, 64, 457-458
    measurement
Core-loss limited design, 496-497
Co-spray roasting, 321
Creepage, 485
Crystal structure, ferrites, 407-427
Curie law, 26-27
Curie point, 26
Curie-Weiss law, 32, 35

DC applications, 4
Deflection yoke, 425-430
Delay lines,. 571
Demagnetization curves, 78, 80
    intrinsic, 80
    normal, 80
Demagnetizing field, 99
Design, converters, 503, 524,
Design, power transformers, 459, 484, 488, 498
Densities, 581-582
Diamagnetism, 26,
Differential-mode filter, 407
Digitizing oscilloscope, 610-613
Disaccommodation, 66-68, 250
Domains, 41-42, 45, 47
    domain wall, 45
    domain wall motion, 46-47
Double exchange, 38
Duplex structure, 271, 273

E-cores, 471
    E-C cores, 472
    ETD cores, 474
Eddy current losses, 59-61, 251
Electrodes, 570
Electromagnetism, 21
EMI applications, 12, 391-422
    limits, 393-395
    Test setup, 396
Energy product, 78
Entertainment applications, 9, 425-439
Expansion coefficient, 583

Faraday equation, 361
Faraday rotator, 71
Ferrimagnetic resonance, 70
Ferrimagnetism, 39
Ferromagnetic resonance, 70-71
Ferromagnetism, 29
Ferrofluids, 570
Ferroresonant transformers, 508
Ferrous ferrite, 211, 214, 236

644 INDEX

Ferrous ion content, 234
Ferroxplana, 221
Films, ferrite, 343-350
Filters, channel, 8, 367,
Flux density, 50, 379, 456
Flux magnetic, 49-50
Flyback transformers, 431-433, 522
Forward converters, 449, 451
Freeze drying, 322
Fused salt synthesis, 322

g factor, 23
Gallium oxide, 233
Gamma iron oxide, 214-215, 262,
Gapped cores, 478
Gapped inductors, 476, 515-516
Garnets, rare earth, 222-225, 260, 291
GFI applications, 7
Grain boundaries, 281, 293-299
Grain growth, 268-271
Grain size, 266-268, 273-280, 286
Gyromagnetic ratio, 64, 563

Hall probe, 590
Hanna curves, 512, 516-516-518
Hexagonal ferrites, 220. 260, 293, 342
High permeability alloys, 153-179
Hiperco, 140
Hysteresigraph, 594-595
Hysteresis loops, 59, 44451-52
Hysteresis losses, 66, 250-251

IEC publications, 617-620
Impedance, 398, 401-402, 405
Inductance, 361-362
Induction, 50
Inductors, 378, 510-512, 520-522, 527
Integrated magnetics, 481-482
Intrinsic properties, 229
Inverters, 447-455, 524-533
Inverse spinel, 214
Ionic radius, 209-212
Iron content, 234-238, 248
Iron oxide, 306-307
Iron powder cores, 183-190, 411, 414-415
Iron, soft, 108-110
Iron, sintered, 128-132
ISDN, 387-388
Isolator, microwave, 556, 559-561

Kiln, ferrite, 339-341

Lamination steel, see Low-carbon steels
Lattice sites,
    garnet, 220-225
    spinel, 208-220
Leakage flux, 104
Linewidth, microwave, 713

Lithium ferrite, 214-215, 220, 236
Loading coils, 9, 368
Longitudinal field devices, 560-562
Loss coefficients, 66
Losses, , 66, 444
Loss factor, 65, 66
Loss tangent, 65
Low carbon steels, 113
Low frequency power,106, 133
Low profile ferrite cores, 406

Magnesium ferrite, 211, 214-215, 271
Magnesium zinc ferrite427
Magnetic balance, 592
Magnetic domains, 41-42, 45, 47
Magnetic fields, 17-18
Magnetic field strength, 17-19, 583
Magnetic induction, 51
    saturation, 51,52
Magnetic moment, 22
    ferromagnetic metals, 32-34
Magnetic polarization
    see Magnetization
Magnetic poles, 18-21,
Magnetic recording, 13, 57, 72, 535-557
    magneto-optical, 553-555
    magneto-resistive, 545-553
    oxide-coated tape, 13, 73
    recording heads, 14, 544-545
Magnetic shielding, 43
Magnetic susceptibility, 26-27
Magnetization, 47,232
Magnetization curve, 47-49
Magnetocrystalline anisotropy, 43, 54-55
Magnetometer, pendulum, 592-593
Magnetomotive force, 103-104
Magnetoplumbite structure, 220-221
Magnetostriction, 56
    magnetostriction contsant, 44, 236
    measurement, 397
Magnetostrictive transducers, 565
Magnons, 72
Manganese ferrite, 211,214-215,236-238, 240
Manganese-zinc ferrite, 236, 239, 244, 246,
    249, 253-255, 256-259, 283, 328-339,
    398, 427,
Maximum energy product , 79
Measurements magnetic, 589-621
Mechanical properties, 575
Melting points, 581-587
Memory cores, 73, 533-536
Metallic magnetic materials, 105-202
Metal powder cores, 181, 202
    EMI applications, 408
    RF applications, 183-184
    Processing, 184
Microstructures, ferrite, 265-300, 325
Microwave Faraday rotation, 561-563

Microwave ferrite applications, 12, 69-72, 261-263, 559-570
Microwave ferrites, 69-72, 260-261, 341-342
Minor loops, 52
MKSA units, 51
Moly-perm powder cores,
    Processing, 91-92
    Properties, 196-205
    EMI applications, 412, 416-418
Multiplying voltmeter, 608
Mumetal, 157, 161, 165-166

Nanocrystalline materials, 128, 176-181, 421-422, 469
Néel theory, 34-36
Néel temperature, 35
Neodymium-iron-boron permanent magnets, 77, 91-94
Neutron diffraction, 215
Nickel-cobalt ferrites
Nickel ferrites, 211, 213-214, 236
Nickel-zinc ferrites, 217-218, 232, 248, 269, 271, 286, 454, 504, 506
Nickel-iron alloys, 154-162
    with Mo, 163-166
    with Cu, 157-158
Normal spinel, 214

Octahedral sites, 208-210
Orbital magnetic moment, 22-23
Organic precursors, 320-321
Orthoferrites, 225
Output power, 461
Oxygen parameter, 211, 241-243
Oxygen stoichiometry, 240-243

Paramagnetic susceptibility, 26-27
    of Conduction electrons, 28
    of Transition metal and RE ions, 28
    above Curie Temperature, 39
Paramagnetism, 26
Permanent magnet ferrites, 75-105
Permanent magnet
    Steels, 75, 81
    Ductile, 95
    Design, 96-105
    Corrosion, 98
    Costs, 101-102
    Leakage, 104
Permeability, 53-54, 62-65, 365, 458
Permeance coefficient, 103
Permeance, magnetic, 80, 103-104
Perminvar, 156
Phase relation, inductor, 362, 364
Phase shifter, microwave, 558
    digital, 562
Phase transformation, 299-300
Planar technology, 481-483

Poles, see Magnetic poles,
Porosity 271, 273-280,
Powder cores, see Metal powder cores
Pot cores, 370-377, 468
Power applications, 5
Power ferrites, 256-259, 444, 465
Power supplies, 11
PQ cores, 478
Precession, electron spin, 70-72
Prepolarized cores, 480
Processing, conventional, 305-317
Processing, non-conventional, 317-326
Pulse measurements, 386, 615
Pulse permeability, 385
Pulse width modulation, 449
Purity, raw materials, 305-308
Push-pull converters, 449, 452

Q-factor, 367

Radar-absorbing ferrites, 564
Radiation resistance, 582
Rare-earth cobalt materials, 77, 90-91
Raw materials, 305-308
Reactance, capacitive, 362-363
Ractance, inductive, 362-363
Recoil lines, 105
Recording magnetic, 262-273
Regulation, 492, 494-495
Regulation-limited transformer, 493
Reluctance, 104-105
Remanence, 53
Residual losses,
    see anomalous losses
Resistivity, 61, 293-300
Resonant frequency, 368
Ripple current, 516
RM cores, 470

Saturation induction, 50-51, 232
Saturation-limited design, 495-496
Saturation magnetization, 47-50
Scanning electron microscopy, 282-284
Sensors, 566-570
Shielding applications, 143-152
    Measurements, 147

Shearing line, 78
Shielding effect, 372
Silica, effect of, 256-247
Silicon steels, 113-120
    Non-oriented, 115-117
    Oriented, 117-120
    6.5% Si, 125-128
Single crystal ferrites, 350-353
Sintering, 326-343
Skin depth 61-62
Snoeks's limit, 400

646 INDEX

Sol-gel synthesis, 322-323
Specific heat, 575
Spinel ferrites, 208-220
Spin moment, 22-25
Spin-orbit coupling, 28, 44
Spin waves, 72
Spray firing, 321
Squareness ratio, 73
Square loop ferrites, 73
Square wave drive, 447
Steinmetz coefficient, 458
Stress, effect of, 585-587
    encapsulation plastic, 587
    impact shock, 587
    machining, 585
    pressure, 585
Strontium ferrite, 260
Superexchange, 38
Supermalloy, 165, 170, 173
Supermendur, 141
Surface-mount design, 467
Susceptibility, magnetic, 26-27
Switched-mode power supplies, 11
Switching regulator, 11

Telecommunications applications, 8
Television applications, see Entertainment
Temperature coefficient, 378
Temperature factor,, 66, 218, 378
Temperature rise, 455, 457
Tensile strength, 581, 583
Thermal characteristics, 583-584
Thermal conductivity, 292-293, 583-584
Thermal properties of ferrites, 583-584

Thermal resistance, 486
Thermal shock resistance, 584
Thin films, ferrite, 343-350
Tin oxide, 249, 251, 257
Titanium dioxide, 251-254, 257, 289
Titanium cobalt additions, 252-255
Toroids. 475
Transformer, 6, 10
    Power, 442, 452
    Wide-band, 9
Transmission electron microscopy, 280, 283-284
Transverse field devices, 559
Tuned circuits, 366

Ultrasonic generators, 441
Unipolar drive, 445-446
Units, conversion, 640

Vanadium pentoxide, 270
Very high frequency operation, 506-508
Vibrating sample magnetometer, 594

Wattmeter, electronic, 608
Wavelength, microwave, 559-560
Winding, 504
Winding-loss limited design, 492-493

X-Ray diffraction, 312-313

Yttrium-iron garnet, 260-261

Zinc ferrite, 211, 213, 215-216, 218, 232, 239
Zinc loss, 271

Dr. Alex Goldman
is awarded the TAKESHI TAKEI Prize for 1996

During the opening of the ICF '97 banquet in Saint-Emilion, the TAKESHI TAKEI PRIZE for 1996 was awarded to Dr. ALEX GOLDMAN in recognition of his contribution in the field of ferrite technology and materials.

****

Dr. Alex Goldman received his M.A. and Ph.D. degrees in physical chemistry from Columbia University.

Since 1987, he is president of Ferrite Technology Worldwide (Pittsburgh, Pennsylvania, USA), an international consultant firm in ferrite research, manufacturing and applications.

Previous to that time, he was Corporate Director of Research of Spang and Co. and its Magnetics Division. His major technical involvement there was in high permeability and low loss ferrites, coprecipitation and spray roasting of ferrites, low-loss Sendust-type powder cores and other new magnetic materials.

Dr. Goldman is the author of "Modern Ferrite Technology" (Van Nostrand Reinhold, 1992) and is completing "Handbook of Modern Ferromagnetic Materials" (Kluwer Academic Publishers).

He has contributed chapters on ferrites and other magnetic materials in various books: "Modern Electrical and Electronic Engineering" (John Wiley, 1986), "Electronic Ceramics, Properties, Devices and Applications" (Marcel Dekker, 1988), "Engineered Materials Handbook" Vol.4 (A.S.M., 1991), "Encyclopedia of Advanced Materials" (Pergamon, 1995), and with B.B. Ghate, "Material Science and Technology" (VCH, 1992).

He has authored numerous technical papers and is the holder of 5 patents on ferrites.

He was co-chairman of the 4th ICF held in San Francisco in 1984 and is an honorary member of the ICF International Advisory Committee. He is a life member of the Institute of Electrical and Electronic Engineers, of the American Chemical Society and the American Ceramic Society of which he is a Fellow.

He has been an invited Lecturer in several universities (China, Korea, USA) and scientific institutions (IEEE, AcerS…).

****

At a short ceremony, Dr. Patrick Beuzelin of Siemens Matsushita Components asked Professor Sugimoto of Teikyo Heisei University, the recipient of the 1992 Takeshi Takei Prize, to make the presentation of the 1996 ICF 7 prize.

Acting on behalf of the ICF International Committee and of the ICF 7 Organizing Committee, Professor Sugimoto presented Dr. Alex Goldman with a brass and oak commemorative plaque as a token of sincere recognition of his much valued contributions in the field of ferrites.